ALGEBRA

ABSTRACT AND CONCRETE

STRESSING SYMMETRY

2ND EDITION

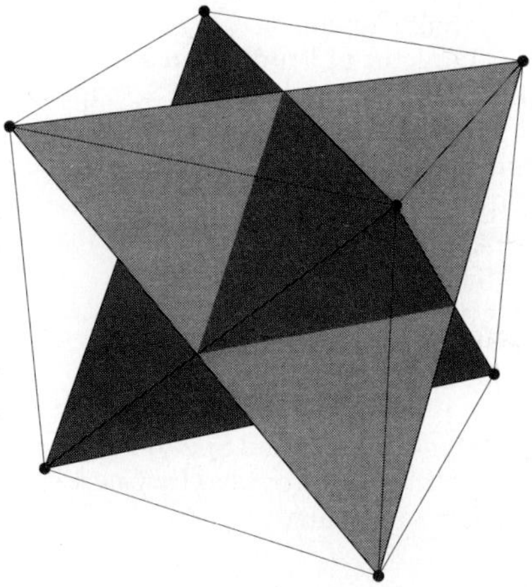

FREDERICK M. GOODMAN

UNIVERSITY OF IOWA

Prentice Hall

Pearson Education, Inc.
Upper Saddle River, NJ 07458

Library of Congress Cataloging-in-Publication Data

Goodman, Frederick M.
Algebra: abstract and concrete / Frederick M. Goodman—2nd ed.
p. cm.
Includes index.
ISBN 0-13-067342-0
1.Algebra. I. Title

QA155 .R64 2003
512–dc21

2003123456

Acquisitions Editor: George Lobell
Editor-in-Chief: Sally Yagan
Vice President/Director of Production and Manufacturing: David W. Riccardi
Executive Managing Editor: Kathleen Schiaparelli
Senior Managing Editor: Linda Mihatov Behrens
Assistant Managing Editor: Bayani Mendoza de Leon
Production Editor: Steven S. Pawlowski
Manufacturing Buyer: Michael Bell
Manufacturing Manager: Trudy Pisciotti
Marketing Assistant: Rachel Beckman
Editorial Assistant: Jennifer Brady
Art Director: Jayne Conte
Cover Designer: Bruce Kenselaar
Cover Image: Uzbek kilim called Suzani.
Cover Image Credit: Slide provided by Geert Keppens, Keppens Gallery.
Photo Researcher: Kim Marsden

Prentice Hall

Pearson Education, Inc.
Upper Saddle River, NJ 07458

Printed in the United States of America

10 9 8 7 6 5 4 3 2 1

ISBN 0-13-067342-0

Pearson Education LTD., *London*
Pearson Education Australia PTY, Limited, *Sydney*
Pearson Education Singapore, Pte. Ltd
Pearson Education North Asia Ltd, *Hong Kong*
Pearson Education Canada, Ltd., *Toronto*
Pearson Educacion de Mexico, S.A. de C.V.
Pearson Education Japan, *Tokyo*
Pearson Education Malaysia, Pte. Ltd

Contents

Preface vii

A Note to the Reader xi

Part I. Basics 1

Chapter 1. Algebraic Themes 1

1.1. What Is Symmetry? 1
1.2. Symmetries of the Rectangle and the Square 3
1.3. Multiplication Tables 7
1.4. Symmetries and Matrices 11
1.5. Permutations 16
1.6. Divisibility in the Integers 26
1.7. Modular Arithmetic 36
1.8. Polynomials 44
1.9. Counting 56
1.10. Groups 69
1.11. Rings and Fields 75
1.12. An Application to Cryptography 80

Chapter 2. Basic Theory of Groups 84

2.1. First Results 84
2.2. Subgroups and Cyclic Groups 91
2.3. The Dihedral Groups 104
2.4. Homomorphisms and Isomorphisms 109
2.5. Cosets and Lagrange's Theorem 119
2.6. Equivalence Relations and Set Partitions 125
2.7. Quotient Groups and Homomorphism Theorems 132

Chapter 3. Products of Groups 145

3.1. Direct Products 145
3.2. Semidirect Products 152
3.3. Finite Abelian Groups 155
3.4. Vector Spaces 165

Chapter 4. Symmetries of Polyhedra 176
4.1. Rotations of Regular Polyhedra 176
4.2. Rotations of the Dodecahedron and Icosahedron 185
4.3. What about Reflections? 189
4.4. Linear Isometries 195
4.5. The Full Symmetry Group and Chirality 200

Chapter 5. Actions of Groups 203
5.1. Group Actions on Sets 203
5.2. Group Actions—Counting Orbits 210
5.3. Symmetries of Groups 214
5.4. Group Actions and Group Structure 216
5.5. Application: Transitive Subgroups of S_5 224
5.6. Additional Exercises for Chapter 5 225

Chapter 6. Rings 228
6.1. A Recollection of Rings 228
6.2. Homomorphisms and Ideals 234
6.3. Quotient Rings 242
6.4. Integral Domains 248
6.5. Euclidean Domains, Principal Ideal Domains, and Unique Factorization 252
6.6. Unique Factorization Domains 261
6.7. Noetherian Rings 266
6.8. Irreducibility Criteria 268

Chapter 7. Field Extensions – First Look 272
7.1. A Brief History 272
7.2. Solving the Cubic Equation 274
7.3. Adjoining Algebraic Elements to a Field 277
7.4. Splitting Field of a Cubic Polynomial 284
7.5. Splitting Fields of Polynomials in $\mathbb{C}[x]$ 292

Part II. Topics 300

Chapter 8. Field Extensions – Second Look 300
8.1. Finite and Algebraic Extensions 300
8.2. Splitting Fields 302
8.3. The Derivative and Multiple Roots 305
8.4. Splitting Fields and Automorphisms 307
8.5. The Galois Correspondence 314
8.6. Symmetric Functions 319
8.7. The General Equation of Degree n 327
8.8. Quartic Polynomials 335
8.9. Galois Groups of Higher Degree Polynomials 342

Chapter 9. Solvability 348
9.1. Composition Series and Solvable Groups 348
9.2. Commutators and Solvability 351
9.3. Simplicity of the Alternating Groups 352
9.4. Cyclotomic Polynomials 356
9.5. The Equation $x^n - b = 0$ 359
9.6. Solvability by Radicals 360
9.7. Radical Extensions 363

Chapter 10. Isometry Groups 367
10.1. More on Isometries of Euclidean Space 367
10.2. Euler's Theorem 374
10.3. Finite Rotation Groups 377
10.4. Crystals 382

Appendices 400

Appendix A. Almost Enough about Logic 400
A.1. Statements 400
A.2. Logical Connectives 401
A.3. Quantifiers 405
A.4. Deductions 407

Appendix B. Almost Enough about Sets 409
B.1. Families of Sets; Unions and Intersections 413
B.2. Finite and Infinite Sets 414

Appendix C. Induction 416
C.1. Proof by Induction 416
C.2. Definitions by Induction 417
C.3. Multiple Induction 418

Appendix D. Complex Numbers 421

Appendix E. Review of Linear Algebra 423
E.1. Linear algebra in K^n 423
E.2. Bases and Dimension 428
E.3. Inner Product and Orthonormal Bases 433

Appendix F. Models of Regular Polyhedra 435

Appendix G. Suggestions for Further Study 443

Index 445

Preface

This text is an introduction to "modern" or "abstract" algebra for undergraduate students. The book addresses the conventional topics: groups, rings, and fields, with symmetry as a unifying theme. This subject matter is central and ubiquitous in modern mathematics and in applications ranging from quantum physics to digital communications.

The most important goal of this book is to engage students in the active practice of mathematics. Students are given the opportunity to participate and investigate, starting on the first page. Exercises are plentiful, and working exercises should be the heart of the course.

This text provides a thorough introduction to abstract algebra at a level suitable for upper-level undergraduates. The text would also be useful for an undergraduate topics course (for example, on geometric aspects of group theory or on Galois theory).

The required background for using this text is a standard first course in linear algebra. I have included a brief summary of linear algebra in an appendix to help students review. I have also provided appendices on sets, logic, mathematical induction, and complex numbers. The instructor may wish to go through this "corequisite" material systematically, or to dip into it from time to time as needed. It might also be useful to recommend a short supplementary text on set theory, logic, and proofs to be used as a reference and aid; several such texts are currently available.

The text is adaptable to different teaching styles. My own preference is increasingly to lecture little, and to use class time for discussing problems. But those who wish to present the material in systematic lectures will find the subject matter cleanly organized and presented in the text. I offer one piece of pedagogical advice both to instructors and students: This stuff takes time, give it the time it needs. Students have an enormous amount to learn "between the lines" of the text. They not only have to learn the mathematics, but they need to learn how mathematics is done, how to read and write mathematics, and how to approach solving problems. They need to learn to tinker, to try examples, to formulate and solve a simpler problem, to try a special case, to think about analogies, to guess at intermediate results, and so on. It is important for the instructor to slow the pace of the course in order to encourage exploration, and it is important for the student to devote the (many) hours which are actually needed to absorb the material and to solve the problems.

The subject treated in this text is usually called abstract algebra. In common language, abstract means both "difficult" and "impractical," and it is a little unfortunate to start out by labeling the subject as hard but useless! It won't be out of place here to make some (encouraging) remarks about abstraction. It takes some effort to remember that even the counting numbers were once (and in principle still are) an enormous abstraction. But they are familiar, they no longer seem difficult, and no one would doubt their usefulness. Abstractions with which we have become familiar eventually lose their aura of abstractness, but those with which we are not yet familiar seem abstract indeed. So it is with the ideas of this course. They may seem abstract today, but as they become familiar they will seem more concrete.

Mathematics involves a continual interplay between the abstract and the concrete: Abstraction is necessary in order to understand concrete phenomena, and concrete phenomena are necessary in order to understand the abstractions. Meanwhile, as one continues to study mathematics the perceived boundary between the abstract and concrete inevitably shifts.

Organization.

The text has been substantially revised for this second edition. The most important innovation is the long introductory chapter, entitled "Algebraic Themes," which treats many of the more elementary topics – symmetries, the integers, modular arithmetic, polynomials, and permutations – and introduces the important algebraic objects – groups, rings and fields – that are the subject of the rest of the book. The chapter also contains new sections on counting and cryptography. Chapter 6, on rings has been expanded, with a more complete discussion of Euclidean, principal ideal, and unique factorization domains. The classification of finite abelian groups has been moved to Chapter 3, on products of groups, and a section on vector spaces and bases has, appropriately, been added to this chapter. There is a new appendix containing a review of linear algebra (Appendix E).

Acknowledgements.

I am grateful to the following reviewers (of the first and second editions) for their exceedingly helpful comments: Manfred Dugas, Joseph Fisher, Daniel King, Jonh Labute, Michael Neuhauer, Emma Previato, David Schmidt, Jo Ann Turisco, Nicholas Vaughn, David Weinberg, and Carolyn Yackel.

I would like to thank George Lobell, Steven S. Pawlowski, and the staff at Prentice Hall for expert guidance and assistance.

Supplements.

I maintain a World Wide Web site with electronic supplements to the text, at `http://www.math.uiowa.edu/~goodman`. Materials available at this site may include

- Additional exercises
- Additional chapters or sections
- Color versions of graphics from the text
- Manipulable three-dimensional graphics
- Programs for algebraic computations
- Errata

I would be grateful for any comments on the text, reports of errors, and suggestions for improvements. I thank those students and instructors who wrote to me about the first edition.

Frederick M. Goodman
`goodman@math.uiowa.edu`

A Note to the Reader

I would like to show you a passage from one of my favorite books, *A River Runs Through It*, by Norman Maclean. The narrator Norman is fishing with his brother Paul on a mountain river near their home in Montana. The brothers have been fishing a "hole" blessed with sunlight and a hatch of yellow stone flies, on which the fish are vigorously feeding. They descend to the next hole downstream, where the fish will not bite. After a while Paul, who is fishing the opposite side of the river, makes some adjustment to his equipment and begins to haul in one fish after another. Norman watches in frustration and admiration, until Paul wades over to his side of the river to hand him a fly:

> He gave me a pat on the back and one of George's No. 2 Yellow Hackles with a feather wing. He said, "They are feeding on drowned yellow stone flies."
>
> I asked him, "How did you think that out?"
>
> He thought back on what had happened like a reporter. He started to answer, shook his head when he found he was wrong, and then started out again. "All there is to thinking," he said, "is seeing something noticeable which makes you see something you weren't noticing which makes you see something that isn't even visible."
>
> I said to my brother, "Give me a cigarette and say what you mean."
>
> "Well," he said, "the first thing I noticed about this hole was that my brother wasn't catching any. There's nothing more noticeable to a fisherman than that his partner isn't catching any.
>
> "This made me see that I hadn't seen any stone flies flying around this hole."
>
> Then he asked me, "What's more obvious on earth than sunshine and shadow, but until I really saw that there were no stone flies hatching here I didn't notice that the upper hole where they were hatching was mostly in sunshine and this hole was in shadow."
>
> I was thirsty to start with, and the cigarette made my mouth drier, so I flipped the cigarette into the water.

> "Then I knew," he said, "if there were flies in this hole they had to come from the hole above that's in the sunlight where there's enough heat to make them hatch.
>
> "After that, I should have seen them dead in the water. Since I couldn't see them dead in the water, I knew they had to be at least six or seven inches under the water where I couldn't see them. So that's where I fished."
>
> He leaned against the rock with his hands behind his head to make the rock soft. "Wade out there and try George's No. 2," he said, pointing at the fly he had given me. [1]

In mathematical practice the typical experience is to be faced by a problem whose solution is an mystery: The fish won't bite. Even if you have a toolbox full of methods and rules, the problem doesn't come labeled with the applicable method, and the rules don't seem to fit. There is no other way but to think things through for yourself.

The purpose of this course is to introduce you to the practice of mathematics; to help you learn to think things through for yourself; to teach you to see "something noticeable which makes you see something you weren't noticing which makes you see something that isn't even visible." And then to explain accurately what you have understood.

Not incidentally, the course aims to show you some algebraic and geometric ideas that are interesting and important and worth thinking about.

It's not at all easy to learn to work things out for yourself, and it's not at all easy to explain clearly what you have worked out. These arts have to be learned by thoughtful practice.

You must have patience, or learn patience, and you must have time. You can't learn these things without getting frustrated, and you can't learn them in a hurry. If you can get someone else to explain how to do the problems, you will learn something, but not patience, and not persistence, and not vision. So rely on yourself as far as possible.

But rely on your teacher as well. Your teacher will give you hints, suggestions, and insights that can help you see for yourself. A book alone cannot do this, because it cannot listen to you and respond.

I wish you success, and I hope you will someday fish in waters not yet dreamed of. Meanwhile, I have arranged a tour of some well known but interesting streams.

---

[1] From Norman Maclean, *A River Runs Through It*, University of Chicago Press, 1976. Reprinted by permission.

PART I: BASICS

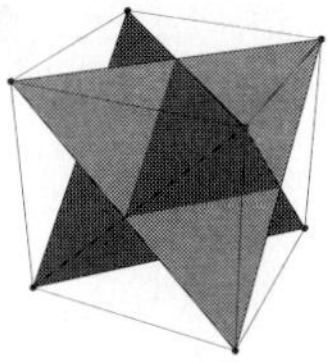

CHAPTER 1

Algebraic Themes

The first task of mathematics is to understand the "found" objects in the mathematical landscape. We have to try to understand the integers, the rational numbers, polynomials, matrices, and so forth, because they are "there." In this chapter we will examine objects that are familiar and concrete, but we will sometimes pose questions about them that are not so easy to answer. Our purpose is to introduce the algebraic themes that will be studied in the rest of the text, but also to begin the practice of looking closely and exactly at concrete situations.

We begin by looking into the idea of symmetry. What is more familiar to us than the symmetry of faces and flowers, of balls and boxes, of virtually everything in our biological and manufactured world? And yet, if we ask ourselves what we actually mean by symmetry, we may find it quite hard to give an adequate answer. We will soon see that symmetry can be given an operational definition, which will lead us to associate an algebraic structure with each symmetric object.

1.1. What Is Symmetry?

What is symmetry? Imagine some symmetric objects and some nonsymmetric objects. What makes a symmetric object symmetric? Are different symmetric objects symmetric in different ways?

The goal of this book is to encourage you to think things through for yourself. Take some time, and consider these questions for yourself. Start by making a list of symmetric objects: a sphere, a circle, a cube, a square, a rectangle, a rectangular box, etc. What do we mean when we say that these objects are symmetric? How is symmetry a common feature of these objects? How do the symmetries of the different objects differ? Close the book and take some time to think about these questions before going on with your reading.

As an example of a symmetric object, let us take a (nonsquare, blank, undecorated) rectangular card. What makes the card symmetric? The rather subtle answer adopted by mathematicians is that the card admits *motions* that leave its appearance unchanged. For example, if I left the room, you could secretly rotate the card by π radians (180 degrees) about the axis through two opposite edges, as shown in the figure, and when I returned, I could not tell that you had moved the card. *A symmetry is an undetectable motion. An object is symmetric if it has symmetries.* [1]

In order to examine this idea, work through the following exercises before continuing with your reading.

---

[1]We have to choose whether to idealize the card as a two–dimensional object (which can only be moved around in a plane) or to think of it as a thin three–dimensional object (which can be moved around in space). I choose to regard it as three–dimensional.

A related notion of symmetry involves *reflections* of the card rather than motions. I propose to ignore reflections for the moment in order to simplify matters, but to bring reflection symmetry into the picture later. You can see, by the way, that reflection symmetry and motion symmetry are different notions by considering a human face; a face has left-right symmetry, but there is no actual motion of a face that is a symmetry.

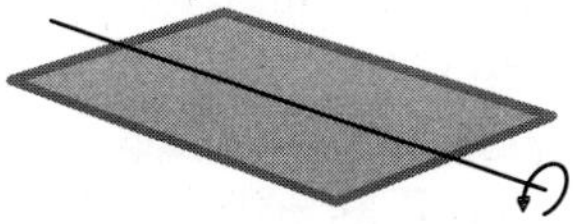

Figure 1.1.1. A symmetry

Exercises 1.1

1.1.1. Catalog all the symmetries of a (nonsquare) rectangular card. Get a card and look at it. Turn it about. Mark its parts as you need. Write out your observations and conclusions.

1.1.2. Do the same for a square card.

1.1.3. Do the same for a brick (i.e., a rectangular solid with three unequal edges). Are the symmetries the same as those of a rectangular card?

1.2. Symmetries of the Rectangle and the Square

What symmetries did you find for the rectangular card? Perhaps you found exactly three motions: two rotations of π (that is, 180 degrees) about axes through centers of opposite edges, and one rotation of π about an axis perpendicular to the faces of the card and passing through their centroids (Figure 1.2.1).

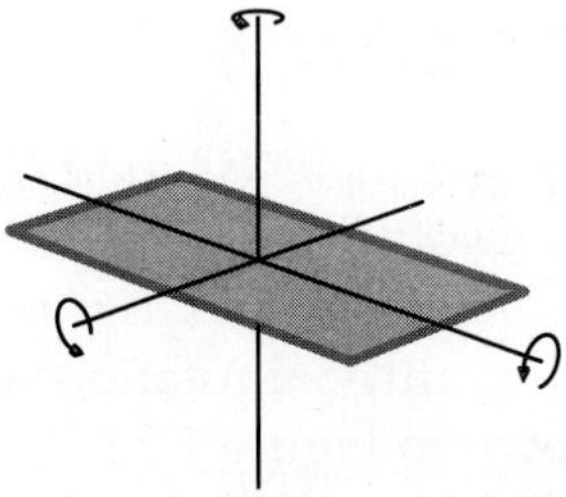

Figure 1.2.1. Symmetries of the rectangle

It turns out to be essential to include the *nonmotion* as well, that is, the rotation through 0 radians about any axis of your choice. One of the things that you could do to the card while I am out of the room is *nothing*. When I returned I could not tell that you had done nothing rather than something; nothing is also undetectable.

Including the nonmotion, we can readily detect four different symmetries of the rectangular card.[2]

However, another sensible answer is that there are infinitely many symmetries. As well as rotating by π about one of the axes, you could rotate by $-\pi$, $\pm 2\pi$, $\pm 3\pi, \dots$. (Rotating by $-\pi$ means rotating by π in the opposite sense.)

Which is it? Are there four symmetries of the rectangular card, or are there infinitely many symmetries? Since the world doesn't come equipped with a solutions manual, we have to make our own choice and see what the consequences are.

To distinguish only the four symmetries does turn out to lead to a rich and useful theory, and this is the choice that I will make here. With this choice, we have to consider rotation by 2π about one of the axes the same as the nonmotion, and rotation by -3π the same as rotation by π. Essentially, our choice is to disregard the path of the motion and to take into account only the final position of the parts of the card. When we rotate by -3π or by π, all the parts of the card end up in the same place.

Another issue is whether to include reflection symmetries as well as rotation symmetries, and as I mentioned previously, I propose to exclude reflection symmetries temporarily, for the sake of simplicity.

Making the same choices regarding the square card (to include the nonmotion, to distinguish only finitely many symmetries, and to exclude reflections), you find that there are eight symmetries of the square: the nonmotion, two rotations by π about axes through centers of opposite edges, two rotations by π about axes through opposite corners, and rotation by $\pi/2$, π, or $3\pi/2$ about the axis perpendicular to the faces and passing through their centroids.[3] (See Figure 1.2.2.)

Here is an essential observation: If I leave the room and you perform two undetectable motions one after the other, I will not be able to detect the result. The result of two symmetries one after the other is also a symmetry.

Let's label the three nontrivial rotations of the rectangular card by r_1, r_2, and r_3, as shown in Figure 1.2.3, and let's call the nonmotion e. If you perform first r_1, and then r_2, the result must be one of r_1, r_2, r_3, or e (because these are all of the symmetries of the card). Which is it? I claim that it is r_3. Likewise, if you perform first r_2 and then r_3, the result is r_1. Take your rectangular card in your hands and verify these assertions.

---

[2]Later we will take up the issue of why there are exactly four.

[3]Again, we will consider later why these are all the symmetries.

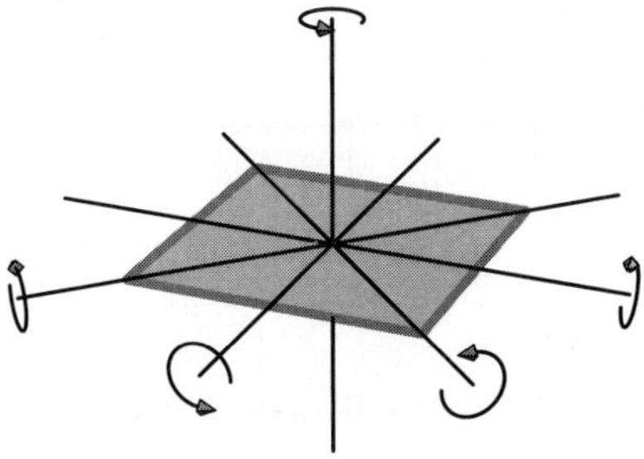

Figure 1.2.2. Symmetries of the square

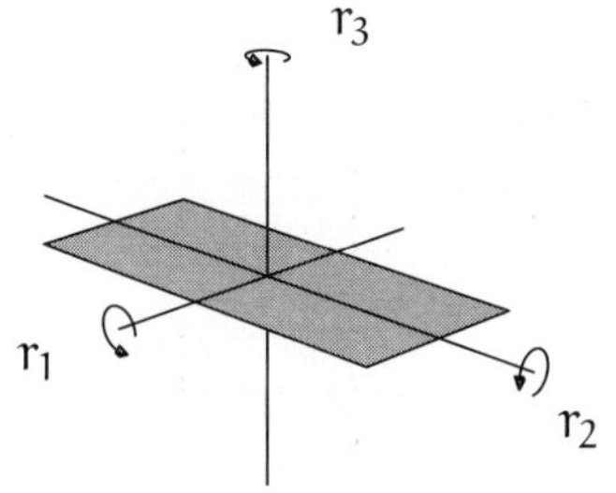

Figure 1.2.3. Labeling symmetries of the rectangle.

So we have a "multiplication" of symmetries by composition: The product xy of symmetries x and y is the symmetry "first do y and then do x." (The order is a matter of convention; the other convention is also possible.)

Your next investigation is to work out all the products of all the symmetries of the rectangular card and of the square card. A good way to record the results of your investigation is in a multiplication table: Label rows and columns of a square table by the various symmetries; for the rectangle you will have four rows and columns, for the square eight rows and columns. In the cell of the table in row x and column y record the product xy. For example, in the cell in row r_1 and column r_2 in the table for the rectangle, you will record the symmetry r_3; see Figure 1.2.4. Your job is to fill in the rest of the table.

When you are finished with the multiplication table for symmetries of the rectangular card, continue with the table for the square

	e	r_1	r_2	r_3
e				
r_1			r_3	
r_2				
r_3				

Figure 1.2.4. Beginning of the multiplication table for symmetries of the rectangle.

card. You will have to choose some labeling for the eight symmetries of the square card in order to begin to work out the multiplication table. In order to compare our results, it will be helpful if we agree on a labeling beforehand.

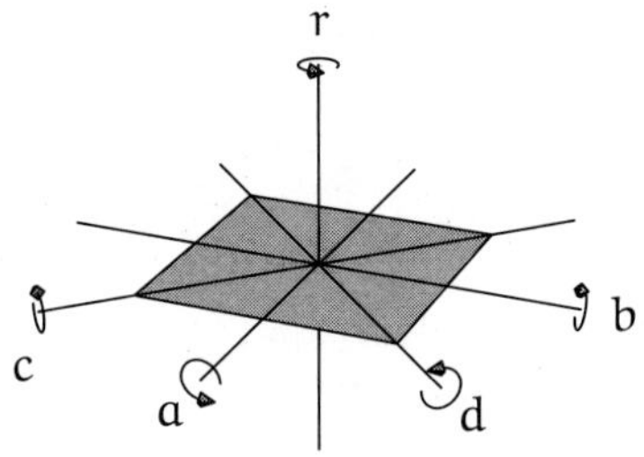

Figure 1.2.5. Labeling symmetries of the square.

Call the rotation by $\pi/2$ around the axis through the centroid of the faces r. The other rotations around this same axis are then r^2 and r^3; we don't need other names for them. Call the nonmotion e. Call the rotations by π about axes through centers of opposite edges a and b, and the rotations by π about axes through opposite vertices c and d. Also, to make comparing our results easier, let's agree to list the symmetries in the order $e, r, r^2, r^3, a, b, c, d$ in our tables (Figure 1.2.5). I have filled in a few entries of the table to help you get going (Figure 1.2.6). Your job is to complete the table.

Before going on with your reading, stop here and finish working out the multiplication tables for the symmetries of the rectangular and square cards. For learning mathematics, it is essential to work things out for yourself.

	e	r	r^2	r^3	a	b	c	d
e				r^3				
r	r						a	
r^2						a		
r^3								
a								
b								
c							e	
d							r^2	

Figure 1.2.6. Beginning of the multiplication table for symmetries of the square.

1.3. Multiplication Tables

In computing the multiplication tables for symmetries of the rectangle and square, we have to devise some sort of bookkeeping device to keep track of the results. One possibility is to "break the symmetry" by labeling parts of the rectangle or square. At the risk of overdoing it somewhat, I'm going to number both the *locations* of the four corners of the rectangle or square and the corners themselves. The numbers on the corners will travel with the symmetries of the card; those on the locations stay put. See Figure **??**, where the labeling for the square card is shown.

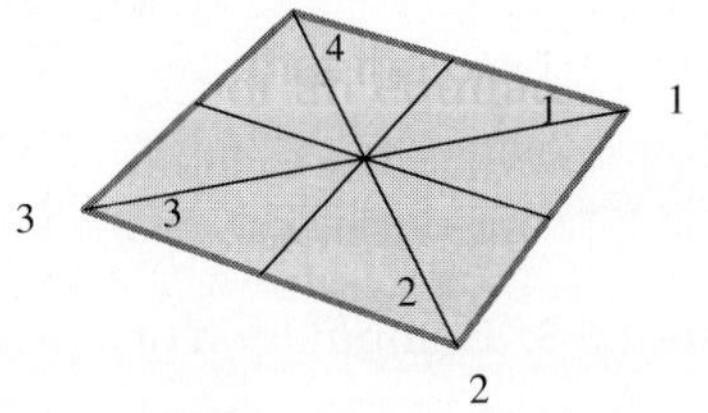

Figure 1.3.1. Breaking of symmetry

The various symmetries can be distinguished by the location of the numbered corners after performing the symmetry, as in Figure 1.3.2.

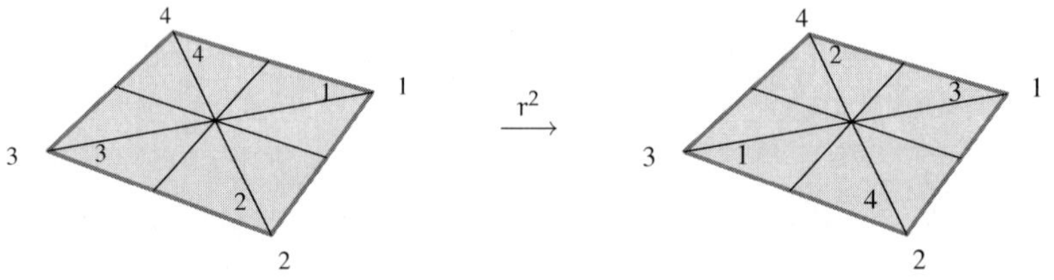

Figure 1.3.2. The symmetry r^2.

You can make a list of where each of the eight symmetries send the numbered vertices, and then you can compute products by diagrams as in Figure 1.3.3. Comparing Figures 1.3.3 and 1.3.2, you see that $cd = r^2$.

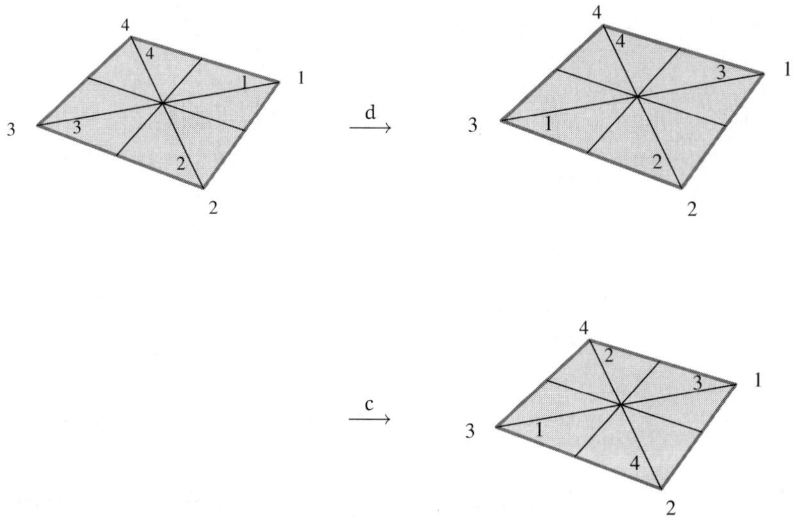

Figure 1.3.3. Computation of a product.

The multiplication table for the symmetries of the rectangle is shown in Figure 1.3.4.

There is a straightforward rule for computing all of the products: The square of any element is the nonmotion e. The product of any two elements other than e is the third such element.

	e	r_1	r_2	r_3
e	e	r_1	r_2	r_3
r_1	r_1	e	r_3	r_2
r_2	r_2	r_3	e	r_1
r_3	r_3	r_2	r_1	e

Figure 1.3.4. Multiplication table for symmetries of the rectangle.

Note that in *this* multiplication table, it doesn't matter in which order the elements are multiplied. The product of two elements in either order is the same. This is actually unusual; generally order *does* matter.

The multiplication table for the symmetries of the square card is shown in Figure 1.3.5.

	e	r	r^2	r^3	a	b	c	d
e	e	r	r^2	r^3	a	b	c	d
r	r	r^2	r^3	e	d	c	a	b
r^2	r^2	r^3	e	r	b	a	d	c
r^3	r^3	e	r	r^2	c	d	b	a
a	a	c	b	d	e	r^2	r	r^3
b	b	d	a	c	r^2	e	r^3	r
c	c	b	d	a	r^3	r	e	r^2
d	d	a	c	b	r	r^3	r^2	e

Figure 1.3.5. Multiplication table for symmetries of the square.

This table has the following properties, which I have emphasized by choosing the order in which to write the symmetries: The product of two powers of r (i.e., of two rotations around the axis through the centroid of the faces) is again a power of r. The square of any of the elements $\{a, b, c, d\}$ is the nonmotion e. The product of any two of $\{a, b, c, d\}$ is a power of r, while the product of a power of r and one of $\{a, b, c, d\}$ (in either order) is again one of $\{a, b, c, d\}$.

Actually this last property is obvious, without doing any close computation of the products, if we think as follows: The symmetries $\{a, b, c, d\}$ exchange the two faces (i.e., top and bottom) of the square card, while the powers of r do not. So, for example, the product of two symmetries that exchange the faces leaves the upper face above and the lower face below, so it has to be a power of r.

Notice that in this table, order in which symmetries are multiplied does matter. For example, $ra = d$, whereas $ar = c$.

We end this section by observing the following more or less obvious properties of the set of symmetries of a geometric figure (such as a square or rectangular card):

1. The product of three symmetries is independent of how the three are associated: The product of two symmetries followed by a third gives the same result as the first symmetry followed by the product of the second and third. This is the associative law for multiplication. In notation, the law is expressed as $s(tu) = (st)u$ for any three symmetries s, t, u.
2. The nonmotion e composed with any other symmetry (in either order) is the second symmetry. In notation, $eu = ue = u$ for any symmetry u.
3. For each symmetry there is an inverse, such that the composition of the symmetry with its inverse (in either order) is the nonmotion e. (The inverse is just the reversed motion; the inverse of a rotation about a certain axis is the rotation about the same axis by the same angle but in the opposite sense.) One can denote the inverse of a symmetry u by u^{-1}. Then the relation satisfied by u and u^{-1} is $uu^{-1} = u^{-1}u = e$.

Later we will pay a great deal of attention to consequences of these apparently modest observations.

Exercises 1.3

1.3.1. List the symmetries of an equilateral triangular plate (there are six) and work out the multiplication table for the symmetries. (See Figure 1.3.6.)

1.3.2. Consider the symmetries of the square card.

(a) Show that any positive power of r must be one of $\{e, r, r^2, r^3\}$. First work out some examples, say through r^{10}. Show that for any natural number k, $r^k = r^m$, where m is the nonnegative remainder after division of k by 4.

(b) Observe that r^3 is the same symmetry as the rotation by $\pi/2$ about the axis through the centroid of the faces of the square,

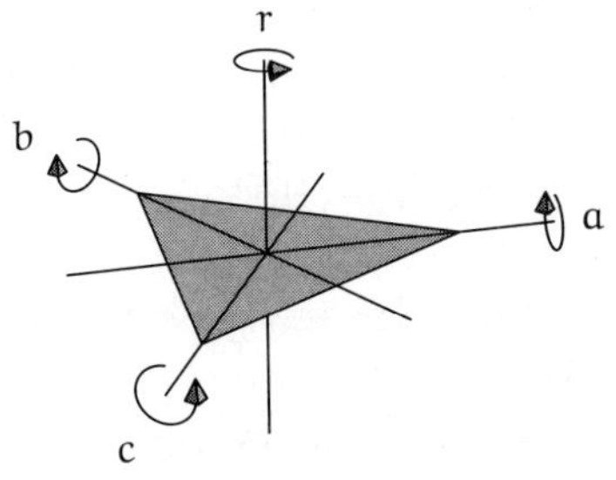

Figure 1.3.6. Symmetries of an equilateral triangle

in the clockwise sense, looking from the top of the square; that is, r^3 is the opposite motion to r, so $r^3 = r^{-1}$.

Define $r^{-k} = (r^{-1})^k$ for any positive integer k. Show that $r^{-k} = r^{3k} = r^m$, where m is the unique element of $\{0, 1, 2, 3\}$ such that $m + k$ is divisible by 4.

1.3.3. Here is another way to list the symmetries of the square card that makes it easy to compute the products of symmetries quickly.

(a) Verify that the four symmetries a, b, c, and d that exchange the top and bottom faces of the card are a, ra, r^2a, and r^3a, in some order. Which is which? Thus a complete list of the symmetries is

$$\{e, r, r^2, r^3, a, ra, r^2a, r^3a\}.$$

(b) Verify that $ar = r^{-1}a = r^3a$.

(c) Conclude that $ar^k = r^{-k}a$ for all integers k.

(d) Show that these relations suffice to compute any product.

1.4. Symmetries and Matrices

While looking at some examples, we have also been gradually refining our notion of a symmetry of a geometric figure. In fact, we are developing a mathematical model for a physical phenomenon — the symmetry of a physical object such as a ball or a brick or a card. So far, we have decided to pay attention only to the final position of the parts of an object, and to ignore the path by which they arrived at this position. This means that a symmetry of a figure R is a transformation or map from R to R. We have also implicitly assumed that the symmetries are *rigid* motions; that is, we don't allow our objects to be distorted by a symmetry.

We can formalize the idea that a transformation is rigid or nondistorting by the requirement that it be *distance preserving* or *isometric*. A transformation $\tau : R \to R$ is called an isometry if for all points $\mathbf{a}, \mathbf{b} \in R$, we have $d(\tau(\mathbf{a}), \tau(\mathbf{b})) = d(\mathbf{a}, \mathbf{b})$, where d denotes the usual Euclidean distance function.

We can show that an isometry $\tau : R \to \mathbb{R}^3$ defined on a subset R of $\mathbb{R}^3$ always extends to an *affine* isometry of $\mathbb{R}^3$. That is, there is a vector $\mathbf{b}$ and a linear isometry $T : \mathbb{R}^3 \to \mathbb{R}^3$ such that $\tau(\mathbf{x}) = \mathbf{b} + T(\mathbf{x})$ for all $\mathbf{x} \in R$. Moreover, if R is not contained in any two–dimensional plane, then the affine extension is uniquely determined by τ. [Note that if $\mathbf{0} \in R$ and $\tau(\mathbf{0}) = \mathbf{0}$, then we must have $\mathbf{b} = \mathbf{0}$, so τ extends to a linear isometry of $\mathbb{R}^3$.] These facts are established in Section 10.1; for now we will just assume them.

Now suppose that R is a (square or a nonsquare) rectangle, which we suppose lies in the (x, y)–plane, in three–dimensional space. Consider an isometry $\tau : R \to R$. We can show that τ must map the set of vertices of R to itself. (It would certainly be surprising if a symmetry did not map vertices to vertices!) Now there is exactly one point in R that is equidistant from the four vertices; this is the centroid of the figure, which is the intersection of the two diagonals of R. Denote the centroid C. What is $\tau(C)$? Since τ is an isometry and maps the set of vertices to itself, $\tau(C)$ is still equidistant from the four vertices, so $\tau(C) = C$. We can assume without loss of generality that the figure is located with its centroid at $\mathbf{0}$, the origin of coordinates. It follows from the results quoted in the previous paragraph that τ extends to a *linear* isometry of $\mathbb{R}^3$.

The same argument and the same conclusion are valid for many other geometric figures (for example, polygons in the plane, or polyhedra in space). For such figures, there is (at least) one point that is mapped to itself by every symmetry of the figure. If we place such a point at the origin of coordinates, then every symmetry of the figure extends to a linear isometry of $\mathbb{R}^3$.

Let's summarize with a proposition:

Proposition 1.4.1. *Let R denote a polygon or a polyhedron in three–dimensional space, located with its centroid at the origin of coordinates. Then every symmetry of R is the restriction to R of a linear isometry of $\mathbb{R}^3$.*

Since our symmetries extend to linear transformations of space, they are implemented by 3-by-3 *matrices*. That is, for each symmetry σ of one of our figures, there is an (invertible) matrix A such that for all points $\mathbf{x}$ in our figure, $\sigma(\mathbf{x}) = A\mathbf{x}$.[4]

[4] A brief review of elementary linear algebra is provided in Appendix E.

Here is an important observation: Let τ_1 and τ_2 be two symmetries of a three-dimensional object R. Let T_1 and T_2 be the (uniquely determined) linear transformations of $\mathbb{R}^3$, extending τ_1 and τ_2. The composed linear transformation T_1T_2 is then the unique linear extension of the composed symmetry $\tau_1\tau_2$. Moreover, if A_1 and A_2 are the matrices implementing T_1 and T_2, then the matrix product A_1A_2 implements T_1T_2. Consequently, we can compute the composition of symmetries by computing the product of the corresponding matrices.

This observation gives us an alternative, and more or less automatic, way to do the bookkeeping for composing symmetries.

Let us proceed to find, for each symmetry of the square or rectangle, the matrix that implements the symmetry.

We can arrange that the figure (square or rectangle) lies in the (x, y)–plane with sides parallel to the coordinate axes and centroid at the origin of coordinates. Then certain axes of symmetry will coincide with the coordinate axes. For example, we can orient the rectangle in the plane so that the axis of rotation for r_1 coincides with the x–axis, the axis of rotation for r_2 coincides with the y–axis, and the axis of rotation for r_3 coincides with the z–axis.

The rotation r_1 leaves the x–coordinate of a point in space unchanged and changes the sign of the y– and z–coordinates. We want to compute the matrix that implements the rotation r_1, so let us recall how the standard matrix of a linear transformation is determined. Consider the standard basis of $\mathbb{R}^3$:

$$\hat{\mathbf{e}}_1 = \begin{bmatrix} 1 \\ 0 \\ 0 \end{bmatrix} \quad \hat{\mathbf{e}}_2 = \begin{bmatrix} 0 \\ 1 \\ 0 \end{bmatrix} \quad \hat{\mathbf{e}}_3 = \begin{bmatrix} 0 \\ 0 \\ 1 \end{bmatrix}.$$

If T is any linear transformation of $\mathbb{R}^3$, then the 3-by-3 matrix M_T with columns $T(\hat{\mathbf{e}}_1), T(\hat{\mathbf{e}}_2)$, and $T(\hat{\mathbf{e}}_3)$ satisfies $M_T\mathbf{x} = T(\mathbf{x})$ for all $\mathbf{x} \in R^3$. Now we have

$$r_1(\hat{\mathbf{e}}_1) = \hat{\mathbf{e}}_1, \;\; r_1(\hat{\mathbf{e}}_2) = -\hat{\mathbf{e}}_2, \;\; \text{and } r_1(\hat{\mathbf{e}}_3) = -\hat{\mathbf{e}}_3,$$

so the matrix R_1 implementing the rotation r_1 is

$$R_1 = \begin{bmatrix} 1 & 0 & 0 \\ 0 & -1 & 0 \\ 0 & 0 & -1 \end{bmatrix}.$$

We still have a slight problem with nonuniqueness of the linear transformation implementing a symmetry of a two–dimensional object such as the rectangle or the square. However, if we insist on implementing our symmetries by *rotational* transformations of space, then the linear transformation implementing each symmetry is unique.

Similarly, we can trace through what the rotations r_2 and r_3 do in terms of coordinates. The result is that the matrices

$$R_2 = \begin{bmatrix} -1 & 0 & 0 \\ 0 & 1 & 0 \\ 0 & 0 & -1 \end{bmatrix} \quad \text{and} \quad R_3 = \begin{bmatrix} -1 & 0 & 0 \\ 0 & -1 & 0 \\ 0 & 0 & 1 \end{bmatrix}$$

implement the rotations r_2 and r_3. Of course, the identity matrix

$$E = \begin{bmatrix} 1 & 0 & 0 \\ 0 & 1 & 0 \\ 0 & 0 & 1 \end{bmatrix}$$

implements the nonmotion. Now you can check that the square of any of the R_i's is E and the product of any two of the R_i's is the third. Thus the matrices R_1, R_2, R_3, and E have the same multiplication table (using matrix multiplication) as do the symmetries r_1, r_2, r_3, and e of the rectangle, as expected.

Let us similarly work out the matrices for the symmetries of the square: Choose the orientation of the square in space so that the axes of symmetry for the rotations a, b, and r coincide with the x–, y–, and z–axes, respectively.

Then the symmetries a and b are implemented by the matrices

$$A = \begin{bmatrix} 1 & 0 & 0 \\ 0 & -1 & 0 \\ 0 & 0 & -1 \end{bmatrix} \quad B = \begin{bmatrix} -1 & 0 & 0 \\ 0 & 1 & 0 \\ 0 & 0 & -1 \end{bmatrix}.$$

The rotation r is implemented by the matrix

$$R = \begin{bmatrix} 0 & -1 & 0 \\ 1 & 0 & 0 \\ 0 & 0 & 1 \end{bmatrix},$$

and powers of r by powers of this matrix

$$R^2 = \begin{bmatrix} -1 & 0 & 0 \\ 0 & -1 & 0 \\ 0 & 0 & 1 \end{bmatrix} \quad \text{and} \quad R^3 = \begin{bmatrix} 0 & 1 & 0 \\ -1 & 0 & 0 \\ 0 & 0 & 1 \end{bmatrix}.$$

The symmetries c and d are implemented by matrices

$$C = \begin{bmatrix} 0 & -1 & 0 \\ -1 & 0 & 0 \\ 0 & 0 & -1 \end{bmatrix} \quad \text{and} \quad D = \begin{bmatrix} 0 & 1 & 0 \\ 1 & 0 & 0 \\ 0 & 0 & -1 \end{bmatrix}.$$

Therefore, the set of matrices $\{E, R, R^2, R^3, A, B, C, D\}$ necessarily has the same multiplication table (under matrix multiplication) as does the corresponding set of symmetries $\{e, r, r^2, r^3, a, b, c, d\}$. So we could have worked out the multiplication table for the symmetries of the square by computing products of the corresponding matrices. For example, we compute that $CD = R^2$ and can conclude that $cd = r^2$.

We can now return to the question of whether we have found *all* the symmetries of the rectangle and the square. We suppose, as before, that the figure (square or rectangle) lies in the (x, y)–plane with sides parallel to the coordinate axes and centroid at the origin of coordinates. Any symmetry takes vertices to vertices, and line segments to line segments (see Exercise 1.4.4), and so takes edges to edges. Since a symmetry is an isometry, it must take each edge to an edge of the same length, and it must take the midpoints of edges to midpoints of edges. Let 2ℓ and $2w$ denote the lengths of the edges; for the rectangle $\ell \neq w$, and for the square $\ell = w$. The midpoints of the edges are at $\pm\ell\hat{\mathbf{e}}_1$ and $\pm w\hat{\mathbf{e}}_2$. A symmetry τ is determined by $\tau(\ell\hat{\mathbf{e}}_1)$ and $\tau(w\hat{\mathbf{e}}_2)$, since the symmetry is linear and these two vectors are a basis of the plane, which contains the figure R.

For the rectangle, $\tau(\ell\hat{\mathbf{e}}_1)$ must be $\pm\ell\hat{\mathbf{e}}_1$, since these are the only two midpoints of edges length $2w$. Likewise, $\tau(w\hat{\mathbf{e}}_2)$ must be $\pm w\hat{\mathbf{e}}_2$, since these are the only two midpoints of edges length 2ℓ. Thus there are at most four possible symmetries of the rectangle. Since we have already found four distinct symmetries, there are exactly four.

For the square (with sides of length $2w$), $\tau(w\hat{\mathbf{e}}_1)$ and $\tau(w\hat{\mathbf{e}}_2)$ must be contained in the set $\{\pm w\hat{\mathbf{e}}_1, \pm w\hat{\mathbf{e}}_2\}$. Furthermore if $\tau(w\hat{\mathbf{e}}_1)$ is $\pm w\hat{\mathbf{e}}_1$, then $\tau(w\hat{\mathbf{e}}_2)$ is $\pm w\hat{\mathbf{e}}_2$; and if $\tau(w\hat{\mathbf{e}}_1)$ is $\pm w\hat{\mathbf{e}}_2$, then $\tau(w\hat{\mathbf{e}}_2)$ is $\pm w\hat{\mathbf{e}}_1$. Thus there are at most eight possible symmetries of the square. As we have already found eight distinct symmetries, there are exactly eight.

Exercises 1.4

1.4.1. Work out the products of the matrices E, R, R^2, R^3, A, B, C, D, and verify that these products reproduce the multiplication table for the symmetries of the square, as expected.

1.4.2. Find matrices implementing the six symmetries of the equilateral triangle. (Compare Exercise 1.3.1.) In order to standardize our notation and our coordinates, let's agree to put the vertices of the triangle at $(1, 0, 0)$, $(-1/2, \sqrt{3}/2, 0)$, and $(-1/2, -\sqrt{3}/2, 0)$. (You may have to review some linear algebra in order to compute the matrices of the symmetries; review how to get the matrix of a linear transformation, given the way the transformation acts on a basis.) Verify that the products of the matrices reproduce the multiplication table for the symmetries of the equilateral triangle.

1.4.3. Let $T(\mathbf{x}) = A\mathbf{x} + \mathbf{b}$ be an invertible affine transformation of $\mathbb{R}^3$. Show that T^{-1} is also affine.

The next four exercises outline an approach to showing that a symmetry of a rectangle or square sends vertices to vertices. The approach is based on the notion of *convexity*.

1.4.4. A line segment $[\mathbf{a}_1, \mathbf{a}_2]$ in $\mathbb{R}^3$ is the set

$$[\mathbf{a}_1, \mathbf{a}_2] = \{s\mathbf{a}_1 + (1-s)\mathbf{a}_2 : 0 \le s \le 1\}.$$

Show that if $T(\mathbf{x}) = A\mathbf{x} + \mathbf{b}$ is an affine transformation of $\mathbb{R}^3$, then $T([\mathbf{a}_1, \mathbf{a}_2]) = [T(\mathbf{a}_1), T(\mathbf{a}_2)]$.

1.4.5. A subset $R \subseteq \mathbb{R}^3$ is convex if for all $\mathbf{a}_1, \mathbf{a}_2 \in R$, the segment $[\mathbf{a}_1, \mathbf{a}_2]$ is a subset of R. Show that if R is convex and $T(\mathbf{x}) = A\mathbf{x} + \mathbf{b}$ is an affine transformation of $\mathbb{R}^3$, then $T(R)$ is convex.

1.4.6. A vertex $\mathbf{v}$ of a convex set R can be characterized by the following property: If $\mathbf{a}_1$ and $\mathbf{a}_2$ are elements of R and $\mathbf{v} \in [\mathbf{a}_1, \mathbf{a}_2]$, then $\mathbf{a}_1 = \mathbf{a}_2$. That is, $\mathbf{v}$ is not contained in any nontrivial line segment in R. Show that if $\mathbf{v}$ is a vertex of a convex set R, and $T(\mathbf{x}) = A\mathbf{x} + b$ is an invertible affine transformation of $\mathbb{R}^3$, then $T(\mathbf{v})$ is a vertex of the convex set $T(R)$.

1.4.7. Let τ be symmetry of a convex subset of $\mathbb{R}^3$. Show that τ maps vertices of R to vertices. (In particular, a symmetry of a rectangle or square maps vertices to vertices.)

1.4.8. Show that there are exactly six symmetries of an equilateral triangle.

1.5. Permutations

Suppose that I put three identical objects in front of you on the table:

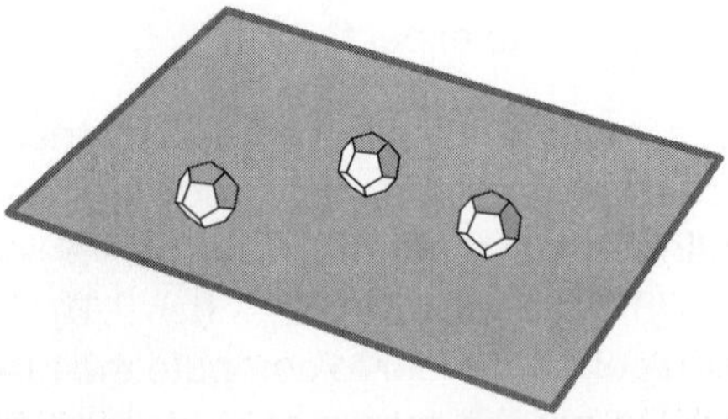

This configuration has symmetry, regardless of the nature of the objects or their relative position, just because the objects are identical. If you glance away, I could switch the objects around, and when you look back you could not tell whether I had moved them. There is a remarkable insight here: *Symmetry is not intrinsically a geometric concept.*

What are all the symmetries of the configuration of three objects? Any two objects can be switched while the third is left in place; there are three such symmetries. One object can be put in the place of a second, the second in the place of the third, and the third in the place of the first; There are two possibilities for such a rearrangement (corresponding to the two ways to traverse the vertices of a triangle). And there is the nonrearrangement, in that all the objects are left in place. So there are six symmetries in all.

The symmetries of a configuration of identical objects are called *permutations.*

What is the multiplication table for the set of six permutations of three objects? Before we can work this out, we have to devise some sort of bookkeeping system. Let's number not the objects but the three positions they occupy. Then we can describe each symmetry by recording for each i, $1 \leq i \leq 3$, the final position of the object that starts in position i. For example, the permutation that switches the objects in positions 1 and 3 and leaves the object in position 2 in place will be described by

$$\begin{pmatrix} 1 & 2 & 3 \\ 3 & 2 & 1 \end{pmatrix}.$$

The permutation that moves the object in position 1 to position 2, that in position 2 to position 3, and that in position 3 to position 1 is denoted by

$$\begin{pmatrix} 1 & 2 & 3 \\ 2 & 3 & 1 \end{pmatrix}.$$

With this notation, the six permutations of three objects are

$$\begin{pmatrix} 1 & 2 & 3 \\ 1 & 2 & 3 \end{pmatrix} \qquad \begin{pmatrix} 1 & 2 & 3 \\ 2 & 3 & 1 \end{pmatrix} \qquad \begin{pmatrix} 1 & 2 & 3 \\ 3 & 1 & 2 \end{pmatrix}$$

$$\begin{pmatrix} 1 & 2 & 3 \\ 2 & 1 & 3 \end{pmatrix} \qquad \begin{pmatrix} 1 & 2 & 3 \\ 1 & 3 & 2 \end{pmatrix} \qquad \begin{pmatrix} 1 & 2 & 3 \\ 3 & 2 & 1 \end{pmatrix}.$$

The product of permutations is computed by following each object as it is moved by the two permutations. If the first permutation moves an object from position i to position j and the second moves an object from position j to position k, then the composition moves an object from i to k. For example,

$$\begin{pmatrix} 1 & 2 & 3 \\ 2 & 3 & 1 \end{pmatrix} \begin{pmatrix} 1 & 2 & 3 \\ 1 & 3 & 2 \end{pmatrix} = \begin{pmatrix} 1 & 2 & 3 \\ 2 & 1 & 3 \end{pmatrix}.$$

Recall our convention that the element on the right in the product is the first permutation and that on the left is the second. Given this bookkeeping system, you can now write out the multiplication table for the six permutations of three objects (Exercise 1.5.1.) We can match up permutations of three objects with symmetries of an equilateral triangle, so that the multiplication tables match up as well; see Exercise 1.5.2.

Notice that the order in which permutations are multiplied matters in general. For example,

$$\begin{pmatrix} 1 & 2 & 3 \\ 2 & 1 & 3 \end{pmatrix} \begin{pmatrix} 1 & 2 & 3 \\ 2 & 3 & 1 \end{pmatrix} = \begin{pmatrix} 1 & 2 & 3 \\ 1 & 3 & 2 \end{pmatrix},$$

but

$$\begin{pmatrix} 1 & 2 & 3 \\ 2 & 3 & 1 \end{pmatrix} \begin{pmatrix} 1 & 2 & 3 \\ 2 & 1 & 3 \end{pmatrix} = \begin{pmatrix} 1 & 2 & 3 \\ 3 & 2 & 1 \end{pmatrix}.$$

For any natural number n, the permutations of n identical objects can be denoted by two-line arrays of numbers, containing the numbers 1 through n in each row. If a permutation π moves an object from position i to position j, then the corresponding two line array has j positioned below i. The numbers in the first row are generally arranged in increasing order, but this is not essential. Permutations are multiplied or composed according to the same rule as given previously for permutations of three objects. For example, we have the following product of permutations of seven objects:

$$\begin{pmatrix} 1 & 2 & 3 & 4 & 5 & 6 & 7 \\ 3 & 2 & 1 & 7 & 4 & 5 & 6 \end{pmatrix} \begin{pmatrix} 1 & 2 & 3 & 4 & 5 & 6 & 7 \\ 4 & 3 & 1 & 2 & 6 & 5 & 7 \end{pmatrix}$$
$$= \begin{pmatrix} 1 & 2 & 3 & 4 & 5 & 6 & 7 \\ 7 & 1 & 3 & 2 & 5 & 4 & 6 \end{pmatrix}.$$

The set of permutations of n identical objects shares the following properties with the set of symmetries of a geometric figure:

1. The multiplication of permutations is associative.
2. There is an identity permutation e, which leaves each object in its original position. The product of e with any other permutation σ, in either order, is σ.
3. For each permutation σ, there is an inverse permutation σ^{-1}, which "undoes" σ. For all i, j, if σ moves an object from position i to position j, then σ^{-1} moves an object from position

j to position i. The product of σ with σ^{-1}, in either order, is e.

A slightly different point of view makes these properties even more evident.

Recall that a function (or map) $f : X \to Y$ is *one to one* (or *injective*) if $f(x_1) \neq f(x_2)$ whenever $x_1 \neq x_2$ are distinct elements of X.[5] A map $f : X \to Y$ is said to be *onto* (or *surjective*) if the range of f is all of Y; that is, for each $y \in Y$, there is an $x \in X$ such that $f(x) = y$. A map $f : X \to Y$ is called *invertible* (or *bijective*) if it is both injective and surjective. For each invertible map $f : X \to Y$, there is a map $f^{-1} : Y \to X$ called the inverse of f and satisfying $f \circ f^{-1} = \mathrm{id}_Y$ and $f^{-1} \circ f = \mathrm{id}_X$. Here id_X denotes the identity map $\mathrm{id}_X : x \mapsto x$ on X and similarly id_Y denotes the identity map on Y. For $y \in Y$, $f^{-1}(y)$ is the unique element of X such that$f(x) = y$.

Now consider maps from a set X to itself. Maps from X to itself can be composed, and composition of maps is associative. If f and g are bijective maps on X, then the composition $f \circ g$ is also bijective (with inverse $g^{-1} \circ f^{-1}$). Denote the set of bijective maps from X to X by $\mathrm{Sym}(X)$, short for "symmetries of X." $\mathrm{Sym}(X)$ satisfies the following properties:

1. Composition of maps defines an associative product on the set $\mathrm{Sym}(X)$.
2. The identity map id_X is the identity element for this product; that is, for any $f \in \mathrm{Sym}(X)$, the composition of id_X with f, in either order, is f.
3. The inverse of a map is the inverse for this product; that is, for any $f \in \mathrm{Sym}(X)$, the composition of f and f^{-1}, in either order, is id_X.

Now a permutation of n objects can be identified with a bijective function on the set $\{1, 2, \ldots, n\}$; the permutation moves an object from position i to position j if the function maps i to j. For example, the permutation

$$\pi = \begin{pmatrix} 1 & 2 & 3 & 4 & 5 & 6 & 7 \\ 4 & 3 & 1 & 2 & 6 & 5 & 7 \end{pmatrix}$$

in S_7 is identified with the bijective map of $\{1, 2, \ldots, 7\}$ that sends 1 to 4, 2, to 3, 3 to 1, and so on. It should be clear, upon reflection, that the multiplication of permutations is the same as the composition of bijective maps. Thus the three properties listed for permutations follow immediately from the corresponding properties of bijective maps.

---

[5]Sets and functions are discussed in Appendix B.

We generally write S_n for the permutations of a set of n elements rather than $\mathrm{Sym}(\{1, 2, \dots, n\})$. It is not difficult to see that the size of S_n is $n! = n(n-1)\cdots(2)(1)$. In fact, the image of 1 under an invertible map can be any of the n numbers $1, 2, \dots, n$; for each of these possibilities, there are $n-1$ possible images for 2, and so forth. When n is moderately large, the set S_n of permutations is enormous; for example, the number of permutations of a deck of 52 cards is $(52)(51)\dots(2)(1) =$

$$80658175170943878571660636856403766975289505440883277824000000000000.$$

We couldn't begin to write out the multiplication table for S_{52} or even to list the elements of S_{52}, and yet S_{52} is not in principle a different sort of object than S_5 (for example), which has 120 elements.

There is an alternative notation for permutations that is convenient for many purposes. I explain it by example. Consider the permutation

$$\pi = \begin{pmatrix} 1 & 2 & 3 & 4 & 5 & 6 & 7 \\ 4 & 3 & 1 & 2 & 6 & 5 & 7 \end{pmatrix}$$

in S_7. This permutation takes 1 to 4, 4 to 2, 2 to 3, and 3 back to 1; it takes 5 to 6 and 6 back to 5; and it fixes (doesn't move) 7. Correspondingly, we write $\pi = (1423)(56)$.

A permutation such as $(1\ 4\ 2\ 3)$ that permutes several numbers cyclically (1 to 4, 4 to 2, 2 to 3, and 3 to 1) and leaves all other numbers fixed is called a *cycle*. Note that $(1\ 4\ 2\ 3) = (4\ 2\ 3\ 1) = (2\ 3\ 1\ 4) = (3\ 1\ 4\ 2)$. There is no preferred first entry in the notation for a cycle; only the cyclic order of the entries matters.

Two cycles are called *disjoint* if each leaves fixed the numbers moved by the other. The expression $\pi = (1\ 4\ 2\ 3)\ (5\ 6)$ for π as a product of disjoint cycles is called *cycle notation*.

Let's check by example how permutations written in cycle notation are multiplied.

Example 1.5.1. We compute the product

$$\left[(1\ 3)\ (4\ 7\ 6\ 5)\right]\left[(1\ 4\ 2\ 3)\ (5\ 6)\right].$$

Remember that in multiplying permutations, the permutation on the right is taken first. The first of the permutations takes 1 to 4 and the second takes 4 to 7, so the product takes 1 to 7. The first leaves 7 fixed and the second takes 7 to 6, so the product takes 7 to 6. The first takes 6 to 5 and the second takes 5 to 4, so the product takes 6 to 4. The first takes 4 to 2 and the second leaves 2 fixed, so the product takes 4 to 2. The first takes 2 to 3 and the second takes 3 to 1, so the product takes 2 to 1. This "closes the cycle" $(1\ 7\ 6\ 4\ 2)$. The first permutation takes 5 to 6 and the second takes 6 to 5, so the product fixes 5. The first takes 3 to 1 and the second takes 1 to 3, so the product fixes 3. Thus the product is

$$[(1\ 3)\ (4\ 7\ 6\ 5)]\ [(1\ 4\ 2\ 3)\ (5\ 6)] = (1\ 7\ 6\ 4\ 2)\,.$$

Notice that the permutation $\pi = (1\ 4\ 2\ 3)\ (5\ 6)$ is the product of the cycles $(1\ 4\ 2\ 3)$ and $(5\ 6)$. Disjoint cycles commute; their product is independent of the order in which they are multiplied. For example,

$$[(1\ 4\ 2\ 3)][(5\ 6)] = [(5\ 6)][(1\ 4\ 2\ 3)] = \begin{pmatrix} 1 & 2 & 3 & 4 & 5 & 6 & 7 \\ 4 & 3 & 1 & 2 & 6 & 5 & 7 \end{pmatrix}.$$

A permutation π is said to have *order* k if the kth power of π is the identity and no lower power of π is the identity. A k-cycle (that is, a cycle of length k) has order k. For example, $(2\ 4\ 3\ 5)$ has order 4. A product of disjoint cycles has order equal to the least common multiple of the lengths of the cycles. For example, $(2\ 4\ 3\ 5)\ (1\ 6)\ (7\ 9\ 10)$ has order 12, the least common multiple of 4, 2, and 3.

Example 1.5.2. Consider the "perfect shuffle" of a deck of cards containing $2n$ cards. The deck is first cut perfectly, with the half deck consisting of the first n cards placed on the left and the half deck consisting of the last n cards placed on the right. The two half decks are then interlaced, the first card from the right going on top, then the first card from the left, then the second card from the right, and so forth. For example, with a deck of 10 cards, the perfect shuffle is the permutation

$$\begin{pmatrix} 1 & 2 & 3 & 4 & 5 & 6 & 7 & 8 & 9 & 10 \\ 2 & 4 & 6 & 8 & 10 & 1 & 3 & 5 & 7 & 9 \end{pmatrix} = (1\ 2\ 4\ 8\ 5\ 10\ 9\ 7\ 3\ 6)\,.$$

Thus the order of the perfect shuffle of a deck of 10 cards is 10.

The perfect shuffle of a deck with 8 cards is the permutation

$$\begin{pmatrix} 1 & 2 & 3 & 4 & 5 & 6 & 7 & 8 \\ 2 & 4 & 6 & 8 & 1 & 3 & 5 & 7 \end{pmatrix} = (1\ 2\ 4\ 8\ 7\ 5)\ (3\ 6)\,.$$

The order of the perfect shuffle of a deck of eight cards is 6.

It is an interesting project to investigate the order of the perfect shuffle of decks of different sizes, as well as the decomposition of the perfect shuffle into a product of disjoint cycles. (We cannot help but be curious about the order of the perfect shuffle of a standard deck of 52 cards.)[6]

---

[6] A sophisticated analysis of the mathematics of card shuffling is carried out in D. Aldous and P. Diaconis, "Shuffling cards and stopping times," *Amer. Math. Monthly*, **93** (1986), no. 5, 333–348.

Here is (the outline of) an algorithm for writing a permutation $\pi \in S_n$ in cycle notation. Let a_1 be the first number $(1 \le a_1 \le n)$ which is not fixed by π. Write

$$\begin{aligned} a_2 &= \pi(a_1) \\ a_3 &= \pi(a_2) = \pi(\pi(a_1)) \\ a_4 &= \pi(a_3) = \pi(\pi(\pi(a_1))), \end{aligned}$$

and so forth. The numbers

$$a_1, a_2, \dots$$

cannot be all distinct since each is in $\{1, 2, \dots, n\}$. It follows that there is a number k such that $a_1, a_2, \dots, a_k$ are all distinct, and $\pi(a_k) = a_1$. (Exercise 1.5.13). The permutation π permutes the numbers $\{a_1, a_2, \dots, a_k\}$ among themselves, and the remaining numbers

$$\{1, 2, \dots, n\} \setminus \{a_1, a_2, \dots, a_k\}$$

among themselves, and the restriction of π to $\{a_1, a_2, \dots, a_k\}$ is the cycle $(a_1, a_2, \dots, a_k)$ (Exercise 1.5.12). If π fixes all numbers in $\{1, 2, \dots, n\} \setminus \{a_1, a_2, \dots, a_k\}$, then

$$\pi = (a_1, a_2, \dots, a_k).$$

Otherwise, consider the first number $b_1 \notin \{a_1, a_2, \dots, a_k\}$ that is not fixed by π. Write

$$\begin{aligned} b_2 &= \pi(b_1) \\ b_3 &= \pi(b_2) = \pi(\pi(b_1)) \\ b_4 &= \pi(b_3) = \pi(\pi(\pi(b_1))), \end{aligned}$$

and so forth; as before, there is an integer l such that $b_1, \dots, b_l$ are all distinct and $\pi(b_l) = b_1$. Now π permutes the numbers

$$\{a_1, a_2, \dots, a_k\} \cup \{b_1, \dots, b_l\}$$

among themselves, and the remaining numbers

$$\{1, 2, \dots, n\} \setminus (\{a_1, a_2, \dots, a_k\} \cup \{b_1, \dots, b_l\})$$

among themselves; furthermore, the restriction of π to

$$\{a_1, a_2, \dots, a_k\} \cup \{b_1, \dots, b_l\}$$

is the product of disjoint cycles

$$(a_1, a_2, \dots, a_k)(b_1, \dots, b_l).$$

Continue in this way until π has been written as a product of disjoint cycles.

Let me show you how to express the idea of the algorithm a little more formally and also more concisely, using mathematical induction.[7] In the preceding explanation, the phrase "continue in this way" is a signal that to formalize the argument it is necessary to use induction.

Because disjoint cycles π_1 and π_2 commute ($\pi_1\pi_2 = \pi_2\pi_1$, Exercise 1.5.12), uniqueness in the following statement means uniqueness up to order; the factors are unique, and the order in which the factors are written is irrelevant. Also note that $(a_1, a_2, \dots, a_k)$ is the same cyclic permutation as $(a_2, \dots, a_k, a_1)$, and there is no preferred first entry in the cycle notation. Finally, in order not to have to make an exception for the identity element e, we regard e as the product of the empty collection of cycles.

Theorem 1.5.3. *Every permutation of a finite set can be written uniquely as a product of disjoint cycles.*

Proof. We prove by induction on the cardinality of a finite set X that every permutation in $\mathrm{Sym}(X)$ can be written uniquely as a product of disjoint cycles.[8] If $|X| = 1$, there is nothing to do, since the only permutation of X is the identity e. Suppose, therefore, that the result holds for all finite sets of cardinality less than $|X|$. Let π be a nonidentity permutation of X. Choose $x_0 \in X$ such that $\pi(x_0) \neq x_0$. Denote $x_1 = \pi(x_0)$, $x_2 = \pi(x_1)$, and so forth. Since $|X|$ is finite, there is a number k such that $x_0, x_1, \dots, x_k$ are all distinct and $\pi(x_k) = x_0$. See Exercise 1.5.13. The sets $X_1 = \{x_0, x_1, \dots, x_k\}$ and $X_2 = X \setminus X_1$ are each invariant under π; that is, $\pi(X_i) = X_i$ for $i = 1, 2$, and therefore π is the product of $\pi_1 = \pi_{|X_1}$ and $\pi_2 = \pi_{|X_2}$. See Exercise 1.5.12. But π_1 is the cycle $(x_0, x_1, \dots, x_k)$, and by the induction hypothesis π_2 is a product of disjoint cycles. Hence π is also a product of disjoint cycles.

The uniqueness statement follows from a variation on the same argument: The cycle containing x_0 is uniquely determined by the list $x_0, x_1, \dots$. Any expression of π as a product of disjoint cycles must contain this cycle. The product of the remaining cycles in the expression yields π_2; but by the induction hypothesis, the decomposition of π_2 as a product of disjoint cycles is unique. Hence, the cycle decomposition of π is unique. ■

---

[7] Mathematical induction is discussed in Appendix C.

[8] The cardinality of a finite set is just the number of elements in the set; we denote the cardinality of X by $|X|$.

Exercises 1.5

1.5.1. Work out the full multiplication table for the set of permutations of three objects.

1.5.2. Compare the multiplication table of S_3 with that for the set of symmetries of an equilateral triangular card. (See Figure 1.3.6 and compare Exercise 1.3.1.) Find an appropriate matching identification or matching of elements of S_3 with symmetries of the triangle that makes the two multiplication tables agree.

1.5.3. Work out the decomposition in disjoint cycles for the following:

(a) $\begin{pmatrix} 1 & 2 & 3 & 4 & 5 & 6 & 7 \\ 2 & 5 & 6 & 3 & 7 & 4 & 1 \end{pmatrix}$

(b) $(12)(1234)$

(c) $(12)(234)$

(d) $(12)(23)(34)$

(e) $(13)(1234)(13)$

(f) $(12)(13)(14)$

1.5.4. On the basis of your computations in Exercise 1.5.3, make some conjectures about patterns for certain products of 2-cycles, and for certain products of two-cycles and other cycles.

1.5.5. Show that any k-cycle $(a_1, \ldots, a_k)$ can be written as a product of $(k-1)$ 2-cycles. Conclude that any permutation can be written as a product of some number of 2-cycles. Hint: For the first part, look at your computations in Exercise 1.5.3 to discover the right pattern. Then do a proper proof by induction.

1.5.6. Explain how to compute the inverse of a permutation that is given in two-line notation. Compute the inverse of

$$\begin{pmatrix} 1 & 2 & 3 & 4 & 5 & 6 & 7 \\ 2 & 5 & 6 & 3 & 7 & 4 & 1 \end{pmatrix}.$$

1.5.7. Explain how to compute the inverse of a permutation that is given as a product of cycles (disjoint or not). One trick of problem solving is to simplify the problem by considering special cases. First you should consider the case of a single cycle, and it will probably be helpful to begin with a short cycle. A 2-cycle is its own inverse, so the first interesting case is that of a 3-cycle. Once you have figured out the inverse for a 3-cycle and a 4-cycle, you will probably be able to guess the general pattern. Now you can begin work on a product of several cycles.

1.5.8. Let σ_n denote the perfect shuffle of a deck of $2n$ cards. Regard σ_n as a bijective function of the set $\{1, 2, \dots, 2n\}$. Find a formula for $\sigma_n(j)$, when $1 \le j \le n$, and another formula for $\sigma_n(j)$, when $n + 1 \le j \le 2n$.

1.5.9. Explain why a cycle of length k has order k. Explain why the order of a product of disjoint cycles is the least common multiple of the lengths of the cycles. Use examples to clarify the phenomena for yourself and to illustrate your explanation.

1.5.10. Find the cycle decomposition for the perfect shuffle for decks of size 2, 4, 6, 12, 14, 16, 52. What is the order of each of these shuffles?

1.5.11. Find the inverse, in two–line notation, for the perfect shuffle for decks of size 2, 4, 6, 8, 10, 12, 14, 16. Can you find a rule describing the inverse of the perfect shuffle in general?

The following two exercises supply important details for the proof of the existence and uniqueness of the disjoint cycle decomposition for a permutation of a finite set:

1.5.12. Suppose X is the union of disjoint sets X_1 and X_2, $X = X_1 \cup X_2$ and $X_1 \cap X_2 = \emptyset$. Suppose X_1 and X_2 are invariant for a permutation $\pi \in \mathrm{Sym}(X)$. Write π_i for the permutation $\pi_{|X_i} \in \mathrm{Sym}(X_i)$ for $i = 1, 2$, and (noticing the abuse of notation) also write π_i for the permutation of X that is π_i on X_i and the identity on $X \setminus X_i$. Show that $\pi = \pi_1 \pi_2 = \pi_2 \pi_1$.

1.5.13.

(a) Let π be a nonidentity permutation in $\mathrm{Sym}(X)$, where X is a finite set. Let x_0 be some element of X that is not fixed by π. Denote $x_1 = \pi(x_0)$, $x_2 = \pi(x_1)$, and so forth. Show that there is a number k such that $x_0, x_1, \dots, x_k$ are all distinct and $\pi(x_k) = x_0$. Hint: Let k be the least integer such that $\pi(x_k) = x_{k+1} \in \{x_0, x_1, \dots, x_k\}$. Show that $\pi(x_k) = x_0$. To do this, show that the assumption $\pi(x_k) = x_l$ for some l, $1 \le l \le k$ leads to a contradiction.

(b) Show that $X_1 = \{x_0, x_1, \dots, x_k\}$ and $X_2 = X \setminus X_1$ are both invariant under π.

1.5.14. Show that the multiplication in S_n is noncommutative for all $n \ge 3$. Hint: Find a pair of 2–cycles that do not commute.

1.6. Divisibility in the Integers

So far we have found algebraic structures in some possibly unexpected places; we have seen that the set of symmetries of a geometric figure or the set of permutations of a collection of identical objects has an algebraic structure. We can do computations in these algebraic systems in order to answer natural (or unnatural) questions, for example, to find out the order of a perfect shuffle of a deck of cards.

In this section, we return to more familiar mathematical territory. We study the set of integers, probably most familiar algebraic system. The integers have two operations, addition and multiplication, but as you learned in elementary school, multiplication in the integers can be interpreted in terms of repeated addition: For integers a and n, with $n > 0$, we have $na = a + \cdots + a$ (n times), and $(-n)a = n(-a)$. Finally, $0a = 0$.

We denote the set of integers $\{0, \pm1, \pm2, \ldots\}$ by $\mathbb{Z}$ and the set of natural numbers $\{1, 2, 3, \ldots\}$ by $\mathbb{N}$.

The integers, with addition and multiplication, have the following properties, which we take to be known.

Proposition 1.6.1.

(a) *Addition on $\mathbb{Z}$ is commutative and associative.*

(b) *0 is an identity element for addition; that is, for all $a \in \mathbb{Z}$, $0 + a = a$.*

(c) *Every element a of $\mathbb{Z}$ has an additive inverse $-a$, satisfying $a + (-a) = 0$.*

(d) *Multiplication on $\mathbb{Z}$ is commutative and associative.*

(e) *1 is an identity element for multiplication; that is, for all $a \in \mathbb{Z}$, $1a = a$.*

(f) *The distributive law holds: For all $a, b, c \in \mathbb{Z}$,*

$$a(b + c) = ab + ac.$$

Although the multiplicative structure of the integers is subordinate to the additive structure, many of the most interesting properties of the integers have to do with *divisibility*, factorization, and prime numbers. Of course, these concepts are already familiar to you from school mathematics, so the emphasis in this section will be more on a systematic, logical development of the material, rather than on exploration of unknown territory. The main goal will be to demonstrate that every natural numbers has a unique factorization as a product of prime numbers; this is trickier than one might expect, the uniqueness being the difficult part. On the way, we will, of course, be practicing with logical argument, and we will have an introduction to computational issues: How do we actually compute some abstractly defined quantity?

Let's begin with a definition of divisibility. We say that an integer a *divides* an integer b (or that b is *divisible* by a) if there is an integer q (the quotient) such that $aq = b$; that is, b can be divided by a *without remainder.* We write $a|b$ for "a divides b."

Here are some elementary properties of divisibility:

Proposition 1.6.2. *Let a, b, c, u, and v denote integers.*

(a) *If $uv = 1$, then $u = v = 1$ or $u = v = -1$.*
(b) *If $a|b$ and $b|a$, then $a = \pm b$.*
(c) *Divisibility is transitive: If $a|b$ and $b|c$, then $a|c$.*
(d) *If $a|b$ and $a|c$, then a divides all integers that can be expressed in the form $sb + tc$, where s and t are integers.*

Proof. For part (a), note that neither u nor v can be zero. Suppose first that both are positive. We have $uv \geq \max\{u, v\} \geq 1$. If equality holds, then $u = v = 1$. In general, if $uv = 1$, then also $|u||v| = 1$, so $|u| = |v| = 1$. Thus both u and v are ± 1. Since their product is positive, both have the same sign.

For (b), let u, v be integers such that $b = ua$ and $a = vb$. Then $b = uvb$. Thus $0 = (uv - 1)b$. Now the product of nonzero integers is nonzero, so either $b = 0$ or $uv = 1$. In the former case, $a = b = 0$, and in the latter case, $v = \pm 1$ by part (a), so $a = \pm b$.

The proofs of parts (c) and (d) are left to the reader. ■

As you know, some natural numbers like 2 and 13 are special in that they cannot be expressed as a product of strictly smaller natural numbers; these numbers are called *prime.* Natural numbers that are not prime are called *composite*; they can be written as a product of prime numbers, for example, $42 = 2 \times 3 \times 7$.

Definition 1.6.3. A natural number is *prime* if it is greater than 1 and not divisible by any natural number other than 1 and itself.

To show formally that every natural number is a product of prime numbers, we have to use mathematical induction.

Proposition 1.6.4. *Any natural number other than* 1 *can be written as a product of prime numbers.*

Proof. We have to show that for all natural numbers $n \geq 2$, n can be written as a product of prime numbers. We prove this statement using the second form of mathematical induction. (See Appendix C.) The natural number 2 is a prime, so it is a product of primes (with only one factor). Now suppose that $n > 2$ and that for all natural numbers r satisfying $2 \leq r < n$, the number r is a product of primes. If n happens to be prime, then it is a product of primes. Otherwise, n can be written as a product $n = ab$, where $1 < a < n$ and $1 < b < n$ (by the definition of prime number). According to the induction hypothesis, each of a and b can be written as a product of prime numbers; therefore, $n = ab$ is also a product of prime numbers. ■

Remark 1.6.5. It is a usual convention in mathematics to consider 0 to be the sum of an *empty* collection of numbers and 1 to be the product of an *empty* collection of numbers. This convention saves a lot of circumlocution and argument by cases. So we will consider 1 to have a prime factorization as well; it is the product of an empty collection of primes.

A fundamental question is whether there exist infinitely many prime numbers or only finitely many. This question was considered and resolved by Greek mathematicians before 300 B.C. The solution is attributed to Euclid and appears as Proposition 20 of Book IX of his *Elements*:

Theorem 1.6.6. *There are infinitely many prime numbers.*

Proof. We show than for all natural numbers n, there exist at least n prime numbers. There exists at least one prime number, because 2 is prime. Let k be a natural number and suppose that there exist at least k prime numbers. We will show that there exist at least $k+1$ prime numbers. Let $\{p_1, p_2, \dots, p_k\}$ be a collection of k (distinct) prime numbers. Consider the natural number $M = p_1 p_2 \cdots p_k + 1$. M is not divisible by any of the primes $p_1, p_2, \dots, p_k$, but M is a product of prime numbers, according to the previous proposition. If p is any prime dividing M, then $\{p_1, p_2, \dots, p_k, p\}$ is a collection of $k + 1$ (distinct) prime numbers. ■

The fundamental fact about divisibility is the following familiar result (division with remainder): It is always possible to divide an integer a by any divisor $d \geq 1$, to get a quotient q and a remainder r, with the remainder being strictly smaller than the divisor.

I would like to emphasize how the proof of this fact is related to the algorithm for long division which you learned in school. How does this algorithm work (for a a positive number)? In case $a < d$, we just put $q = 0$ and $r = a$. If $a \geq d$, first we guess an approximation q_1 for the quotient such that $0 < q_1 d \leq a$, and compute $r_1 = a - q_1 d$, which is less than a. Now we have

$$\begin{aligned} a &= q_1 d + r_1, \\ 0 &\leq r_1 < a. \end{aligned}$$

If $r_1 < d$, we are finished; we put $q = q_1$ and $r = r_1$. Otherwise, we try to divide r_1 by d, guessing a q_2 such that $0 < q_2 d < r_1$, and compute $r_2 = r_1 - q_2 d$. Now we have

$$\begin{aligned} a &= q_1 d + r_1, \\ r_1 &= q_2 d + r_2, \\ 0 &\leq r_2 < r_1 < a. \end{aligned}$$

Combining the two equations, we get

$$\begin{aligned} a &= (q_1 + q_2) d + r_2, \\ 0 &\leq r_2 < r_1 < a. \end{aligned}$$

If $r_2 < d$, then we put $q = q_1 + q_2$ and $r = r_2$, and we are done. Otherwise, we continue by trying to divide r_2 by d. If the process continues to the nth stage, we obtain partial quotients $q_1, q_2, \dots, q_n$, and successive remainders $r_1, r_2, \dots, r_n$, such that

$$\begin{aligned} a &= (q_1 + q_2 + \cdots + q_n) d + r_n, \\ 0 &\leq r_n < \cdots < r_2 < r_1 < a. \end{aligned}$$

Eventually, it must be that $0 \leq r_n < d$, because the r_n are non-negative and strictly decreasing. (Can you justify this "obvious" conclusion by reference to some version of the principle of mathematical induction?) Then we stop, and put $q = q_1 + q_2 + \cdots + q_n$ and $r = r_n$.

The procedure for negative a is similar.

The description of this algorithm, made a little more formal by reference to the principle of mathematical induction, constitutes a proof of the existence part of the following assertion. In the following proof, I'm going to arrange the inductive argument in a more streamlined way, but the essential idea is the same.

Proposition 1.6.7. *Given integers a and d, with $d \geq 1$, there exist unique integers q and r such $a = qd + r$ and $0 \leq r < d$.*

Proof. First consider the case $a \geq 0$. If $a < d$, take $q = 0$ and $r = a$. Now suppose that $a \geq d$. Assume inductively that the existence assertion holds for all nonnegative integers that are strictly smaller than a. Then in particular it holds for $a - d$, so there exist integers q' and r such that $(a - d) = q'd + r$ and $0 \leq r < d$. Then $a = (q' + 1)d + r$, and we are done.

We deal with the case $a < 0$ by induction on $|a|$. If $-d < a < 0$, take $q = -1$ and $r = a + d$. Suppose that $a \leq -d$, and assume inductively that the existence assertion holds for all nonpositive integers whose absolute values are strictly smaller than $|a|$. Then in particular it holds for $a + d$, so there exist integers q' and r such that $(a + d) = q'd + r$ and $0 \leq r < d$. Then $a = (q' - 1)d + r$.

So far, we have shown the existence of q and r with the desired properties. For uniqueness, suppose that $a = qd + r$, and $a = q'd + r'$, with $0 \leq r, r' < d$. It follows that $r - r' = (q' - q)d$, so $r - r'$ is divisible by d. But $|r - r'| \leq \max\{r, r'\} < d$, so the only possibility is $r - r' = 0$. But then $(q' - q)d = 0$, so $q' - q = 0$. ■

We have shown the existence of a prime factorization of any natural number, but we have not shown that the prime factorization is unique. This is a more subtle issue, which is addressed in the following discussion. The key idea is that the greatest common divisor of two integers can be computed *without knowing their prime factorizations.*

Definition 1.6.8. A natural number α is the *greatest common divisor* of nonzero integers m and n if

(a) α divides m and n and
(b) whenever $\beta \in \mathbb{N}$ divides m and n, then β also divides α.

Notice that if the greatest common divisor is unique, if it exists at all, by Proposition 1.6.2. Next, we show that the greatest common divisor of two nonzero integers m and n does indeed exist, can be found by an algorithm involving repeated use of division with remainder, and is an element of the set

$$I(m, n) = \{am + bn : a, b \in \mathbb{Z}\}.$$

This set has several important properties, which we record in the following proposition.

Proposition 1.6.9. *For integers* n *and* m, *let*

$$I(m, n) = \{am + bn : a, b \in \mathbb{Z}\}.$$

(a) *For* $x, y \in I(m, n)$, $x + y \in I(m, n)$ *and* $-x \in I(m, n)$.
(b) *For all* $x \in \mathbb{Z}$, $xI(m, n) \subseteq I(m, n)$.
(c) *If* $b \in \mathbb{Z}$ *divides* m *and* n, *then* b *divides all elements of* $I(m, n)$.

Proof. Exercise 1.6.2. ■

Now we proceed to the algorithm for the greatest common divisor of nonzero integers m and n. Suppose without loss of generality that $|m| \geq |n|$. Define sequences $|n| > n_1 > n_2 \cdots \geq 0$ and $q_1, q_2, \ldots$ by induction, as follows. Define q_1 and n_1 as the quotient and remainder upon dividing m by n:

$$m = q_1 n + n_1 \quad \text{and} \quad 0 \leq n_1 < |n|.$$

If $n_1 > 0$, define q_2 and n_2 as the quotient and remainder upon dividing n by n_1:

$$n = q_2 n_1 + n_2 \quad \text{and} \quad 0 \leq n_2 < n_1.$$

In general, if $n_1, \ldots, n_{k-1}$ and $q_1, \ldots, q_{k-1}$ have been defined and $n_{k-1} > 0$, then define q_k and n_k as the quotient and remainder upon dividing n_{k-2} by n_{k-1}:

$$n_{k-2} = q_k n_{k-1} + n_k \quad \text{and} \quad 0 \leq n_k < n_{k-1}.$$

This process must stop after no more than n steps with some remainder $n_{r+1} = 0$. Then we have the following system of relations:

$$\begin{aligned}
m &= q_1 n + n_1 \\
n &= q_2 n_1 + n_2 \\
&\ldots \\
n_{k-2} &= q_k n_{k-1} + n_k \\
&\ldots \\
n_{r-1} &= q_{r+1} n_r.
\end{aligned}$$

Proposition 1.6.10. *The natural number* n_r *is the greatest common divisor of* m *and* n, *and furthermore* $n_r \in I(m, n)$.

Proof. Write $m = n_{-1}$ and $n = n_0$.

First, we show by induction on k that all the n_k $(-1 \le k \le r)$ are elements of $I(m, n)$. This is clear for n_{-1} and n_0. Now if $n_j = s_j m + t_j n$ for $-1 \le j \le k$, then we have

$$\begin{aligned} n_{k+1} &= n_{k-1} - q_{k+1} n_k \\ &= (s_{k-1} - q_{k+1} s_k) m + (t_{k-1} - q_{k+1} t_k) n. \end{aligned}$$

Thus $n_{k+1} \in I(m, n)$ as well. It follows by induction that $n_k \in I(m, n)$ for all natural numbers k.

Next, we show by induction that all the n_j are divisible by n_r; the induction goes "from the top down" rather than "from the bottom up." Clearly, n_r and n_{r-1} are divisible by n_r. Fix some k with $1 \le k \le r$ and suppose that $n_r, n_{r-1}, \dots, n_{r-k}$ are all divisible by n_r. Then we have

$$n_{r-(k+1)} = q_{r-(k-1)} n_{r-k} + n_{r-(k-1)}.$$

Therefore, $n_{r-(k+1)}$ is divisible by n_r as well. It follows by induction that n_{r-k} is divisible by n_r for all k $(0 \le k \le r+1)$. That is, n_j is divisible by n_r for all j $(-1 \le j \le r)$. In particular, n_r divides both $m = n_{-1}$ and $n = n_0$.

Finally, since $n_r \in I(m, n)$, it is divisible by any common divisor of m and n, by Proposition 1.6.9 (c). On the other hand, n_r is a common divisor of m and n. Hence n_r is the greatest common divisor of m and n. ■

We denote the greatest common divisor by g.c.d.(m, n).

Example 1.6.11. Find the greatest common divisor of 1734282 and 452376. Successive divisions with remainder give

$$\begin{aligned} 1734282 &= 3 \times 452376 + 377154 \\ 452376 &= 377154 + 75222 \\ 377154 &= 5 \times 75222 + 1044 \\ 75222 &= 72 \times 1044 + 54 \\ 1044 &= 19 \times 54 + 18 \\ 54 &= 3 \times 18. \end{aligned}$$

Thus $18 =$ g.c.d.$(1734282, 452376)$.

We can find the coefficients s_i, t_i such that $n_i = s_i m + t_i n$ using the recursion relation

$$s_i = s_{i-2} - q_i s_{i-1}, \quad t_i = t_{i-2} - q_i t_{i-1},$$

which was obtained in the proof of the proposition, together with

$$s_{-1} = 1, \quad t_{-1} = 0, \quad s_0 = 0, \quad t_0 = 1.$$

Thus,

$$\begin{aligned} s_1 &= 1, & t_1 &= -3 \\ s_2 &= -1, & t_2 &= 4 \\ s_3 &= 6, & t_3 &= -23 \\ s_4 &= -43, & t_4 &= 1660 \\ s_5 &= 8233, & t_5 &= -31563. \end{aligned}$$

In particular,

$$18 = \text{g.c.d.}(1734282, 452376) = n_5 = 8233 \times 1734282 - 31563 \times 452376.$$

Definition 1.6.12. Nonzero integers m and n are *relatively prime* if g.c.d.$(m, n) = 1$.

Proposition 1.6.13. *Two integers m and n are relatively prime if, and only if, $1 \in I(m, n)$.*

Proof. Exercise 1.6.8. ■

Example 1.6.14. The integers 21 and 16 are relatively prime and $1 = -3 \times 21 + 4 \times 16$.

Proposition 1.6.15. *If p is a prime number and a is any nonzero integer, then either p divides a or p and a are relatively prime.*

Proof. Exercise 1.6.9. ■

From here, it is a only a short way to the proof of uniqueness of prime factorizations. The key observation is the following:

Proposition 1.6.16. *Let p be a prime number, and a and b nonzero integers. If p divides ab, then p divides a or p divides b.*

Proof. If p does not divide a, then a and p are relatively prime, so $1 = \alpha a + \beta p$ for some integers α and β, by Proposition 1.6.13. Multiplying by b gives $b = \alpha ab + \beta pb$, which shows that b is divisible by p. ■

Corollary 1.6.17. *Suppose that a prime number p divides a product $a_1 a_2 \dots a_r$ of nonzero integers. Then p divides one of the factors.*

Proof. Exercise 1.6.12. ■

Theorem 1.6.18. *The prime factorization of a natural number is unique.*

Proof. We have to show that for all natural numbers n, if n has factorizations

$$n = q_1 q_2 \dots q_r,$$
$$n = p_1 p_2 \dots p_s,$$

where the q_i's and p_j's are prime and $q_1 \leq q_2 \leq \cdots \leq q_r$ and $p_1 \leq p_2 \leq \cdots \leq p_s$, then $r = s$ and $q_i = p_i$ for all i. We do this by induction on n. First check the case $n = 1$; 1 cannot be written as the product of any nonempty collection of prime numbers. So consider a natural number $n \geq 2$ and assume inductively that the assertion of unique factorization holds for all natural numbers less than n. Consider two factorizations of n as before, and assume without loss of generality that $q_1 \leq p_1$. Since q_1 divides $n = p_1 p_2 \dots p_s$, it follows from Proposition 1.6.17 that q_1 divides, and hence is equal to, one of the p_i. Since also $q_1 \leq p_1 \leq p_k$ for all k, it follows that $p_1 = q_1$. Now dividing both sides by q_1, we get

$$n/q_1 = q_2 \dots q_r,$$
$$n/q_1 = p_2 \dots p_s.$$

(Note that n/q_1 could be 1 and one or both of $r - 1$ and $s - 1$ could be 0.) Since $n/q_1 < n_1$, it follows from the induction hypothesis that $r = s$ and $q_i = p_i$ for all $i \geq 2$. ■

How do we actually compute the prime factorization of a natural number? The conceptually clear but computationally difficult method that you learned in school for factoring a natural number n is to test all natural numbers no larger than $\sqrt{n}$ to see if any divides n. If no factor is found, then n must be prime. If a factor a is found, then we can write $n = a \times (n/a)$ and proceed to search for factors of a and n/a. We continue this procedure until only prime factors appear. Unfortunately, this procedure is very inefficient. Better methods are known, but no truly efficient methods are available for factoring very large natural numbers.

Exercises 1.6

1.6.1. Complete the following sketch of a proof of Proposition 1.6.7 using the well–ordering principle.

(a) If $a \neq 0$, consider the set S of nonnegative integers that can be written in the form $a - sd$, where s is an integer. Show that S is nonempty.

(b) By the well–ordering principle, S has a least element, which we write as $r = a - qd$. Then we have $a = qd + r$. Show that that $r < d$.

1.6.2. Prove Proposition 1.6.9. Hint: For part (b), you have to show that if $y \in I(m, n)$, then $xy \in I(m.n)$. For part (c), you have to show that if $y \in I(m, n)$, then b divides y.

1.6.3. Suppose that a natural number p has the property that for all nonzero integers a and b, if p divides the product ab, then p divides a or p divides b. Show that p is prime. This is the converse of Proposition 1.6.16.

1.6.4. For each of the following pairs of numbers m, n, compute $\text{g.c.d.}(m, n)$ and write $\text{g.c.d.}(m, n)$ explicitly as an integer linear combination of m and n.

(a) $m = 60$ and $n = 8$

(b) $m = 32242$ and $n = 42$

1.6.5. Show that for nonzero integers m and n, $\text{g.c.d.}(m, n) = \text{g.c.d.}(|m|, |n|)$.

1.6.6. Show that for nonzero integers m and n, $\text{g.c.d.}(m, n)$ is the largest natural number dividing m and n.

1.6.7. Show that for nonzero integers m and n, $\text{g.c.d.}(m, n)$ is the smallest element of $I(m, n) \cap \mathbb{N}$. [The existence of a smallest element of $I(m, n) \cap \mathbb{N}$ follows from the well–ordering principle for the natural numbers; see Appendix C.]

1.6.8. Show that two nonzero integers m and n are relatively prime if, and only if, $1 \in I(m, n)$.

1.6.9. Show that if p is a prime number and a is any nonzero integer, then either p divides a or p and a are relatively prime.

1.6.10. Suppose that a and b are relatively prime integers and that x is an integer. Show that if a divides the product bx, then a divides x. Hint: Use the existence of s, t such that $sa + tb = 1$.

1.6.11. Suppose that a and b are relatively prime integers and that x is an integer. Show that if a divides x and b divides x, then ab divides x.

1.6.12. Show that if a prime number p divides a product $a_1 a_2 \ldots a_r$ of nonzero integers, then p divides one of the factors.

1.6.13.

(a) Write a program in your favorite programming language to compute the greatest common divisor of two nonzero integers, using the approach of repeated division with remainders. Get your program to explicitly give the greatest common divisor as an integer linear combination of the given nonzero integers.

(b) Another method of finding the greatest common divisor would be to compute the prime factorizations of the two integers and then to take the largest collection of prime factors common to the two factorizations. This method is often taught in school mathematics. How do the two methods compare in computational efficiency?

The next three exercises explore the idea of the greatest common divisor of *several* integers, $n_1, n_2, \ldots, n_k$.

1.6.14.

(a) Make a reasonable definition of $\text{g.c.d}(n_1, n_2, \ldots, n_k)$.

(b) Show that $\text{g.c.d}(n_1, n_2, \ldots, n_k) = \text{g.c.d}(n_1, \text{g.c.d}(n_2, \ldots, n_k))$.

1.6.15.

(a) Let $I = I(n_1, n_2, \ldots, n_k) =$

$$\{m_1 n_1 + m_2 n_2 + \ldots m_k n_k : m_1, \ldots, m_k \in \mathbb{Z}\}.$$

Show that if $x, y \in I$, then $x + y \in I$ and $-x \in I$. Show that if $x \in \mathbb{Z}$ and $a \in I$, then $xa \in I$.

(b) Show that $\text{g.c.d}(n_1, n_2, \ldots, n_k)$ is the smallest element of $I \cap \mathbb{N}$.

1.6.16.

(a) Develop an algorithm to compute $\text{g.c.d}(n_1, n_2, \ldots, n_k)$.

(b) Develop a computer program to compute the greatest common divisor of any finite collection of nonzero integers.

1.7. Modular Arithmetic

We are all familiar with the arithmetic appropriate to the hours of a clock: If it is now 9 o'clock, then in 7 hours it will be 4 o'clock. Thus in clock arithmetic, $9 + 7 = 4$. The clock number 12 is the identity for clock addition: Whatever time the clock now shows, in 12 hours it will show the same time. Multiplication of hours is not quite so familiar an operation, but it does make sense: If it is now 12 o'clock, then after 5 periods of seven hours it will be 11 o'clock, so $5 \times 7 = 11$ in clock arithmetic. Clock arithmetic is an arithmetic system with only 12 numbers, in which all the usual laws of arithmetic hold, except that division is not generally possible.

The goal of this section is to examine clock arithmetic for a clock face with n hours, for any natural number n. (We don't do this because we want to build crazy clocks with strange numbers of hours, but because the resulting algebraic systems are important in algebra and its applications.)

Fix a natural number $n > 1$. Think of a clock face with n hours (labeled $0, 1, 2, \dots, n-1$) and of circumference n. Imagine taking a number line, with the integer points marked, and wrapping it around the circumference of the clock, with 0 on the number line coinciding with 0 on the clock face. Then the numbers

$$\dots, -3n, -2n, -n, 0, n, 2n, 3n, \dots$$

on the number line all overlay 0 on the clock face. The numbers

$$\dots, -3n+1, -2n+1, -n+1, 1, n+1, 2n+1, 3n+1, \dots$$

on the number line all overlay 1 on the clock face. In general, for $0 \le k \le n-1$, the numbers

$$\dots, -3n+k, -2n+k, -n+k, k, n+k, 2n+k, 3n+k, \dots$$

on the number line all overlay k on the clock face.

When do two integers on the number line land on the same spot on the clock face? This happens precisely when the distance between the two numbers on the number line is some multiple of n, so that the interval between the two numbers on the number line wraps some integral number of times around the clock face. Since the distance between two numbers a and b on the number line is $|a - b|$, this suggests the following definition:

Definition 1.7.1. Given integers a and b, and a natural number n, we say that "a is congruent to b modulo n" and we write $a \equiv b \pmod{n}$ if $a - b$ is divisible by n.

The relation $a \equiv b \pmod{n}$ has the following properties:

Lemma 1.7.2.

(a) *For all* $a \in \mathbb{Z}$, $a \equiv a \pmod{n}$.

(b) *For all* $a, b \in \mathbb{Z}$, $a \equiv b \pmod{n}$ *if, and only if,* $b \equiv a \pmod{n}$.

(c) *For all* $a, b, c \in \mathbb{Z}$, *if* $a \equiv b \pmod{n}$ *and* $b \equiv c \pmod{n}$, *then* $a \equiv c \pmod{n}$.

Proof. For (a), $a - a = 0$ is divisible by n. For (b), $a - b$ is divisible by n if, and only if, $b - a$ is divisible by n. Finally, if $a - b$ and $b - c$ are both divisible by n, then also $a - c = (a - b) + (b - c)$ is divisible by n. ■

For each integer a, write

$$[a] = \{b \in \mathbb{Z} : a \equiv b \pmod{n}\} = \{a + kn : k \in \mathbb{Z}\}.$$

Note that this is just the set of all integer points that land at the same place as a when the number line is wrapped around the clock face. The set $[a]$ is called the *residue class* or *congruence class* of a modulo n.

Also denote by $\text{rem}_n(a)$ the unique number r such that $0 \le r < n$ and $a - r$ is divisible by n. (Proposition 1.6.7). Thus $\text{rem}_n(a)$ is the unique element of $[a]$ that lies in the interval $\{0, 1, \dots, n-1\}$. Equivalently, $\text{rem}_n(a)$ is the label on the circumference of the clock face at the point where a falls when the number line is wrapped around the clock face.

Lemma 1.7.3. *For* $a, b \in Z$, *the following are equivalent:*

(a) $a \equiv b \pmod{n}$.

(b) $[a] = [b]$.

(c) $\text{rem}_n(a) = \text{rem}_n(b)$.

(d) $[a] \cap [b] \neq \emptyset$.

Proof. Suppose that $a \equiv b \pmod{n}$. For any $c \in \mathbb{Z}$, if $c \equiv a \pmod{n}$, then $c \equiv b \pmod{n}$, by the previous lemma, part (c). This shows that $[a] \subseteq [b]$. Likewise, if $c \equiv b \pmod{n}$, then $c \equiv a \pmod{n}$, so $[b] \subseteq [a]$. Thus $[a] = [b]$. This shows that (a) implies (b).

Since for all integers x, $\text{rem}_n(x)$ is characterized as the unique element of $[x] \cap \{0, 1, \dots, n-1\}$, we have (b) implies (c), and also (c) implies (d).

Finally, if $[a] \cap [b] \neq \emptyset$, let $c \in [a] \cap [b]$. Then $a \equiv c \pmod{n}$ and $c \equiv b \pmod{n}$, so $a \equiv b \pmod{n}$. This shows that (c) implies (a). ■

Corollary 1.7.4. *There exist exactly n distinct residue classes modulo n, namely* $[0], [1], \ldots, [n-1]$. *These classes are mutually disjoint.*

Congruence respects addition and multiplication, in the following sense:

Lemma 1.7.5. *Let a, a', b, b' be integers with $a \equiv a' \pmod{n}$ and $b \equiv b' \pmod{n}$. Then $a + b \equiv a' + b' \pmod{n}$ and $ab \equiv a'b' \pmod{n}$.*

Proof. By hypothesis, $a - a'$ and $b - b'$ are divisible by n. Hence

$$(a + b) - (a' + b') = (a - a') + (b - b')$$

is divisible by n, and

$$\begin{aligned} ab - a'b' &= (ab - a'b) + (a'b - a'b') \\ &= (a - a')b + a'(b - b') \end{aligned}$$

is divisible by n. ■

We denote by $\mathbb{Z}_n$ the set of residue classes modulo n. The set $\mathbb{Z}_n$ has a natural algebraic structure which we now describe. Let A and B be elements of $\mathbb{Z}_n$, and let $a \in A$ and $b \in B$; we say that a is a *representative* of the residue class A, and b a representative of the residue class B.

The class $[a+b]$ and the class $[ab]$ are independent of the choice of representatives. For if a' is another representative of A and b' another representative of B, then $a \equiv a' \pmod{n}$ and $b \equiv b' \pmod{n}$; therefore $a + b \equiv a' + b' \pmod{n}$ and $ab \equiv a'b' \pmod{n}$ according to Lemma 1.7.5. Thus $[a + b] = [a' + b']$ and $[ab] = [a'b']$. This means that it makes sense to define $A + B = [a + b]$ and $AB = [ab]$. Another way to write these definitions is

$$[a] + [b] = [a + b], \qquad [a][b] = [ab]. \tag{1.7.1}$$

Example 1.7.6. Let us look at another example in which we *cannot* define operations on classes of numbers in the same way (in order to see what the issue is in the preceding discussion). Let N be the set of all negative integers and P the set of all nonnegative integers. Every integer belongs to exactly one of these two classes. Write $s(a)$ for the class of a [i.e., $s(a) = N$ if $a < 0$ and $s(a) = P$ if $a \geq 0$]. If $n \in N$ and $p \in P$, then, depending on the choice of n and p, the sum $n + p$ can be in either of N or P; that is, the sum of a positive number and a negative number can be either positive or negative. So it does *not* make sense to define $N + P = s(n + p)$.

Once we have cleared the hurdle of defining sensible operations on $\mathbb{Z}_n$, it is easy to check that these operations satisfy most of the usual rules of arithmetic, as recorded in the following proposition.

Each of the assertions in the proposition follows from the corresponding property of $\mathbb{Z}$, together with Equation (1.7.1). For example, the commutativity of multiplication on $\mathbb{Z}_n$ is shown as follows. For $a, b \in \mathbb{Z}$,

$$[a][b] = [ab] = [ba] = [b][a].$$

The proof of associativity of multiplication proceeds similarly; let $a, b, c \in \mathbb{Z}$. Then

$$([a][b])[c] = [ab][c] = [(ab)c] = [a(bc)] = [a][bc] = [a]([b][c]).$$

Checking the details for the other assertions is left to the reader.

Proposition 1.7.7.

(a) *Addition on $\mathbb{Z}_n$ is commutative and associative; that is, for all $[a], [b], [c] \in \mathbb{Z}_n$,*

$$[a] + [b] = [b] + [a],$$

and

$$([a] + [b]) + [c] = [a] + ([b] + [c]).$$

(b) *$[0]$ is an identity element for addition; that is, for all $[a] \in \mathbb{Z}_n$,*

$$[0] + [a] = [a].$$

(c) *Every element $[a]$ of $\mathbb{Z}_n$ has an additive inverse $[-a]$, satisfying*

$$[a] + [-a] = [0].$$

(d) *Multiplication on $\mathbb{Z}_n$ is commutative and associative; that is, for all $[a], [b], [c] \in \mathbb{Z}_n$,*

$$[a][b] = [b][a],$$

and

$$([a][b])[c] = [a]([b][c]).$$

(e) *$[1]$ is an identity for multiplication; that is, for all $[a] \in \mathbb{Z}_n$,*

$$[1][a] = [a].$$

(f) *The distributive law hold; that is, for all $[a], [b], [c] \in \mathbb{Z}_n$,*

$$[a]([b] + [c]) = [a][b] + [a][c].$$

Multiplication in $\mathbb{Z}_n$ has features that you might not expect. On the one hand, nonzero elements can sometimes have a zero product. For example, in $\mathbb{Z}_6$, $[4][3] = [12] = [0]$. We call a nonzero element $[a]$

a *zero divisor* if there exists a nonzero element $[b]$ such that $[a][b] = [0]$. Thus, in $\mathbb{Z}_6$, $[4]$ and $[3]$ are zero divisors.

On the other hand, many elements have *multiplicative inverses*; an element $[a]$ is said to *have a multiplicative inverse* or to *be invertible* if there exists an element $[b]$ such that $[a][b] = [1]$. For example, in $\mathbb{Z}_{14}$, $[1][1] = [1]$, $[3][5] = [15] = [1]$, $[9][11] = [-5][-3] = [15] = [1]$, and $[13][13] = [-1][-1] = [1]$. Thus, in $\mathbb{Z}_{14}$, $[1], [3], [5], [9], [11]$, and $[13]$ have multiplicative inverses. You can check that the remaining nonzero elements $[2], [4], [6], [7], [8], [10]$, and $[12]$ are zero divisors. Thus in $\mathbb{Z}_{14}$, every nonzero element is either a zero divisor or is invertible, and there are just about as many zero divisors as invertible elements.

In $\mathbb{Z}_7$, on the other hand, *every* nonzero element is invertible: $[1][1] = [1]$, $[2][4] = [8] = [1]$, $[3][5] = [15] = [1]$, and $[6][6] = [-1][-1] = [1]$.

You should compare these phenomena with the behavior of multiplication in $\mathbb{Z}$, where no nonzero element is a zero divisor, and only ± 1 are invertible. In the Exercises for this section, you are asked to investigate further the zero divisors and invertible elements in $\mathbb{Z}_n$, and solutions to congruences of the form $ax \equiv b \pmod{n}$.

Let's look at a few examples of computations with congruences, or, equivalently, computations in $\mathbb{Z}_n$. The main principle to remember is that the congruence $a \equiv b \pmod{n}$ is equivalent to $[a] = [b]$ in $\mathbb{Z}_n$.

Example 1.7.8.

(a) Compute the congruence class modulo 5 of 4^{237}. This is easy because $4 \equiv -1 \pmod{5}$, so $4^{237} \equiv (-1)^{237} \equiv -1 \equiv 4 \pmod{5}$. Thus in $\mathbb{Z}_5$, $[4^{237}] = [4]$.

(b) Compute the congruence class modulo 9 of 4^{237}. As a strategy, let's compute a few powers of 4 modulo 9. We have $4^2 \equiv 7 \pmod{9}$ and $4^3 \equiv 1 \pmod{9}$. It follows that in $\mathbb{Z}_9$, $[4^{3k}] = [4^3]^k = [1]^k = [1]$ for all natural numbers k; likewise, $[4^{3k+1}] = [4]^{3k}[4] = [4]$, and $[4^{3k+2}] = [4]^{3k}[4]^2 = [7]$. So to compute 4^{237} modulo 9, we only have to find the conjugacy class of 237 modulo 3; since 237 is divisible by 3, we have $[4^{237}] = [1]$ in $\mathbb{Z}_9$.

(c) Show that every integer a satisfies $a^7 \equiv a \pmod{7}$. The assertion is equivalent to $[a]^7 = [a]$ for all $[a] \in \mathbb{Z}_7$. Now we only have to check this for each of the 7 elements in $\mathbb{Z}_7$, which is straightforward. This is a special case of Fermat's little theorem; see Proposition 1.9.10.

(d) You have to be careful about canceling in congruences. If $[a][b] = [a][c]$ in $\mathbb{Z}_n$, it is not necessarily true that $[b] = [c]$. For example, in $\mathbb{Z}_{12}$, $[4][2] = [4][5] = [8]$.

We now introduce the problem of simultaneous solution of congruences: given positive integers a and b, and integers α and β, when can we find an integer x such that $x \equiv \alpha \pmod{a}$ and $x \equiv \beta \pmod{b}$? For example, can we find an integer x that is congruent to 3 modulo 4 and also congruent to 12 modulo 15? (Try it!) The famous Chinese remainder theorem says that this is always possible if a and b are relatively prime.

Proposition 1.7.9. *(Chinese remainder theorem). Suppose a and b are relatively prime natural numbers, and α and β are integers. There exists an integer x such that $x \equiv \alpha \pmod{a}$ and $x \equiv \beta \pmod{b}$.*

Proof. Exercise 1.7.15 guides you to a proof of this result. (If you don't do it now, that's OK, because the theorem will be discussed again later. If you want to try it, I suggest that you begin by looking at examples, so you can see what is involved in finding a solution to two simultaneous congruences.) ■

What are the objects $\mathbb{Z}_n$ actually good for? First, they are devices for studying the integers. Congruence modulo n is a fundamental relation in the integers, and any statement concerning congruence modulo n is equivalent to a statement about $\mathbb{Z}_n$. Sometimes it is easier, or it provides better insight, to study a statement about congruence in the integers in terms of $\mathbb{Z}_n$.

Second, the objects $\mathbb{Z}_n$ are fundamental building blocks in several general algebraic theories, as we shall see later. For example, all finite fields are constructed using $\mathbb{Z}_p$ for some prime p.

Third, although the algebraic systems $\mathbb{Z}_n$ were first studied in the nineteenth century without any view toward practical applications, simply because they had a natural and necessary role to play in the development of algebra, they are now absolutely fundamental in modern digital engineering. Algorithms for digital communication, for error detection and correction, for cryptography, and for digital signal processing all employ $\mathbb{Z}_n$.

Exercises 1.7

Exercises 1.7.1 through 1.7.3 ask you to prove parts of Proposition 1.7.7.

1.7.1. Prove that addition in $\mathbb{Z}_n$ is commutative and associative.

1.7.2. Prove that $[0]$ is an identity element for addition in $\mathbb{Z}_n$, and that $[a] + [-a] = [0]$ for all $[a] \in \mathbb{Z}_n$.

1.7.3. Prove that multiplication in $\mathbb{Z}_n$ is commutative and associative, and that $[1]$ is an identity element for multiplication.

1.7.4. Compute the congruence class modulo 12 of 4^{237}.

Exercises 1.7.5 through 1.7.10 constitute an experimental investigation of zero divisors and invertible elements in $\mathbb{Z}_n$.

1.7.5. Can an element of $\mathbb{Z}_n$ be both invertible and a zero divisor?

1.7.6. If an element of $\mathbb{Z}_n$ is invertible, is its multiplicative inverse unique? That is, if $[a]$ is invertible, can there be two distinct elements $[b]$ and $[c]$ such that $[a][b] = [1]$ and $[a][c] = [1]$?

1.7.7. If a nonzero element $[a]$ of $\mathbb{Z}_n$ is a zero divisor, can there be two distinct nonzero elements $[b]$ and $[c]$ such that $[a][b] = [0]$ and $[a][c] = [0]$?

1.7.8. Write out multiplication tables for $\mathbb{Z}_n$ for $n \leq 10$.

1.7.9. Using your multiplication tables from the previous exercise, catalog the invertible elements and the zero divisors in $\mathbb{Z}_n$ for $n \leq 10$. Is it true (for $n \leq 10$) that every nonzero element in $\mathbb{Z}_n$ is either invertible or a zero divisor?

1.7.10. Based on your data for $\mathbb{Z}_n$ with $n \leq 10$, make a conjecture (guess) about which elements in Z_n are invertible and which are zero divisors. Does your conjecture imply that every nonzero element is either invertible or a zero divisor?

The next three exercises provide a guide to a more analytical approach to invertibility and zero divisors in $\mathbb{Z}_n$.

1.7.11. Suppose a is relatively prime to n. Then there exist integers s and t such that $as + nt = 1$. What does this say about the invertibility of $[a]$ in $\mathbb{Z}_n$?

1.7.12. Suppose a is not relatively prime to n. Then there do not exist integers s and t such that $as + nt = 1$. What does this say about the invertibility of $[a]$ in $\mathbb{Z}_n$?

1.7.13. Suppose that $[a]$ is not invertible in $\mathbb{Z}_n$. Consider the left multiplication map $L_{[a]} : \mathbb{Z}_n \to \mathbb{Z}_n$ defined by $L_{[a]}([b]) = [a][b] = [ab]$. Since $[a]$ is not invertible, $[1]$ is not in the range of $L_{[a]}$, so $L_{[a]}$ is not surjective. Conclude that $L_{[a]}$ is not injective, and use this to show that there exists $[b] \neq [0]$ such that $[a][b] = [0]$ in $\mathbb{Z}_n$.

1.7.14. Suppose a is relatively prime to n.

(a) Show that for all $b \in \mathbb{Z}$, the congruence $ax \equiv b \pmod{n}$ has a solution.

(b) Can you find an algorithm for solving congruences of this type? Hint: Consider Exercise 1.7.11.

(c) Solve the congruence $8x = 12 \pmod{125}$.

1.7.15. This exercise guides you to a proof of the Chinese remainder theorem.

(a) To prove the Chinese remainder theorem, show that it suffices to find m and n such that $\alpha + ma = \beta + nb$.

(b) To find m and n as in part (a), show that it suffices to find s and t such that $as + bt = (\alpha - \beta)$.

(c) Show that the existence of s and t as in part (b) follows from a and b being relatively prime.

1.8. Polynomials

Let K denote the set $\mathbb{Q}$ of rational numbers, the set $\mathbb{R}$ of real numbers, or the set $\mathbb{C}$ of complex numbers. (K could actually be any *field*; fields are algebraic systems that generalize the examples $\mathbb{Q}$, $\mathbb{R}$, and $\mathbb{C}$; we will give a careful definition of fields later.)

Polynomials with coefficients in K are expressions of the form $a_n x^n + a_{n-1}x^{n-1} + \cdots + a_0$, where the a_i are elements in K. The set of all polynomials with coefficients in K is denoted by $K[x]$. Addition and multiplication of polynomials are defined according to the familiar rules:

$$(\sum_j a_j x^j) + (\sum_j b_j x^j) = \sum_j (a_j + b_j)x^j,$$

and

$$(\sum_i a_i x^i)(\sum_j b_j x^j) = \sum_i \sum_j (a_i b_j) x^{i+j}$$
$$= \sum_k (\sum_{i,j:i+j=k} a_i b_j) x^k = \sum_k (\sum_i a_i b_{k-i}) x^k.$$

I trust that few readers will be disturbed by the informality of defining polynomials as "an expression of the form ..." . However, it is possible to formalize the concept by defining a polynomial to be a function $\alpha : n \mapsto \alpha_n$ from the nonnegative integers into K such that α_n is zero for all but finitely many n. Thus $7x^2 + 2x + 3$ would be interpreted as the function whose value at 2 is 7, whose value at 1 is 2, whose value at 0 is 3, and whose value at all other nonnegative integers is 0. The operations of addition and multiplication of polynomials can be recast as operations on such functions. It is straightforward (but not especially enlightening) to carry this out.

Example 1.8.1.

$$(2x^3 + 4x + 5) + (5x^7 + 9x^3 + x^2 + 2x + 8) = 5x^7 + 11x^3 + x^2 + 6x + 13,$$

and

$$(2x^3 + 4x + 5)(5x^7 + 9x^3 + x^2 + 2x + 8) = \\ 10\,x^{10} + 20\,x^8 + 25\,x^7 + 18\,x^6 + 2\,x^5 + 40\,x^4 + 65\,x^3 + 13\,x^2 + 42\,x + 40.$$

K can be regarded as subset of $K[X]$, and the addition and multiplication operations on $K[x]$ extend those on K; that is, for any two elements in K, their sum and product *as elements of* K agree with their sum and product *as elements of* $K[x]$.

The operations of addition and multiplication of polynomials satisfy properties exactly analogous to those listed for the integers in Proposition 1.6.1; see Proposition 1.8.2 on the next page.

All of these properties can be verified by straightforward computations, using the definitions of the operations and the corresponding properties of the operations in K.

As an example of the sort of computations needed, let us verify the distributive law: Let $f(x) = \sum_{i=0}^{\ell} a_i x^i$, $g(x) = \sum_{j=0}^{n} b_j x^j$, and $h(x) = \sum_{j=0}^{n} c_j x^j$. Then

$$\begin{aligned} f(x)(g(x) + h(x)) &= (\sum_{i=0}^{\ell} a_i x^i)(\sum_{j=0}^{n} b_j x^j + \sum_{j=0}^{n} c_j x^j) \\ &= \sum_{i=0}^{\ell} a_i x^i (\sum_{j=0}^{n} (b_j + c_j) x^j) \\ &= \sum_{i=0}^{\ell} \sum_{j=0}^{n} a_i (b_j + c_j) x^{i+j} \end{aligned}$$

$$
\begin{aligned}
&= \sum_{i=0}^{\ell}\sum_{j=0}^{n}(a_i b_j + a_i c_j)x^{i+j} \\
&= \sum_{i=0}^{\ell}\sum_{j=0}^{n}(a_i b_j)x^{i+j} + \sum_{i=0}^{\ell}\sum_{j=0}^{n}(a_i c_j)x^{i+j} \\
&= (\sum_{i=0}^{\ell} a_i x^i)(\sum_{j=0}^{n} b_j x^j) + (\sum_{i=0}^{\ell} a_i x^i)(\sum_{j=0}^{n} c_j x^j) \\
&= f(x)g(x) + f(x)h(x).
\end{aligned}
$$

Verification of the remaining properties listed in the following proposition is left to the reader.

Proposition 1.8.2.

(a) *Addition in* $K[x]$ *is commutative and associative; that is, for all* $f, g, h \in K[x]$,
$$f + g = g + f,$$
and
$$f + (g + h) = (f + g) + h.$$

(b) 0 *is an identity element for addition; that is, for all* $f \in K[x]$,
$$0 + f = f.$$

(c) *Every element* f *of* $K[x]$ *has an additive inverse* $-f$, *satisfying*
$$f + (-f) = 0.$$

(d) *Multiplication in* $K[x]$ *is commutative and associative; that is, for all* $f, g, h \in K[x]$,
$$fg = gf,$$
and
$$f(gh) = (fg)h.$$

(e) 1 *is an identity for multiplication; that is, for all* $f \in K[x]$,
$$1f = f.$$

(f) *The distributive law holds: For all* $f, g, h \in K[x]$,
$$f(g + h) = fg + fh.$$

Definition 1.8.3. The *degree* of a polynomial $\sum_k a_k x^k$ is the largest k such that $a_k \neq 0$. (The degree of a constant polynomial c is zero, unless $c = 0$. By convention, the degree of the constant polynomial 0 is $-\infty$.) The degree of $p \in K[x]$ is denoted $\deg(p)$.

If $p = \sum_j a_j x^j$ is a nonzero polynomial of degree k, the *leading coefficient* of p is a_k and the *leading term* of p is $a_k x^k$. A polynomial is said to be *monic* if its leading coefficient is 1.

Example 1.8.4. The degree of $p = (\pi/2)x^7 + ix^4 - \sqrt{2}x^3$ is 7; the leading coefficient is $\pi/2$; $(2/\pi)p$ is a monic polynomial.

Proposition 1.8.5. *Let* $f, g \in K[x]$.

(a) $\deg(fg) = \deg(f) + \deg(g)$; *in particular, if* f *and* g *are both nonzero, then* $fg \neq 0$.

(b) $\deg(f + g) \leq \max\{\deg(f), \deg(g)\}$.

Proof. Exercise 1.8.3. ■

We say that a polynomial f *divides* a polynomial g (or that g is *divisible* by f) if there is a polynomial q such that $fq = g$. We write $f|g$ for "f divides g."

The goal of this section is to show that $K[x]$ has a theory of divisibility, or factorization, that exactly parallels the theory of divisibility for the integers, which was presented in Section 1.6. In fact, all of the results of this section are analogues of results of Section 1.6, with the proofs also following a nearly identical course. In this discussion, the degree of a polynomial plays the role that absolute value plays for integers.

Proposition 1.8.6. *Let* f, g, h, u, *and* v *denote polynomials in* $K[x]$.

(a) *If* $uv = 1$, *then* $u, v \in K$.

(b) *If* $f|g$ *and* $g|f$, *then there is a* $k \in K$ *such that* $g = kf$.

(c) *Divisibility is transitive: If* $f|g$ *and* $g|h$, *then* $f|h$.

(d) *If* $f|g$ *and* $f|h$, *then for all polynomials* s, t, $f|(sg + th)$.

Proof. For part (a), if $uv = 1$, then both of u, v must be nonzero. If either of u or v had positive degree, then uv would also have positive degree. Hence both u and v must be elements of K.

For part (b), if $g = vf$ and $f = ug$, then $g = uvg$, or $g(1-uv) = 0$. If $g = 0$, then $k = 0$ meets the requirement. Otherwise, $1 - uv = 0$, so both u and v are elements of K, by part (a), and $k = v$ satisfies the requirement.

The remaining parts are left to the reader. ■

What polynomials should be considered the analogues of prime numbers? The polynomial analogue of a prime number should be a polynomial that does not admit any nontrivial factorization. It is always possible to "factor out" an arbitrary nonzero element of K from a polynomial $f \in K[x]$, $f(x) = c(c^{-1}f(x))$, but this should not count as a genuine factorization. A nontrivial factorization $f(x) = g(x)h(x)$ is one in which both of the factors have positive degree [or, equivalently, each factor has degree less than $\deg(f)$] and the polynomial analogue of a prime number is a polynomial for which such a factorization is is not possible. Such polynomials are called *irreducible* rather than *prime.*

Definition 1.8.7. We say that a polynomial in $K[x]$ is *irreducible* if its degree is positive and it cannot be written as a product of two polynomials each of strictly smaller (positive) degree.

The analogue of the existence of prime factorizations for integers is the following statement.

Proposition 1.8.8. *Any polynomial in* $K[x]$ *of positive degree can be written as a product of irreducible polynomials.*

Proof. The proof is by induction on the degree. Every polynomial of degree 1 is irreducible, by the definition of irreducibility. So let f be a polynomial of degree greater than 1, and make the inductive hypothesis that every polynomial whose degree is positive but less than the degree of f can be written as a product of irreducible polynomials. If f is not itself irreducible, then it can be written as a product, $f = g_1 g_2$, where $1 \le \deg(g_i) < \deg(f)$. By the inductive hypothesis, each g_i is a product of irreducible polynomials, and thus so is f. ■

Proposition 1.8.9. $K[x]$ *contains infinitely many irreducible polynomials.*

Proof. If K is an field with infinitely many elements like $\mathbb{Q}$, $\mathbb{R}$ or $\mathbb{C}$, then $\{x - k : k \in K\}$ is already an infinite set of irreducible polynomials. However, there also exist fields with only finitely many elements, as we will see later. For such fields, we can apply the same proof as for Theorem 1.6.6 (replacing prime numbers by irreducible polynomials). ■

Remark 1.8.10. It is not true in general that $K[x]$ has irreducible polynomials of arbitrarily large degree. In fact, in $\mathbb{C}[x]$, every irreducible polynomial has degree 1, and in $\mathbb{R}[x]$, every irreducible polynomial has degree ≤ 2. But in $\mathbb{Q}[x]$, there do exist irreducible polynomials of arbitrarily large degree. If K is a finite field, then for each $n \in \mathbb{N}$, $K[x]$ contains only finitely many polynomials of degree $\leq n$. Therefore, since $K[x]$ has infinitely many irreducible polynomials, it has irreducible polynomials of arbitrarily large degree.

Example 1.8.11. Let's recall the process of long division of polynomials by working out an example. Let $p = 7x^5 + 3x^2 + x + 2$ and $d = 5x^3+2x^2+3x+4$. We wish to find polynomials q and r (quotient and remainder) such that $p = qd + r$ and $\deg(r) < \deg(d)$. The first contribution to q is $\frac{7}{5}x^2$, and

$$p - \frac{7}{5}x^2 d = -\frac{14}{5}x^4 - \frac{21}{5}x^3 - \frac{13}{5}x^2 + x + 2.$$

The next contribution to q is $-\frac{14}{25}x$, and

$$p - (\frac{7}{5}x^2 - \frac{14}{25}x)\, d = -\frac{77}{25}x^3 - \frac{23}{25}x^2 + \frac{81}{25}x + 2.$$

The next contribution to q is $-\frac{77}{125}$, and

$$p - (\frac{7}{5}x^2 - \frac{14}{25}x - \frac{77}{125})\, d = \frac{39}{125}x^2 + \frac{636}{125}x + \frac{558}{125}.$$

Thus,

$$q = \frac{7}{5}x^2 - \frac{14}{25}x - \frac{77}{125}$$

and

$$r = \frac{39}{125}x^2 + \frac{636}{125}x + \frac{558}{125}.$$

Lemma 1.8.12. *Let p and d be elements of $K[x]$, with $\deg(p) \geq \deg(d) \geq 0$. Then there is a monomial $m = bx^k \in K[x]$ and a polynomial $p' \in K[x]$ such that $p = md + p'$, and $\deg(p') < \deg(p)$.*

Proof. Write $p = a_n x^n + a_{n-1}x^{d-1} + \cdots + a_0$ and $d = b_s x^s + b_{s-1}x^{s-1} + \cdots + b_0$, where $n = \deg(p)$ and $s = \deg(d)$, and $s \leq n$. [Note that $d \neq 0$, because we required $\deg(d) \geq 0$.] Put $m = (a_n/b_s)x^{n-s}$ and $p' = p - md$. Then both p and md have leading term equal to $a_n x^n$, so $\deg(p') < \deg(p)$. ■

Proposition 1.8.13. *Let* p *and* d *be elements of* $K[x]$, *with* $\deg(d) \geq 0$. *Then there exist polynomials* q *and* r *in* $K[x]$ *such that* $p = dq + r$ *and* $\deg(r) < \deg(d)$.

Proof. The idea of the proof is illustrated by the preceding example : We divide p by d, obtaining a monomial quotient and a remainder p' of degree strictly less than the degree of p. We then divide p' by d, obtaining a remainder p'' of still smaller degree. We continue in this fashion until we finally get a remainder of degree less than $\deg(d)$. The formal proof goes by induction on the degree of p.

If $\deg(p) < \deg(d)$, then put $q = 0$ and $r = p$. So assume now that $\deg(p) \geq \deg(d) \geq 0$, and that the result is true when p is replaced by any polynomial of lower degree.

According to the lemma, we can write $p = md + p'$, where $\deg(p') < \deg(p)$. By the induction hypothesis, there exist polynomials q' and r with $\deg(r) < \deg(d)$ such that $p' = q'd + r$. Putting $q = q' + m$, we have $p = qd + r$. ■

Definition 1.8.14. A polynomial $f \in K[x]$ is a *greatest common divisor* of nonzero polynomials $p, q \in K[x]$ if

(a) f divides p and q in $K[x]$ and

(b) whenever $g \in K[x]$ divides p and q, then g also divides f.

We are about to show that two nonzero polynomials in $K[x]$ always have a greatest common divisor. Notice that a greatest common divisor is unique up to multiplication by a nonzero element of K, by Proposition 1.8.6 (b). There is a *unique* greatest common divisor that is *monic* (i.e., whose leading coefficient is 1). When we need to refer to *the* greatest common divisor, we will mean the one that is monic. We denote the monic greatest common divisor of p and q by g.c.d.(p, q).

The following results (1.8.15 through 1.8.22) are analogues of results for the integers, and the proofs are *virtually identical* to those for the integers, with Proposition 1.8.13 playing the role of Proposition 1.6.7. For each of these results, *you* should write out a complete proof modeled on the proof of the analogous result for the integers. You will end up understanding the proofs for the integers better, as well as understanding how they have to be modified to apply to polynomials.

For integers m, n, we studied the set $I(m, n) = \{am + bn : a, b \in \mathbb{Z}\}$, which played an important role in our discussion of the greatest common divisor. Here we introduce the analogous set for polynomials.

Proposition 1.8.15. *For polynomials* $f, g \in K[x]$, *let*

$$I(f, g) = \{af + bg : a, b \in K[x]\}.$$

(a) *For all* $p, q \in I(f, g)$, $p + q \in I(f, g)$ *and* $-p \in I(f, g)$
(b) *For all* $p \in K[x]$, $pI(f, g) \subseteq I(f, g)$.
(c) *If* $p \in K[x]$ *divides* f *and* g, *then* p *divides all elements of* $I(f, g)$.

Proof. Exercise 1.8.4. ■

Theorem 1.8.16. *Any two nonzero polynomials* $f, g \in K[x]$ *have a greatest common divisor, which is an element of* $I(f, g)$.

Proof. Mimic the proof of Proposition 1.6.10, with Proposition 1.8.13 replacing 1.6.7. Namely (assuming $\deg(f) \leq \deg(g)$), we do repeated division with remainder, each time obtaining a remainder of smaller degree. The process must terminate with a zero remainder after at most $\deg(f)$ steps:

$$\begin{aligned} g &= q_1 f + f_1 \\ f &= q_2 f_1 + f_2 \\ &\vdots \\ f_{k-2} &= q_k f_{k-1} + f_k \\ &\vdots \\ f_{r-1} &= q_{r+1} f_r. \end{aligned}$$

Here $\deg f > \deg(f_1) > \deg(f_2) > \dots$. By the argument of Proposition 1.6.10, the final nonzero remainder f_r is an element of $I(f, g)$ and is a greatest common divisor of f and g. ■

Remark 1.8.17. The procedure described in the proof can be modified to product an algorithm for computing the *monic* greatest common divisor of f and g, as follows. [Denote the leading coefficient of a polynomial q by $LC(q)$.]

$$\begin{aligned} g &= q_1 f + f_1 \\ \rho_1 &= LC(f_1) \\ \bar{f}_1 &= \rho_1^{-1} f_1 \end{aligned}$$

$$\begin{aligned}
f &= q_2\bar{f}_1 + f_2 \\
&\quad \rho_2 = LC(f_2) \\
&\quad \bar{f}_2 = \rho_2^{-1} f_2 \\
&\quad \vdots \\
\bar{f}_{k-2} &= q_k \bar{f}_{k-1} + f_k \\
&\quad \rho_k = LC(f_k) \\
&\quad \bar{f}_k = \rho_k^{-1} f_k \\
&\quad \vdots \\
\bar{f}_{r-1} &= q_{r+1}\bar{f}_r
\end{aligned}$$

We can also find the (polynomial) coefficients s_i, t_i such that

$$\bar{f}_i = s_i f + t_i g$$

by using the recursion relation

$$s_i = \rho_i^{-1}(s_{i-2} - q_i s_{i-1}), \quad t_i = \rho_i^{-1}(t_{i-2} - q_i t_{i-1}),$$

together with the initial conditions

$$s_{-1} = 1, \quad t_{-1} = 0, \quad s_0 = 0, \quad t_0 = 1.$$

Example 1.8.18. Compute the (monic) greatest common divisor of

$$f(x) = -4 + 9x - 3x^2 - 6x^3 + 6x^4 - 3x^5 + x^6$$

and

$$g(x) = 3 - 6x + 4x^2 - 2x^3 + x^4.$$

Repeated division with remainder gives

$$\begin{aligned}
&(-4 + 9x - 3x^2 - 6x^3 + 6x^4 - 3x^5 + x^6) \\
&\qquad = (-x + x^2)(3 - 6x + 4x^2 - 2x^3 + x^4) + (-4 + 12x - 12x^2 + 4x^3), \\
&(3 - 6x + 4x^2 - 2x^3 + x^4) = (1 + x)(-1 + 3x - 3x^2 + x^3) + (4 - 8x + 4x^2), \\
&(-1 + 3x - 3x^2 + x^3) = (-1 + x)(1 - 2x + x^2) + 0.
\end{aligned}$$

(The remainder at each step has been divided by its leading coefficient before being used as the divisor in the next step.) Thus the greatest common divisor is $1-2x+x^2$. Furthermore, the successive coefficients s_i, t_i are given by

$$\begin{aligned}
s_1 &= \frac{1}{4}, & t_1 &= \frac{x}{4} - \frac{x^2}{4} \\
s_2 &= -\frac{1}{16} - \frac{x}{16}, & t_2 &= \frac{1}{4} - \frac{x}{16} + \frac{x^3}{16}.
\end{aligned}$$

Thus, $1 - 2x + x^2 = (-\frac{1}{16} - \frac{x}{16})f + (\frac{1}{4} - \frac{x}{16} + \frac{x^3}{16})g.$

Definition 1.8.19. Two polynomials $f, g \in K[x]$ are *relatively prime* if $\text{g.c.d.}(f, g) = 1$.

Proposition 1.8.20. *Two polynomials* $f, g \in K[x]$ *are relatively prime if, and only if,* $1 \in I(f, g)$.

Proof. Exercise 1.8.5. ∎

Proposition 1.8.21.

(a) *Let* p *be an irreducible polynomial in* $K[x]$ *and* $f, g \in K[x]$ *nonzero polynomials. If* p *divides the product* fg*, then* p *divides* f *or* p *divides* g.

(b) *Suppose that an irreducible polynomial* $p \in K[x]$ *divides a product* $f_1 f_2 \cdots f_s$ *of nonzero polynomials. Then* p *divides one of the factors.*

Proof. Exercise 1.8.8. ∎

Theorem 1.8.22. *The factorization of a polynomial in* $K[x]$ *into irreducible factors is essentially unique. That is, the irreducible factors appearing are unique up to multiplication by nonzero elements in* K.

Proof. Exercise 1.8.9. ∎

This completes our treatment of unique factorization of polynomials. Before we leave the topic, let us notice that you haven't yet learned any general methods for recognizing irreducible polynomials, or for carrying out the factorization of a polynomial by irreducible polynomials. In the integers, you could, at least in principle, test whether a number n is prime, and find its prime factors if it is composite, by searching for divisors among the natural numbers $\leq \sqrt{n}$. For an infinite field such as $\mathbb{Q}$, we cannot factor polynomials in $\mathbb{Q}[x]$ by exhaustive search, as there are infinitely many polynomials of each degree.

We finish this section with some elementary but important results relating *roots* of polynomials to divisibility.

Proposition 1.8.23. *Let* $p \in K[x]$ *and* $a \in K$. *Then there is a polynomial* q *such that* $p(x) = q(x)(x - a) + p(a)$. *Consequently,* $p(a) = 0$ *if, and only if,* $x - a$ *divides* p.

Proof. Write $p(x) = q(x)(x - a) + r$, where the remainder r is a constant. Substituting a for x gives $p(a) = r$. ■

Definition 1.8.24. Say an element $\alpha \in K$ is a *root* of a polynomial $p \in K[x]$ if $p(\alpha) = 0$. Say the *multiplicity of the root* α is k if $x - \alpha$ appears exactly k times in the irreducible factorization of p.

Corollary 1.8.25. *A polynomial* $p \in K[x]$ *of degree* n *has at most* n *roots in* K, *counting with multiplicities. That is, the sum of multiplicities of all roots is at most* n.

Proof. If $p = (x - \alpha_1)^{m_1}(x - \alpha_2)^{m_2} \cdots (x - \alpha_k)^{m_k} q_1 \cdots q_s$, where the q_i are irreducible, then evidently $m_1 + m_2 + \cdots + m_k \le \deg(p)$. ■

Exercises 1.8

Exercises 1.8.1 through 1.8.2 ask you to prove parts of Proposition 1.8.2.

1.8.1. Prove that addition in $K[x]$ is commutative and associative, that 0 is an identity element for addition in $K[x]$, and that $f + -f = 0$ for all $f \in K[x]$.

1.8.2. Prove that multiplication in $K[x]$ is commutative and associative, and that 1 is an identity element for multiplication.

1.8.3. Prove Proposition 1.8.5.

1.8.4. Prove Proposition 1.8.15.

1.8.5. Let h be an element of $I(f, g)$ of least degree. Show that h is a greatest common divisor of f and g. Hint: Apply division with remainder.

1.8.6. Show that two polynomials $f, g \in K[x]$ are relatively prime if, and only if, $1 \in I(f, g)$.

1.8.7. Show that if $p \in K[x]$ is irreducible and $f \in K[x]$, then either p divides f, or p and f are relatively prime.

1.8.8. Let $p \in K[x]$ be irreducible. Prove the following statements.

(a) If p divides a product fg of elements of $K[x]$, then p divides f or p divides g.
(b) If p divides a product $f_1 f_2 \dots f_r$ of several elements of $K[x]$, then p divides one of the f_i.

Hint: Mimic the arguments of Proposition 1.6.16 and Corollary 1.6.17.

1.8.9. Prove Theorem 1.8.22. (Mimic the proof of Theorem 1.6.18.)

1.8.10. For each of the following pairs of polynomials f, g, find the greatest common divisor and find polynomials r, s such that $rf + sg =$ g.c.d.(f, g).

(a) $x^3 - 3x + 3$, $x^2 - 4$
(b) $-4 + 6x - 4x^2 + x^3$, $x^2 - 4$

1.8.11. Write a computer program to compute the greatest common divisor of two polynomials f, g with real coefficients. Make your program find polynomials r, s such that $rf + sg = \text{g.c.d.}(f, g)$.

The next three exercises explore the idea of the greatest common divisor of *several* nonzero polynomials, $f_1, f_2, \dots, f_k \in K[x]$.

1.8.12. Make a reasonable definition of $\text{g.c.d}(f_1, f_2, \dots, f_k)$, and show that $\text{g.c.d}(f_1, f_2, \dots, f_k) = \text{g.c.d.}(f_1, \text{g.c.d}(f_2, \dots, f_k)))$.

1.8.13.

(a) Let $I = I(f_1, f_2, \dots, f_k) =$

$$\{m_1 f_1 + m_2 f_2 + \cdots + m_k f_k : m_1, \dots, m_k \in \mathbb{Z}\}.$$

Show that I has all the properties of $I(f, g)$ listed in Proposition 1.8.15.

(b) Show that $f = \text{g.c.d}(f_1, f_2, \dots, f_k)$ is an element of I of smallest degree and that $I = fK[x]$.

1.8.14.

(a) Develop an algorithm to compute $\text{g.c.d}(f_1, f_2, \dots, f_k)$.
(b) Develop a computer program to compute the greatest common divisor of any finite collection of nonzero polynomials with real coefficients.

1.8.15.

(a) Suppose that $p(x) = a_n x^n + \cdots + a_1 x + a_0 \in Z[x]$. Suppose that $r/s \in \mathbb{Q}$ is a root of p, where r and s are relatively prime integers. Show that s divides a_n and r divides a_0. Hint: Start with the equation $p(r/s) = 0$, multiply by s^n, and note, for example, that all the terms except $a_n r^n$ are divisible by s.

(b) Conclude that any rational root of a monic polynomial in $\mathbb{Z}[x]$ is an integer.

(c) Conclude that $x^2 - 2$ has no rational root, and therefore $\sqrt{2}$ is irrational.

1.8.16.

(a) Show that a quadratic or cubic polynomial $f(x)$ in $K[x]$ is irreducible if, and only if, $f(x)$ has no root in K.

(b) A monic quadratic or cubic polynomial $f(x) \in \mathbb{Z}[x]$ is irreducible if, and only if, it has no integer root.

(c) $x^3 - 3x + 1$ is irreducible in $\mathbb{Q}[x]$.

1.9. Counting

Counting is a fundamental and pervasive technique in algebra. In this section we will discuss two basic counting tools, the binomial coefficients and the method of inclusion-exclusion. These tools will be used to establish some not at all obvious results in number theory: First, the binomial coefficients and the binomial theorem are used to prove Fermat's little theorem (Proposition 1.9.10); then inclusion –exclusion is used to obtain a formula for the Euler φ function (Proposition 1.9.18). Finally, we use the formula for the φ function to obtain Euler's generalization of the little Fermat theorem (Theorem 1.9.20).

Let's begin with some problems on counting subsets of a set. How many subsets are there of the set $\{1, 2, 3\}$? There are $8 = 2^3$ subsets, namely $\emptyset$, $\{1\}$, $\{2\}$, $\{3\}$, $\{1, 2\}$, $\{1, 3\}$, $\{2, 3\}$, and $\{1, 2, 3\}$.

How many subsets are there of a set with n elements? For each of the n elements we have two possibilities: The element is in the subset, or not. The in/out choices for the n elements are independent, so there are 2^n possibilities, that is, 2^n subsets of a set with n elements.

Here is another way to see this: Define a map θ from the set of subsets of $\{1, 2, \dots, n\}$ to the set of sequences $(s_1, s_2, \dots, s_n)$ with each $s_i \in \{0, 1\}$. Given a subset A of $\{1, 2, \dots, n\}$, form the corresponding sequence $\theta(A)$ by putting a 1 in the jth position if $j \in A$

and a 0 in the jth position otherwise. It's not hard to check that θ is a bijection. Therefore, there are just as many subsets of $\{1, 2, \ldots, n\}$ as there are sequences of n 0's and 1's, namely 2^n.

Proposition 1.9.1. *A set with n elements has 2^n subsets.*

How many two–element subsets are there of a set with five elements? You can list all the two–element subsets of $\{1, 2, \ldots, 5\}$ and find that there are 10 of them.

How many two–element subsets are there of a set with n elements? Let's denote the number of two–element subsets by $\binom{n}{2}$. A two–element subset of $\{1, 2, \ldots, n\}$ either includes n or not. There are $n-1$ two–element subsets that *do* include n, since there are $n-1$ choices for the second element, and the number of two–element subsets that *do not* include n is $\binom{n-1}{2}$. Thus, we have the recursive relation

$$\binom{n}{2} = (n-1) + \binom{n-1}{2}.$$

For example,

$$\begin{aligned}\binom{5}{2} &= 4 + \binom{4}{2} = 4 + 3 + \binom{3}{2} \\ &= 4 + 3 + 2 + \binom{2}{2} = 4 + 3 + 2 + 1 = 10.\end{aligned}$$

In general,

$$\binom{n}{2} = (n-1) + (n-2) + \cdots + 2 + 1.$$

This sum is well known and equal to $n(n-1)/2$. [You can find an inductive proof of the formula

$$(n-1) + (n-2) + \cdots + 2 + 1 = n(n-1)/2$$

in Appendix C.1.]

Here is another argument for the formula

$$\binom{n}{2} = n(n-1)/2$$

that is better because it generalizes. Think of building the $n!$ permutations of $\{1, 2, \ldots, n\}$ in the following way. First choose two elements to be the first two (leaving $n-2$ to be the last $n-2$). This can be done in $\binom{n}{2}$ ways. Then arrange the first 2 (in $2! = 2$ ways) and the last $n-2$ (in $(n-2)!$ ways). This strange process for building permutations gives the formula

$$n! = \binom{n}{2} 2!\,(n-2)!\,.$$

Now dividing by $2!\,(n-2)!$ gives

$$\binom{n}{2} = \frac{n!}{2!(n-2)!} = \frac{n(n-1)}{2}\,.$$

The virtue of this argument is that it remains valid if 2 is replaced by any k, $0 \le k \le n$. So we have the following:

Proposition 1.9.2. *Let n be a natural number and let k be an integer in the range $0 \le k \le n$. Let $\binom{n}{k}$ denote the number of k-element subsets of an n-element set. Then*

$$\binom{n}{k} = \frac{n!}{k!(n-k)!}\,.$$

We extend the definition by declaring $\binom{n}{k} = 0$ if k is negative or greater than n. Also, we declare $\binom{0}{0} = 1$ and $\binom{0}{k} = 0$ if $k \neq 0$. Note that $\binom{n}{0} = \binom{n}{n} = 1$ for all $n \ge 0$. The expression $\binom{n}{k}$ is generally read as "n choose k."

Here are some elementary properties of the numbers $\binom{n}{k}$.

Lemma 1.9.3. *Let n be a natural number and $k \in \mathbb{Z}$.*

(a) $\binom{n}{k}$ *is a nonnegative integer.*

(b) $\binom{n}{k} = \binom{n}{n-k}$.

(c) $\binom{n}{k} = \binom{n-1}{k} + \binom{n-1}{k-1}$.

Proof. Part (a) is evident from the definition of $\binom{n}{k}$.

The formula $\binom{n}{k} = \dfrac{n!}{k!(n-k)!}$ implies (b) when $0 \le k \le n$. But when k is not in this range, neither is $n-k$, so both sides of (b) are zero.

When k is 0, both sides of (c) are equal to 1. When k is negative or greater than n, both sides of (c) are zero. For $0 < k \leq n$, (c) follows from a combinatorial argument: To choose k elements out of $\{1, 2, \ldots, n\}$, we can either choose n, together with $k - 1$ elements out of $\{1, 2, \ldots, n-1\}$, which can be done in $\binom{n-1}{k-1}$ ways, or we can choose k elements out of $\{1, 2, \ldots, n-1\}$, which can be done in $\binom{n-1}{k}$ ways. ■

Example 1.9.4. The coefficients $\binom{n}{k}$ have an interpretation in terms of paths. Consider paths in the (x, y)–plane from $(0, 0)$ to a point (a, b) with nonnegative integer coordinates. We admit only paths of $a + b$ "steps," in which each step goes one unit to the right or one unit up; that is, each step is either a horizontal segment from an integer point (x, y) to $(x + 1, y)$, or a vertical segment from (x, y) to $(x, y + 1)$. How many such paths are there? Each path has exactly a steps to the right and b steps up, so a path can be specified by a sequence with a R's and b U's. Such a sequence is determined by choosing the positions of the a R's, so the number of sequences (and the number of paths) is $\binom{a+b}{a} = \binom{a+b}{b}$.

The numbers $\binom{n}{k}$ are called *binomial coefficients*, because of the following proposition:

Proposition 1.9.5. *(Binomial theorem). Let x and y be numbers (or variables). For $n \geq 0$ we have*

$$(x + y)^n = \sum_{k=0}^{n} \binom{n}{k} x^k y^{n-k}.$$

Proof. $(x+y)^n$ is a sum of 2^n monomials, each obtained by choosing x from some of the n factors, and choosing y from the remaining factors. For fixed k, the number of monomials $x^k y^{n-k}$ in the sum is the number of ways of choosing k objects out of n, which is $\binom{n}{k}$. Hence the coefficient of $x^k y^{n-k}$ in the product is $\binom{n}{k}$. ■

Corollary 1.9.6.

(a) $2^n = \sum_{k=0}^{n} \binom{n}{k}$.

(b) $0 = \sum_{k=0}^{n} (-1)^k \binom{n}{k}$.

(c) $2^{n-1} = \sum_{\substack{k=0 \\ k \text{ odd}}}^{n} \binom{n}{k} = \sum_{\substack{k=0 \\ k \text{ even}}}^{n} \binom{n}{k}$.

Proof. Part (a) follows from the combinatorial interpretation of the two sides: The total number of subsets of an n element set is the sum over k of the number of subsets with k elements. Part (a) also follows from the binomial theorem by putting $x = y = 1$. Part (b) follows from the binomial theorem by putting $x = -1, y = 1$. The two sums in part (c) are equal by part (b), and they add up to 2^n by part (a); hence each is equal to 2^{n-1}. ■

Example 1.9.7. We can obtain many identities for the binomial coefficients by starting with the special case of the binomial theorem:

$$(1+x)^n = \sum_{k=0}^{n} \binom{n}{k} x^k,$$

regarding both sides as functions in a real variable x, manipulating the functions (for example, by differentiating, integrating, multiplying by x, etc.), and finally evaluating for a specific value of x. For example, differentiating the basic formula gives

$$n(1+x)^{n-1} = \sum_{k=1}^{n} k \binom{n}{k} x^{k-1}.$$

Evaluating at $x = 1$ gives

$$n2^{n-1} = \sum_{k=1}^{n} k \binom{n}{k},$$

while evaluating at $x = -1$ gives

$$0 = \sum_{k=1}^{n} (-1)^{k-1} k \binom{n}{k}.$$

Lemma 1.9.8. *Let* p *be a prime number.*

(a) *If* $0 < k < p$*, then* $\binom{p}{k}$ *is divisible by* p.

(b) *For all integers* a *and* b, $(a+b)^p \equiv a^p + b^p \pmod{p}$.

Proof. For part (a), we have $p! = \binom{p}{k}\, k!\,(p-k)!$. Now p divides $p!$, but p does not divide $k!$ or $(n-k)!$. Consequently, p divides $\binom{p}{k}$.

The binomial theorem gives $(a+b)^p = \sum_{k=0}^{p} \binom{p}{k} a^k b^{n-k}$. By part (a), all the terms for $0 < k < p$ are divisible by p, so $(a+b)^p$ is congruent modulo p to the sum of the terms for $k = 0$ and $k = p$, namely to $a^p + b^p$. ■

I hope you already discovered the first part of the next lemma while doing the exercises for Section 1.7.

Proposition 1.9.9.

(a) *Let* $n \geq 2$ *be a natural number. An element* $[a] \in \mathbb{Z}_n$ *has a multiplicative inverse if, and only if,* a *is relatively prime to* n.

(b) *If* p *is a prime, then every nonzero element of* $\mathbb{Z}_p$ *is invertible.*

Proof. If a is relatively prime to n, there exist integers s, t such that $as + nt = 1$. But then $as \equiv 1 \pmod{n}$, or $[a][s] = [1]$ in $\mathbb{Z}_n$. On the other hand, if a is not relatively prime to n, then a and n have a common divisor $k > 1$. Say $kt = a$ and $ks = n$. Then $as = kts = nt$. Reducing modulo n gives $[a][s] = [0]$, so $[a]$ is a zero divisor in $\mathbb{Z}_n$, and therefore not invertible.

If p is a prime, and $[a] \neq 0$ in $\mathbb{Z}_p$, then a is relatively prime to p, so $[a]$ is invertible in Z_p by part (a). ■

Proposition 1.9.10. *(Fermat's little theorem). Let* p *be a prime number.*

(a) *For all integers* a*, we have* $a^p \equiv a \pmod{p}$.

(b) *If* a *is not divisible by* p*, then* $a^{p-1} \equiv 1 \pmod{p}$.

Proof. It suffices to show this for a a natural number (since the case $a = 0$ is trivial and the case $a < 0$ follows from the case $a > 0$). The proof goes by induction on a. For $a = 1$, the assertion is obvious.

So assume that $a > 1$ and that $(a-1)^p \equiv a - 1 \pmod{p}$. Then

$$\begin{aligned} a^p &= ((a-1)+1)^p \\ &\equiv (a-1)^p + 1 \pmod{p} \\ &\equiv (a-1) + 1 \pmod{p} = a, \end{aligned}$$

where the first congruence follows from Lemma 1.9.8, and the second from the induction assumption.

The conclusion of part (a) is equivalent to $[a]^p = [a]$ in $\mathbb{Z}_p$. If a is not divisible by p, then by the previous proposition, $[a]$ has a multiplicative inverse $[a]^{-1}$ in $\mathbb{Z}_p$. Multiplying both sides of the equation by $[a]^{-1}$ gives $[a]^{p-1} = [1]$. But this is equivalent to $a^{p-1} \equiv 1 \pmod{p}$. ■

Inclusion–Exclusion

Suppose you have three subsets A, B and C of a finite set U, and you want to count $A \cup B \cup C$. If you just add $|A| + |B| + |C|$, then you might have too large a result, because any element that is in more than one of the sets has been counted multiple times. So you can try to correct this problem by subtracting off the sizes of the intersections of pairs of sets, obtaining a new estimate for $|A \cup B \cup C|$, namely $|A| + |B| + |C| - |A \cap B| - |A \cap C| - |B \cap C|$. But this might be too small! If an element lies in $A \cap B \cap C$, then it has been counted three times in $|A| + |B| + |C|$, but uncounted three times in $-|A \cap B| - |A \cap C| - |B \cap C|$, so altogether it hasn't been counted at all. To fix this, we had better add back $|A \cap B \cap C|$. So our next (and final) estimate is

$$\begin{aligned} |A \cup B \cup C| &= |A| + |B| + |C| \\ &\quad - |A \cap B| - |A \cap C| - |B \cap C| + |A \cap B \cap C|. \quad (1.9.1) \end{aligned}$$

This is correct, because if an element is in just one of the sets, it is counted just once; if it is in exactly two of the sets, then it is counted twice in $|A| + |B| + |C|$, but then uncounted once in $|A \cap B| - |A \cap C| - |B \cap C|$; if it is in all three sets, then it is counted three times in $|A| + |B| + |C|$, uncounted three times in $-|A \cap B| - |A \cap C| - |B \cap C|$, and again counted once in $|A \cap B \cap C|$.

We want to obtain a generalization of Formula (1.9.1) for any number of subsets.

Let U be any set. For a subset $X \subseteq U$, the *characteristic function* of X is the function $\mathbf{1}_X : U \to \{0, 1\}$ defined by $\mathbf{1}_X(u) = 1$ if $u \in X$ and $\mathbf{1}_X(u) = 0$ otherwise. Evidently, the characteristic function of the relative complement of X is $\mathbf{1}_{U \setminus X} = 1 - \mathbf{1}_X$, and the characteristic function of the intersection of two sets is the product of the characteristic functions: $\mathbf{1}_{X \cap Y} = \mathbf{1}_X \mathbf{1}_Y$.

Let X' denote the relative complement of a subset $X \subseteq U$; that is, $X' = U \setminus X$.

Proposition 1.9.11. *Let* $A_1, A_2, \dots, A_n$ *be subsets of* U. *Then*

(a)

$$\mathbf{1}_{A_1' \cap A_2' \cap \cdots \cap A_n'} = 1 - \sum_i \mathbf{1}_{A_i} + \sum_{i<j} \mathbf{1}_{A_i \cap A_j} - \sum_{i<j<k} \mathbf{1}_{A_i \cap A_j \cap A_k} + \cdots + (-1)^n \mathbf{1}_{A_1 \cap \cdots A_n}.$$

(b)

$$\mathbf{1}_{A_1 \cup A_2 \cup \cdots \cup A_n} = \sum_i \mathbf{1}_{A_i} - \sum_{i<j} \mathbf{1}_{A_i \cap A_j} + \sum_{i<j<k} \mathbf{1}_{A_i \cap A_j \cap A_k} - \cdots + (-1)^{n-1} \mathbf{1}_{A_1 \cap \cdots A_n}.$$

Proof. For part (a), evaluate both sides of the equation at some $u \in U$. If u is in none of the A_i, then both the left and right sides of the equation, evaluated at u, give 1. On the other hand, if u is in exactly k of the A_i, then the left side of the equation, evaluated at u is zero, and the right side is

$$1 - k + \binom{k}{2} - \binom{k}{3} + \cdots + (-1)^k \binom{k}{k},$$

which is equal to zero by Corollary 1.9.6(b). Part (b) follows from part (a), because $A_1' \cap A_2' \cap \cdots \cap A_n' = (A_1 \cup A_2 \cup \cdots \cup A_n)'$. ■

Corollary 1.9.12. *Suppose that* U *is a finite set and that* $A_1, A_2, \dots, A_n$ *are subsets of* U. *Then*

(a)

$$|A_1' \cap A_2' \cap \cdots \cap A_n'| = |U| - \sum_i |A_i| + \sum_{i<j} |A_i \cap A_j| - \sum_{i<j<k} |A_i \cap A_j \cap A_k| + \cdots + (-1)^n |A_1 \cap \cdots A_n|.$$

(b)

$$|A_1 \cup A_2 \cup \cdots \cup A_n| = \sum_i |A_i| - \sum_{i<j} |A_i \cap A_j| + \sum_{i<j<k} |A_i \cap A_j \cap A_k| - \cdots + (-1)^{n-1} |A_1 \cap \cdots A_n|.$$

Proof. For any subset X of U, $|X| = \sum_{u \in U} \mathbf{1}_X(u)$. The desired equalities are obtained by starting with the identities for characteristic functions given in the proposition, evaluating both sides at $u \in U$, and summing over u. ■

The formulas given in the corollary are called the inclusion-exclusion formulas.

Example 1.9.13. Find a formula for the number of permutations π of n with no fixed points. That is, π is required to satisfy $\pi(j) \neq j$ for all $1 \leq j \leq n$. Such permutations are sometimes called *derangements*. Take U to be the set of all permutations of $\{1, 2, \dots, n\}$, and let A_i be the set of permutations π of $\{1, 2, \dots, n\}$ such that $\pi(i) = i$. Thus each A_i is the set of permutations of $n - 1$ objects, and so has size $(n - 1)!$. In general, the intersection of any k of the A_i is the set of permutations of $n - k$ objects, and so has cardinality $(n - k)!$. The situation is ideal for application of the inclusion-exclusion formula because the size of the intersection of k of the A_i does not depend on the choice of the k subsets. The set of derangements is $A_1' \cap A_2 \cap \cdots \cap A_n'$, and its cardinality is

$$D_n = |A_1' \cap A_2' \cap \cdots \cap A_n'| = |U| - \sum_i |A_i| + \sum_{i<j} |A_i \cap A_j|$$
$$- \sum_{i<j<k} |A_i \cap A_j \cap A_k| + \cdots + (-1)^n |A_1 \cap \cdots A_n|.$$

As each k-fold intersection has cardinality $(n - k)!$ and there are $\binom{n}{k}$ such intersections, D_n evaluates to

$$D_n = n! - n(n-1)! + \binom{n}{2}(n-2)! - \ldots + (-1)^k \binom{n}{k}(n-k)! + \cdots + (-1)^n.$$

This sum can be simplified as follows:

$$D_n = = \sum_{k=0}^{n} (-1)^k \binom{n}{k}(n-k)! = n! \sum_{k=0}^{n} (-1)^k \frac{1}{k!}.$$

Since $\sum_{k=0}^{\infty} (-1)^k \frac{1}{k!}$ is an alternating series with limit $1/e$, we have

$$|1/e - \sum_{k=0}^{n} (-1)^k \frac{1}{k!}| \leq 1/(n+1)!,$$

so

$$|D_n - n!/e| \leq n!/(n+1)! = 1/(n+1).$$

Therefore, D_n is the integer closest to $n!/e$.

Example 1.9.14. Ten diners leave coats in the wardrobe of a restaurant. In how many ways can the coats be returned so that no customer gets his own coat back? The number of ways in which the coats can be returned, each to the wrong customer, is the number of derangements of 10 objects, $D_{10} = 1,333,961$.

The primary goal of our discussion of inclusion-exclusion is to obtain a formula for the *Euler φ-function*:

Definition 1.9.15. For each natural number n, $\varphi(n)$ is defined to be the the cardinality of the set of natural numbers $k < n$ such that k is relatively prime to n.

Lemma 1.9.16. *Let k and n be be natural numbers, with k dividing n. The number of natural numbers $j \le n$ such that k divides j is n/k.*

Proof. Say $kd = n$. The set of natural numbers that are no greater than n and divisible by k is $\{k, 2k, 3k, \dots, dk = n\}$, so the size of this set is $d = n/k$. ■

Corollary 1.9.17. *If p is a prime number, then for all $k \ge 1$, $\varphi(p^k) = p^{k-1}(p-1)$.*

Proof. A natural number j less than p^k is relatively prime to p^k if, and only if, p does not divide j. The number of natural numbers $j \le p^k$ such that p *does* divide j is p^{k-1}, so the number of natural numbers $j \le n$ such that p *does not* divide j is $p^k - p^{k-1}$. ■

Let n be a natural number with prime factorization $n = p_1^{k_1} \cdots p_s^{k_s}$. A natural number is relatively prime to n if it is not divisible by any of the p_i appearing in the prime factorization of n. Let us take our "universal set" to be the set of natural numbers less than or equal to n. For each i $(1 \le i \le s)$, let A_i be the set of natural numbers less than or equal to n that are divisible by p_i. Then $\varphi(n) = |A_1' \cap A_2' \cap \cdots \cap A_s'|$. In order to use the inclusion-exclusion formula, we need to know $|A_{i_1} \cap \cdots \cap A_{i_r}|$ for each choice of r and of $\{i_1, \dots, i_r\}$. In fact, $A_{i_1} \cap \cdots \cap A_{i_r}$ is the set of natural numbers less than or equal to n that are divisible by each of $p_{i_1}, p_{i_2}, \dots, p_{i_r}$ and thus by $p_{i_1} p_{i_2} \cdots p_{i_r}$. Since $p_{i_1} p_{i_2} \cdots p_{i_r}$ divides n, the number of natural numbers $a \le n$ such that $p_{i_1} p_{i_2} \cdots p_{i_r}$ divides a is

$\dfrac{n}{p_{i_1}p_{i_2}\cdots p_{i_r}}$. Thus we have the formula

$$\begin{aligned}\varphi(n) &= |A_1' \cap A_2' \cap \cdots \cap A_n'| \\ &= |U| - \sum_i |A_i| + \sum_{i<j} |A_i \cap A_j| - \sum_{i<j<k} |A_i \cap A_j \cap A_k| \\ &\quad + \cdots + (-1)^s |A_1 \cap \cdots \cap A_s| \\ &= n - \sum_i \frac{n}{p_i} + \sum_{i<j} \frac{n}{p_i p_j} - \sum_{i<j<k} \frac{n}{p_i p_j p_k} + \cdots + (-1)^s \frac{n}{p_1 p_2 \cdots p_s}.\end{aligned}$$

The very nice feature of this formula is that the right side factors,

$$\varphi(n) = n(1 - \frac{1}{p_1})(1 - \frac{1}{p_2}) \cdots (1 - \frac{1}{p_s}).$$

Proposition 1.9.18. *Let n be a natural number with prime factorization $n = p_1^{k_1} \cdots p_s^{k_s}$. Then*

(a)
$$\varphi(n) = n(1 - \frac{1}{p_1})(1 - \frac{1}{p_2}) \cdots (1 - \frac{1}{p_s}).$$

(b)
$$\varphi(n) = \varphi(p_1^{k_1}) \cdots \varphi(p_s^{k_s}).$$

Proof. The formula in part (a) was obtained previously. The result in part (b) is obtained by rewriting

$$\begin{aligned}n(1 - \frac{1}{p_1})&(1 - \frac{1}{p_2}) \cdots (1 - \frac{1}{p_s}) \\ &= p_1^{k_1}(1 - \frac{1}{p_1}) p_2^{k_2}(1 - \frac{1}{p_2}) \cdots p_s^{k_s}(1 - \frac{1}{p_s}) \\ &= \varphi(p_1^{k_1}) \cdots \varphi(p_s^{k_s}).\end{aligned}$$

■

Corollary 1.9.19. *If m and n are relatively prime, then $\varphi(mn) = \varphi(m)\varphi(n)$.*

The following theorem of Euler is a substantial generalization of Fermat's little theorem:

Theorem 1.9.20. *(Euler's theorem). Fix a natural number n. If $a \in \mathbb{Z}$ is relatively prime to n, then*

$$a^{\varphi(n)} \equiv 1 \pmod{n}.$$

One proof of this theorem is outlined in the Exercises. Another proof is given in the next section (based on group theory).

Example 1.9.21. Take $n = 7$ and $a = 4$. $\varphi(7) = 6$, and 4 is relatively prime to 7, so $4^6 \equiv 1 \pmod 7$. In fact, $4^6 - 1 = 4095 = 7 \times 585$.

Take $n = 16$ and $a = 7$. $\varphi(16) = 8$, and 7 is relatively prime to 16, so $7^8 \equiv 1 \pmod{16}$. In fact, $7^8 - 1 = 5764801 = 16 \times 360300$.

Exercises 1.9

1.9.1. Prove the binomial theorem by induction on n.

1.9.2. Prove that $3^n = \sum_{k=0}^{n} \binom{n}{k} 2^k$.

1.9.3. Prove that $3^n = \sum_{k=0}^{n} (-1)^k \binom{n}{k} 4^{n-k}$.

1.9.4. Prove that

$$n(n-1)2^{n-2} = \sum_{k=0}^{n} k(k-1) \binom{n}{k}.$$

Use this to find a formula for $\sum_{k=0}^{n} k^2 \binom{n}{k}$.

1.9.5. Bernice lives in a city whose streets are arranged in a grid, with streets running north-south and east-west. How many shortest paths are there from her home to her business, that lies 4 blocks east and 10 blocks north of her home? How many paths are there which avoid the pastry shop located 3 blocks east and 6 blocks north of her home?

1.9.6. Bernice travels from home to a restaurant located n blocks east and n blocks north. On the way, she must pass through one of $n + 1$ intersections located n blocks from home: The intersections have coordinates $(0, n)$, $(1, n-1)$, ... , $(n, 0)$. Show that the number of paths to the restaurant that pass through the intersection with coordinates $(k, n-k)$ is $\binom{n}{k}^2$. Conclude that

$$\sum_{k=0}^{n} \binom{n}{k}^2 = \binom{2n}{n}.$$

1.9.7. How many natural numbers ≤ 1000 are there that are divisible by 7? How many are divisible by both 7 and 6? How many are not divisible by any of 7, 6, 5?

1.9.8. A party is attended by 10 married couples (one man and one woman per couple).

(a) In how many ways can the men and women form pairs for a dance so that no man dances with his wife?

(b) In how many ways the men and women form pairs for a dance so that exactly three married couples dance together?

(c) In how many ways can the 20 people sit around a circular table, with men and women sitting in alternate seats, so that no man sits opposite his wife? Two seating arrangements are regarded as being the same if they differ only by a rotation of the guests around the table.

1.9.9. Show that Fermat's little theorem is a special case of Euler's theorem.

The following three exercises outline a proof of Euler's theorem.

1.9.10. Let p be a prime number. Show that for all integers k and for all nonnegative integers s,

$$(1 + kp)^{p^s} \equiv 1 \pmod{p^{s+1}}.$$

Moreover, if p is odd and k is not divisible by p, then

$$(1 + kp)^{p^s} \not\equiv 1 \pmod{p^{s+2}}.$$

Hint: Use induction on s.

1.9.11.

(a) Suppose that the integer a is relatively prime to the prime p. Then for all integers $s \geq 0$,

$$a^{p^s(p-1)} \equiv 1 \pmod{p^{s+1}}.$$

Hint: By Fermat's little theorem, $a^{p-1} = 1 + kp$ for some integer k. Thus $a^{p^s(p-1)} = (1 + kp)^{p^s}$. Now use the previous exercise.

(b) Show that the statement in part (a) is the special case of Euler's theorem, for n a power of a prime.

1.9.12. Let n be a natural number with prime factorization $n = p_1^{k_1} \cdots p_s^{k_s}$, and let a be an integer that is relatively prime to n.

(a) Fix an index i $(1 \leq i \leq s)$ and put $b = a^{\prod_{j \neq i} \varphi(p_j^{k_j})}$. Show b is also relatively prime to n.

(b) Show that $a^{\varphi(n)} = b^{\varphi(p_i^{k_i})}$, and apply the previous exercise to show that $a^{\varphi(n)} \equiv 1 \pmod{p_i^{k_i}}$.

(c) Observe that $a^{\varphi(n)} - 1$ is divisible by $p_i^{k_i}$ for each i. Conclude that $a^{\varphi(n)} - 1$ is divisible by n. This is the conclusion of Euler's theorem.

1.10. Groups

We have seen several examples of sets with an operation (or product, or multiplication) satisfying the following three properties:

(a) The product is associative.
(b) There is an identity element e with the property that the product of e with any other element a (in either order) is a.
(c) For each element a there is an inverse element a^{-1} satisfying $aa^{-1} = a^{-1}a = e$.

Examples we have considered so far are as follows:

1. The set of symmetries of a geometric figure with composition of symmetries as the product.
2. The set of permutations of a (finite) set, with composition of permutations as the product.
3. The set of integers with addition as the operation.
4. $\mathbb{Z}_n$ with addition as the operation. Indeed, Proposition 1.7.7, parts (a), (b), and (c), says that addition in $\mathbb{Z}_n$ is associative and has an identity [0], and that all elements of $\mathbb{Z}_n$ have an additive inverse.
5. $K[x]$ with addition as the operation. In fact, Proposition 1.8.2, parts (a), (b), and (c), says that addition in $K[x]$ is associative, that 0 is an identity element for addition, and that all elements of $K[x]$ have an additive inverse.

It is convenient and fruitful to make a *concept* out of the common characteristics of these several examples. So we make the following definition:

Definition 1.10.1. A *group* is a (nonempty) set G with a product, denoted here simply by juxtaposition, satisfying the following properties:

(a) The product is associative: For all $a, b, c \in G$, we have $(ab)c = a(bc)$.
(b) There is an identity element $e \in G$ with the property that for all $a \in G$, $ea = ae = a$.
(c) For each element $a \in G$ there is an element $a^{-1} \in G$ satisfying $aa^{-1} = a^{-1}a = e$.

A few additional examples of groups are as follows:

Example 1.10.2.

(a) Any of the familiar number systems $\mathbb{Q}$, $\mathbb{R}$, or $\mathbb{C}$, with addition as the operation.
(b) The positive real numbers, with multiplication as the operation.
(c) The nonzero complex numbers, with multiplication as the operation.
(d) The set of invertible n-by-n matrices with entries in $\mathbb{R}$, with matrix multiplication as the product. Indeed, the product AB of invertible n-by-n matrices A and B is invertible with inverse $B^{-1}A^{-1}$. The product of matrices is associative, so matrix multiplication defines an associative product on the set of invertible n-by-n matrices. The identity matrix E, with 1's on the diagonal and 0's off the diagonal, is the identity element for matrix multiplication, and the inverse of a matrix is the inverse for matrix multiplication.

Abstraction, the process of elevating recurring phenomena to a concept, has several purposes and advantages. First, it is a tool for organizing our knowledge of phenomena. To understand phenomena, it already helps to classify them according to their common features. Second, abstraction yields efficiency by eliminating the need for redundant arguments; certain results are valid for all groups, or for all finite groups, or for all groups satisfying some additional hypothesis. Instead of of proving these results again and again for symmetry groups, for permutation groups, for groups of invertible matrices, and so on, we can prove them once and for all for all groups (or for all finite groups, or for all groups satisfying property xyz). Indeed, finding more or less the same arguments employed to establish analogous properties of different objects is a clue that we should identify some abstract class of objects, such that the arguments apply to all members of the class.

The most important advantage of abstraction, however, is that it allows us to see different objects of a class *in relation to one another.* This gives us a deeper understanding of the individual objects.

The first way that two groups may be related is that they are essentially the same group. Let's consider an example:

Inside the symmetry group of the square card, consider the set $\mathcal{R} = \{e, r, r^2, r^3\}$. Verify that this subset of the symmetry group is a group under composition of symmetries (Exercise 1.10.1).

On the other hand, the set $C_4 = \{i, -1, -i, 1\}$ of fourth roots of 1 in $\mathbb{C}$ is a group under multiplication of complex numbers.

The group $\mathcal{R}$ is essentially the same as the group H . Define a bijection between these two groups by

$$\begin{aligned} e &\leftrightarrow 1 \\ r &\leftrightarrow i \\ r^2 &\leftrightarrow -1 \\ r^3 &\leftrightarrow -i. \end{aligned}$$

Under this bijection, the multiplication tables of the two groups match up: If we apply the bijection to each entry in the multiplication table of $\mathcal{R}$, we obtain the multiplication table for H. Verify this statement; see Exercise 1.10.3. So although the groups seem to come from different contexts, they really are essentially the same and differ only in the names given to the elements.

Two groups G and H are said to be *isomorphic* if there is a *bijective* map $f : H \to G$ between them that makes the multiplication table of one group match up with the multiplication table of the other. The map f is called an *isomorphism*. [The requirement on f is this: Given a, b, c in H, we have $c = ab$ if and only if $f(c) = f(a)f(b)$.]

For another example, the permutation group S_3 is isomorphic to the group of symmetries of an equilateral triangular card, as is shown in Exercise 1.5.2.

Another simple way in which two groups may be related is that one may be *contained* in the other. For example, the set of symmetries of the square card that do not exchange top and bottom is $\{e, r, r^2, r^3\}$; this is a group in its own right. So the group of symmetries of the square card contains the group $\{e, r, r^2, r^3\}$ *as a subgroup*. Another example: The set of eight invertible 3-by-3 matrices $\{E, A, B, C, D, R, R^2, R^3\}$ introduced in Section 1.4 is a group under the operation of matrix multiplication; so it is a *subgroup* of the group of all invertible 3-by-3 matrices.

A third way in which two groups may be related to one another is more subtle. A map $f : H \to G$ between two groups is said to be a *homomorphism* if f take products to products, identity to identity, and inverses to inverses. (An isomorphism is a bijective homomorphism.) Actually these requirements are redundant, as we shall see later; it is enough to require $f(ab) = f(a)f(b)$ for all $a, b \in H$, as the other properties follow from this.

For example, the map $f : \mathbb{Z} \to \mathbb{Z}_n$ defined by $f(a) = [a]$ is a homomorphism of groups, because

$$f(a + b) = [a + b] = [a] + [b] = f(a) + f(b).$$

For another example, the map $x \mapsto e^x$ is a homomorphism of groups between the group $\mathbb{R}$, with addition, to the group of nonzero real numbers, with multiplication, because $e^{x+y} = e^x e^y$.

Let us use the group concept to obtain a new proof of Euler's theorem, from the previous section. Recall that an element $[a]$ of $\mathbb{Z}_n$ is invertible if, and only if a is relatively prime to n (Proposition 1.9.9). The number of invertible elements is therefore $\varphi(n)$.

Lemma 1.10.3. *The set $\Phi(n)$ of elements in $\mathbb{Z}_n$ possessing a multiplicative inverse forms a group (of cardinality $\varphi(n)$) under multiplication, with identity element* $[1]$.

Proof. The main point is to show that if $[a]$ and $[b]$ are elements of $\Phi(n)$, then their product $[a][b] = [ab]$ is also an element of $\Phi(n)$. We can see this in two different ways.

First, by hypothesis $[a]$ has a multiplicative inverse $[x]$ and $[b]$ has a multiplicative inverse $[y]$. Then $[a][b]$ has multiplicative inverse $[y][x]$, for

$$\begin{aligned}([a][b])([y][x]) &= [a](([b]([y][x])) \\ &= [a](([b][y])[x])) = [a]([1][x]) = [a][x] = [1].\end{aligned}$$

A second way to see that $[ab] \in \Phi(n)$ is that the invertibility of $[a]$ and $[b]$ implies that a and b are relatively prime to n, and it follows that ab is also relatively prime to n. Hence $[ab] \in \Phi(n)$.

It is clear that $[1] \in \Phi(n)$.

Now since multiplication is associative on $\mathbb{Z}_n$, it is also associative on $\Phi(n)$, and since $[1]$ is a multiplicative identity for $\mathbb{Z}_n$, it is also a multiplicative identity for $\Phi(n)$. Finally, every element in $\Phi(n)$ has a multiplicative inverse, by definition of $\Phi(n)$. This proves that $\Phi(n)$ is a group. ■

Now we state a basic theorem about finite groups, whose proof will have to wait until the next chapter (Theorem 2.5.6):

If G *is a finite group of size* $|G|$, *then for every element* $g \in G$, *we have* $g^{|G|} = e$.

Now we observe that Euler's theorem is an immediate consequence of this principle of group theory:

New proof of Euler's theorem. Since $\Phi(n)$ is a group of size $|\Phi(n)| = \varphi(n)$, we have $[a]^{\varphi(n)} = [1]$, for all $[a] \in \Phi(n)$, by the theorem of group theory just mentioned. But $[a] \in \Phi(n)$ if, and only if, a is relatively prime to n, and $[a]^{\varphi(n)} = [1]$ translates to $a^{\varphi(n)} \equiv 1 \pmod{n}$. ■

This proof is meant to impress the reader with the power of abstraction. Our first proof of Euler's theorem (in the Exercises of the previous section) was entirely elementary, but one has to admit that it was somewhat involved, and it required detailed information about the Euler φ function. The proof here requires only that we recognize that $\Phi(n)$ is a group of size $\varphi(n)$ and apply a general principle about finite groups.

Exercises 1.10

1.10.1. Show that set of symmetries $\mathcal{R} = \{e, r, r^2, r^3\}$ of the square card is a group under composition of symmetries.

1.10.2. Show that $C_4 = \{i, -1, -i, 1\}$ is a group under complex multiplication, with 1 the identity element.

1.10.3. . Consider the group $C_4 = \{i, -1, -i, 1\}$ of fourth roots of unity in the complex numbers and the group $\mathcal{R} = \{e, r, r^2, r^3\}$ contained in the group of rotations of the square card. Show that the bijection

$$\begin{aligned} e &\leftrightarrow 1 \\ r &\leftrightarrow i \\ r^2 &\leftrightarrow -1 \\ r^3 &\leftrightarrow -i \end{aligned}$$

produces a matching of the multiplication tables of the two groups. That is, if we apply the bijection to each entry of the multiplication table of H, we produce the multiplication table of $\mathcal{R}$. Thus, the two groups are isomorphic.

1.10.4. Show that the group $C_4 = \{i, -1, -i, 1\}$ of fourth roots of unity in the complex numbers is isomorphic to $\mathbb{Z}_4$.

The next several exercises give examples of groups coming from various areas of mathematics and require some topology or real and complex analysis. Skip the exercises for which you do not have the appropriate background.

1.10.5. An isometry of $\mathbb{R}^3$ is a bijective map $T : \mathbb{R}^3 \to \mathbb{R}^3$ satisfying $d(T(x), T(y)) = d(x, y)$ for all $x, y \in \mathbb{R}^3$. Show that the set of isometries of $\mathbb{R}^3$ forms a group. (You can replace $\mathbb{R}^3$ with any metric space.)

1.10.6. A homeomorphism of $\mathbb{R}^3$ is a bijective map $T : \mathbb{R}^3 \to \mathbb{R}^3$ such that both T and its inverse are continuous. Show that the set of homeomorphisms of $\mathbb{R}^3$ forms a group under composition of maps. (You can replace $\mathbb{R}^3$ with any metric space or, more generally, with any topological space. Do not confuse the similar words homomorphism and homeomorphism.)

1.10.7. A C^1 diffeomorphism of $\mathbb{R}^3$ is a bijective map $T : \mathbb{R}^3 \to \mathbb{R}^3$ having continuous first–order partial derivatives. Show that the set of C^1 diffeomorphisms of $\mathbb{R}^3$ forms a group under composition of maps. (You can replace $\mathbb{R}^3$ with any open subset of $\mathbb{R}^n$.)

1.10.8. Show that the set of bijective holomorphic (complex differentiable) maps from an open subset U of $\mathbb{C}$ onto itself forms a group under composition of maps.

1.10.9. Show that the set of affine maps transformations of $\mathbb{R}^n$, $\mathbf{x} \mapsto S(\mathbf{x}) + \mathbf{b}$, where S is an invertible linear transformation and $\mathbf{b}$ is a vector, forms a group, under composition of maps.

1.10.10. A fractional linear transformation of $\mathbb{C}$ is a transformation of the form $z \mapsto \dfrac{az+b}{cz+d}$, where a, b, c, d are complex numbers. Actually such a transformation should be regarded as a transformation of $\mathbb{C} \cup \{\infty\}$; for example, $z \mapsto \dfrac{2z+1}{3z-1/2}$ maps $1/6$ to ∞ and ∞ to $2/3$. Show that the set of fractional linear transformations is closed under composition. Find the condition for a fractional linear transformation to be invertible, and show that the set of invertible fractional linear transformations forms a group.

1.11. Rings and Fields

You are familiar with several algebraic systems having two operations, addition and multiplication, satisfying several of the usual laws of arithmetic:

1. The set $\mathbb{Z}$ of integers
2. The set $K[x]$ of polynomials over a field K
3. The set $\mathbb{Z}_n$ of integers modulo n
4. The set $\text{Mat}_n(\mathbb{R})$ of n-by-n matrices with real entries, with entry-by-entry addition of matrices, and matrix multiplication

In each of these examples, the set is group under the operation of addition, multiplication is associative, and there are distributive laws relating multiplication and addition: $x(a+b) = xa + xb$, and $(a+b)x = ax + bx$. In the first three examples, multiplication is commutative, but in the fourth example, it is not.

Again is fruitful to make a *concept* out of the common characteristics of these examples, so we make the following definition:

Definition 1.11.1. A *ring* is a (nonempty) set R with two operations: addition, denoted here by $+$, and multiplication, denoted by juxtaposition, satisfying the following requirements:

(a) Under addition, R is an abelian group.
(b) Multiplication is associative.
(c) Multiplication distributes over addition: $a(b+c) = ab+ac$, and $(b+c)a = ba+ca$ for all $a, b, c \in R$.

Multiplication need not be commutative in a ring, in general. If multiplication *is* commutative, the ring is called a *commutative ring*.

Some additional examples of rings are as follows:

Example 1.11.2.

(a) The familiar number systems $\mathbb{R}$, $\mathbb{Q}$, $\mathbb{C}$ are rings. The set of natural numbers $\mathbb{N}$ is *not* a ring, because it is not a group under addition.
(b) The set $K[X, Y]$ of polynomials in two variables over a field K is a commutative ring.
(c) The set of real–valued functions on a set X, with pointwise addition and multiplication of functions, $(f+g)(x) = f(x) + g(x)$, and $(fg)(x) = f(x)g(x)$, for functions f and g and $x \in X$, is a commutative ring.
(d) The set of n-by-n matrices with integer entries is a noncommutative ring.

There are many, many variations on these examples: matrices with complex entries or with with polynomial entries; functions with complex values, or integer values; continuous or differentiable functions; polynomials with any number of variables.

Many rings have an *identity element for multiplication*, usually denoted by 1. (The identity element satisfies $1a = a1 = a$ for all elements a of the ring. Some rings do not have an identity element!)

Example 1.11.3.

(a) The identity matrix (diagonal matrix with diagonal entries equal to 1) is the multiplicative identity in the n-by-n matrices.
(b) The constant polynomial 1 is the multiplicative identity in the ring $\mathbb{R}[x]$.
(c) The constant function 1 is the multiplicative identity in the ring of real–valued functions on a set X.

In a ring with a multiplicative identity, nonzero elements may or may not have *multiplicative inverses*; A multiplicative inverse for an element a is an element b such that $ab = ba = 1$. An element with a multiplicative inverse is called a *unit* or an *invertible element*.

Example 1.11.4.

(a) Some nonzero square matrices have multiplicative inverses, and some do not. The condition for a square matrix to be invertible is that the rows (or columns) are linearly independent or, equivalently, that the determinant is nonzero.
(b) In the ring of integers, the only units are ± 1. In the real numbers, every nonzero element is a unit.
(c) In the ring $\mathbb{Z}_n$, the units are the classes $[a]$ where a is relatively prime to n, by Proposition 1.9.9.
(d) In the ring $\mathbb{R}[x]$, the units are nonzero constant polynomials.
(e) In the ring of real–valued functions defined on a set, the units are the functions that never take the value zero.

Just as with groups, rings may be related to one another, and understanding their relations helps us to understand their structure. For example, one ring may be contained in another as a *subring*.

Example 1.11.5.

(a) $\mathbb{Z}$ is a subring of $\mathbb{Q}$.
(b) The ring of 3-by-3 matrices with *rational* entries is a subring of the ring of 3-by-3 matrices with *real* entries.

(c) The set of *upper triangular* 3-by-3 matrices with real entries is a subring of the ring of *all* 3-by-3 matrices with real entries.
(d) The set of *continuous* functions from $\mathbb{R}$ to $\mathbb{R}$ is a subring of the ring of *all* functions from $\mathbb{R}$ to $\mathbb{R}$.
(e) The set of *rational–valued* functions on a set X is a subring of the ring of *real–valued* functions on X.

A second way in which two rings may be related to one another is by a *homomorphism*. A map $f : R \to S$ between two rings is said to be a *homomorphism* if it "respects" the ring structures. More explicitly, f must take sums to sums and products to products. In notation, $f(a+b) = f(a)+f(b)$ and $f(ab) = f(a)f(b)$ for all $a, b \in R$. A bijective homomorphism is called an *isomorphism*.

Example 1.11.6.

(a) The map $f : \mathbb{Z} \to \mathbb{Z}_n$ defined by $f(a) = [a]$ is a homomorphism of rings, because

$$f(a + b) = [a + b] = [a] + [b] = f(a) + f(b),$$

and

$$f(ab) = [ab] = [a][b] = f(a)f(b).$$

(b) Let T be any n-by-n matrix with real entries. We can define a ring homomorphism from $\mathbb{R}[x]$ to $\mathrm{Mat}_n(\mathbb{R})$ by

$$a_0 + a_1x + \cdots + a_sx^s \mapsto a_0 + a_1T + \cdots + a_sT^s.$$

Let us observe that we can interpret, and prove, the Chinese remainder theorem (Proposition 1.7.9) in terms of a certain ring homomorphism.

First we need the idea of the *direct sum of rings*. Given rings R and S, we define a ring structure on the Cartesian product $R \times S$, by $(r, s) + (r', s') = (r + r', s + s')$ and $(r, s)(r', s') = (rr', ss')$. It is straightforward to check that these operations make the Cartesian product into a ring. The Cartesian product of R and S, endowed with these operations, is called the *direct sum* of R and S and is usually denoted $R \oplus S$.

Proof of the Chinese remainder theorem. Given a and b relatively prime, we want to define a map from $\mathbb{Z}_{ab}$ to $\mathbb{Z}_a \oplus \mathbb{Z}_b$ by $[x]_{ab} \mapsto ([x]_a, [x]_b)$. (Here, as we have several rings $\mathbb{Z}_n$ in the picture, we denote the class of x in $\mathbb{Z}_n$ by $[x]_n$.) We want to check two things:

1. The map is well defined.
2. The map is one to one.

The first assertion is this: If $x \equiv y \pmod{ab}$, then $x \equiv y \pmod{a}$ and $x \equiv y \pmod{b}$. But this is actually evident, as it just says that if $x - y$ is divisible by ab, then it is divisible by both a and b.

The second assertion is similar but slightly more subtle. We have to show if $[x]_{ab} \neq [y]_{ab}$, then $([x]_a, [x]_b) \neq ([y]_a, [y]_b)$; that is, $[x]_a \neq [y]_a$ or $[x]_b \neq [y]_b$. Equivalently, if $x - y$ is not divisible by ab, then $x - y$ is not divisible by a or $x - y$ is not divisible by b. The (equally valid) contrapositive statement is: If $x - y$ is divisible by both a and b, then it is divisible by ab. This was shown in Exercise 1.6.11.

Now, since both $\mathbb{Z}_{ab}$ and $\mathbb{Z}_a \oplus \mathbb{Z}_b$ have ab elements, a one-to-one map between them must also be onto. This means that, given integers α and β, there exists an integer x such that

$$([x]_a, [x]_b) = ([\alpha]_a, [\beta]_b),$$

or $[x]_a = [\alpha]_a$ and $[x]_b = [\beta]_b$. But this means that $x \equiv \alpha \pmod{a}$ and $x \equiv \beta \pmod{b}$. ■

We didn't need it here (in order to prove the Chinese remainder theorem), but the (bijective) map from $\mathbb{Z}_{ab}$ to $\mathbb{Z}_a \oplus \mathbb{Z}_b$ defined by $[x]_{ab} \mapsto ([x]_a, [x]_b)$ is a ring isomorphism (exercise).

A *field* is a special sort of ring:

Definition 1.11.7. A *field* is a commutative ring with multiplicative identity element $1 \neq 0$ (in which *every nonzero element is a unit*.

Example 1.11.8.

(a) $\mathbb{R}$, $\mathbb{Q}$, and $\mathbb{C}$ are fields.

(b) $\mathbb{Z}$ is not a field. $\mathbb{R}[x]$ is not a field.

(c) If p is a prime, then $\mathbb{Z}_p$ is a field. This follows at once from Proposition 1.9.9.

Exercises 1.11

1.11.1. Show that the only units in the ring of integers are ± 1.

1.11.2. Let K be any field. (If you prefer, you may take $K = \mathbb{R}$.) Show that the set $K[x]$ of polynomials with coefficients in K is a commutative ring with the usual addition and multiplication of polynomials. Show that the constant polynomial 1 is the multiplicative identity, and the only units are the constant polynomials.

1.11.3. A Laurent polynomial is a "polynomial" in which negative as well as positive powers of the variable x are allowed, for example, $p(x) = 7x^{-3} + 4x^{-2} + 4 + 2x$. Show that the set of Laurent polynomials with coefficients in a field K forms a ring with identity. This ring is denoted by $K[x, x^{-1}]$. (If you prefer, you may take $K = \mathbb{R}$.) What are the units?

1.11.4. A trigonometric polynomial is a finite linear combination of the functions $t \mapsto e^{int}$, where n is an integer; for example, $f(t) = 3e^{-i2t} + 4e^{it} + i\sqrt{3}e^{i7t}$. Show that the set of trigonometric polynomials is a subring of the ring of continuous complex–valued functions on $\mathbb{R}$. Show that the ring of trigonometric polynomials is isomorphic to the ring of Laurent polynomials with complex coefficients.

1.11.5. Show that the set of polynomials with real coefficients in three variables, $\mathbb{R}[x, y, z]$ is a ring with identity. What are the units?

1.11.6.

(a) Let X be any set. Show that the set of functions from X to $\mathbb{R}$ is a ring. What is the multiplicative identity? What are the units?

(b) Let $S \subseteq X$. Show that the set of functions from X to $\mathbb{R}$ whose restriction to S is zero is a ring. Does this ring have a multiplicative identity element?

1.11.7. Show that the set of continuous functions from $\mathbb{R}$ into $\mathbb{R}$ is a ring, with the operations of pointwise addition and multiplication of functions. What is the multiplicative identity? What are the units?

1.11.8. Show that the set of continuous functions $f : \mathbb{R} \to \mathbb{R}$ such that $\lim_{x \to \pm\infty} f(x) = 0$, with the operations of pointwise addition and multiplication of functions is a ring without multiplicative identity.

1.11.9. Let R be a ring. Show that the set of 2-by-2 matrices with entries in R is a ring.

1.11.10. Show that the (bijective) map from $\mathbb{Z}_{ab}$ to $\mathbb{Z}_a \oplus \mathbb{Z}_b$ defined by $[x]_{ab} \mapsto ([x]_a, [x]_b)$ is a ring homomorphism.

1.11.11. Show that the set of real numbers of the form $a + b\sqrt{2}$, where a and b are rational is a field. Show that $a + b\sqrt{2} \mapsto a - b\sqrt{2}$ is an isomorphism of this field onto itself.

1.12. An Application to Cryptography

Cryptography is the science of secret or secure communication. The sender of a message, who wishes to keep it secret from all but the intended recipient, *encrypts* the message by applying some transformation rule to it. The recipient *decrypts* the message by applying the inverse transformation.

All sorts of encryption/decryption rules have been used, from simple letter by letter substitutions to rules that transform the message as a whole. But one feature traditional methods of cryptography have in common is that if one knows the encryption rule or *key*, then one can also derive the decryption key. Therefore, the sender and recipient must somehow manage to transmit the key secretly in order to keep their communications secure.

A remarkable *public key* cryptography system, based on properties of congruences of integers, circumvents the problem of transmitting the key. The trick is that it is impractical to derive the decryption key from the encryption key, so that the recipient can simply publish the encryption key; anybody can encrypt messages with this key, but only the recipient has the key to decrypt them.

The RSA public key cryptography method was discovered in 1977 by R. Rivest, A. Shamir, and L. Adleman[9] and is very widely used. It's a good bet that you are using this method when you use an automatic teller machine to communicate with your bank, or when you use a Web browser to transmit personal or confidential information over the internet by a secure connection.

The RSA method is based on the following number theoretic observation:

Let p and q be distinct prime numbers (in practice, very large prime numbers). Let $n = pq$. Recall that

$$\varphi(n) = \varphi(p)\varphi(q) = (p-1)(q-1),$$

by Proposition 1.9.18. Consider the *least common multiple* m of $p-1$ and $q-1$,

$$m = \text{l.c.m.}(p-1, q-1) = \varphi(n)/\text{g.c.d.}(p-1, q-1).$$

If a is relatively prime to n, then $a^{\varphi(n)} \equiv 1 \pmod{n}$, by Euler's theorem (Theorem 1.9.20). In fact, we have the following:

Lemma 1.12.1. *If a is relatively prime to n, then $a^m \equiv 1 \pmod{n}$.*

[9] R. L. Rivest, A. Shamir, L. Adleman, "Public key cryptography," *Communications of the ACM*, 21, 120–126, 1978.

Proof. If a is relatively prime to n, then a is also relatively prime to p and $a^m \equiv 1 \pmod{p}$, by Fermat's little theorem. For the same reason, $a^m \equiv 1 \pmod{q}$. But then $a^m - 1$ is divisible by both p and q, and hence by $n = pq = \text{l.c.m.}(p, q)$. ■

Let r be any natural number relatively prime to m, and let s be an inverse of r modulo m, $rs \equiv 1 \pmod{m}$. We can encrypt and decrypt natural numbers a that are relatively prime to n by first raising them to the rth power modulo n, and then raising the result to the sth power modulo n.

Lemma 1.12.2. *Suppose a is relatively prime to n. If*

$$b \equiv a^r \pmod{n},$$

then

$$b^s \equiv a \pmod{n}.$$

Proof. Write $rs = 1 + tm$ Then $b^s \equiv a^{rs} = a^1 a^{tm} \equiv a \pmod{n}$. ■

Here is how these observations can be used to encrypt and decrypt information: I pick two very large primes p and q, and I compute the quantities n, m, r, and s. I publish n and r (or I send them to you privately, but I don't worry very much if some snoop intercepts them). I keep p, q, m, and s secret.

Any message that you might like to send me is first encoded as a natural number by a standard procedure; for example, a text message is converted into a list of ASCII character codes (which are in the range $0 \le d \le 255$).[10] The list is then converted into an integer by the rule $a = \sum_i d_i (256)^i$, where the d_i are the ascii codes read in reversed order. We suppose that $a < n$; if not, you break your message into several pieces, which are treated separately.

Now a is almost certainly relatively prime to n, since otherwise it would be equal to either p or q, and it is extremely unlikely to just guess p or q by chance. To encrypt your message, you raise a to the rth power and reduce modulo n. (To make the computation practical, you compute the sequence $a_1, a_2, \dots a_r$, where $a_1 = a$, and $a_j =$ the remainder of $a_{j-1}a$ upon division by n for $2 \le j \le r$.) You transmit the result b to me, and I decrypt it by raising b to the sth power and reducing modulo n. According to the previous lemma, I recover the natural number a, from which I extract the base 256 digits, which are the character codes of your message.

---

[10] ASCII codes are the computer industry's standard encoding of textual characters by integers.

Now we understand why this procedure succeeds in transmitting your message, but perhaps not why the transmission cannot be intercepted and decrypted by an unfriendly snoop. If the snoop just factors n, he recovers p and q, and therefore can compute m and s, which enables him to decrypt the message. What foils the snoop and makes the RSA method useful for secure communications is that it is at present computationally difficult (which means either too time consuming or requiring too expensive computer resources) to factor very large integers.[11]

On the other hand, it is computationally easy to find large prime numbers and to do the computations necessary for encryption and decryption. So if my primes p and q are sufficiently large, you and I will still be able to do our computations quickly and inexpensively, but the snoop will not be able to factor n to find p and q and decrypt our secret message.

Example 1.12.3. I find two (randomly chosen) 50–digit prime numbers:

$$p = 4588423984\ 0513596008\ 9179371668\ 8547296304\ 3161712479$$

and

$$q = 8303066083\ 0407235737\ 6288737707\ 9465758615\ 4960341401.$$

Their product is

$$\begin{aligned} n =& 38097987556587433854770986078646810108958511558183\\ &98481072459510812271047829671161055819764204 3079. \end{aligned}$$

I choose a small prime $r = 55589$, and I send you n and r by email. I secretly compute

$$\begin{aligned} m =& 19048993778293716927385493039323405054479255779091\\ &27534955026837138188081833679515740004499759994600, \end{aligned}$$

check that r and m are relatively prime, and compute

$$\begin{aligned} s =& 13006252295510587094189792734309302898824590352325\\ &26859436543985385236803072501862237346791422594309. \end{aligned}$$

Your secret text is "ALGEBRA IS REALLY INTERESTING". The ASCII codes for the characters are 65, 76, 71, 69, 66, 82, 65, 32, 73, 83,

[11]No algorithm is known for factoring integers which has the property that the time required to factor n is bounded by n^k for some k, nor is it known whether such an algorithm is possible in principle. At present, it is practically impossible to factor 300 digit integers. It is possible, however, that the U.S. National Security Agency has factorization algorithms more capable than those which are publicly known.

32, 82, 69, 65, 76, 76, 89, 32, 73, 78, 84, 69, 82, 69, 83, 84, 73, 78, 71. You convert this list of character codes into the integer

$$a = \sum_i d_i (256)^{i-1} =$$

$$1394756013\ 7671806612\ 0093742431\ 8737275883\ 4636804720$$
$$235524153\ 9381439874\ 7285049884\ 8199313352\ 6469320736,$$

where d_i are the character codes, read in reversed order. Now you compute $b =$ the remainder of $a^r \pmod n =$

$$1402590192\ 4491156271\ 5456170360\ 6218336917\ 7495217553$$
$$3838307479\ 8636900168\ 3433148116\ 7995123149\ 9324473812.$$

You send me b in an e–mail message. I recover a by raising b to the sth power and reducing modulo n. I then recover the list of character codes by extracting the base 256 digits of a and finally I use an ASCII character code table to restore your message "ALGEBRA IS REALLY INTERESTING".

It goes without saying that one does not do such computations by hand. The Mathematica notebook **RSA.nb** on my Web site contains programs to automate all the steps of finding large primes, and encoding, encrypting, decrypting, and decoding a text message.

Exercises 1.12

1.12.1.

(a) Let $G \cong H \times K$ be a direct product of finite groups. Show that every element in G has order dividing $\text{l.c.m.}(|H|, |K|)$.

(b) Let $n = pq$ the product of two primes, and let $m = \text{l.c.m.}(p - 1, q-1)$. Use the isomorphism $\Phi(pq) \cong \Phi(p) \times \Phi(q)$ to show that $a^m \equiv 1 \pmod n$ whenever a is relatively prime to n.

1.12.2. Show that if a snoop were able to find $\varphi(n)$, then he could also find p and q.

1.12.3. Suppose that André needs to send a message to Bernice in such a way that Bernice will know that the message comes from André and not from some impostor. The issue here is not the secrecy of the message but rather its authenticity. How can the RSA method be adapted to solve the problem of message authentification?

1.12.4. How can André and Bernice adapt the RSA method so that they can exchange messages that are both secure and authenticated?

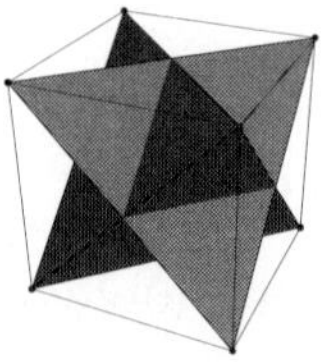

CHAPTER 2

Basic Theory of Groups

2.1. First Results

In the previous chapter, we saw many examples of groups and finally arrived at a definition, or collection of axioms, for groups. In this section we will try our hand at obtaining some first theorems about groups. For many students, this will be the first experience with constructing proofs concerning an algebraic object described by axioms. I would like to urge both students and instructors to take time with this material and not to go on before mastering it.

Our first results concern the *uniqueness of the identity element in a group.*

Proposition 2.1.1. *(Uniqueness of the identity). Let* G *be a group and suppose* e *and* e' *are both identity elements in* G*; that is, for all* $g \in G$, $eg = ge = e'g = ge' = g$. *Then* $e = e'$.

Proof. Since e' is an identity element, we have $e = ee'$. And since e is an identity element, we have $ee' = e'$. Putting these two equations together gives $e = e'$. ■

Likewise, *inverses in a group are unique*:

Proposition 2.1.2. *(Uniqueness of inverses). Let* G *be a group and* $h, g \in G$. *If* $hg = e$, *then* $h = g^{-1}$. *Likewise, if* $gh = e$, *then* $h = g^{-1}$.

Proof. Assume $hg = e$. Then $h = he = h(gg^{-1}) = (hg)g^{-1} = eg^{-1} = g^{-1}$. The proof when $gh = e$ is similar. ■

Corollary 2.1.3. *Let* g *be an element of a group* G*. We have* $g = (g^{-1})^{-1}$.

Proof. Since $gg^{-1} = e$, it follows from the proposition that g is the inverse of g^{-1}. ■

Proposition 2.1.4. *Let* G *be a group and let* $a, b \in G$*. Then* $(ab)^{-1} = b^{-1}a^{-1}$

Proof. It suffices to show that $ab(b^{-1}a^{-1}) = e$. But by associativity, $ab(b^{-1}a^{-1}) = a(bb^{-1})a^{-1} = aea^{-1} = aa^{-1} = e$. ■

Let G be a group and $a \in G$. We define a map $L_a : G \longrightarrow G$ by $L_a(x) = ax$. L_a stands for left multiplication by a. Likewise, we define $R_a : G \longrightarrow G$ by $R_a(x) = xa$. R_a stands for right multiplication by a.

Proposition 2.1.5. *Let* G *be a group and* $a \in G$*. The map* $L_a : G \longrightarrow G$ *defined by* $L_a(x) = ax$ *is a bijection. Similarly, the map* $R_a : G \longrightarrow G$ *defined by* $R_a(x) = xa$ *is a bijection.*

Proof. The assertion is that the map L_a has an inverse map. What could the inverse map possibly be except left multiplication by a^{-1}? So let's try that.

We have

$$L_{a^{-1}}(L_a(x)) = a^{-1}(ax) = (a^{-1}a)x = ex = x,$$

so

$L_{a^{-1}} \circ L_a = \mathrm{id}_G$. A similar computation shows that $L_a \circ L_{a^{-1}} = \mathrm{id}_G$. This shows that L_a and $L_{a^{-1}}$ are inverse maps, so both are bijective.

The proof for R_a is similar. ■

Corollary 2.1.6. *Let* G *be a group and let* a *and* b *be elements of* G*. The equation* $ax = b$ *has a unique solution in* G*, and likewise the equation* $xa = b$ *has a unique solution in* G.

Proof. The *existence* of a solution to $ax = b$ for all b is equivalent to the *surjectivity* of the map L_a. The *uniqueness* of the solution for all b is equivalent to the *injectivity* of L_a.

Likewise, the existence and uniqueness of solutions to $xa = b$ for all b is equivalent to the surjectivity and injectivity of R_a. ■

Corollary 2.1.7. *(Cancellation). Suppose* a, x, y *are elements of a group* G. *If* $ax = ay$, *then* $x = y$. *Similarly, if* $xa = ya$, *then* $x = y$.

Proof. The first assertion is equivalent to the injectivity of L_a, the second to the injectivity of R_a. ■

Perhaps you have already noticed in the group multiplication tables that you have computed that each row and column contains each group element exactly once. This is always true.

Corollary 2.1.8. *If* G *is a finite group, each row and each column of the multiplication table of* G *contains each element of* G *exactly once.*

Proof. You are asked to prove this in Exercise 2.1.5. ■

Example 2.1.9. The conclusions of Proposition 2.1.5 and its corollaries can be false if we are not working in a group. For example, let G be the set of nonzero integers. The equation $2x = 3$ has no solution in G; we cannot, in general, divide in the integers. Or consider $\mathbb{Z}_{12}$ with multiplication. We have $[2][8] = [4] = [2][2]$, so cancellation fails.

Definition 2.1.10. The *order* of a group is its size or cardinality. We will denote the order of a group G by $|G|$.

Groups of Small Order

Let's produce some examples of groups of small order.

Example 2.1.11. For any natural number n, $\mathbb{Z}_n$ (with the operation $+$) is a group of order n. This gives us one example of a group of each order $1, 2, 3, 4, \dots$.

Example 2.1.12. *Another group of order 4:* The set of rotational symmetries of the rectangular card is a group of order 4.

Recall that we have a notion of two groups being essentially the same:

Definition 2.1.13. We say that two groups G and H are *isomorphic* if there is a bijection $\varphi : G \longrightarrow H$ such that for all $g_1, g_2 \in G$ $\varphi(g_1 g_2) = \varphi(g_1)\varphi(g_2)$. The map φ is called an *isomorphism*.

You are asked to show in Exercise 2.1.6 that $\mathbb{Z}_4$ is not isomorphic to the group of rotational symmetries of a rectangular card. Thus there exist at least two nonisomorphic groups of order 4.

Definition 2.1.14. A group G is called *abelian* (or *commutative*) if for all elements $a, b \in G$, the products in the two orders are equal: $ab = ba$.

Example 2.1.15. For any natural number n, the symmetric group S_n is a group of order $n!$. According to Exercise 1.5.14, for all $n \geq 3$, S_n is nonabelian.

If two groups are isomorphic, then either they are both abelian or both nonabelian. That is, if one of two groups is abelian and the other is nonabelian, then the two groups are not isomorphic (exercise).

Example 2.1.16. S_3 is nonabelian and $\mathbb{Z}_6$ is abelian. Thus these are two nonisomorphic groups of order 6.

So far we have one example each of groups of order 1, 2, 3, and 5 and two examples each of groups of order 4 and 6. In fact, we can classify all groups of order no more than 5:

Proposition 2.1.17.

(a) *Up to isomorphism, $\mathbb{Z}_1$ is the unique group of order 1.*
(b) *Up to isomorphism, $\mathbb{Z}_2$ is the unique group of order 2.*
(c) *Up to isomorphism, $\mathbb{Z}_3$ is the unique group of order 3.*
(d) *Up to isomorphism, there are exactly two groups of order 4, namely $\mathbb{Z}_4$, and the group of rotational symmetries of the rectangular card.*
(e) *Up to isomorphism, $\mathbb{Z}_5$ is the unique group of order 5.*
(f) *All groups of order no more than 5 are abelian.*
(g) *There are at least two nonisomorphic groups of order 6, one abelian and one nonabelian.*

The statement (c) means, for example, that any group of order 3 is isomorphic to $\mathbb{Z}_3$. Statement (d) means that there are two distinct (nonisomorphic) groups of order 4, and any group of order 4 must be isomorphic to one of them.

Proof. The reader is guided through the proof of statements (a) through (e) in the exercises. The idea is to try to write down the group multiplication table, observing the constraint that each group element must appear exactly once in each row and column.

Statements (a)–(e) give us the complete list, up to isomorphism, of groups of order no more than 5, and we can see by going through the list that all of them are abelian. Finally, we have already seen that $\mathbb{Z}_6$ and S_3 are two nonisomorphic groups of order 6, and S_3 is nonabelian. ■

The definition of isomorphism says that under the bijection, the multiplication tables of the two groups match up, so the two groups differ only by a renaming of elements. Since the multiplication tables match up, one could also expect that the identity elements and inverses of elements match up, and in fact this is so:

Proposition 2.1.18. *If* $\varphi : G \to H$ *is an isomorphism, then* $\varphi(e_G) = e_H$*, and for each* $g \in G$*,* $\varphi(g^{-1}) = \varphi(g)^{-1}$.

Proof. For any $h \in H$, there is a $g \in G$ such that $\varphi(g) = h$. Then $\varphi(e_G)h = \varphi(e_G)\varphi(g) = \varphi(e_G g) = \varphi(g) = h$. Hence by the uniqueness of the identity in H, $\varphi(e_G) = e_H$. Likewise, $\varphi(g^{-1})\varphi(g) = \varphi(gg^{-1}) = \varphi(e_G) = e_H$. This shows that $\varphi(g^{-1}) = \varphi(g)^{-1}$. ■

Exercises 2.1

2.1.1. Determine the symmetry group of a nonsquare rhombus. That is, describe all the symmetries, find the size of the group, and determine whether it is isomorphic to a known group of the same size. If it is an entirely new group, determine its multiplication table.

2.1.2. Consider a square with one pair of opposite vertices painted red and the other pair of vertices painted blue. Determine the symmetry group of the painted square. That is, describe all the (color-preserving) symmetries, find the size of the group and determine whether it is isomorphic to a known group of the same size. If it is an entirely new group, determine its multiplication table.

2.1.3. Prove the following refinement of the uniqueness of the identity in a group: Let G be a group with identity element e, and let $e', g \in G$. Suppose e' and g are elements of G. If $e'g = g$, then $e' = e$. Similarly, if $ge' = g$, then $e' = e$. (This result says that if a group element acts like the identity when multiplied by one element on one side, then it is the identity.)

2.1.4. Prove Proposition 2.1.6

2.1.5. Show that each row and each column of the multiplication table of a finite group contains each group element exactly once. Use Proposition 2.1.6.

2.1.6. Show that the groups $\mathbb{Z}_4$ and the group of rotational symmetries of the rectangle are not isomorphic, although each group has four elements. Hint: In one of the groups, but not the other, each element has square equal to the identity. Show that if two groups G and H are isomorphic, and G has the property that each element has square equal to the identity, then H also has this property.

2.1.7. Suppose that $\varphi : G \to H$ is an isomorphism of groups. Show that for all $g \in G$ and $n \in \mathbb{N}$, $\varphi(g^n) = (\varphi(g))^n$. Show that if $g^n = e$, then also $(\varphi(g))^n = e$.

2.1.8. Suppose that $\varphi : G \to H$ is an isomorphism of groups. Show that G is abelian if, and only if, H is abelian.

The following several exercises investigate groups with a small number of elements by means of their multiplication tables. The requirements $ea = a$ and $ae = a$ for all a determine one row and one column of the multiplication table. The other constraint on the multiplication table that we know is that each row and each column must contain every group element exactly once. When the size of the group is small, these constraints suffice to determine the possible tables.

2.1.9. Show that there is up to isomorphism only one group of order 2. Hint: Call the elements $\{e, a\}$. Show that there is only one possible multiplication table. Since the row and the column labeled by e are known, there is only one entry of the table that is not known. But that entry is determined by the requirement that each row and column contain each group element.

2.1.10. Show that there is up to isomorphism only one group of order 3. Hint: Call the elements $\{e, a, b\}$. Show that there is only one possible multiplication table. Since the row and column labeled by e are known, there are four table entries left to determine. Show that there is only one way to fill in these entries that is consistent with the requirement that each row and column contain each group element exactly once.

2.1.11. Show that any group with four elements must have a nonidentity element whose square is the identity. That is, some nonidentity element must be its own inverse.

Hint: If you already knew that there were exactly two groups of order four, up to isomorphism, namely $\mathbb{Z}_4$ and the group of rotational symmetries of the rectangle, then you could verify the statement by checking that these two groups have the desired property. But our purpose is to prove that these are the only two groups of order 4, so we are not allowed to use this information in the proof!

So you must start by assuming that you have a group G with four elements; you may not assume anything else about G except that it is a group and that it has four elements. You must show that one of the three nonidentity elements has square equal to e. Call the elements of the group $\{e, a, b, c\}$. There are two possibilities: Each nonidentity element has square equal to e, in which case there is nothing more to show, or some element does not have square equal to e. Suppose that $a^2 \neq e$. Thus, $a \neq a^{-1}$. Without loss of generality, write $b = a^{-1}$. Then also $a = b^{-1}$. Then what is the inverse of c?

2.1.12. Show that there are only two distinct groups with four elements, as follows. Call the elements of the group e, a, b, c. Let a denote a nonidentity element whose square is the identity, which exists by Exercise 2.1.11. The row and column labeled by e are known. Show that the row labeled by a is determined by the requirement that each group element must appear exactly once in each row and column; similarly, the column labeled by a is determined. There are now four table entries left to determine. Show that there are exactly two possible ways to complete the multiplication table that are consistent with the constraints on multiplication tables. Show that these two ways of completing the table yield the multiplication tables of the two groups with four elements that we have already encountered.

2.1.13. The group $\Phi(10)$ of invertible elements in the ring $\mathbb{Z}_{10}$ has four elements, $\Phi(10) = \{[1], [3], [7], [9]\}$. Is this group isomorphic to $\mathbb{Z}_4$ or to the rotation group of the rectangle?

The group $\Phi(8)$ also has four elements, $\Phi(8) = \{[1], [3], [5], [7]\}$. Is this group isomorphic to $\mathbb{Z}_4$ or to the rotation group of the rectangle?

2.1.14. Generalizing Exercise 2.1.11, show that any group with an even number of elements must have a nonidentity element whose square is the identity, that is, a nonidentity element that is its own inverse. Hint: Show that there is an even number of nonidentity elements that are not equal to their own inverses.

2.1.15. It is possible to show the uniqueness of the group of order 5 with the techniques that we have at hand. Try it. Hint: Suppose G is a group of order 5. First show it is not possible for a nonidentity element $a \in G$ to satisfy $a^2 = e$, because there is no way to complete the multiplication table that respects the constraints on group multiplication tables.

Next, show that there is no nonidentity element a such that e, a, a^2 are distinct elements but $a^3 = e$. Finally, show that there is no nonidentity element a such that e, a, a^2, a^3 are distinct elements but $a^4 = e$. Consequently, for any nonidentity element a, the elements e, a, a^2, a^3, a^4 are distinct, and necessarily $a^5 = e$.

In Section 2.5, we will be able to obtain the uniqueness of the groups of order 2, 3, and 5 as an immediate corollary of a general result.

2.1.16. Show that the following conditions are equivalent for a group G:

(a) G is abelian.
(b) For all $a, b \in G$, $(ab)^{-1} = a^{-1}b^{-1}$.
(c) For all $a, b \in G$, $aba^{-1}b^{-1} = e$.
(d) For all $a, b \in G$, $(ab)^2 = a^2b^2$.
(e) For all $a, b \in G$ and natural numbers n, $(ab)^n = a^nb^n$. (Use induction.)

2.2. Subgroups and Cyclic Groups

Definition 2.2.1. A nonempty subset H of a group G is called a *subgroup* if H is itself a group *with the group operation inherited from* G.

For a nonempty subset H of G to be a subgroup of G, it is necessary that

1. For all elements h_1 and h_2 of H, the product h_1h_2 is also an element of H.
2. For all $h \in H$, the inverse h^{-1} is an element of H.

These conditions also suffice for H to be a subgroup. Associativity of the product is inherited from G, so it need not be checked. Also, if conditions (1) and (2) are satisfied, then e is automatically in H; indeed, H is nonempty, so contains some element h; according to (2), $h^{-1} \in H$ as well, and then according to (1), $e = hh^{-1} \in H$.

These observations are a great labor–saving device. Very often when we need to check that some set H with an operation is a group, H is already contained in some known group, so we need only check points (1) and (2).

We say that a subset H of a group G is *closed under multiplication* if condition (1) is satisfied. We say that H is *closed under inverses* if condition (2) is satisfied.

Example 2.2.2. An n-by-n matrix A is said to be *orthogonal* if $A^tA = E$. Show that the set $O(n, \mathbb{R})$ of n-by-n real–valued orthogonal matrices is a group.

Proof. If $A \in O(n, \mathbb{R})$, then A has a left inverse A^t, so A is invertible with inverse A^t. Thus $O(n, \mathbb{R}) \subseteq GL(n, \mathbb{R})$. Therefore, it suffices to check that the product of orthogonal matrices is orthogonal and that the inverse of an orthogonal matrix is orthogonal. But if A and B are orthogonal, then $(AB)^t = B^tA^t = B^{-1}A^{-1} = (AB)^{-1}$; hence AB is orthogonal. If $A \in O(n, \mathbb{R})$, then $(A^{-1})^t = (A^t)^t = A = (A^{-1})^{-1}$, so $A^{-1} \in O(n, \mathbb{R})$. ■

Here are some additional examples of subgroups:

Example 2.2.3. In any group G, G itself and $\{e\}$ are subgroups.

Example 2.2.4. The set of all complex numbers of modulus (absolute value) equal to 1 is a subgroup of the group of all nonzero complex numbers, with multiplication as the group operation. See Appendix D.

Proof. For any nonzero complex numbers a and b, $|ab| = |a||b|$, and $|a^{-1}| = |a|^{-1}$. It follows that the set of complex number of modulus 1 is closed under multiplication and under inverses. ■

Example 2.2.5. In the group of symmetries of the square, the subset $\{e, r, r^2, r^3\}$ is a subgroup. Also, the subset $\{e, r^2, a, b\}$ is a subgroup; the latter subgroup is isomorphic to the symmetry group of the rectangle, since each nonidentity element has square equal to the identity, and the product of any two nonidentity elements is the third.

Example 2.2.6. In the permutation group S_4, the set of permutations π satisfying $\pi(4) = 4$ is a subgroup. This subgroup, since it permutes the numbers $\{1, 2, 3\}$ and leaves 4 fixed, is isomorphic to S_3.

Example 2.2.7. In S_4, there are eight 3–cycles. There are three elements that are products of disjoint 2–cycles, namely $(12)(34)$, $(13)(24)$, and $(14)(23)$. These eleven elements, together with the identity, form a subgroup of S_4.

Proof. At the moment we have no theory to explain this fact, so we have to verify by computation that the set is closed under multiplication. The amount of computation required can be reduced substantially by observing some patterns in products of cycles, as in Exercise 2.2.4. The set is clearly closed under inverses.

Eventually we will have a theory that will make this result transparent. ■

We conclude this subsection with some very general observations about subgroups.

Proposition 2.2.8. *Let* G *be a group and let* $H_1, H_2, \dots, H_n$ *be subgroups of* G*. Then* $H_1 \cap H_2 \cap \cdots \cap H_n$ *is a subgroup of* G*. More generally, if* $\{H_\alpha\}$ *is any collection of subgroups, then* $\cap_\alpha H_\alpha$ *is a subgroup.*

Proof. Exercise 2.2.7. ■

For any group G and any subset $S \subseteq G$, there is a smallest subgroup of G that contains S, which is called the *subgroup generated by* S and denoted $\langle S \rangle$. When $S = \{a\}$ is a singleton, the subgroup generated by S is denoted by $\langle a \rangle$. We say that G *is generated by* S or that S *generates* G if $G = \langle S \rangle$.

A "constructive" view of $\langle S \rangle$ is that it consists of all possible products $g_1 g_2 \cdots g_n$, where $g_i \in S$ or $g_i^{-1} \in S$. Another view of $\langle S \rangle$, which is sometimes useful, is that it is the intersection of the family of all subgroups of G that contain S; this family is nonempty since G itself is such a subgroup.

The family of subgroups of a group G are partially ordered by set inclusion.[1] In fact, the family of subgroups forms what is called a *lattice*.[2] This means that, given two subgroups A and B of G there is a unique smallest subgroup C such that $C \supseteq A$ and $C \supseteq B$. In fact, the subgroup C is $\langle A \cup B \rangle$. Furthermore, there is a unique largest subgroup D such that $D \subseteq A$ and $D \subseteq B$; in fact, $D = A \cap B$.

Example 2.2.9. Determine all subgroups of S_3 and all containment relations between these subgroups. The smallest subgroup of S_3 is $\{e\}$. The next smallest subgroups are those generated by single element. There are three distinct subgroups of order 2 generated by the 2–cycles, and there is one subgroup of order 3 generated by either of the 3–cycles. Now we can compute that if a subgroup contains two distinct 2–cycles, or a 2–cycle and a 3–cycle, then it is equal to S_3. Therefore, the only subgroups of S_3 are $\langle (12) \rangle$, $\langle (13) \rangle$, $\langle (23) \rangle$, $\langle (123) \rangle$, $\{e\}$, and S_3. The inclusion relations among these subgroups are shown in Figure 2.2.1.

---

[1]A partial order on a set X is a relation $\le$ satisfying *transitivity* ($x \le y$ and $y \le z$ implies $x \le z$) and *asymmetry* ($x \le y$ and $y \le x$ implies $x = y$). Evidently, the family of subgroups of a group, with the relation $\subseteq$, is a partially ordered set.

[2]The word *lattice* is used in several completely different senses in mathematics. Here we mean a lattice in the context of partially ordered sets.

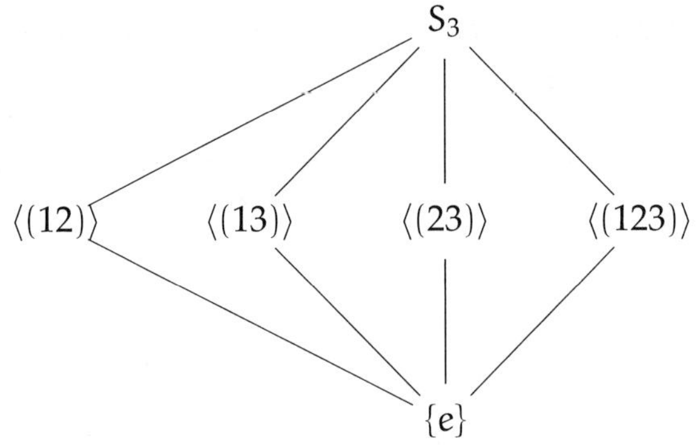

Figure 2.2.1. Lattice of subgroups of S_3.

Cyclic Groups and Cyclic Subgroups

I now discuss a certain type of subgroup that appears in all groups. Take any group G and any element $a \in G$. Consider all powers of a: Define $a^0 = e$, $a^1 = a$, and for $k > 1$, define a^k to be the product of k factors of a. (A little more properly, a^k is defined inductively by declaring $a^k = aa^{k-1}$.) For $k > 1$ define $a^{-k} = (a^{-1})^k$.

We now have a^k defined for all integers k, and it is a fact that $a^k a^l = a^{k+l}$ for all integers k and l. Likewise, we can show that for all integers k, $(a^k)^{-1} = a^{-k}$. Finally, $a^{kl} = (a^k)^l$ for all integers k and l. All of these assertions can be proved by induction, and you are asked to write the proofs in Exercise 2.2.8.

Proposition 2.2.10. *Let a be an element of a group G. The subgroup $\langle a \rangle$ generated by a is $\{a^k : k \in \mathbb{Z}\}$.*

Proof. The formulas $a^k a^l = a^{k+l}$ and $(a^k)^{-1} = a^{-k}$ show that $\{a^k : k \in \mathbb{Z}\}$ is a subgroup of G containing $a = a^1$. Therefore, $\langle a \rangle \subseteq \{a^k : k \in \mathbb{Z}\}$. On the other hand, a subgroup is always is closed under taking integer powers, so $\langle a \rangle \supseteq \{a^k : k \in \mathbb{Z}\}$. ■

Definition 2.2.11. Let a be an element of a group G. The set $\langle a \rangle = \{a^k : k \in \mathbb{Z}\}$ of powers of a is called the *cyclic subgroup generated by* a. If there is an element $a \in G$ such that $\langle a \rangle = G$, we say that G is a *cyclic group*. We say that a is a *generator* of the cyclic group.

Example 2.2.12. Take $G = \mathbb{Z}$, with addition as the group operation, and take any element $d \in \mathbb{Z}$. Because the group operation is addition, the set of powers of d with respect to this operation is the set of integer multiples of d, in the ordinary sense. For example, the third power of d is $d + d + d = 3d$.

Thus, $\langle d \rangle = d\mathbb{Z} = \{nd : n \in \mathbb{Z}\}$ is a cyclic subgroup of $\mathbb{Z}$. Note that $\langle d \rangle = \langle -d \rangle$.

$\mathbb{Z}$ itself is cyclic, $\langle 1 \rangle = \langle -1 \rangle = \mathbb{Z}$.

Example 2.2.13. In $\mathbb{Z}_n$, the cyclic subgroup generated by an element $[d]$ is $\langle [d] \rangle = \{[kd] : k \in \mathbb{Z}\} = \{[k][d] : [k] \in \mathbb{Z}_n\}$.

$\mathbb{Z}_n$ itself is cyclic, since $\langle [1] \rangle = \mathbb{Z}_n$.

Example 2.2.14. Let C_n denote the set of nth roots of unity in the complex numbers.

(a) $C_n = \{e^{2\pi i k/n} : 0 \le k \le n-1\}$.

(b) C_n is a cyclic group of order n with generator $\xi = e^{2\pi i/n}$.

Proof. C_n is a subgroup of the multiplicative group of nonzero complex numbers, as the product of two n^{th} roots of unity is again an nth root of unity, and the inverse of an nth root is also an nth root or unity.

If z is an nth root of unity in $\mathbb{C}$, then $|z|^n = |z^n| = 1$; thus $|z|$ is a positive n^{th} root of 1, so $|z| = 1$. Therefore, z has the form $z = e^{i\theta}$, where $\theta \in \mathbb{R}$. Then $1 = z^n = e^{in\theta}$. Hence $n\theta$ is an integer multiple of 2π, and $z = e^{2\pi i k/n}$ for some $k \in \mathbb{Z}$. These numbers are exactly the powers of $\xi = e^{2\pi i/n}$, which has order n. ∎

Example 2.2.15. The set of all powers of r in the symmetries of the square is $\{e, r, r^2, r^3\}$.

There are two possibilities for a cyclic group $\langle a \rangle$, as we are reminded by these examples. One possibility is that all the powers a^k are distinct, in which case, of course, the subgroup $\langle a \rangle$ is infinite; if this is so, we say that a has infinite order.

The other possibility is that two powers of a coincide. Suppose $k < l$ and $a^k = a^l$. Then $e = (a^k)^{-1}a^l = a^{l-k}$, so some positive power of a is the identity. Let n be the least positive integer such that $a^n = e$. Then $e, a, a^2, \ldots, a^{n-1}$ are all distinct (Exercise 2.2.9) and $a^n = e$. Now any integer k (positive or negative) can be written as $k = mn + r$, where the remainder r satisfies $0 \le r \le n-1$. Hence $a^k = a^{mn+r} = a^{mn}a^r = e^m a^r = ea^r = a^r$. Thus $\langle a \rangle =$

$\{e, a, a^2, \ldots, a^{n-1}\}$. Furthermore, $a^k = a^l$ if, and only if, k and l have the same remainder upon division by n, if, and only if, $k \equiv l \pmod{n}$.

Definition 2.2.16. The *order* of the cyclic subgroup generated by a is called *the order of* a. We denote the order of a by $o(a)$.

The discussion just before the definition establishes the following assertion:

Proposition 2.2.17. *If the order of a is finite, then it is the least positive integer n such that $a^n = e$. Furthermore, $\langle a \rangle = \{a^k : 0 \le k < o(a)\}$.*

Example 2.2.18. What is the order of [4] in $\mathbb{Z}_{14}$? Since the operation in the group $\mathbb{Z}_{14}$ is addition, powers of an element are multiples. We have $2[4] = [8]$, $3[4] = [12]$, $4[4] = [2]$, $5[4] = [6]$, $6[4] = [10]$, $7[4] = [0]$. So the order of [4] is 7.

Example 2.2.19. What is the order of [5] in $\Phi(14)$? We have $[5]^2 = [11]$, $[5]^3 = [13]$, $[5]^4 = [9]$, $[5]^5 = [3]$, $[5]^6 = [1]$, so the order of [5] is 6. Note that $|\Phi(14)| = \varphi(14) = 6$, so this computation shows that $\Phi(14)$ is cyclic, with generator [5].

The next result says that cyclic groups are completely classified by their order. That is, any two cyclic groups of the same order are isomorphic.

Proposition 2.2.20. *Let a be an element of a group G.*
(a) *If a has infinite order, then $\langle a \rangle$ is isomorphic to $\mathbb{Z}$.*
(b) *If a has finite order n, then $\langle a \rangle$ is isomorphic to the group $\mathbb{Z}_n$.*

Proof. For part (a), define a map $\varphi : \mathbb{Z} \longrightarrow \langle a \rangle$ by $\varphi(k) = a^k$. This map is surjective by definition of $\langle a \rangle$, and it is injective because all powers of a are distinct. Furthermore, $\varphi(k + l) = a^{k+l} = a^k a^l$. So φ is an isomorphism between $\mathbb{Z}$ and $\langle a \rangle$.

For part (b), since Z_n has n elements $[0], [1], [2], \ldots, [n-1]$ and $\langle a \rangle$ has n elements $e, a, a^2, \ldots, a^{n-1}$, we can define a bijection $\varphi : \mathbb{Z}_n \longrightarrow \langle a \rangle$ by $\varphi([k]) = a^k$ for $0 \le k \le n-1$. The product (addition) in $\mathbb{Z}_n$ is given by $[k]+[l] = [r]$, where r is the remainder after division of $k + l$ by n. The multiplication in $\langle a \rangle$ is given by the analogous rule: $a^k a^l = a^{k+l} = a^r$, where r is the remainder after division of $k + l$ by n. Therefore, φ is an isomorphism. ■

Subgroups of Cyclic Groups

In this subsection, we determine all subgroups of cyclic groups. Since every cyclic group is isomorphic either to $\mathbb{Z}$ or to $\mathbb{Z}_n$ for some n, it suffices to determine the subgroups of $\mathbb{Z}$ and of $\mathbb{Z}_n$.

Proposition 2.2.21.

(a) *Let H be a subgroup of $\mathbb{Z}$. Then either $H = \{0\}$, or there is a unique $d \in \mathbb{N}$ such that $H = \langle d \rangle = d\mathbb{Z}$.*

(b) *If $d \in \mathbb{N}$, then $d\mathbb{Z} \cong \mathbb{Z}$.*

(c) *If $a, b \in \mathbb{N}$, then $a\mathbb{Z} \subseteq b\mathbb{Z}$ if, and only if, b divides a.*

Proof. Let's check part (c) first. If $a\mathbb{Z} \subseteq b\mathbb{Z}$, then $a \in b\mathbb{Z}$, so b divides a. On the other hand, if $b|a$, then $a \in b\mathbb{Z}$, so $a\mathbb{Z} \subseteq b\mathbb{Z}$.

Next we verify part (a). Let H be a subgroup of $\mathbb{Z}$. If $H \neq \{0\}$, then H contains a nonzero integer; since H contains, together with any integer a, its opposite $-a$, it follows that H contains a positive integer. Let d be the smallest element of $H \cap \mathbb{N}$. I claim that $H = d\mathbb{Z}$. Since $d \in H$, it follows that $\langle d \rangle = d\mathbb{Z} \subseteq H$. On the other hand, let $h \in H$, and write $h = qd + r$, where $0 \leq r < d$. Since $h \in H$ and $qd \in H$, we have $r = h - qd \in H$. But since d is the least positive element of H and $r < d$, we must have $r = 0$. Hence $h = qd \in d\mathbb{Z}$. This shows that $H \subseteq d\mathbb{Z}$.

So far we have shown that there is a $d \in \mathbb{N}$ such that $d\mathbb{Z} = H$. If also $d' \in H$ and $d'\mathbb{Z} = H$, then by part (a), d and d' divide one another. Since they are both positive, they are equal. This proves the uniqueness of d in part (a).

Finally, for part (b), we can check that $a \mapsto da$ is an isomorphism from $\mathbb{Z}$ onto $d\mathbb{Z}$. ■

Corollary 2.2.22. *Every subgroup of $\mathbb{Z}$ other than $\{0\}$ is isomorphic to $\mathbb{Z}$.*

Lemma 2.2.23. *Let $n \geq 2$ and let d be a positive divisor of n. The cyclic subgroup $\langle [d] \rangle$ generated by $[d]$ in $\mathbb{Z}_n$ has cardinality $|\langle [d] \rangle| = n/d$.*

Proof. The order of $[d]$ is the least positive integer s such that $s[d] = [0]$, by Proposition 2.2.17, that is, the least positive integer s such that n divides sd. But this is just n/d, since d divides n. ■

Proposition 2.2.24. *Let* H *be a subgroup of* $\mathbb{Z}_n$.

(a) *Either* $H = \{[0]\}$, *or there is a* $d > 0$ *such that* $H = \langle [d] \rangle$.

(b) *If* d *is the smallest of positive integers* s *such that* $H = \langle [s] \rangle$, *then* $d|H| = n$.

Proof. Let H be a subgroup of $\mathbb{Z}_n$. If $H \neq \{[0]\}$, let d denote the smallest of positive integers s such that $[s] \in H$. An argument identical to that used for Proposition 2.2.21 (a) shows that $\langle [d] \rangle = H$.

Clearly, d is then also the smallest of positive integers s such that $\langle [s] \rangle = H$.

Write $n = qd + r$, where $0 \leq r < d$. Then $[r] = -q[d] \in \langle [d] \rangle$. Since $r < d$ and d is the least of positive integers s such that $[s] \in \langle [d] \rangle$, it follows that $r = 0$. Thus n is divisible by d. It now follows from the previous lemma that $|H| = n/d$. ■

Corollary 2.2.25. *Fix a natural number* $n \geq 2$.

(a) *Any subgroup of* $\mathbb{Z}_n$ *is cyclic.*

(b) *Any subgroup of* $\mathbb{Z}_n$ *has cardinality dividing* n.

Proof. Both parts are immediate from the proposition. ■

Corollary 2.2.26. *Fix a natural number* $n \geq 2$.

(a) *For any positive divisor* q *of* n, *there is a unique subgroup of* $\mathbb{Z}_n$ *of cardinality* q, *namely* $\langle [n/q] \rangle$.

(b) *For any two subgroups* H *and* H′ *of* $\mathbb{Z}_n$, *we have* $H \subseteq H' \Leftrightarrow |H|$ *divides* $|H'|$.

Proof. If q is a positive divisor of n, then the cardinality of $\langle [n/q] \rangle$ is q, by Lemma 2.2.23. On the other hand, if H is a subgroup of cardinality q, then by Proposition 2.2.24 (b), n/q is the least of positive integers s such that $s \in H$, and $H = \langle [n/q] \rangle$. So $\langle [n/q] \rangle$ is the unique subgroup of $\mathbb{Z}_n$ of cardinality q.

Part (b) is left as an exercise for the reader. ■

Example 2.2.27. Determine all subgroups of $\mathbb{Z}_{12}$ and all containments between these subgroups. We know that $\mathbb{Z}_{12}$ has exactly one subgroup of cardinality q for each positive divisor of 12. The sizes of subgroups are $1, 2, 3, 4, 6$, and 12. The canonical generators of these subgroups are [0], [6], [4], [3], [2], and [1], respectively. The inclusion relations among the subgroups of $\mathbb{Z}_{12}$ is pictured in Figure 2.2.2.

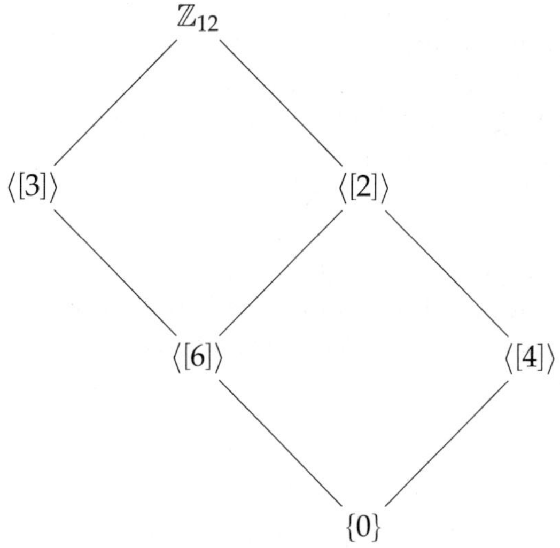

Figure 2.2.2. Lattice of subgroups of $\mathbb{Z}_{12}$.

Corollary 2.2.28. *Let* $b \in \mathbb{Z}$.

(a) *The cyclic subgroup* $\langle[b]\rangle$ *of* $\mathbb{Z}_n$ *generated by* $[b]$ *is equal to the cyclic subgroup generated by* $[d]$, *where* $d = \text{g.c.d.}(b, n)$.
(b) *The order of* $[b]$ *in* $\mathbb{Z}_n$ *is* $n/\text{g.c.d.}(b, n)$.
(c) *In particular,* $\langle[b]\rangle = \mathbb{Z}_n$ *if, and only if,* b *is relatively prime to* n.

Proof. One characterization of $d = \text{g.c.d.}(b, n)$ is as the smallest positive integer in $\{\beta b + \nu n : \beta, \nu \in \mathbb{Z}\}$. But then d is also the smallest of positive integers s such that s is congruent modulo n to an integer multiple of b, or, equivalently, the smallest of positive integers s such that $[s] \in \langle[b]\rangle$. By the proof of Proposition 2.2.24, $\langle[d]\rangle = \langle[b]\rangle$. The order of $[b]$ is the order of $\langle[b]\rangle$, which is n/d, by Proposition 2.2.24 (b). Part (c) is left as an exercise. ■

Example 2.2.29. Find all generators of $\mathbb{Z}_{12}$. Find all $[b] \in \mathbb{Z}_{12}$ such that $\langle[b]\rangle = \langle[3]\rangle$, the unique subgroup of order 4. The generators of $\mathbb{Z}_{12}$ are exactly those $[a]$ such that $1 \le a \le 11$ and a is relatively prime to 12. Thus the generators are $[1], [5], [7], [11]$. The generators of $\langle[3]\rangle$ are those elements $[b]$ satisfying $\text{g.c.d.}(b, 12) = \text{g.c.d.}(3, 12) = 3$; the complete list of these is $[3], [9]$.

The results of this section carry over to arbitrary cyclic groups, since each cyclic group is isomorphic to $\mathbb{Z}$ or to $\mathbb{Z}_n$ for some n. For example, we have

Proposition 2.2.30. *Every subgroup of a cyclic group is cyclic.*

Proposition 2.2.31. *Let a be an element of finite order n in a group. Then $\langle a^k \rangle = \langle a \rangle$, if, and only if, k is relatively prime to n. The number of generators of $\langle a \rangle$ is $\varphi(n)$.*

Proposition 2.2.32. *Let a be an element of finite order n in a group. For each positive integer q dividing n, $\langle a \rangle$ has a unique subgroup of order q.*

Proposition 2.2.33. *Let a be an element of finite order n in a group. For each nonzero integer s, a^s has order $n/\text{g.c.d.}(n, s)$.*

Example 2.2.34. The group $\Phi(2^n)$ has order $\varphi(2^n) = 2^{n-1}$. $\Phi(2)$ has order 1, and $\Phi(4)$ has order 2, so these are cyclic groups. However, for $n \geq 3$, $\Phi(2^n)$ is not cyclic. In fact, the three elements $[2^n - 1]$ and $[2^{n-1} \pm 1]$ are distinct and each has order 2. But if $\Phi(2^n)$ were cyclic, it would have a unique subgroup of order 2, and hence a unique element of order 2.

Exercises 2.2

2.2.1. Verify the assertions made about the subgroups of S_3 in Example 2.2.9.

2.2.2. Determine the subgroup lattice of the group of symmetries of the square card.

2.2.3. Determine the subgroup lattice of the group of symmetries of the rectangular card.

2.2.4. Let H be the subset of S_4 consisting of all 3–cycles, all products of disjoint 2–cycles, and the identity. The purpose of this exercise is to show that H is a subgroup of S_4.

(a) Show that $\{e, (12)(34), (13)(24), (14)(23)\}$ is a subgroup of S_4.

(b) Now examine products of two 3–cycles in S_4. Notice that the two 3–cycles have either all three digits in common, or they have two out of three digits in common. If they have three digits in common, they are either the same or inverses. If they have two digits in common, then they can be written as $(a_1 a_2 a_3)$ and $(a_1 a_2 a_4)$, or as $(a_1 a_2 a_3)$ and $(a_2 a_1 a_4)$. Show that in all cases the product is either the identity, another 3–cycle, or a product of two disjoint 2–cycles.

(c) Show that the product of a 3–cycle and an element of the form $(ab)(cd)$ is again a 3–cycle.

(d) Show that H is a subgroup.

2.2.5.

(a) Let R_θ denote the rotation matrix

$$R_\theta = \begin{bmatrix} \cos\theta & -\sin\theta \\ \sin\theta & \cos\theta \end{bmatrix}.$$

Show that the set of R_θ, where θ varies through the real numbers, forms a group under matrix multiplication. In particular, $R_\theta R_\mu = R_{\theta+\mu}$, and $R_\theta^{-1} = R_{-\theta}$.

(b) Let J denote the matrix of reflection in the x–axis,

$$J = \begin{bmatrix} 1 & 0 \\ 0 & -1 \end{bmatrix}.$$

Show $JR_\theta = R_{-\theta}J$.

(c) Let J_θ be the matrix of reflection in the line containing the origin and the point $(\cos\theta, \sin\theta)$. Compute J_θ and show that

$$J_\theta = R_\theta J R_{-\theta} = R_{2\theta} J.$$

(d) Let $R = R_{\pi/2}$. Show that the eight matrices

$$\{R^k J^l : 0 \le k \le 3 \text{ and } 0 \le l \le 1\}$$

form a subgroup of $GL(s)$, isomorphic to the group of symmetries of the square.

2.2.6. Let S be a subset of a group G, and let S^{-1} denote $\{s^{-1} : s \in S\}$. Show that $\langle S^{-1}\rangle = \langle S\rangle$. In particular, for $a \in G$, $\langle a\rangle = \langle a^{-1}\rangle$, so also $o(a) = o(a^{-1})$.

2.2.7. Prove Proposition 2.2.8. (Refer to Appendix B for a discussion of intersections of arbitrary collections of sets.)

2.2.8. Prove by induction the following facts about powers of elements in a group. For all integers k and l,

(a) $a^k a^l = a^{k+l}$.

(b) $(a^k)^{-1} = a^{-k}$.

(c) $(a^k)^l = a^{kl}$.

(Refer to Appendix C for a discussion of induction and multiple induction.)

2.2.9. Let a be an element of a group. Let n be the least positive integer such that $a^n = e$. Show that $e, a, a^2, \ldots, a^{n-1}$ are all distinct. Conclude that the order of the subgroup generated by a is n.

2.2.10. $\Phi(14)$ is cyclic of order 6. Which elements of $\Phi(14)$ are generators? What is the order of each element of $\Phi(14)$?

2.2.11. Show that the order of a cycle in S_n is the length of the cycle. For example, the order of (1234) is 4. What is the order of a product of two disjoint cycles? Begin (of course!) by considering some examples. Note, for instance, that the product of a 2–cycle and a 3–cycle (disjoint) is 6,while the order of the product of two disjoint 2–cycles is 2.

2.2.12. Can an abelian group have exactly two elements of order 2?

2.2.13. Suppose an abelian group has an element a of order 4 and an element b of order 3. Show that it must also have elements of order 2 and 6.

2.2.14. Suppose that a group G contains two elements a and b such that $ab = ba$ and the orders of $o(a)$ and $o(b)$ of a and b are relatively prime. Show that the order of ab is $o(a)o(b)$.

2.2.15. How must the conclusion of the previous exercise be altered if $o(a)$ and $o(b)$ are not relatively prime?

2.2.16. Show that the symmetric group S_n (for $n \geq 2$) is generated by the 2–cycles $(12), (23), \dots, (n-1\ n)$.

2.2.17. Show that the symmetric group S_n (for $n \geq 2$) is generated by the 2–cycle (12) and the n–cycle $(12\dots n)$.

2.2.18. Show that the subgroup of $\mathbb{Z}$ generated by any finite set of nonzero integers $n_1, \dots, n_k$ is $\mathbb{Z}d$, where d is the greatest common divisor of $\{n_1, \dots, n_k\}$.

2.2.19. Determine the subgroup lattice of $\mathbb{Z}_{24}$.

2.2.20. Suppose a is an element of order 24 in a group G. Determine the subgroup lattice of $\langle a \rangle$. Determine all the generators of $\langle a \rangle$.

2.2.21. Determine the subgroup lattice of $\mathbb{Z}_{30}$.

2.2.22. Suppose a is an element of order 30 in a group G. Determine the subgroup lattice of $\langle a \rangle$. Determine all the generators of $\langle a \rangle$.

2.2.23. How many elements of order 10 are there in $\mathbb{Z}_{30}$? How many elements of order 10 are therein $\mathbb{Z}_{200000}$? How many elements of order 10 are there in $\mathbb{Z}_{15}$?

2.2.24. Suppose that a group G of order 20 has at least three elements of order 4. Can G be cyclic? What if G has exactly two elements of order 4?

2.2.25. Suppose k divides n. How many elements are there of order k in $\mathbb{Z}_n$? Suppose k does not divide n. How many elements are there of order k in $\mathbb{Z}_n$?

2.2.26. Show that for any two subgroups H and H' of $\mathbb{Z}_n$, we have $H \subseteq H' \Leftrightarrow |H|$ divides $|H'|$. This is the assertion of Corollary 2.2.26(b).

2.2.27. Let $a, b \in \mathbb{Z}$. Show that the subgroups $\langle [a] \rangle$ and $\langle [b] \rangle$ generated by $[a]$ and $[b]$ in $\mathbb{Z}_n$ are equal, if, and only if, $\text{g.c.d.}(a, n) = \text{g.c.d.}(b, n)$. Show that $\langle [b] \rangle = \mathbb{Z}_n$ if, and only if, $\text{g.c.d.}(b, n) = 1$. This is the assertion of Corollary 2.2.28 (c).

2.2.28. Let $n \geq 3$. Verify that $[2^n - 1]$, $[2^{n-1} + 1]$, and $[2^{n-1} - 1]$ are three distinct elements of order 2 in $\Phi(2^n)$. Show, moreover, that these are the only elements of order 2 in $\Phi(2^n)$. (These facts are used in Example 2.2.34.)

2.2.29. Verify that $[3]$ has order 2^{n-2} in $\Phi(2^n)$ for $n = 3, 4, 5$.

Remark. Using a short computer program, you can quickly verify that $[3]$ has order 2^{n-2} in $\Phi(2^n)$ for $n = 6, 7, 8$, and so on as well. For example, use the following *Mathematica* program or its equivalent in your favorite computer language:

```
n = 8;
For[j = 1, j ≤ 2^(n-2), j++, Print[Mod[3^j, 2^n]]]
```

2.2.30.

(a) Show that for all natural numbers k, $3^{2^k} - 1$ is divisible by 2^{k+2} but not by 2^{k+3}. (Compare Exercise 1.9.10.)

(b) Lagrange's theorem (Theorem 2.5.6) states that the order of any subgroup of a finite group divides the order of the group. Consequently, the order of any element of a finite group divides the order of the group. Applying this to the group $\Phi(2^n)$, we see that the order of any element is a power of 2. Assuming this, conclude from part (a) that the order of $[3]$ in $\Phi(2^n)$ is 2^{n-2} for all $n \geq 3$.

2.3. The Dihedral Groups

In this section, we will work out the symmetry groups of regular polygons[3] and of the disk, which might be thought of as a "limit" of regular polygons as the number of sides increases. We regard these figures as thin plates, capable of rotations in three dimensions. Their symmetry groups are known collectively as the *dihedral groups*.

We have already found the symmetry group of the equilateral triangle (regular 3-gon) in Exercise 1.3.1 and of the square (regular 4-gon) in Sections 1.2 and 1.3. For now, it will be convenient to work first with the disk

$$\{\begin{bmatrix} x \\ y \\ 0 \end{bmatrix} : x^2 + y^2 \leq 1\},$$

whose symmetry group we denote by D.

Observe that the rotation r_t through any angle t around the z–axis is a symmetry of the disk. Such rotations satisfy $r_t r_s = r_{t+s}$ and in particular $r_t r_{-t} = r_0 = e$, where e is the nonmotion. It follows that $N = \{r_t : t \in \mathbb{R}\}$ is a subgroup of D.

For any line passing through the origin in the (x, y)–plane , the rotation by π about that line is a symmetry of the disk (which interchanges the top and bottom faces of the disk). Denote by j_t the rotation about the line ℓ_t which passes through the origin and the point $\begin{bmatrix} \cos(t) \\ \sin(t) \\ 0 \end{bmatrix}$, and write $j = j_0$ for the rotation about the x–axis. Each j_t generates a subgroup of D of order 2. Symmetries of the disk are illustrated in Figure 2.3.1.

Next, we observe that each j_t can be expressed in terms of j and the rotation r_t. To perform the rotation j_t about the line ℓ_t, we can rotate the disk until the line ℓ_t overlays the x–axis, then perform the rotation j about the x–axis, and finally rotate the disk so that ℓ_t is returned to its original position. Thus $j_t = r_t j r_{-t}$, or $j_t r_t = r_t j$. Therefore, we need only work out how to compute products involving j and the rotations r_t.

[3]A regular polygon is one all of whose sides are congruent and all of whose internal angles are congruent.

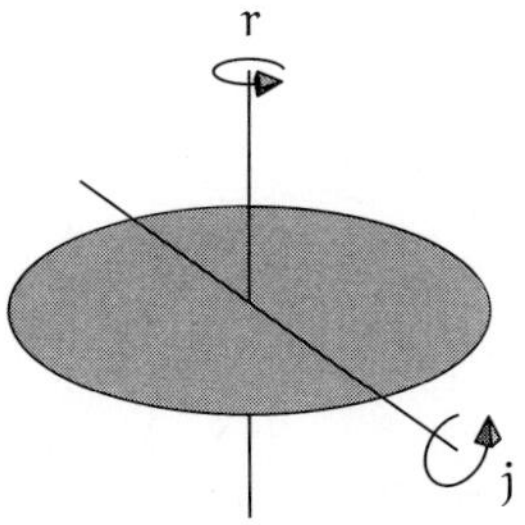

Figure 2.3.1. Symmetries of the disk.

Note that j applied to a point $\begin{bmatrix} \rho\cos(s) \\ \rho\sin(s) \\ 0 \end{bmatrix}$ in the disk is $\begin{bmatrix} \rho\cos(-s) \\ \rho\sin(-s) \\ 0 \end{bmatrix}$, and r_t applied to $\begin{bmatrix} \rho\cos(s) \\ \rho\sin(s) \\ 0 \end{bmatrix}$ is $\begin{bmatrix} \rho\cos(s+t) \\ \rho\sin(s+t) \\ 0 \end{bmatrix}$.

In the Exercises, you are asked to verify the following facts about the group D:

1. $jr_t = r_{-t}j$, and $j_t = r_{2t}j = jr_{-2t}$.
2. All products in D can be computed using these relations.
3. The symmetry group D of the disk consists of the rotations r_t for $t \in \mathbb{R}$ and the "flips" $j_t = r_{2t}j$. Writing $N = \{r_t : t \in \mathbb{R}\}$, we have $D = N \cup Nj$.
4. The subgroup N of D satisfies $aNa^{-1} = N$ for all $a \in D$.

Next, we turn to the symmetries of the regular polygons. Consider a regular n-gon with vertices at $\begin{bmatrix} \cos(k\pi/n) \\ \sin(k\pi/n) \\ 0 \end{bmatrix}$ for $k = 0, 1, \dots, n-1$. Denote the symmetry group of the n-gon by D_n. In the exercises, you are asked to verify the following facts about the symmetries of the n-gon:

1. The rotation $r = r_{2\pi/n}$ through an angle of $2\pi/n$ about the z–axis generates a cyclic subgroup of D_n of order n.
2. The "flips" $j_{k\pi/n} = r_{k2\pi/n}\, j = r^k\, j$, for $k \in \mathbb{Z}$, are symmetries of the n-gon.
3. The distinct flip symmetries of the n-gon are $r^k\, j$ for $k = 0, 1, \dots, n-1$.

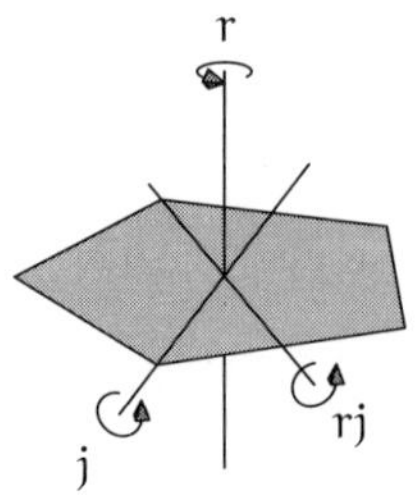

Figure 2.3.2. Symmetries of the pentagon.

4. If n is odd, then the axis of each of the "flips" passes through a vertex of the n-gon and the midpoint of the opposite edge. See Figure 2.3.2 for the case $n = 5$.
5. If n is even and k is even, then $j_{k\pi/n} = r^k j$ is a rotation about an axis passing through a pair of opposite vertices of the n-gon.
6. If n is even and k is odd, then $j_{k\pi/n} = r^k j$ is a rotation about an axis passing through the midpoints of a pair of opposite edges of the n-gon. See Figure 2.3.3 for the case $n = 6$.

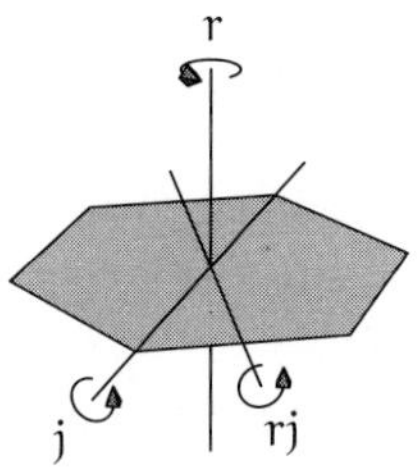

Figure 2.3.3. Symmetries of the hexagon.

The symmetry group D_n consists of the $2n$ symmetries r^k and $r^k j$, for $0 \le k \le n-1$. It follows from our computations for the symmetries of the disk that $jr = r^{-1}j$, so $jr^k = r^{-k}j$ for all k. This relation allows the computation of all products in D_n.

The group D_n can appear as the symmetry group of a geometric figure, or of a physical object, in a slightly different form. Think, for example, of a five-petalled flower, or a star-fish, which look quite different from the top and from the bottom. Or think of a pentagonal card with its two faces of different colors. Such an object or figure does not admit rotational symmetries that exchange the top and bottom faces. However, a starfish or a flower does have reflection symmetries, as well as rotational symmetries that do not exchange top and bottom.

Consider a regular n-gon in the plane. The reflections in the lines passing through the centroid of the n-gon and a vertex, or through the centroid and the midpoint of an edge, are symmetries; there are n such reflection symmetries. We can show that the n rotational symmetries in the plane together with the n reflection symmetries form a group that is isomorphic to D_n. See Exercise 2.3.10.

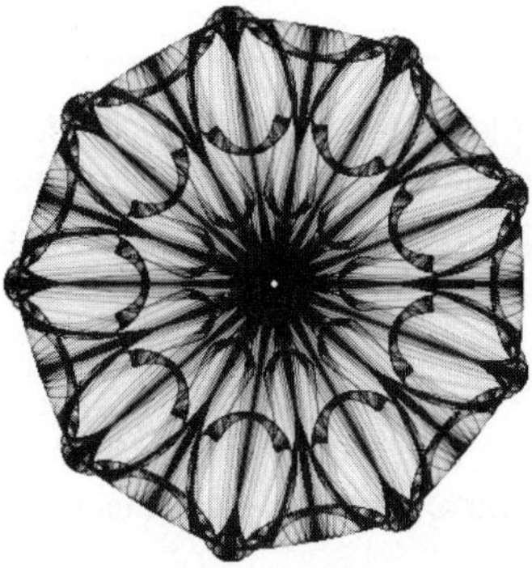

Figure 2.3.4. Object with D_9 symmetry.

Figure 2.3.5. Object with Z_5 symmetry.

The plane figure shown as Figure 2.3.4 has D_9 symmetry.

Is it possible for a plane figure to have Z_n symmetry but not D_n symmetry? Certainly. Figure 2.3.5 possesses Z_5 symmetry, but not D_5 symmetry.

Figure 2.3.4 looks as if it might be the product of an art glass studio, and Figure 2.3.5 looks as if it might be of biological origin. In fact, each figure was generated by several million iterations of a discrete dynamical system exhibiting "chaotic" behavior; the figures are shaded according to the probability of the moving "particle" entering a region of the diagram — the darker regions are visited more frequently. A beautiful book by M. Field and M. Golubitsky, *Symmetry in Chaos* (Oxford University Press, 1992), discusses symmetric figures arising from chaotic dynamical systems and displays many beautiful figures produced by such systems. Figures 2.3.4 and 2.3.5 were produced using algorithms from the book of Field and Golubitsky.

Exercises 2.3

2.3.1. Show that the elements j and r_t of the symmetry group D of the disk satisfy the relations $jr_t = r_{-t}j$, and $j_t = r_{2t}j = jr_{-2t}$.

2.3.2. The symmetry group D of the disk consists of the rotations r_t for $t \in \mathbb{R}$ and the "flips" $j_t = r_{2t}j$.

(a) Writing $N = \{r_t : t \in \mathbb{R}\}$, show that $D = N \cup Nj$.

(b) Show that all products in D can be computed using the relation $jr_t = r_{-t}j$.

(c) Show that the subgroup N of D satisfies $aNa^{-1} = N$ for all $a \in D$.

2.3.3. The symmetries of the disk are implemented by linear transformations of $\mathbb{R}^3$. Write the matrices of the symmetries r_t and j with respect to the standard basis of $\mathbb{R}^3$. Denote these matrices by R_t and J, respectively. Confirm the relation $JR_t = R_{-t}J$.

2.3.4. Consider the group D_n of symmetries of the n-gon.

(a) Show that the rotation $r = r_{2\pi/n}$ through an angle of $2\pi/n$ about the z–axis generates a cyclic subgroup of D_n of order n.

(b) Show that the "flips" $j_{k\pi/n} = r_{k2\pi/n}\, j = r^k\, j$, for $k \in \mathbb{Z}$, are symmetries of the n-gon.

(c) Show that the distinct flip symmetries of the n-gon are $r^k\, j$ for $k = 0, 1, \ldots, n-1$.

2.3.5.

(a) Show that if n is odd, then the axis of each of the "flips" passes through a vertex of the n-gon and the midpoint of the opposite edge. See Figure 2.3.2 for the case $n = 5$.

(b) If n is even and k is even, show that $j_{k\pi/n} = r^k\, j$ is a rotation about an axis passing through a pair of opposite vertices of the n-gon.

(c) Show that if n is even and k is odd, then $j_{k\pi/n} = r^k\, j$ is a rotation about an axis passing through the midpoints of a pair of opposite edges of the n-gon. See Figure 2.3.3 for the case $n = 6$.

2.3.6. Find a subgroup of D_6 that is isomorphic to D_3.

2.3.7. Find a subgroup of D_6 that is isomorphic to the symmetry group of the rectangle.

2.3.8. Consider a regular 10-gon in which alternate vertices are painted red and blue. Show that the symmetry group of the painted 10-gon is isomorphic to D_5.

2.3.9. Consider a regular 15-gon in which every third vertex is painted red. Show that the symmetry group of the painted 15-gon is isomorphic to D_5. However, if the vertices are painted with the pattern red, green, blue, red, green, blue, ..., red, green, blue, then the symmetry group is reduced to Z_5.

2.3.10. Consider a card in the shape of an n-gon, whose two faces (top and bottom) are painted red and green. Show that the symmetry group of the card (including reflections) is isomorphic to D_n.

2.3.11. Consider a card in the shape of an n-gon, whose two faces (top and bottom) are indistinguishable. Determine the group of symmetries of the card (including both rotations and reflections).

2.4. Homomorphisms and Isomorphisms

We have already introduced the concept of an *isomorphism* between two groups: An isomorphism $\varphi : G \longrightarrow H$ is a bijection that preserves group multiplication [i.e., $\varphi(g_1 g_2) = \varphi(g_1)\varphi(g_2)$ for all $g_1, g_2 \in G$].

For example, the set of eight 3-by-3 matrices

$$\{E, R, R^2, R^3, A, RA, R^2A, R^3A\},$$

where E is the 3-by-3 identity matrix, and

$$A = \begin{bmatrix} 1 & 0 & 0 \\ 0 & -1 & 0 \\ 0 & 0 & -1 \end{bmatrix} \quad R = \begin{bmatrix} 0 & -1 & 0 \\ 1 & 0 & 0 \\ 0 & 0 & 1 \end{bmatrix},$$

given in Section 1.4, is a subgroup of $GL(3, \mathbb{R})$. The map $\varphi : r^k a^l \mapsto R^k A^l$ $(0 \le k \le 3, 0 \le l \le 1)$ is an isomorphism from the group of symmetries of the square to this group of matrices.

Similarly, the set of eight 2-by-2 matrices

$$\{E, R, R^2, R^3, J, RJ, R^2J, R^3J\},$$

where now E is the 2-by-2 identity matrix and

$$J = \begin{bmatrix} 1 & 0 \\ 0 & -1 \end{bmatrix} \quad R = \begin{bmatrix} 0 & -1 \\ 1 & 0 \end{bmatrix}$$

is a subgroup of $GL(2, \mathbb{R})$, and the map $\psi : r^k a^l \mapsto R^k J^l$ $(0 \le k \le 3, 0 \le l \le 1)$ is an isomorphism from the group of symmetries of the square to this group of matrices.

There is a more general concept that is very useful:

Definition 2.4.1. A map between groups $\varphi : G \longrightarrow H$ is called a *homomorphism* if it preserves group multiplication, $\varphi(g_1 g_2) = \varphi(g_1)\varphi(g_1)$ for all $g_1, g_2 \in G$.

There is no requirement here that φ be either injective or surjective.

Example 2.4.2. We consider some homomorphisms of the symmetry group of the square into permutation groups. Place the square card in the $(x-y)$–plane so that the axes of symmetry for the rotations a, b, and r coincide with the x–, y–, and z–axes, respectively. Each symmetry of the card induces a bijective map of the space $S = \{(x, y, 0) : |x| \le 1, |y| \le 1\}$ occupied by the card. For example, the symmetry a induces the map

$$\begin{bmatrix} x \\ y \\ 0 \end{bmatrix} \mapsto \begin{bmatrix} x \\ -y \\ 0 \end{bmatrix}.$$

The map associated to each symmetry sends the set V of four vertices of S onto itself. So for each symmetry σ of the square, we get an element $\pi(\sigma)$ of $\mathrm{Sym}(V)$. Composition of symmetries corresponds to composition of maps of S and of V, so the assignment $\sigma \mapsto \pi(\sigma)$ is a homomorphism from the symmetry group of the square to $\mathrm{Sym}(V)$. This homomorphism is injective, since a symmetry of the square is entirely determined by what it does to the vertices, but it cannot be surjective, since the square has only eight symmetries while $|\mathrm{Sym}(V)| = 24$.

Example 2.4.3. To make these observations more concrete and computationally useful, we number the vertices of S. It should be emphasized that we are not numbering the corners of the card, which move along with the card, but rather *the locations* of these corners, which stay put. (This is a point about which one can easily be confused.) See Figure 2.4.1.

Numbering the vertices gives us a homomorphism φ from the group of symmetries of the square into S_4. Observe, for example, that $\varphi(r) = (1432)$, $\varphi(a) = (14)(23)$, and $\varphi(c) = (24)$. Now you can compute that

$$\varphi(a)\varphi(r) = (14)(23)(1432) = (24) = \varphi(c) = \varphi(ar).$$

You are asked in Exercise 2.4.1 to complete the tabulation of the map φ from the symmetry group of the square into S_4 and to verify the homomorphism property.

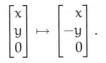

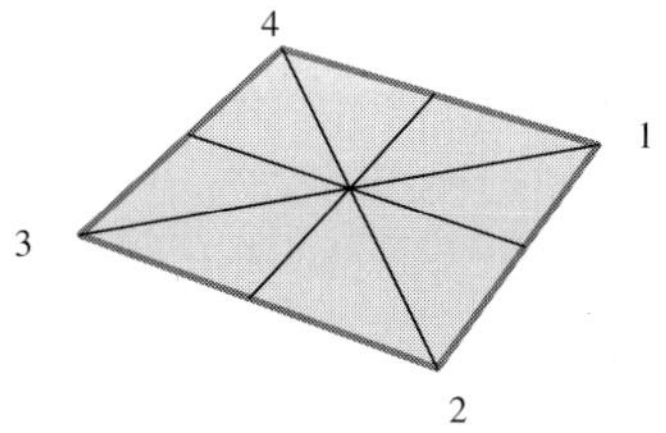

Figure 2.4.1. Labeling the vertices of the square.

Note that all of this is a formalization of the computation by pictures that was done in Section 1.3.

Example 2.4.4. There are other sets of geometric objects associated with the square that are permuted by symmetries of the square: the set of edges, the set of diagonals, the set of pairs of opposite edges. Let's consider the diagonals (Figure 2.4.2). Numbering the diagonals gives a homomorphism ψ from the group of symmetries of the square into S_2.

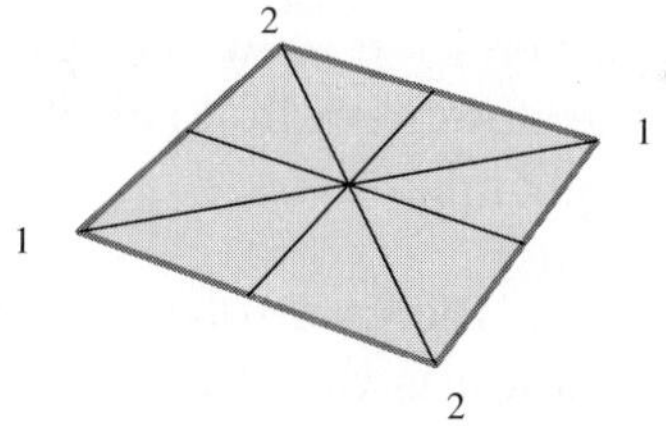

Figure 2.4.2. Labeling the diagonals of the square.

You can compute, for example, that $\psi(r) = \psi(a) = (12)$, while $\psi(c) = e$. You are asked in Exercise 2.4.2 to complete the tabulation of the map ψ and to verify its homomorphism property.

Example 2.4.5. It is well known that the determinant of an invertible matrix is nonzero, and that the determinant satisfies the identity $\det(AB) = \det(A)\det(B)$. Therefore, $\det : GL(n, \mathbb{R}) \to \mathbb{R}^*$ is a homomorphism from the group of invertible matrices to the group of nonzero real numbers under multiplication.

Example 2.4.6. Recall that a linear transformation $T : \mathbb{R}^n \to \mathbb{R}^n$ has the property that $T(\mathbf{a} + \mathbf{b}) = T(\mathbf{a}) + T(\mathbf{b})$. Thus T is a group homomorphism from the additive group $\mathbb{R}^n$ to itself. More concretely, for any n-by-n matrix M, we have $M(\mathbf{a} + \mathbf{b}) = M\mathbf{a} + M\mathbf{b}$. Thus multiplication by M is a group homomorphism from the additive group $\mathbb{R}^n$ to itself.

Example 2.4.7. Let G be any group and a and element of G. The the map from $\mathbb{Z}$ to G given by $k \mapsto a^k$ is a group homomorphism. This is equivalent to the statement that $a^{k+\ell} = a^k a^\ell$ for all integers k and ℓ. The image of this homomorphism is the cyclic subgroup of G generated by g.

Example 2.4.8. There is a homomorphism from $\mathbb{Z}$ to $\mathbb{Z}_n$ defined by $k \mapsto [k]$. This follows directly from the definition of the addition in $\mathbb{Z}_n$: $[a] + [b] = [a + b]$. (But recall that we had to check that this definition makes sense; see the discussion following Lemma 1.7.5.)

This example is also a special case of the previous example, with $G = \mathbb{Z}_n$ and the chosen element $a = [1] \in G$. The map is given by $k \mapsto k[1] = [k]$.

Example 2.4.9. Let G be an *abelian* group and n a fixed integer. Then the map from G to G given by $g \mapsto g^n$ is a group homomorphism. This is equivalent to the statement that $(ab)^n = a^n b^n$ when a, b are elements in an abelian group.

Let us now turn from the examples to some general observations. Our first observation is that the composition of homomorphisms is a homomorphism.

Proposition 2.4.10. *Let* $\varphi : G \longrightarrow H$ *and* $\psi : H \longrightarrow K$ *be homomorphisms of groups. Then the composition* $\psi \circ \varphi : G \longrightarrow K$ *is also a homomorphism.*

Proof. Exercise 2.4.3. ■

Next we check that homomorphisms preserve the group identity and inverses.

Proposition 2.4.11. *Let* $\varphi : G \longrightarrow H$ *be a homomorphism of groups.*

(a) $\varphi(e_G) = e_H$.

(b) *For each* $g \in G$, $\varphi(g^{-1}) = (\varphi(g))^{-1}$.

Proof. For any $g \in G$,

$$\varphi(e_G)\varphi(g) = \varphi(e_G g) = \varphi(g).$$

It follows from Proposition 2.1.1(a) that $\varphi(e_G) = e_H$. Similarly, for any $g \in G$,

$$\varphi(g^{-1})\varphi(g) = \varphi(g^{-1}g) = \varphi(e_G) = e_H,$$

so Proposition 2.1.1(b) implies that $\varphi(g^{-1}) = (\varphi(g))^{-1}$. ■

The next observation is that homomorphisms preserve subgroups.

Proposition 2.4.12. *Let $\varphi : G \longrightarrow H$ be a homomorphism of groups.*

(a) *For each subgroup $A \subseteq G$, $\varphi(A)$ is a subgroup of H.*

(b) *For each subgroup $B \subseteq H$,*

$$\varphi^{-1}(B) = \{g \in G : \varphi(g) \in B\}$$

is a subgroup of G.

Proof. We have to show that $\varphi(A)$ is closed under multiplication and inverses. Let h_1 and h_2 be elements of $\varphi(A)$. There exist elements $a_1, a_2 \in A$ such that $h_i = \varphi(a_i)$ for $i = 1, 2$. Then

$$h_1 h_2 = \varphi(a_1)\varphi(a_2) = \varphi(a_1 a_2) \in \varphi(A),$$

since $a_1 a_2 \in A$. Likewise, for $h \in \varphi(A)$, there is an $a \in A$ such that $\varphi(a) = h$. Then, using Proposition 2.4.11(b) and closure of A under inverses, we compute

$$h^{-1} = (\varphi(a))^{-1} = \varphi(a^{-1}) \in \varphi(A).$$

The proof of part (b) is left as an exercise. ■

The Kernel of a Homomorphism

You might think at first that it is not very worthwhile to look at noninjective homomorphisms $\varphi : G \to H$ since such a homomorphism loses information about G. But in fact, such a homomorphism also reveals certain information about the structure of G that otherwise might be missed. For example, consider the homomorphism ψ from the symmetry group G of the square into the symmetric group S_2 induced by the action of G on the two diagonals of the square, as discussed in Example 2.4.4. Let N denote the set of symmetries σ of the square such that $\psi(\sigma) = e$. You can compute that $N = \{e, c, d, r^2\}$. (Do it now!) From the general theory, which I am about to explain, we see that N is a special sort of subgroup G, called a normal subgroup. Understanding such subgroups helps to understand the structure of G. For now, verify for yourself that N is, in fact, a subgroup.

Definition 2.4.13. A subgroup N of a group G is said to be *normal* if for all $g \in G$, $gNg^{-1} = N$. Here gNg^{-1} means $\{gng^{-1} : n \in N\}$.

Definition 2.4.14. Let $\varphi : G \longrightarrow H$ be a homomorphism of groups. The *kernel* of the homomorphism φ, denoted $\ker(\varphi)$, is $\varphi^{-1}(e_H) = \{g \in G : \varphi(g) = e_H\}$.

According to Proposition 2.4.12 (b), $\ker(\varphi)$ is a subgroup of G (since $\{e_H\}$ is a subgroup of H). We now observe that it is a normal subgroup.

Proposition 2.4.15. *Let* $\varphi : G \longrightarrow H$ *be a homomorphism of groups. Then* $\ker(\varphi)$ *is a normal subgroup of* G.

Proof. It suffices to show that $g\ker(\varphi)g^{-1} = \ker(\varphi)$ for all $g \in G$. If $x \in \ker(\varphi)$, then $\varphi(gxg^{-1}) = \varphi(g)\varphi(x)(\varphi(g))^{-1} = \varphi(g)e(\varphi(g))^{-1} = e$. Thus $gxg^{-1} \in \ker(\varphi)$. We have now shown that for all $g \in G$, $g\ker(\varphi)g^{-1} \subseteq \ker(\varphi)$. We still need to show the opposite containment. But if we replace g by g^{-1}, we obtain that for all $g \in G$, $g^{-1}\ker(\varphi)g \subseteq \ker(\varphi)$; this is equivalent to $\ker(\varphi) \subseteq g\ker(\varphi)g^{-1}$. Since we have both $g\ker(\varphi)g^{-1} \subseteq \ker(\varphi)$ and $\ker(\varphi) \subseteq g\ker(\varphi)g^{-1}$, we have equality of the two sets. ■

If a homomorphism $\varphi : G \longrightarrow H$ is injective, then its kernel $\varphi^{-1}(e_H)$ contains only e_G. The converse of this statement is also valid:

Proposition 2.4.16. *A homomorphism* $\varphi : G \longrightarrow H$ *is injective if, and only if,* $\ker(\varphi) = \{e_G\}$.

Proof. If φ is injective, then e_G is the unique preimage of e_H under φ. Conversely, suppose that $\ker(\varphi) = \{e_G\}$. Let $h \in H$ and suppose that $g_1, g_2 \in G$ satisfy $\varphi(g_1) = \varphi(g_2) = h$. Then $\varphi(g_1^{-1}g_2) = \varphi(g_1)^{-1}\varphi(g_2) = h^{-1}h = e_H$, so $g_1^{-1}g_2 \in \ker(\varphi)$. Therefore, $g_1^{-1}g_2 = e_G$, which gives $g_1 = g_2$. ■

Example 2.4.17. The kernel of the determinant $\det : GL(n, \mathbb{R}) \to \mathbb{R}^*$ is the subgroup of matrices with determinant equal to 1. This subgroup is called the *special linear group* and denoted $SL(n, \mathbb{R})$.

Example 2.4.18. Let G be any group and $a \in G$. If the order of a is n, then the kernel of the homomorphism $k \mapsto a^k$ from $\mathbb{Z}$ to G is the set of all multiples of n, $\{kn : k \in \mathbb{Z}\}$. If a is of infinite order, then the kernel of the homomorphism is $\{0\}$.

Example 2.4.19. In particular, the kernel of the homomorphism from $\mathbb{Z}$ to $\mathbb{Z}_n$ defined by $k \mapsto [k]$ is $[0] = \{kn : k \in \mathbb{Z}\}$.

Example 2.4.20. If G is an abelian group and n is a fixed integer, then the kernel of the homomorphism $g \mapsto g^n$ from G to G is the set of elements whose order divides n.

Parity of Permutations

Additional examples of homomorphisms are explored in the Exercises. In particular, it is shown in the Exercises that there is a homomorphisms $\varepsilon : S_n \to \{\pm 1\}$ with the property that $\varepsilon(\tau) = -1$ for any 2–cycle τ. This is an example of a homomorphism that is very far from being injective and that picks out an essential structural feature of the symmetric group.

Definition 2.4.21. The homomorphism ε is called the *sign* (or *parity*) homomorphism. A permutation π is said to be *even* if $\varepsilon(\pi) = 1$, that is, if π is in the kernel of the sign homomorphism. Otherwise, π is said to be *odd*. The subgroup of even permutations (that is, the kernel of ε) is generally denoted A_n. This subgroup is also referred to as the *alternating group*.

The following statement about even and odd permutations is implicit in the Exercises:

Proposition 2.4.22. *A permutation π is even if, and only if, π can be written as a product of an even number of 2–cycles.*

Even and odd permutations have the following property: The product of two even permutations is even; the product of an even and an odd permutation is odd, and the product of two odd permutations is even. Hence

Corollary 2.4.23. *The set of odd permutations in S_n is $(12)A_n$, where A_n denotes the subgroup of even permutations.*

Proof. $(12)A_n$ is contained in the set of odd permutations. But if σ is any odd permutation, then $(12)\sigma$ is even, so $\sigma = (12)((12)\sigma) \in (12)A_n$. ■

Corollary 2.4.24. *A k–cycle is even if k is odd and odd if k is even.*

Proof. According to Exercise 1.5.5, a k cycle can be written as a product of $(k-1)$ 2–cycles. ■

Exercises 2.4

2.4.1. Let φ be the map from symmetries of the square into S_4 induced by the numbering of the vertices of the square in Figure 2.4.1. Complete the tabulation of $\varphi(\sigma)$ for σ in the symmetry group of the square, and verify the homomorphism property of φ by computation.

2.4.2. Let ψ be the map from the symmetry group of the square into S_2 induced by the labeling of the diagonals of the square as in Figure 2.4.2. Complete the tabulation of ψ and verify the homomorphism property by computation. Identify the kernel of ψ.

2.4.3. Prove Proposition 2.4.10.

2.4.4. Prove part (b) of Proposition 2.4.12.

2.4.5. For any subgroup A of a group G, and $g \in G$, show that gAg^{-1} is a subgroup of G.

2.4.6. Show that every subgroup of an abelian group is normal.

2.4.7. Let $\varphi : G \longrightarrow H$ be a homomorphism of groups with kernel N. For $a, x \in G$, show that

$$\varphi(a) = \varphi(x) \Leftrightarrow a^{-1}x \in N \Leftrightarrow aN = xN.$$

Here aN denotes $\{an : n \in N\}$.

2.4.8. Let $\varphi : G \longrightarrow H$ be a homomorphism of G onto H. If A is a normal subgroup of G, show that $\varphi(A)$ is a normal subgroup of H.

2.4.9. Define a map ε from D_n to $C_2 = \{\pm 1\}$ by $\varepsilon(\sigma) = 1$ if σ does not interchange top and bottom of the n-gon, and $\varepsilon(\sigma) = -1$ if σ does interchange top and bottom of the n-gon. Show that ε is a homomorphism.

The following exercises examine an important homomorphism from S_n to $C_2 = \{\pm 1\}$ (for any n).

2.4.10. Let $\{x_1, x_2, \dots, x_n\}$ be variables. For any polynomial p in n variables and for $\sigma \in S_n$, define

$$\sigma(p)(x_1, \dots, x_n) = p(x_{\sigma(1)}, \dots, x_{\sigma(n)}).$$

Check that $\sigma(\tau(p)) = (\sigma\tau)(p)$ for all σ and $\tau \in S_n$

2.4.11. Now fix $n \in \mathbb{N}$, and define

$$\Delta = \prod_{1 \le i < j \le n} (x_i - x_j).$$

For any $\sigma \in S_n$, check that $\sigma(\Delta) = \pm\Delta$. Show that the map $\varepsilon : \sigma \mapsto \sigma(\Delta)/\Delta$ is a homomorphism from S_n to $\{1, -1\}$.

2.4.12. Show that for any 2–cycle (a, b), $\varepsilon((a, b)) = -1$; hence if a permutation π is a product of k 2–cycles , then $\varepsilon(\pi) = (-1)^k$. Now any permutation can be written as a product of 2–cycles (Exercise 1.5.5). If a permutation π can be written as a product of k 2–cycles and also as a product of l 2–cycles, then $\varepsilon(\pi) = (-1)^k = (-1)^l$, so the parity of k and of l is the same. The parity is even if, and only if, $\varepsilon(\pi) = 1$.

2.4.13. For each permutation $\pi \in S_n$, define an n-by-n matrix $T(\pi)$ as follows. Let $\hat{\mathbf{e}}_1, \hat{\mathbf{e}}_2, \dots, \hat{\mathbf{e}}_n$ be the standard basis of $\mathbb{R}^n$; $\hat{\mathbf{e}}_k$ has a 1 in the kth coordinate and zeros elsewhere. Define

$$T(\pi) = [\hat{\mathbf{e}}_{\pi(1)}, \hat{\mathbf{e}}_{\pi(2)}, \dots, \hat{\mathbf{e}}_{\pi(n)}];$$

that is, the kth column of $T(\pi)$ is the basis vector $\hat{\mathbf{e}}_{\pi(k)}$. Show that the map $\pi \mapsto T(\pi)$ is a homomorphism from S_n into $GL(n, \mathbb{R})$. What is the range of T?

Definition 2.4.25. Two elements a and b in a group G are said to be *conjugate* if there is an element $g \in G$ such that $a = gbg^{-1}$.

2.4.14. This exercise determines when two elements of S_n are conjugate.

(a) Show that for any k cycle $(a_1, a_2, \dots, a_k) \in S_n$, and for any permutation $\pi \in S_n$, we have

$$\pi(a_1, a_2, \dots, a_k)\pi^{-1} = (\pi(a_1), \pi(a_2), \dots, \pi(a_k)).$$

Hint: As always, first look at some examples for small n and k. Both sides are permutations (i.e., bijective maps defined on $\{1, 2, \dots, n\}$). Show that they are the same maps by showing that they do the same thing.

(b) Show that for any two k cycles, $(a_1, a_2, \ldots, a_k)$ and $(b_1, b_2, \ldots, b_k)$ in S_n there is a permutation $\pi \in S_n$ such that

$$\pi(a_1, a_2, \ldots, a_k)\pi^{-1} = (b_1, b_2, \ldots, b_k).$$

(c) Suppose that α and β are elements of S_n and that $\beta = g\alpha g^{-1}$ for some $g \in S_n$. Show that when α and β are written as a product of disjoint cycles, they both have exactly the same number of cycles of each length. (For example, if $\alpha \in S_{10}$ is a product of two 3–cycles, one 2–cycle, and four 1–cycles, then so is β.) We say that α and β have the same cycle structure.

(d) Conversely, suppose α and β are elements of S_n and they have the same cycle structure. Show that there is an element $g \in S_n$ such that $\beta = g\alpha g^{-1}$.

The result of this exercise is as follows: Two elements of S_n are conjugate if, and only if, they have the same cycle structure.

2.4.15. Show that ε is the unique homomorphism from S_n onto $\{1, -1\}$. Hint: Let $\varphi : S_n \longrightarrow \{\pm 1\}$ be a homomorphism. If $\varphi((12)) = -1$, show, using the results of Exercise 2.4.14, that $\varphi = \varepsilon$. If $\varphi((12)) = +1$, show that φ is the trivial homomorphism, $\varphi(\pi) = 1$ for all π.

2.4.16. For $m < n$, we can consider S_m as a subgroup of S_n. Namely, S_m is the subgroup of S_n that leaves fixed the numbers from $m + 1$ to n. The parity of an element of S_m can be computed in two ways: as an element of S_m and as an element of S_n. Show that two answers always agree.

The following concept is used in the next exercises:

Definition 2.4.26. An *automorphism* of a group G is an isomorphism from G onto G.

2.4.17. Fix an element g in a group G. Show that the map $c_g : G \to G$ defined by $c_g(a) = gag^{-1}$ is an automorphism of G. (This type of automorphism is called an inner automorphism.)

2.4.18. Show that conjugate elements of S_n have the same parity. More generally, if $\phi : S_n \to S_n$ is an automorphism, then ϕ preserves parity.

2.4.19.

(a) Show that the set of matrices with positive determinant is a normal subgroup of $GL(n, \mathbb{R})$.

(b) Show that $\varepsilon = \det \circ T$, where T is the homomorphism defined in Exercise 2.4.13 and ε is the sign homomorphism. Hint: Determine the range of $\det \circ T$ and use the uniqueness of the sign homomorphism from Exercise 2.4.15.

2.4.20.

(a) For $A \in GL(n, \mathbb{R})$ and $\mathbf{b} \in \mathbb{R}^n$, define the transformation $T_{A,\mathbf{b}} : \mathbb{R}^n \to \mathbb{R}^n$ by $T_{A,\mathbf{b}}(\mathbf{x}) = A\mathbf{x} + \mathbf{b}$. Show that the set of all such transformations forms a group G.

(b) Consider the set of matrices

$$\begin{bmatrix} A & b \\ 0 & 1 \end{bmatrix},$$

where $A \in GL(n, \mathbb{R})$ and $\mathbf{b} \in \mathbb{R}^n$, and where the 0 denotes a 1-by-n row of zeros. Show that this is a subgroup of $GL(n+1, \mathbb{R})$, and that it is isomorphic to the group described in part (a).

(c) Show that the map $T_{A,\mathbf{b}} \mapsto A$ is a homomorphism from G to $GL(n, \mathbb{R})$, and that the kernel K of this homomorphism is isomorphic to $\mathbb{R}^n$, considered as an abelian group under vector addition.

2.4.21. Let G be an abelian group. For any integer $n > 0$ show that the map $\varphi : a \mapsto a^n$ is a homomorphism from G into G. Characterize the kernel of φ. Show that if n is relatively prime to the order of G, then φ is an isomorphism; hence for each element $g \in G$ there is a unique $a \in G$ such that $g = a^n$.

2.5. Cosets and Lagrange's Theorem

Consider the subgroup $H = \{e, (1\ 2)\} \subseteq S_3$. For each of the six elements $\pi \in S_3$ you can compute the set $\pi H = \{\pi\sigma : \sigma \in H\}$. For example, $(23)H = \{(23), (132)\}$. Do the computation now, and check that you get the following results:

$$\begin{aligned} eH &= (1\,2)H &&= H \\ (2\,3)H &= (1\,3\,2)H &&= \{(2\,3), (1\,3\,2)\} \\ (1\,3)H &= (1\,2\,3)H &&= \{(1\,3), (1\,2\,3)\}. \end{aligned}$$

As π varies through S_3, only three different sets πH are obtained, each occurring twice.

Definition 2.5.1. Let H be subgroup of a group G. A subset of the form gH, where $g \in G$, is called a *left coset of* H *in* G. A subset of the form Hg, where $g \in G$, is called a *right coset of* H *in* G.

Example 2.5.2. S_3 may be identified with a subgroup of S_4 consisting of permutations that leave 4 fixed and permute $\{1, 2, 3\}$. For each of the 24 elements $\pi \in S_4$ you can compute the set πS_3.

This computation requires a little labor. If you want, you can get a computer to do some of the repetitive work; for example, programs for computations in the symmetric group are distributed with the symbolic mathematics program *Mathematica*.

With the notation $H = \{\sigma \in S_4 : \sigma(4) = 4\}$, the results are

$$\begin{aligned} eH &= (1\,2)H = (1\,3)H = (2\,3)H = (1\,2\,3)H = (1\,3\,2)H = H \\ (4\,3)H &= (4\,3\,2)H = (2\,1)(4\,3)H \\ &= (2\,4\,3\,1)H = (4\,3\,2\,1)H = (4\,3\,1)H \\ &= \{(4\,3),\ (4\,3\,2),\ (2\,1)(4\,3),\ (2\,4\,3\,1),\ (4\,3\,2\,1),\ (4\,3\,1)\} \end{aligned}$$

$$\begin{aligned} (4\,2)H &= (3\,4\,2)H = (4\,2\,1)H \\ &= (4\,2\,3\,1)H = (3\,4\,2\,1)H = (3\,1)(4\,2)H \\ &= \{(4\,2),\ (3\,4\,2),\ (4\,2\,1),\ (4\,2\,3\,1),\ (3\,4\,2\,1),\ (3\,1)(4\,2)\} \\ (4\,1)H &= (4\,1)\,(3\,2)H = (2\,4\,1)H \\ &= (2\,3\,4\,1)H = (3\,2\,4\,1)H = (3\,4\,1)H \\ &= \{(4\,1),\ (4\,1)\,(3\,2),\ (2\,4\,1),\ (2\,3\,4\,1),\ (3\,2\,4\,1),\ (3\,4\,1)\}. \end{aligned}$$

The regularity of the preceding data for left cosets of subgroups of symmetric groups is striking! Based on these data, can you make any conjectures (guesses!) about properties of cosets of a subgroup H in a group G?

Properties of Cosets

Proposition 2.5.3. *Let* H *be a subgroup of a group* G*, and let* a *and* b *be elements of* G*. The following conditions are equivalent:*

(a) $a \in bH$.
(b) $b \in aH$.
(c) $aH = bH$.
(d) $b^{-1}a \in H$.
(e) $a^{-1}b \in H$.

Proof. If condition (a) is satisfied, then there is an element $h \in H$ such that $a = bh$; but then $b = ah^{-1} \in aH$. Thus, (a) implies (b), and similarly (b) implies (a). Now suppose that (a) holds and choose $h \in H$ such that $a = bh$. Then for all $h_1 \in H$, $ah_1 = bhh_1 \in bH$; thus $aH \subseteq bH$. Similarly, (b) implies that $bH \subseteq aH$. Since (a) is equivalent to (b), each implies (c). Because $a \in aH$ and $b \in bH$, (c) implies (a) and (b). Finally, (d) and (e) are equivalent by taking inverses, and $a = bh \in bH \Leftrightarrow b^{-1}a = h \in H$, so (a) and (d) are equivalent. ■

Proposition 2.5.4. *Let* H *be a subgroup of a group* G.

(a) *Let* a *and* b *be elements of* G. *Either* $aH = bH$ *or* $aH \cap bH = \emptyset$.

(b) *Each left coset* aH *is nonempty and the union of left cosets is* G.

Proof. If $aH \cap bH \neq \emptyset$, let $c \in aH \cap bH$. By the previous proposition $cH = aH$ and $cH = bH$, so $aH = bH$. For each $a \in G$, $a \in aH$; this implies both assertions of part (b). ■

Proposition 2.5.5. *Let* H *be a subgroup of a group* G *and let* a *and* b *be elements of* G. *Then* $x \mapsto ba^{-1}x$ *is a bijection between* aH *and* bH.

Proof. The map $x \mapsto ba^{-1}x$ is a bijection of G (with inverse $y \mapsto ab^{-1}y$). Its restriction to aH is a bijection of aH onto bH. ■

Theorem 2.5.6. *(Lagrange's theorem). Let* G *be a finite group and* H *a subgroup. Then the cardinality of* H *divides the cardinality of* G, *and the quotient* $\dfrac{|G|}{|H|}$ *is the number of left cosets of* H *in* G.

Proof. The distinct left cosets of H are mutually disjoint by Proposition 2.5.4 and each has the same size (namely $|H| = |eH|$) by Proposition 2.5.5. Since the union of the left cosets is G, the cardinality of G is the cardinality of H times the number of distinct left cosets of H. ■

Definition 2.5.7. For a subgroup H of a group G, the *index* of H in G is the number of left cosets of H in G. The index is denoted $[G : H]$.

Index also makes sense for infinite groups. For example, take the larger group to be $\mathbb{Z}$ and the subgroup to be $n\mathbb{Z}$. Then $[\mathbb{Z} : n\mathbb{Z}] = n$, because there are n cosets of $n\mathbb{Z}$ in $\mathbb{Z}$. Every subgroup of $\mathbb{Z}$ (other than the zero subgroup $\{0\}$) has the form $n\mathbb{Z}$ for some n, so every nonzero subgroup has finite index. The zero subgroup has infinite index.

It is also possible for a nontrivial subgroup of an infinite group (i.e., a subgroup that is neither $\{e\}$ nor the entire group) to have infinite index. For example, the cosets of $\mathbb{Z}$ in $\mathbb{R}$ are in one to one correspondence with the elements of the half-open interval $[0, 1)$; there are uncountably many cosets! See Exercise 2.5.11.

Corollary 2.5.8. *Let p be a prime number and suppose G is a group of order p. Then*

(a) *G has no subgroups other than G and $\{e\}$.*
(b) *G is cyclic, and in fact, for any nonidentity element $a \in G$, $G = \langle a \rangle$.*
(c) *Every homomorphism from G into another group is either trivial (i.e., every element of G is sent to the identity) or injective.*

Proof. The first assertion follows immediately from Lagrange's theorem, since the size of a subgroup can only be p or 1. If $a \neq e$, then the subgroup $\langle a \rangle$ is not $\{e\}$, so must be G. The last assertion also follows from the first, since the kernel of a homomorphism is a subgroup. ■

Any two groups of prime order p are isomorphic, since each is cyclic. This generalizes (substantially) the results that we obtained before on the uniqueness of the groups of orders 2, 3, and 5.

Corollary 2.5.9. *Let G be any finite group, and let $a \in G$. Then the order $o(a)$ divides the order of G.*

Proof. The order of a is the cardinality of the subgroup $\langle a \rangle$. ■

The index of subgroups satisfies a multiplicativity property:

Proposition 2.5.10. *Suppose $K \subseteq H \subseteq G$ are subgroups. Then*

$$[G : K] = [G : H][H : K].$$

Proof. If the groups are finite, then by Lagrange's theorem,

$$[G : K] = \frac{|G|}{|K|} = \frac{|G|}{|H|}\frac{|H|}{|K|} = [G : H][H : K].$$

If the groups are infinite, we have to use another approach, which is discussed in the Exercises. ■

Definition 2.5.11. For any group G, the *center* Z(G) of G is the set of elements that commute with all elements of G,

$$Z(G) = \{a \in G : ag = ga \text{ for all } g \in G\}.$$

You are asked in the Exercises to show that the center of a group is a normal subgroup, and to compute the center of several particular groups.

Exercises 2.5

2.5.1. Check that the left cosets of the subgroup

$$K = \{e, (123), (132)\}$$

in S_3 are

$$eK = (123)K = (132)K = K$$
$$(12)K = (13)K = (23)K = \{(12), (13), (23)\}$$

and that each occurs three times in the list $(gK)_{g \in S_3}$. Note that K is the subgroup of even permutations and the other coset of K is the set of odd permutations.

2.5.2. Suppose $K \subseteq H \subseteq G$ are subgroups. Suppose $h_1K, \dots, h_RK$ is a list of the distinct cosets of K in H, and $g_1H, \dots, g_SH$ is a list of the distinct cosets of H in G. Show that $\{g_s h_r H : 1 \le s \le S, 1 \le r \le R\}$ is the set of distinct cosets of H in G. Hint: There are two things to show. First, you have to show that if $(r, s) \ne (r', s')$, then $g_s h_r K \ne g_{s'} h_{r'} K$. Second, you have to show that if $g \in G$, then for some (r, s), $gK = g_s h_r K$.

2.5.3. Try to extend the idea of the previous exercise to the case where at least one of the pairs $K \subseteq H$ and $H \subseteq G$ has infinite index.

2.5.4. Consider the group S_3.

(a) Find all the left cosets and all the right cosets of the subgroup $H = \{e, (12)\}$ of S_3, and observe that not every left coset is also a right coset.

(b) Find all the left cosets and all the right cosets of the subgroup $K = \{e, (123), (132)\}$ of S_3, and observe that every left coset is also a right coset.

2.5.5. What is the analogue of Proposition 2.5.3, with left cosets replaced with right cosets?

2.5.6. Let H be a subgroup of a group G. Show that $aH \mapsto Ha^{-1}$ defines a bijection between left cosets of H in G and right cosets of H in G. (The index of a subgroup was defined in terms of left cosets, but this observation shows that we get the same notion using right cosets instead.)

2.5.7. For a subgroup N of a group G, prove that the following are equivalent:

(a) N is normal.

(b) Each left coset of N is also a right coset. That is, for each $a \in G$, there is a $b \in G$ such that $aN = Nb$.

(c) For each $a \in G$, $aN = Na$.

2.5.8. Suppose N is a subgroup of a group G and $[G : N] = 2$. Show that N is normal using the criterion of the previous exercise.

2.5.9. Show that if G is a finite group and N is a subgroup of index 2, then for elements a and b of G, the product ab is an element of N if, and only if, either both of a and b are in N or neither of a and b is in N.

2.5.10. For two subgroups H and K of a group G and an element $a \in G$, the double coset HaK is the set of all products hak, where $h \in H$ and $k \in K$. Show that two double cosets HaK and HbK are either equal or disjoint.

2.5.11. Consider the additive group $\mathbb{R}$ and its subgroup $\mathbb{Z}$. Describe a coset $t + \mathbb{Z}$ geometrically. Show that the set of all cosets of $\mathbb{Z}$ in $\mathbb{R}$ is $\{t + \mathbb{Z} : 0 \le t < 1\}$. What are the analogous results for $\mathbb{Z}^2 \subseteq \mathbb{R}^2$?

2.5.12. Consider the additive group $\mathbb{Z}$ and its subgroup $n\mathbb{Z}$ consisting of all integers divisible by n. Show that the distinct cosets of $n\mathbb{Z}$ in $\mathbb{Z}$ are $\{n\mathbb{Z}, 1 + n\mathbb{Z}, 2 + n\mathbb{Z}, \dots, n-1 + n\mathbb{Z}\}$.

2.5.13.

(a) Show that the center of a group G is a normal subgroup of G.

(b) What is the center of the group S_3?

2.5.14.

(a) What is the center of the group D_4 of symmetries of the square?

(b) What is the center of the dihedral group D_n?

2.5.15. Find the center of the group $GL(2, \mathbb{R})$ of 2-by-2 invertible real matrices. Do the same for $GL(3, \mathbb{R})$. Hint: It is a good idea to go outside of the realm of invertible matrices in considering this problem. This is because it is useful to exploit linearity, but the group of invertible matrices is not a linear space. So first explore the condition for a matrix A to commute with all matrices B, not just invertible matrices. Note that the condition $AB = BA$ is linear in B, so a matrix A commutes with all matrices if, and only if, it commutes with each member of a linear spanning set of matrices. So now consider the so-called matrix units E_{ij}, which have a 1 in the i, j position and zeros elsewhere. The set of matrix units is a basis of the linear space of matrices. Find the condition for a matrix to commute with all of the E_{ij}'s. It remains to show that if a matrix commutes with all invertible matrices, then it also commutes with all E_{ij}'s. (The results of this exercise hold just as well with the real numbers $\mathbb{R}$ replaced by the complex numbers $\mathbb{C}$ or the rational numbers $\mathbb{Q}$.)

2.5.16. Show that the symmetric group S_n has a unique subgroup of index 2, namely the subgroup A_n of even permutations. Hint: Such subgroup N is normal. Hence if it contains one element of a certain cycle structure, then it contains all elements of that cycle structure, according to Exercises 2.6.6 and 2.4.14. Can N contain any 2–cycles? Show that N contains a product of k 2–cycles if, and only if, k is even, by using Exercise 2.5.9. Conclude $N = A_n$ by Proposition 2.4.22.

2.6. Equivalence Relations and Set Partitions

Associated to the data of a group G and a subgroup H, we can define a binary relation on G by $a \sim b \pmod{H}$ or $a \sim_H b$ if, and only if, $aH = bH$. According to Proposition 2.5.3, $a \sim_H b$ if, and only if, if $b^{-1}a \in H$. This relation has the following properties:

1. $a \sim_H a$.
2. $a \sim_H b \Leftrightarrow b \sim_H a$.
3. If $a \sim_H b$ and $b \sim_H c$, then also $a \sim_H c$.

This is an example of an *equivalence relation.*

Definition 2.6.1. An equivalence relation $\sim$ on a set X is a binary relation with the properties:

(a) *Reflexivity:* For each $x \in X$, $x \sim x$.
(b) *Symmetry:* For $x, y \in X$, $x \sim y \Leftrightarrow y \sim x$.
(c) *Transitivity:* For $x, y, z \in X$, if $x \sim y$ and $y \sim z$, then $x \sim z$.

Associated to the same data (a group G and a subgroup H), we also have the family of left cosets of H in G. Each left coset are nonempty, distinct left cosets are mutually disjoint, and the union of all left cosets is G. This is an example of a *set partition.*

Definition 2.6.2. A *partition* of a set X is a collection of mutually disjoint nonempty subsets whose union is X.

Equivalence relations and set partitions are very common in mathematics. We will soon see that equivalence relations and set partitions are two aspects of one phenomenon.

Example 2.6.3.

(a) For any set X, equality is an equivalence relation on X. Two elements $x, y \in X$ are related if, and only if, $x = y$.

(b) For any set X, declare $x \sim y$ for all $x, y \in X$. This is an equivalence relation on X.

(c) Let n be a natural number. Recall the relation of *congruence modulo* n defined on the set of integers by $a \equiv b \pmod{n}$ if, and only if, $a - b$ is divisible by n. It was shown in Proposition 1.7.2 that congruence modulo n is an equivalence relation on the set of integers. In fact, this is a special case of the coset equivalence relation, with the group $\mathbb{Z}$, and the subgroup $n\mathbb{Z} = \{nd : d \in \mathbb{Z}\}$.

(d) Let X and Y be any sets and $f : X \to Y$ any map. Define a relation on X by $x' \sim_f x''$ if, and only if, $f(x') = f(x'')$. Then $\sim_f$ is an equivalence relation on X.

(e) In Euclidean geometry, *congruence* is an equivalence relation on the set of triangles in the plane. *Similarity* is another equivalence relation on the set of triangles in the plane.

(f) Again in Euclidean geometry, parallelism of lines is an equivalence relation on the set of all lines in the plane.

(g) Let X be any set, and let $T : X \to X$ be an bijective map of X. All integer powers of T are defined and for integers m, n, we have $T^n \circ T^m = T^{n+m}$. In fact, T is an element of the group of all bijective maps of X, and the powers of T are defined as elements of this group. See the discussion preceding Definition 2.2.11.

For $x, y \in X$, declare $x \sim y$ if there is an integer n such that $T^n(x) = y$. Then $\sim$ is an equivalence relation on X. In fact, for all $x \in X$, $x \sim x$ because $T^0(x) = x$. If $x, y \in X$ and $x \sim y$, then $T^n(x) = y$ for some integer n; it follows that $T^{-n}(y) = x$, so $y \sim x$. Finally, suppose that $x, y, z \in X$, $x \sim y$, and $y \sim z$. Then there exist integers n, m such that $T^n(x) = y$ and $T^m(y) = z$. But then $T^{n+m}(x) = T^m(T^n(x)) = T^m(y) = z$, so $x \sim z$.

An equivalence relation on a set X always gives rise to a distinguished family of subsets of X:

Definition 2.6.4. If $\sim$ is an equivalence relation on X, then for each $x \in X$, the *equivalence class* of x is the set

$$[x] = \{y \in X : x \sim y\}.$$

Note that $x \in [x]$ because of reflexivity; in particular, the equivalence classes are nonempty subsets of X.

Proposition 2.6.5. *Let $\sim$ be an equivalence relation on X. For $x, y \in X$, $x \sim y$ if, and only if, $[x] = [y]$.*

Proof. If $[x] = [y]$, then $x \in [x] = [y]$, so $x \sim y$. For the converse, suppose that $x \sim y$ (and hence $y \sim x$, by symmetry of the equivalence relation). If $z \in [x]$, then $z \sim x$. Since $x \sim y$ by assumption, transitivity of the equivalence relation implies that $z \sim y$ (i.e., $z \in [y]$). This shows that $[x] \subseteq [y]$. Similarly, $[y] \subseteq [x]$. Therefore, $[x] = [y]$. ■

Corollary 2.6.6. *Let $\sim$ be an equivalence relation on X, and let $x, y \in X$. Then either $[x] \cap [y] = \emptyset$ or $[x] = [y]$.*

Proof. We have to show that if $[x] \cap [y] \neq \emptyset$, then $[x] = [y]$. But if $z \in [x] \cap [y]$, then $[x] = [z] = [y]$, by the proposition. ■

Consider an equivalence relation $\sim$ on a set X. The equivalence classes of $\sim$ are nonempty and have union equal to X, since for each $x \in X$, $x \in [x]$. Furthermore, the equivalence classes are *mutually disjoint*; this means that any two distinct equivalence classes have empty intersection. So the collection of equivalence classes is a *partition* of the set X.

Any equivalence relation on a set X gives rise to a partition of X by equivalence classes.

On the other hand, given a partition P of X, we can define an relation on X by $x \sim_P y$ if, and only if, x and y are in the same subset of the partition. Let's check that this is an equivalence relation. Write the partition P as $\{X_i : i \in I\}$. We have $X_i \neq \emptyset$ for all $i \in I$, $X_i \cap X_j = \emptyset$ if $i \neq j$, and $\cup_{i \in I} X_i = X$. Our definition of the relation is $x \sim_P y$ if, and only if, there exists $i \in I$ such that both x and y are elements of X_i.

Now for all $x \in X$, $x \sim_P x$ because there is some $i \in I$ such that $x \in X_i$. The definition of $x \sim_P y$ is clearly symmetric in x and y.

Finally, if $x \sim_P y$ and $y \sim_P z$, then there exist $i \in I$ such that both x and y are elements of X_i, and furthermore there exists $j \in I$ such that both y and z are elements of X_j. Now i is necessarily equal to j because y is an element of both X_i and of X_j and $X_i \cap X_j = \emptyset$ if $i \neq j$. But then both x and z are elements of X_i, which gives $x \sim_P z$.

Every partition of a set X *gives rise to an equivalence relation on* X.

Suppose that we start with an equivalence relation on a set X, form the partition of X into equivalence classes, and then build the equivalence relation related to this partition. Then we just get back the equivalence relation we started with! In fact, let $\sim$ be an equivalence relation on X, let $P = \{[x] : x \in X\}$ be the corresponding partition of X into equivalence classes, and let $\sim_P$ denote the equivalence relation derived from P. Then $x \sim y \Leftrightarrow [x] = [y] \Leftrightarrow$ there exists a $[z] \in P$ such that both x and y are elements of $[z] \Leftrightarrow x \sim_P y$.

Suppose, on the other hand, that we start with a partition of a set X, form the associated equivalence relation, and then form the partition of X consisting of equivalence classes for this equivalence relation. Then we end up with exactly the partition we started with. In fact, let P be a partition of X, let $\sim_P$ be the corresponding equivalence relation, and let P' be the family of $\sim_P$ equivalence classes. We have to show that $P = P'$.

Let $a \in X$, let $[a]$ be the unique element of P' containing a, and let A be the unique element of P containing a. We have $b \in [a] \Leftrightarrow b \sim_P a \Leftrightarrow$ there exists a $B \in P$ such that both a and b are elements of B. But since $a \in A$, the last condition holds if, and only if, both a and b are elements of A. That is, $[a] = A$. But this means that P and P' contain exactly the same subsets of X.

The following proposition summarizes this discussion:

Proposition 2.6.7. *Let* X *be any set. There is a one to one correspondence between equivalence relations on* X *and set partitions of* X.

Remark 2.6.8. This proposition, and the discussion preceding it, are still valid (but completely uninteresting) if X is the empty set. There is exactly one equivalence relation on the empty set, namely the empty relation, and there is exactly one partition of the empty set, namely the empty collection of nonempty subsets!

Example 2.6.9. Consider a group G and a subgroup H. Associated to these data, we have the family of left cosets of H in G. According to Proposition 2.5.4, the family of left cosets of H in G forms a partition of G. We defined the equivalence relation $\sim_H$ in terms of this partition, namely $a \sim_H b$ if, and only if, $aH = bH$. The equivalence classes of $\sim_H$ are precisely the left cosets of H in G, since $a \sim_H b \Leftrightarrow aH = bH \Leftrightarrow a \in bH$.

Example 2.6.10.

(a) The equivalence classes for the equivalence relation of equality on a set X are just the singletons $\{x\}$ for $x \in X$.

(b) The equivalence relation $x \sim y$ for all $x, y \in X$ has just one equivalence class, namely X.

(c) The equivalence classes for the relation of congruence modulo n on $\mathbb{Z}$ are $\{[0], [1], \dots, [n-1]\}$.

(d) Let $f : X \to Y$ be any map. Define $x' \sim_f x''$ if, and only if, $f(x') = f(x'')$. The equivalence classes for the equivalence relation $\sim_f$ are the *fibers of* f, namely the sets $f^{-1}(y)$ for y in the range of f.

(e) Let X be any set, and let $T : X \to X$ be an bijective map of X. For $x, y \in X$, declare $x \sim y$ if there is an integer n such that $T^n(x) = y$. The equivalence classes for this relation are the *orbits* of T, namely the sets $O(x) = \{T^n(x) : n \in Z\}$ for $x \in X$.

Equivalence Relations and Surjective Maps

There is a third aspect to the phenomenon of equivalence relations/partitions. We have noted that for any map f from a set X to another set Y, we can define an equivalence relation on X by $x' \sim_f x''$ if $f(x') = f(x'')$. We might as well assume that f is surjective, as we can replace Y by the range of f without changing the equivalence relation. The equivalence classes of $\sim_f$ are the fibers $f^{-1}(y)$ for $y \in Y$. See Exercise 2.6.1.

On the other hand, given an equivalence relation $\sim$ on X, define $X/\sim$ to be the set of equivalence classes of $\sim$ and define a surjection π of X onto $X/\sim$ by $\pi(x) = [x]$. If we now build the equivalence relation $\sim_\pi$ associated with this surjective map, we just recover the original equivalence relation. In fact, for $x', x'' \in X$, we have $x' \sim x'' \Leftrightarrow [x'] = [x''] \Leftrightarrow \pi(x') = \pi(x'') \Leftrightarrow x' \sim_\pi x''$.

We have proved the following result:

Proposition 2.6.11. *Let $\sim$ be an equivalence relation on a set X. Then there exists a set Y and a surjective map $\pi : X \to Y$ such that $\sim$ is equal to the equivalence relation $\sim_\pi$.*

When do two surjective maps $f : X \to Y$ and $f' : X \to Y'$ determine the *same* equivalence relation on X? The condition for this to happen turns out to be the following:

Definition 2.6.12. Two surjective maps $f : X \to Y$ and $f' : X \to Y'$ are *similar* if there exists a bijection $s : Y \to Y'$ such that $f' = s \circ f$.

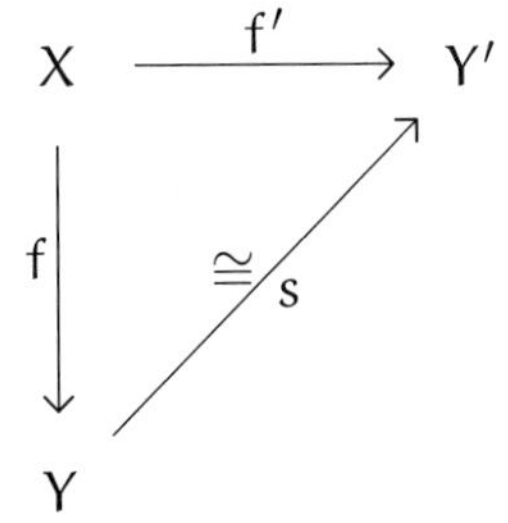

Proposition 2.6.13. *Two surjective maps* $f : X \to Y$ *and* $f' : X \to Y'$ *determine the* same *equivalence relation* X *if, and only if,* f *and* f' *are similar.*

Proof. It is easy to see that if f and f' are similar surjections, then they determine the same equivalence relation on X.

Suppose, on the other hand, that $f : X \to Y$ and $f' : X \to Y'$ are surjective maps that define the same equivalence relation $\sim$ on X. We want to define a map $s : Y \to Y'$ such that $f' = s \circ f$. Let $y \in Y$, and choose any $x \in f^{-1}(y)$. We wish to define $s(y) = f'(x)$, for then $s(f(x)) = s(y) = f'(x)$, as desired. However, we have to be careful to check that this makes sense; that is, we have to check that $s(y)$ really depends only on y and not on the choice of $x \in f^{-1}(y)$! But in fact, if $\bar{x}$ is another element of $f^{-1}(y)$, then $x \sim_f \bar{x}$, so $x \sim_{f'} \bar{x}$, by hypothesis, so $f'(x) = f'(\bar{x})$.

We now have our map $s : Y \to Y'$ such that $f' = s \circ f$. It remains to show that s is a bijection.

In the same way that we defined s, we can also define a map $s' : Y' \to Y$ such that $f = s' \circ f'$. I claim that s and s' are inverse maps, so both are bijective. In fact, we have $f = s' \circ f' = s' \circ s \circ f$, so $f(x) = s'(s(f(x)))$ for all $x \in X$. Let $y \in Y$; choose $x \in X$ such that $y = f(x)$. Substituting y for $f(x)$ gives $y = s'(s(y))$. Similarly, $s(s'(y')) = y'$ for all $y' \in Y'$. This completes the proof that s is bijective. ■

Let us look again at our main example, the equivalence relation on a group G determined by a subgroup H, whose equivalence classes are the left cosets of H in G. We would like to define a canonical surjective map on G whose fibers are the left cosets of H in G.

Definition 2.6.14. The set of left cosets of H in G is denoted G/H. The surjective map $\pi : G \longrightarrow G/H$ defined by $\pi(a) = aH$ is called *the canonical projection* or *quotient map* of G onto G/H.

Proposition 2.6.15. *The fibers of the canonical projection* $\pi : G \to G/H$ *are the left cosets of* H *in* G*. The equivalence relation* $\sim_\pi$ *on* G *determined by* π *is the equivalence relation* $\sim_H$.

Proof. We have $\pi^{-1}(aH) = \{b \in G : bH = aH\} = aH$. Furthermore, $a \sim_\pi b \Leftrightarrow aH = bH \Leftrightarrow a \sim_H b$. ■

Conjugacy

We close this section by introducing another equivalence relation that is extremely useful for studying the structure of groups:

Definition 2.6.16. Let a and b be elements of a group G. We say that b is *conjugate* to a if there is a $g \in G$ such that $b = gag^{-1}$.

You are asked to show in the Exercises that conjugacy is an equivalence relation and to find all the conjugacy equivalence classes in several groups of small order.

Definition 2.6.17. The equivalence classes for conjugacy are called *conjugacy classes.*

Note that the center of a group is related to the notion of conjugacy in the following way: The center consists of all elements whose conjugacy class is a singleton. That is, $g \in Z(G) \Leftrightarrow$ the conjugacy class of g is $\{g\}$.

Exercises 2.6

2.6.1. Consider any surjective map f from a set X onto another set Y. We can define a relation on X by $x_1 \sim x_2$ if $f(x_1) = f(x_2)$. Check that this is an equivalence relation. Show that the associated partition of X is the partition into "fibers" $f^{-1}(y)$ for $y \in Y$.

The next several exercises concern conjugacy classes in a group.

2.6.2. Show that conjugacy of group elements is an equivalence relation.

2.6.3. What are the conjugacy classes in S_3?

2.6.4. What are the conjugacy classes in the symmetry group of the square D_4?

2.6.5. What are the conjugacy classes in the dihedral group D_5?

2.6.6. Show that a subgroup is normal if, and only if, it is a union of conjugacy classes.

2.7. Quotient Groups and Homomorphism Theorems

Consider the permutation group S_n with its normal subgroup of even permutations. For the moment write $\mathcal{E}$ for the subgroup of even permutations and $\mathcal{O}$ for the coset $\mathcal{O} = (12)\mathcal{E} = \mathcal{E}(12)$ consisting of odd permutations. The subgroup $\mathcal{E}$ is the kernel of the sign homomorphism $\varepsilon : S_n \longrightarrow \{1, -1\}$.

Since the product of two permutations is even if, and only if, both are even or both are odd, we have the following multiplication table for the two cosets of $\mathcal{E}$:

	$\mathcal{E}$	$\mathcal{O}$
$\mathcal{E}$	$\mathcal{E}$	$\mathcal{O}$
$\mathcal{O}$	$\mathcal{O}$	$\mathcal{E}$

The products are taken in the sense mentioned previously; namely the product of two even permutations or two odd permutations is even, and the product of an even permutation with an odd permutation is odd. Thus the multiplication on the cosets of $\mathcal{E}$ reproduces the multiplication on the group $\{1, -1\}$.

This is a general phenomenon: If N is a *normal* subgroup of a group G, then the set G/N of left cosets of a N in G has the structure of a group.

The Quotient Group Construction

Theorem 2.7.1. *Let* N *be a normal subgroup of a group* G*. The set of cosets* G/N *has a unique product that makes* G/N *a group and that makes the quotient map* $\pi : G \longrightarrow G/N$ *a group homomorphism.*

Proof. Let A and B be elements of G/N (i.e., A and B are left cosets of N in G). Let $a \in A$ and $b \in B$ (so $A = aN$ and $B = bN$). We would like to define the product AB to be the left coset containing ab, that is,

$$(aN)(bN) = abN.$$

But we have to check that this makes sense (i.e., that the result is independent of the choice of $a \in A$ and of $b \in B$). So let a' be another element of aN and b' another element of bN. We need to check that $abN = a'b'N$, or, equivalently, that $(ab)^{-1}(a'b') \in N$. We have

$$\begin{aligned}(ab)^{-1}(a'b') &= b^{-1}a^{-1}a'b' \\ &= b^{-1}a^{-1}a'(bb^{-1})b' = (b^{-1}a^{-1}a'b)(b^{-1}b').\end{aligned}$$

Since $aN = a'N$, and $bN = b'N$, we have $a^{-1}a' \in N$ and $b^{-1}b' \in N$. Since N is normal, $b^{-1}(a^{-1}a')b \in N$. Therefore, the final expression is a product of two elements of N, so is in N. This completes the verification that the definition of the product on G/H makes sense.

The associativity of the product on G/N follows from repeated use of the definition of the product, and the associativity of the product on G; namely

$$\begin{aligned}(aNbN)cN = abNcN = (ab)cN &= a(bc)N \\ &= aNbcN = aN(bNcN).\end{aligned}$$

It is clear that N itself serves as the identity for this multiplication and that $a^{-1}N$ is the inverse of aN. Thus G/N with this multiplication is a group. Furthermore, π is a homomorphism because

$$\pi(ab) = abN = aNbN = \pi(a)\pi(b).$$

The uniqueness of the product follows simply from the surjectivity of π: in order for π to be a homomorphism, it is necessary that $aNbN = abN$. ■

The group G/N is called the *quotient group* of G by N. The map $\pi : G \to G/N$ is called the quotient homomorphism. Another approach to defining the product in G/N is developed in Exercise 2.7.3.

Example 2.7.2. (Finite cyclic groups as quotients of $\mathbb{Z}$). The construction of $\mathbb{Z}_n$ in Section 1.7 is an example of the quotient group construction. The (normal) subgroup in the construction is $n\mathbb{Z} = \{\ell n : \ell \in \mathbb{Z}\}$. The cosets of $n\mathbb{Z}$ in $\mathbb{Z}$ are of the form $k + n\mathbb{Z} = [k]$; the distinct cosets are $[0] = n\mathbb{Z}, [1] = 1+n\mathbb{Z}, \dots, [n-1] = n-1+n\mathbb{Z}$. The product (sum) of two cosets is $[a] + [b] = [a + b]$. So the group we called $\mathbb{Z}_n$ is precisely $\mathbb{Z}/n\mathbb{Z}$. The quotient homomorphism $\mathbb{Z} \to \mathbb{Z}_n$ is given by $k \mapsto [k]$.

Example 2.7.3. Now consider a cyclic group G of order n with generator a. There is a homomorphism $\varphi : \mathbb{Z} \to G$ of $\mathbb{Z}$ *onto* G defined by $\varphi(k) = a^k$. The kernel of this homomorphism is precisely all multiples of n, the order of a; $\ker(\varphi) = n\mathbb{Z}$. I claim that φ "induces" an isomorphism $\tilde{\varphi} : \mathbb{Z}_n \to G$, defined by $\tilde{\varphi}([k]) = a^k = \varphi(k)$. It is necessary to check that this makes sense (i.e., that $\tilde{\varphi}$ is well defined) because we have attempted to define the value of $\tilde{\varphi}$ on a coset $[k]$ in terms of a particular representative of the coset. Would we get the same result if we took another representative, say $k + 17n$ instead of k? In fact, we would get the same answer: If $[a] = [b]$, then $a - b \in n\mathbb{Z} = \ker(\varphi)$, and, therefore, $\varphi(a) - \varphi(b) = \varphi(a - b) = 0$. Thus $\varphi(a) = \varphi(b)$. This shows that the map $\tilde{\varphi}$ is well defined.

Next we have to check the homomorphism property of $\tilde{\varphi}$. This property is valid because $\tilde{\varphi}([a][b]) = \tilde{\varphi}([ab]) = \varphi(ab) = \varphi(a)\varphi(b) = \tilde{\varphi}([a])\tilde{\varphi}([b])$.

The homomorphism $\tilde{\varphi}$ has the same range as φ, so it is surjective. It also has trivial kernel: If $\tilde{\varphi}([k]) = 0$, then $\varphi(k) = 0$, so $k \in n\mathbb{Z} = [0]$, so $[k] = [0]$. Thus $\tilde{\varphi}$ is an isomorphism.

Example 2.7.4. Take the additive abelian group $\mathbb{R}$ as G and the subgroup $\mathbb{Z}$ as N. Since $\mathbb{R}$ is abelian, all of its subgroups are normal, and, in particular, $\mathbb{Z}$ is a normal subgroup.

The cosets of $\mathbb{Z}$ in $\mathbb{R}$ were considered in Exercise 2.5.11, where you were asked to verify that the cosets are parameterized by the set of real numbers t such that $0 \le t < 1$. In fact, two real numbers are in the same coset modulo $\mathbb{Z}$ precisely if they differ by an integer, $s \equiv t \pmod{\mathbb{Z}} \Leftrightarrow s - t \in \mathbb{Z}$. For any real number t, let $[[t]]$ denote the greatest integer less than or equal to t. Then $t - [[t]] \in [0, 1)$ and $t \equiv (t - [[t]]) \pmod{\mathbb{Z}}$. On the other hand, no two real numbers in $[0, 1)$ are congruent modulo $\mathbb{Z}$. Thus we have a bijection between $\mathbb{R}/\mathbb{Z}$ and $[0, 1)$ which is given by $[t] \mapsto t - [[t]]$.

We get a more instructive geometric picture of the set $\mathbb{R}/\mathbb{Z}$ of cosets of $\mathbb{R}$ modulo $\mathbb{Z}$ if we take, instead of the half–open interval $[0, 1)$, the closed interval $[0, 1]$ but *identify* the endpoints 0 and 1: The picture is a circle of circumference 1. Actually we can take a circle of any convenient size, and it is more convenient to take a circle of radius 1, namely

$$\{e^{2\pi i t} : t \in \mathbb{R}\} = \{e^{2\pi i t} : 0 \le t < 1\}.$$

So now we have bijections between set $\mathbb{R}/\mathbb{Z}$ of cosets of $\mathbb{R}$ modulo $\mathbb{Z}$, the set $[0, 1)$, and the unit circle $\mathbb{T}$, given by

$$[t] \mapsto t - [[t]] \mapsto e^{2\pi i t} = e^{2\pi i (t - [[t]])}.$$

Let us write φ for the map $t \mapsto e^{2\pi i t}$ from $\mathbb{R}$ onto the unit circle, and $\tilde{\varphi}$ for the map $[t] \mapsto \varphi(t) = e^{2\pi i t}$. Our discussion shows that $\tilde{\varphi}$ is well defined. We know that the unit circle $\mathbb{T}$ is itself a group, and we recall that that the exponential map $\varphi : \mathbb{R} \to \mathbb{T}$ is a group homomorphism, namely,

$$\varphi(s+t) = e^{2\pi i(s+t)} = e^{2\pi i s} e^{2\pi i t} = \varphi(s)\varphi(t).$$

Furthermore, the kernel of φ is precisely $\mathbb{Z}$.

We now have a good geometric picture of the quotient group $\mathbb{R}/\mathbb{Z}$ *as a set*, but we still have to discuss the group structure of $\mathbb{R}/\mathbb{Z}$. The definition of the product (addition!) on $\mathbb{R}/\mathbb{Z}$ is $[t] + [s] = [t+s]$. But observe that

$$\tilde{\varphi}([s] + [t]) = \tilde{\varphi}([s+t]) = e^{2\pi i(s+t)} = e^{2\pi i s} e^{2\pi i t} = \tilde{\varphi}(s)\tilde{\varphi}(t).$$

Thus $\tilde{\varphi}$ is a group isomorphism from the quotient group $\mathbb{R}/\mathbb{Z}$ to $\mathbb{T}$.

Our work can be summarized in the following diagram, in which all of the maps are group homomorphisms, and the map π is the quotient map from $\mathbb{R}$ to $\mathbb{R}/\mathbb{Z}$.

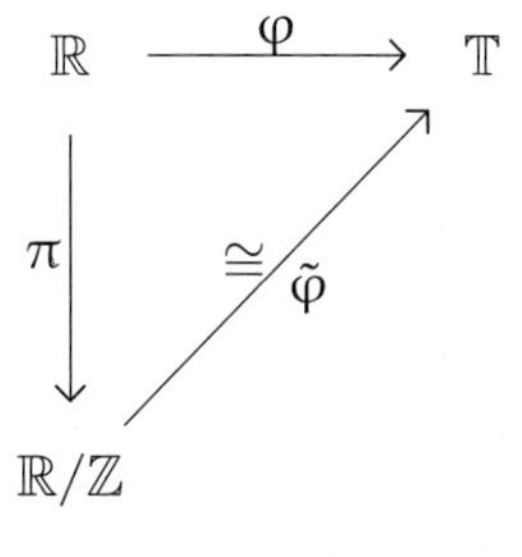

Example 2.7.5. Recall from Exercise 2.4.20 the "$Ax + b$" group or affine group $\mathrm{Aff}(n)$ consisting of transformations of $\mathbb{R}^n$ of the form

$$T_{A,\mathbf{b}}(\mathbf{x}) = A\mathbf{x} + \mathbf{b},$$

where $A \in GL(n, \mathbb{R})$ and $\mathbf{b} \in \mathbb{R}^n$. Let N be the subset consisting of the transformations $T_{E,\mathbf{b}}$, where E is the identity transformation,

$$T_{E,\mathbf{b}}(\mathbf{x}) = \mathbf{x} + \mathbf{b}.$$

The composition rule in $\mathrm{Aff}(n)$ is

$$T_{A,\mathbf{b}} T_{A',\mathbf{b}'} = T_{AA',A\mathbf{b}'+\mathbf{b}}.$$

The inverse of $T_{A,\mathbf{b}}$ is $T_{A^{-1},-A^{-1}\mathbf{b}}$. N is a subgroup isomorphic to the additive group $\mathbb{R}^n$ because

$$T_{E,\mathbf{b}} T_{E,\mathbf{b}'} = T_{E,\mathbf{b}+\mathbf{b}'},$$

and N is normal. In fact,

$$T_{A,\mathbf{b}} T_{E,\mathbf{c}} T_{A,\mathbf{b}}^{-1} = T_{E,A\mathbf{c}}.$$

Let us examine the condition for two elements $T_{A,\mathbf{b}}$ and $T_{A',\mathbf{b}'}$ to be congruent modulo N. The condition is

$$T_{A',\mathbf{b}'}^{-1} T_{A,\mathbf{b}} = T_{A'^{-1},-A'^{-1}\mathbf{b}'} T_{A,\mathbf{b}} = T_{A'^{-1}A,A'^{-1}(\mathbf{b}-\mathbf{b}')} \in N.$$

This is equivalent to $A = A'$. Thus the class of $T_{A,\mathbf{b}}$ modulo N is $[T_{A,\mathbf{b}}] = \{T_{A,\mathbf{b}'} : \mathbf{b}' \in \mathbb{R}^n\}$, and the cosets of N can be parameterized by $A \in GL(n)$. In fact, the map $[T_{A,\mathbf{b}}] \mapsto A$ is a bijection between the set $\mathrm{Aff}(n)/N$ of cosets of $\mathrm{Aff}(n)$ modulo N and $GL(n)$.

Let us write φ for the map $\varphi : T_{A,\mathbf{b}} \mapsto A$ from $\mathrm{Aff}(n)$ to $GL(n)$, and $\tilde{\varphi}$ for the map $\tilde{\varphi} : [T_{A,\mathbf{b}}] \mapsto A$ from $\mathrm{Aff}(n)/N$ to $GL(n)$. The map φ is a (surjective) homomorphism, because

$$\varphi(T_{A,\mathbf{b}} T_{A',\mathbf{b}'}) = \varphi(T_{AA',A\mathbf{b}'+\mathbf{b}}) = AA' = \varphi(T_{A,\mathbf{b}})\varphi(T_{A',\mathbf{b}'}),$$

and furthermore the kernel of φ is N.

The definition of the product in $\mathrm{Aff}(n)/N$ is

$$[T_{A,\mathbf{b}}][T_{A',\mathbf{b}'}] = [T_{A,\mathbf{b}} T_{A',\mathbf{b}'}] = [T_{AA',\mathbf{b}+A\mathbf{b}'}].$$

It follows that

$$\tilde{\varphi}([T_{A,\mathbf{b}}][T_{A',\mathbf{b}'}]) = \tilde{\varphi}([T_{AA',\mathbf{b}+A\mathbf{b}'}]) = AA' = \tilde{\varphi}([T_{A,\mathbf{b}}])\tilde{\varphi}([T_{A',\mathbf{b}'}]),$$

and, therefore, $\tilde{\varphi}$ is an isomorphism of groups.

We can summarize our findings in the diagram:

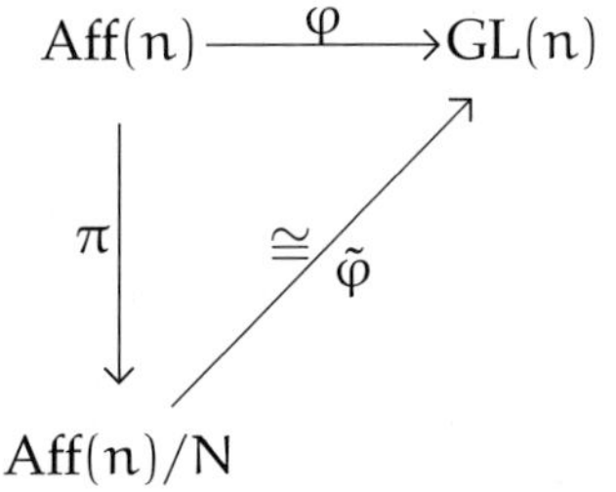

Homomorphism Theorems

The features that we have noticed in the several examples are quite general:

Theorem 2.7.6. *(Homomorphism theorem). Let* $\varphi : G \longrightarrow \bar{G}$ *be a surjective homomorphism with kernel* N. *Let* $\pi : G \longrightarrow G/N$ *be the quotient homomorphism. There is a group isomorphism* $\tilde{\varphi} : G/N \longrightarrow \bar{G}$ *satisfying* $\tilde{\varphi} \circ \pi = \varphi$. *(See the following diagram.)*

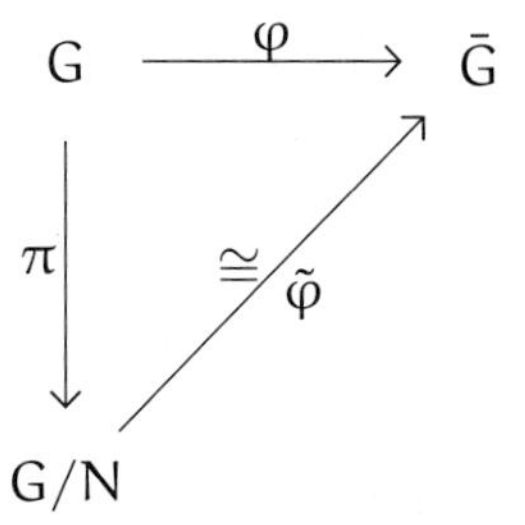

Proof. There is only one possible way to define $\tilde{\varphi}$ so that it will satisfy $\tilde{\varphi} \circ \pi = \varphi$, namely $\tilde{\varphi}(aN) = \varphi(a)$. We can check directly that $\tilde{\varphi}$ is well defined and bijective. (See Exercise 2.7.1.) This assertion also follows from the general discussion of surjective maps and equivalence relations in the previous section. (See Exercise 2.7.2.)

Furthermore,

$$\tilde{\varphi}(aNbN) = \tilde{\varphi}(abN) = \varphi(ab) = \varphi(a)\varphi(b) = \tilde{\varphi}(aN)\tilde{\varphi}(bN).$$

■

A slightly different proof is suggested in Exercise 2.7.2.

The two theorems (Theorems 2.7.1 and 2.7.6) say that normal subgroups and (surjective) homomorphisms are two sides of one coin: Given a normal subgroup N, there is a surjective homomorphism with N as kernel, and, on the other hand, a surjective homomorphism is essentially determined by its kernel.

Theorem 2.7.6 also reveals the best way to understand a quotient group G/N. The best way is to find a natural model, namely some naturally defined group $\bar{G}$ together with a surjective homomorphism $\varphi : G \to \bar{G}$ with kernel N. Then, according to the theorem, $G/N \cong \bar{G}$. With this in mind, we take another look at the examples given above, as well as several more examples.

Example 2.7.7. Let a be an element of order n in a group H. There is a homomorphism $\varphi : \mathbb{Z} \to H$ given by $k \mapsto a^k$. This homomorphism has range $\langle a \rangle$ and kernel $n\mathbb{Z}$. Therefore, by the homomorphism theorem, $\mathbb{Z}/n\mathbb{Z} \cong \langle a \rangle$. In particular, if $\zeta = e^{2\pi i/n}$, then $\varphi(k) = \zeta^k$ induces an isomorphism of $\mathbb{Z}/n\mathbb{Z}$ onto the group C_n of nth roots of unity in $\mathbb{C}$.

Example 2.7.8. The homomorphism $\varphi : \mathbb{R} \to \mathbb{C}$ given by $\varphi(t) = e^{2\pi i t}$ has range $\mathbb{T}$ and kernel $\mathbb{Z}$. Thus by the homomorphism theorem, $\mathbb{R}/\mathbb{Z} \cong \mathbb{T}$.

Example 2.7.9. The map $\varphi : \text{Aff}(n) \to \text{GL}(n)$ defined by $T_{A,\mathbf{b}} \mapsto A$ is a surjective homomorphism with kernel $N = \{T_{E,\mathbf{b}} : \mathbf{b} \in \mathbb{R}^n\}$. Therefore, by the homomorphism theorem, $\text{Aff}(n)/N \cong \text{GL}(n)$.

Example 2.7.10. The set $\text{SL}(n, \mathbb{R})$ of matrices of determinant 1 is a normal subgroup of $\text{GL}(n, \mathbb{R})$. In fact, $\text{SL}(n, \mathbb{R})$ is the kernel of the homomorphism $\det : \text{GL}(n, \mathbb{R}) \to \mathbb{R}^*$, and this implies that $\text{SL}(n, \mathbb{R})$ is a normal subgroup. It also implies that the quotient group $\text{GL}(n, \mathbb{R})/\text{SL}(n, \mathbb{R})$ is naturally isomorphic to $\mathbb{R}^*$.

Example 2.7.11. Consider $G = \text{GL}(n, \mathbb{R})$, the group of n-by-n invertible matrices. Set $Z = G \cap \mathbb{R}E$, the set of invertible scalar matrices. Then Z is evidently a normal subgroup of G, and is, in fact, the center of G. A coset of Z in G has the form $[A] = AZ = \{\lambda A : \lambda \in \mathbb{R}^*\}$, the set of all nonzero multiples of the invertible matrix A; two matrices A and B are equivalent modulo Z precisely if one is a scalar multiple of the other. By our general construction of quotient groups, we can form G/Z, whose elements are cosets of Z in G, with the product $[A][B] = [AB]$. G/Z is called the *projective linear group*.

The rest of this example is fairly difficult, and it might be best to skip it on the first reading. We would like to find some natural realization or model of the quotient group. Now a natural model for a group is generally as a group of transformations of something or the other, so we would have to look for some objects which are naturally transformed not by matrices but rather by matrices modulo scalar multiples.

At least two natural models are available for G/Z. One is as transformations of projective $(n-1)$–dimensional space $\mathbb{P}^{n-1}$, and the other is as transformations of G itself.

Projective $(n-1)$–dimensional space consists of n-vectors modulo scalar multiplication. More precisely, we define an equivalence relation $\sim$ on the set $\mathbb{R}^n \setminus \{\mathbf{0}\}$ of nonzero vectors in $\mathbb{R}^n$ by $\mathbf{x} \sim \mathbf{y}$ if there is a nonzero scalar λ such that $\mathbf{x} = \lambda\mathbf{y}$. Then $\mathbb{P}^{n-1} = (\mathbb{R}^n \setminus \{\mathbf{0}\})/\sim$, the set of equivalence classes of vectors. There is another picture of $\mathbb{P}^{n-1}$ that is a little easier to visualize; every nonzero vector $\mathbf{x}$ is equivalent to the unit vector $\mathbf{x}/||\mathbf{x}||$, and furthermore two unit vectors $\mathbf{a}$ and $\mathbf{b}$ are equivalent if and only if $\mathbf{a} = \pm\mathbf{b}$; therefore, $\mathbb{P}^{n-1}$ is also realized as $S^{n-1}/\pm$, the unit sphere in n–dimensional space, modulo equivalence of antipodal points. Write $[\mathbf{x}]$ for the class of a nonzero vector $\mathbf{x}$.

There is a homomorphism of G into $\text{Sym}(\mathbb{P}^{n-1})$, the group of invertible maps from $\mathbb{P}^{n-1}$ to $\mathbb{P}^{n-1}$, defined by $\varphi(A)([\mathbf{x}]) = [A\mathbf{x}]$; we have to check, as usual, that $\varphi(A)$ is a well-defined transformation of $\mathbb{P}^{n-1}$ and that φ is a homomorphism. I leave this as an exercise. What is the kernel of φ? It is precisely the invertible scalar matrices Z. We have $\varphi(G) \cong G/Z$, by the homomorphism theorem, and thus G/Z has been identified as a group of transformations of projective space.

A second model for G/Z is developed in Exercise 2.7.7, as a group of transformations of G itself.

Everything in this example works in exactly the same way when $\mathbb{R}$ is replaced by $\mathbb{C}$. When moreover $n = 2$, there is a natural realization of $GL(n, \mathbb{C})/Z$ as "fractional linear transformations" of $\mathbb{C}$. For this, see Exercise 2.7.6.

Proposition 2.7.12. *Let* $\varphi : G \longrightarrow \bar{G}$ *be a homomorphism of* G onto $\bar{G}$, *and let* N *denote the kernel of* φ.

(a) *If* $\bar{B}$ *is a subgroup of* $\bar{G}$, *then* $\varphi^{-1}(\bar{B})$ *is a subgroup of* G *containing* N.

(b) *The map* $\bar{B} \mapsto \varphi^{-1}(\bar{B})$ *is a bijection between subgroups of* $\bar{G}$ *and subgroups of* G *containing* N.

(c) $\bar{B} \subseteq \bar{G}$ *is a normal subgroup of* $\bar{G}$ *if, and only if,* $\varphi^{-1}(\bar{B})$ *is a normal subgroup of* G.

Proof. For each subgroup $\bar{B}$ of $\bar{G}$, $\varphi^{-1}(\bar{B})$ is a subgroup of G by Proposition 2.4.12, and furthermore $\varphi^{-1}(\bar{B}) \supseteq \varphi^{-1}\{e\} = N$. This proves part (a).

To prove (b), we show that the map $A \mapsto \varphi(A)$ is the inverse of the map $\bar{B} \mapsto \varphi^{-1}(\bar{B})$. If $\bar{B}$ is a subgroup of G, then $\varphi(\varphi^{-1}(\bar{B}))$ is a subgroup of $\bar{G}$, that a priori is contained in $\bar{B}$. But since φ is surjective, $\bar{B} = \varphi(\varphi^{-1}(\bar{B}))$.

For a subgroup A of G containing $\ker(\varphi)$, $\varphi^{-1}(\varphi(A))$ is a subgroup of G which *a priori* contains A. If x is in that subgroup, then there is an $a \in A$ such that $\varphi(x) = \varphi(a)$. This is equivalent to $a^{-1}x \in \ker(\varphi) \subseteq A$ (exercise). Hence, $x \in aA = A$. This shows that $\varphi^{-1}(\varphi(A)) = A$, which completes the proof of part (b).

Suppose $\bar{B}$ is normal in $\bar{G}$. We have to show that $\varphi^{-1}(\bar{B})$ is normal in G. For all $g \in G$, $g\varphi^{-1}(\bar{B})g^{-1} \supseteq gNg^{-1} = N$, so $g\varphi^{-1}(\bar{B})g^{-1}$ is a subgroup of G containing N. Therefore, by part (b),

$$\begin{aligned} g\varphi^{-1}(\bar{B})g^{-1} &= \varphi^{-1}(\varphi(g\varphi^{-1}(\bar{B})g^{-1})) \\ &= \varphi^{-1}(\varphi(g)\bar{B}\varphi(g)^{-1}) = \varphi^{-1}(\bar{B}), \end{aligned}$$

since $\bar{B}$ is normal. Thus if $\bar{B}$ is normal in $\bar{G}$, then $\varphi^{-1}(\bar{B})$ is normal in G. The converse was shown in Exercise 2.4.8. ∎

Example 2.7.13. What are all the subgroups of $\mathbb{Z}_n$? Since $\mathbb{Z}_n = \mathbb{Z}/n\mathbb{Z}$, the subgroups of $\mathbb{Z}_n$ correspond one to one with subgroups of $\mathbb{Z}$ containing the kernel of the quotient map $\varphi : \mathbb{Z} \to \mathbb{Z}/n\mathbb{Z}$, namely $n\mathbb{Z}$. But the subgroups of $\mathbb{Z}$ are cyclic and of the form $k\mathbb{Z}$ for some $k \in Z$. So when does $k\mathbb{Z}$ contain $n\mathbb{Z}$? Precisely when $n \in k\mathbb{Z}$, or when k divides n. Thus the subgroups of $\mathbb{Z}_n$ correspond one to one with *positive integer divisors of* n. The image of $k\mathbb{Z}$ in $\mathbb{Z}_n$ is cyclic with generator $[k]$ and with order n/k.

The following is a very useful generalization of the homomorphism theorem.

Proposition 2.7.14. *Let $\varphi : G \to \bar{G}$ be a surjective homomorphism of groups with kernel N. Let $H \subseteq N$ be a subgroup that is normal in G, and let $\pi : G \to G/H$ denote the quotient map. Then there is a surjective homomorphism $\tilde{\varphi} : G/H \to \bar{G}$ such that $\tilde{\varphi} \circ \pi = \varphi$. (See the following diagram.) The kernel of $\tilde{\varphi}$ is $N/H \subseteq G/H$.*

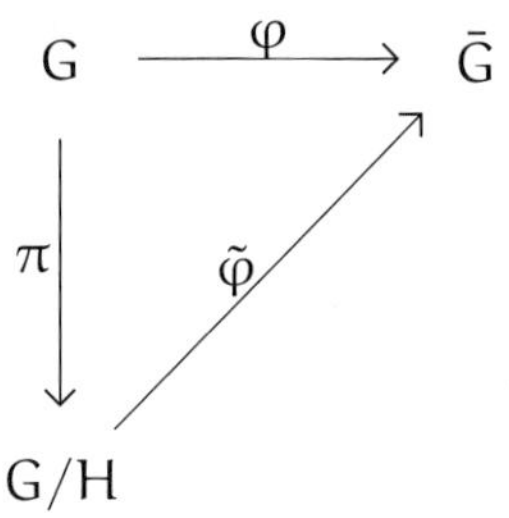

Proof. As in the proof of the homomorphism theorem, there is only one way to define $\tilde{\varphi}$ consistent with the requirement that $\tilde{\varphi} \circ \pi = \varphi$, namely $\tilde{\varphi}(aH) = \varphi(a)$. It is necessary to check that this is well defined and a homomorphism. But if $aH = bH$, then $b^{-1}a \in H \subseteq N = \ker(\varphi)$, so $\varphi(b^{-1}a) = 1$, or $\varphi(a) = \varphi(b)$. This shows that the map $\tilde{\varphi}$ is well defined. The homomorphism property follows as in the proof of the homomorphism theorem. ■

Corollary 2.7.15.

(a) *Let $H \subseteq N \subseteq G$ be subgroups with both H and N normal in G. Then $xH \mapsto xN$ defines a homomorphism of G/H onto G/N with kernel N/H.*

(b) *There is a natural isomorphism $(G/H)/(N/H) \cong G/N$.*

Proof. Statement (a) is the special case of the Proposition with $\bar{G} = G/N$ and $\varphi : G \to G/N$ the quotient map. For part (b), apply the homomorphism theorem to the surjective homomorphism $G/H \to G/N$ with kernel N/H. ■

Example 2.7.16. When is there a surjective homomorphism from one cyclic group $\mathbb{Z}_k$ to another cyclic group $\mathbb{Z}_\ell$?

Suppose first that $\psi : \mathbb{Z}_k \to \mathbb{Z}_\ell$ is a surjective homomorphism such that $\psi[1] = [1]$. Let φ_k and φ_ℓ be the natural quotient maps of $\mathbb{Z}$ onto $\mathbb{Z}_k$ and $\mathbb{Z}_\ell$ respectively. We have maps

$$\mathbb{Z} \xrightarrow{\varphi_k} \mathbb{Z}_k \xrightarrow{\psi} \mathbb{Z}_\ell,$$

and $\psi \circ \varphi_k$ is a surjective homomorphism of $\mathbb{Z}$ onto $\mathbb{Z}_\ell$ such that $\psi \circ \varphi_k(1) = [1]$; therefore, $\psi \circ \varphi_k = \varphi_\ell$. But then the kernel of φ_k is contained in the kernel of φ_ℓ, which is to say that every integer multiple of k is divisible by ℓ. In particular, k is divisible by ℓ.

The assumption that $\psi[1] = [1]$ is not essential and can be eliminated as follows: Suppose that $\psi : \mathbb{Z}_k \to \mathbb{Z}_\ell$ is a surjective homomorphism with $\psi([1]) = [a]$. The cyclic subgroup group generated by $[a]$ is all of $\mathbb{Z}_\ell$, and in particular $[a]$ has order ℓ. Thus there is a surjective homomorphism $\mathbb{Z} \to \mathbb{Z}_\ell$ defined by $n \mapsto [na]$, with kernel $\ell\mathbb{Z}$. It follows from the homomorphism theorem that there is an isomorphism $\theta : \mathbb{Z}_\ell \to \mathbb{Z}_\ell$ such that $\theta([1]) = [a]$. But then $\theta^{-1} \circ \psi : \mathbb{Z}_k \to \mathbb{Z}_\ell$ is a surjective homomorphism such that $\theta^{-1} \circ \psi([1]) = \theta^{-1}([a]) = [1]$. It follows that k is divisible by ℓ.

Conversely, if k is divisible by ℓ, then $k\mathbb{Z} \subseteq \ell\mathbb{Z} \subseteq Z$. Since $\mathbb{Z}$ is abelian, all subgroups are normal, and by the corollary, there is a surjective homomorphism $\mathbb{Z}_k \to \mathbb{Z}_\ell$ such that $[1] \mapsto [1]$.

We conclude that there is a surjective homomorphism from $\mathbb{Z}_k$ to $\mathbb{Z}_l$ if, and only if, ℓ divides k.

Proposition 2.7.17. *Let $\varphi : G \longrightarrow \bar{G}$ be a surjective homomorphism. Let $\bar{K}$ be a normal subgroup of $\bar{G}$ and let $K = \varphi^{-1}(\bar{K})$. Then*

(a) $x \mapsto \varphi(x)\bar{K}$ *is a homomorphism of G onto $\bar{G}/\bar{K}$ with kernel K.*
(b) $xK \mapsto \varphi(x)\bar{K}$ *is an isomorphism of G/K onto $\bar{G}/\bar{K}$.*

Proof. We know from the previous proposition that K is a normal subgroup of G containing $\ker(\varphi)$. Write ψ for the canonical projection $\psi : \bar{G} \longrightarrow \bar{G}/\bar{K}$. Then $\psi \circ \varphi : G \longrightarrow \bar{G}/\bar{K}$ is a surjective homomorphism, because it is a composition of surjective homomorphisms. The kernel of $\psi \circ \varphi$ is the set of $x \in G$ such that $\varphi(x) \in \ker(\psi) = \bar{K}$; that is, $\ker(\psi \circ \varphi) = \varphi^{-1}(\bar{K}) = K$. This proves (a).

Statement (b) now follows from the homomorphism theorem (Theorem 2.7.6). ■

Proposition 2.7.18. *Let* $\varphi : G \longrightarrow \bar{G}$ *be a surjective homomorphism with kernel* N*. Let* A *be a subgroup of* G*. Then*

(a) $\varphi^{-1}(\varphi(A)) = AN = \{an : a \in A \text{ and } n \in N\}$,

(b) AN *is a subgroup of* G *containing* N*.*

(c) $AN/N \cong \varphi(A) \cong A/(A \cap N)$.

Proof. Let $x \in G$. Then

$$\begin{aligned} x \in \varphi^{-1}(\varphi(A)) &\Leftrightarrow \text{there exists } a \in A \text{ such that } \varphi(x) = \varphi(a) \\ &\Leftrightarrow \text{there exists } a \in A \text{ such that } x \in aN \\ &\Leftrightarrow x \in AN. \end{aligned}$$

Thus, $AN = \varphi^{-1}(\varphi(A))$, which, by Proposition 2.7.12, is a subgroup of G containing N. Now applying Theorem 2.7.6 to the restriction of φ to AN gives the isomorphism $AN/N \cong \varphi(AN) = \varphi(A)$. On the other hand, applying the theorem to the restriction of φ to A gives $A/(A \cap N) \cong \varphi(A)$. ■

Example 2.7.19. Let G be the symmetry group of the square, which is generated by elements r and j satisfying $r^4 = e = j^2$ and $jrj = r^{-1}$. Let N be the subgroup $\{e, r^2\}$; then N is normal because $jr^2j = r^{-2} = r^2$. What is G/N? The group G/N has order 4 and is generated by two commuting elements rN and jN each of order 2. (Note that rN and jN commute because $rN = r^{-1}N$, and $jr^{-1} = rj$, so $jrN = jr^{-1}N = rjN$.) Hence, G/N is isomorphic to the group $\mathcal{V}$ of symmetries of the rectangle. Let $A = \{e, j\}$. Then AN is a four–element subgroup of G (also isomorphic to $\mathcal{V}$) and $AN/N = \{N, jN\} \cong \mathbb{Z}_2$. On the other hand, $A \cap N = \{e\}$, so $A/(A \cap N) \cong A \cong \mathbb{Z}_2$.

Example 2.7.20. Let $G = GL(n, \mathbb{C})$, the group of n-by-n invertible complex matrices. Let Z be the subgroup of invertible scalar matrices. G/Z is the complex projective linear group. Let $A = SL(n, \mathbb{C})$. Then $AZ = G$ because for any invertible matrix X, we have $X = \det(X)X'$, where $X' = \det(X)^{-1}X \in A$. On the other hand, $A \cap Z$ is the group of invertible scalar matrices with determinant 1; such a matrix must have the form ζE where ζ is an nth root of unity in $\mathbb{C}$. We have $G/Z = AZ/Z = A/(A \cap Z) = A/\{\zeta E : \zeta \text{ is a root of unity}\}$.

The same holds with $\mathbb{C}$ replaced by $\mathbb{R}$, and here the result is more striking, because $\mathbb{R}$ contains few roots of unity. If n is odd, the only n^{th} root of unity in $\mathbb{R}$ is 1, so we see that the projective linear group is isomorphic to $SL(n, \mathbb{R})$. On the other hand, if n is even, then -1 is also an nth root of unity and the projective linear group is isomorphic to $SL(n, \mathbb{R})/\{\pm 1E\}$.

Exercises 2.7

2.7.1. Let $\varphi : G \longrightarrow \bar{G}$ be a surjective homomorphism with kernel N. Let $\pi : G \longrightarrow G/N$ be the quotient homomorphism. Show directly that the map $\tilde{\varphi} : G/N \longrightarrow \bar{G}$ defined by $\tilde{\varphi}(aN) = \varphi(a)$ is well defined and bijective.

2.7.2. Let $\varphi : G \longrightarrow \bar{G}$ be a surjective homomorphism with kernel N. Let $\pi : G \longrightarrow G/N$ be the quotient homomorphism. Show that for $x, y \in G$, $x \sim_\varphi y \Leftrightarrow x \sim_\pi y \Leftrightarrow x \sim_N y$. Conclude that the map $\tilde{\varphi} : G/N \longrightarrow \bar{G}$ defined by $\tilde{\varphi}(aN) = \varphi(a)$ is well defined and bijective.

2.7.3. Here is a different approach to the definition of the product on G/N, where N is a normal subgroup of G.

(a) Define the product of arbitrary subsets A and B of G to be
$$\{ab : a \in A \text{ and } b \in B\}.$$
Verify that this gives an associative product on subsets.

(b) Take $A = aN$ and $B = bN$. Verify that the product AB in the sense of part (a) is equal to abN. Your verification will use that N is a normal subgroup of G.

(c) Observe that it follows from parts (a) and (b) that $(aN)(bN) = abN$ is a well–defined, associative product on G/N.

2.7.4. Consider the affine group $\mathrm{Aff}(n)$ consisting of transformations of $\mathbb{R}^n$ of the form $T_{A,\mathbf{b}}(\mathbf{x}) = A\mathbf{x} + \mathbf{b}$ ($A \in GL(n, \mathbb{R})$ and $\mathbf{b} \in \mathbb{R}^n$).

(a) Show that the inverse of $T_{A,\mathbf{b}}$ is $T_{A^{-1},-A^{-1}\mathbf{b}}$.

(b) Show that $T_{A,\mathbf{b}} T_{E,\mathbf{c}} T_{A,\mathbf{b}}^{-1} = T_{E,A\mathbf{c}}$. Conclude that $N = \{T_{E,\mathbf{b}} : \mathbf{b} \in \mathbb{R}^n\}$ is a normal subgroup of $\mathrm{Aff}(n)$.

2.7.5. Suppose G is finite. Verify that
$$|AN| = \frac{|A|\,|N|}{|A \cap N|}.$$

2.7.6. Consider the set of fractional linear transformations of the complex plane with ∞ adjoined, $\mathbb{C} \cup \{\infty\}$,
$$T_{a,b;c,d}(z) = \frac{az + b}{cz + d}$$

where $\begin{bmatrix} a & b \\ c & d \end{bmatrix}$ is an invertible 2-by-2 complex matrix. Show that this is a group of transformations and is isomorphic to $\mathrm{GL}(2, \mathbb{C})/Z(\mathrm{GL}(2, \mathbb{C}))$.

2.7.7. Recall that an automorphism of a group G is a group isomorphism from G to G. Denote the set of all automorphisms of G by $\mathrm{Aut}(G)$.

(a) Show that $\mathrm{Aut}(G)$ of G is also a group.
(b) Recall that for each $g \in G$, the map $c_g : G \longrightarrow G$ defined by $c_g(x) = gxg^{-1}$ is an element of $\mathrm{Aut}(G)$. Show that the map $c : g \mapsto c_g$ is a homomorphism from G to $\mathrm{Aut}(G)$.
(c) Show that the kernel of the map c is $Z(G)$.
(d) In general, the map c is not surjective. The image of c is called the group of inner automorphisms and denoted $\mathrm{Int}(G)$. Conclude that $\mathrm{Int}(G) \cong G/Z(G)$.

2.7.8. Let D_4 denote the group of symmetries of the square, and N the subgroup of rotations. Observe that N is normal and check that D_4/N is isomorphic to the cyclic group of order 2.

2.7.9. Find out whether every automorphism of S_3 is inner. Note that any automorphism φ must permute the set of elements of order 2, and an automorphism φ is completely determined by what it does to order 2 elements, since all elements are products of 2–cycles. Hence, there can be at most as many automorphisms of S_3 as there are permutations of the three–element set of 2–cycles, namely 6; that is, $|\mathrm{Aut}(S_3)| \leq 6$. According to Exercises 2.5.13 and 2.7.7, how large is $\mathrm{Int}(S_3)$? What do you conclude?

2.7.10. Let G be a group and let C be the subgroup generated by all elements of the form $xyx^{-1}y^{-1}$ with $x, y \in G$. C is called the commutator subgroup of G. Show that C is a normal subgroup and that G/C is abelian. Show that if H is a normal subgroup of G such that G/H is abelian, then $H \supseteq C$.

2.7.11. Show that any quotient of an abelian group is abelian.

2.7.12. Prove that if $G/Z(G)$ is cyclic, then G is abelian.

2.7.13. Suppose $G/Z(G)$ is abelian. Must G be abelian?

Chapter 3

Products of Groups

3.1. Direct Products

Whenever we have a normal subgroup N of a group G, the group G is in some sense built up from N and the quotient group G/N, both of which are in general smaller and simpler than G itself. So a route to understanding G is to understand N and G/N, and finally to understand the way the two interact.

The simplest way in which a group can be built up from two groups is without any interaction between the two; the resulting group is called the direct product of the two groups.

As a set, the direct product of two groups A and B is the Cartesian product $A \times B$. We define a multiplication on $A \times B$ by declaring the product $(a, b)(a', b')$ to be (aa', bb'). That is, the multiplication is performed coordinate by coordinate. It is straightforward to check that this product makes $A \times B$ a group, with $e = (e_A, e_B)$ serving as the identity element, and (a^{-1}, b^{-1}) the inverse of (a, b). You are asked to check this in Exercise 3.1.1.

Definition 3.1.1. $A \times B$, with this group structure, is called the *direct product* of A and B.

Example 3.1.2. Suppose we have two sets of objects, of different types, say five apples on one table and four bananas on another table. Let A denote the group of permutations of the apples, $A \cong S_5$, and let B denote the group of permutations of the bananas, $B \cong S_4$. The symmetries of the entire configuration of fruit consist of permutations of the apples among themselves and of the bananas among themselves. That is, a symmetry is a pair of permutations (σ, τ), where $\sigma \in A$ and $\tau \in B$. Two such symmetries are composed by separately composing the symmetries of apples and the symmetries of bananas: $(\sigma', \tau')(\sigma, \tau) = (\sigma'\sigma, \tau'\tau)$. Thus the symmetry group is the direct product $A \times B \cong S_5 \times S_4$.

Example 3.1.3. Let a and b be relatively prime natural numbers. We showed in the proof of the Chinese remainder theorem near the end of Section 1.11 that the map $[x]_{ab} \mapsto ([x]_a, [x]_b)$ from $\mathbb{Z}_{ab}$ to $\mathbb{Z}_a \times \mathbb{Z}_b$ is well defined and bijective. It is straightforward to check that this is an isomorphism of groups. Thus $\mathbb{Z}_{ab} \cong \mathbb{Z}_a \times \mathbb{Z}_b$.

The direct product contains a copy of A and a copy of B. Namely, $A \times \{e_B\}$ is a subgroup isomorphic to A, and $\{e_A\} \times B$ is a subgroup isomorphic to B. Both of these subgroups are normal in $A \times B$. Elements of the two subgroups commute:

$$(a, e_B)(e_A, b) = (e_A, b)(a, e_B) = (a, b).$$

The two subgroups have trivial intersection

$$(A \times \{e_B\}) \cap (\{e_A\} \times B) = \{(e_A, e_B)\},$$

and together they generate $A \times B$, namely $(A \times \{e_B\})(\{e_A\} \times B) = A \times B$.

It is useful to be able to recognize when a group is isomorphic to a direct product of groups:

Proposition 3.1.4.

(a) *Suppose* M *and* N *are normal subgroups of* G, *and* $M \cap N = \{e\}$. *Then for all* $m \in M$ *and* $n \in N$, $mn = nm$.

(b) $MN = \{mn : m \in M$ *and* $n \in N\}$ *is a subgroup and* $(m, n) \mapsto mn$ *is an isomorphism of* $M \times N$ *onto* MN.

(c) *If* $MN = G$, *then* $G \cong M \times N$.

Proof. We have $mn = nm \Leftrightarrow mnm^{-1}n^{-1} = e$. Observe that

$$mnm^{-1}n^{-1} = (mnm^{-1})n \in N,$$

since N is normal. Likewise,

$$mnm^{-1}n^{-1} = m(nm^{-1}n^{-1}) \in M,$$

since M is normal. Since the intersection of M and N is trivial, it follows that $mnm^{-1}n^{-1} = e$.

It is now easy to check that MN is a subgroup, and that $(m, n) \mapsto mn$ is a homomorphism of $M \times N$ onto MN. The homomorphism is injective, because if $mn = e$, then $m = n^{-1} \in M \cap N = \{e\}$.

Assertion (c) is evident. ∎

Example 3.1.5. Let's look at Example 3.1.3 again "from the inside" of $\mathbb{Z}_{ab}$. The subgroup $\langle [a] \rangle$ of $\mathbb{Z}_{ab}$ generated by $[a]$ has order b and the subgroup $\langle [b] \rangle$ has order a. These two subgroups have trivial intersection because the order of the intersection must divide the orders of both subgroups, and the orders of the subgroups are relatively prime. The two subgroups are normal since $\mathbb{Z}_{ab}$ is abelian. Therefore, the subgroup of $\mathbb{Z}_{ab}$ generated by $\langle [a] \rangle$ and $\langle [b] \rangle$ is isomorphic to $\langle [b] \rangle \times \langle [a] \rangle \cong \mathbb{Z}_a \times \mathbb{Z}_b$. Since the order of both $\mathbb{Z}_{ab}$ and $\mathbb{Z}_a \times \mathbb{Z}_b$ is ab, it follows that $\mathbb{Z}_{ab} = \langle [b] \rangle + \langle [a] \rangle \cong \langle [b] \rangle \times \langle [a] \rangle \cong \mathbb{Z}_a \times \mathbb{Z}_b$.

Example 3.1.6. Let G be the group of symmetries of the rectangle. Let r be the rotation of order 2 about the axis through the centroid of the faces, and let j be the rotation of order 2 about an axis passing through the centers of two opposite edges. Set $R = \{e, r\}$ and $J = \{e, j\}$. Then R and J are normal (since G is abelian), $R \cap J = \{e\}$, and $RJ = G$. Hence $G \cong R \times J \cong \mathbb{Z}_2 \times \mathbb{Z}_2$.

Remark 3.1.7. Some authors distinguish between *external* direct products and *internal* direct products. For groups A and B, the group constructed previously from the Cartesian product of A and B is the *external* direct product. On the other hand, if a group G has normal subgroups N and M such that $N \cap M = \{e\}$, so that G is *isomorphic* to the direct product $N \times M$, G is said to be the *internal* direct product of N and M. The distinction is more psychological than mathematical.

When the groups involved are abelian and written with additive notation, we often use the terminology *direct sum* instead of direct product and use the notation $A \oplus B$ instead of $A \times B$. In this book, we shall, in general, use the terminology of direct sums and the notation $\oplus$ for abelian groups *with additional structure* (for example, rings, vector spaces, and modules) but we shall stay with the terminology of direct products and with the notation $\times$ when speaking of abelian groups *as groups*.

We can define the direct product of any finite number of groups in the same way as we define the direct product of two groups. As a set, the direct product of groups $A_1, A_2, \dots, A_n$ is the Cartesian product $A_1 \times A_2 \times \cdots \times A_n$. The multiplication on $A_1 \times A_2 \times \cdots \times A_n$ is defined coordinate–by–coordinate:

$$(x_1, x_2, \dots x_n)(y_1, y_2, \dots, y_n) = (x_1 y_1, x_2 y_2, \dots, x_n y_n)$$

for $x_i, y_i \in A_i$. It is again straightforward to check that this makes the Cartesian product into a group. You are asked to verify this in Exercise 3.1.8.

Definition 3.1.8. $A_1 \times A_2 \times \cdots \times A_n$, with the coordinate–by–coordinate multiplication, is called the *direct product* of A_1, $A_2, \dots, A_n$.

The direct product $P = A_1 \times A_2 \times \cdots \times A_n$ contains a copy of each A_i. In fact, for each i, $\tilde{A}_i = \{e\} \times \dots \{e\} \times A_i \times \{e\} \times \dots \{e\}$ is a normal subgroup of P, and $A_i \cong \tilde{A}_i$. (We are writing e for the identity in each of the A_i's as well in P.)

For each $\tilde{A}_i$ there is a "complementary" subgroup $\tilde{A}_i'$ of P,

$$\tilde{A}_i' = A_1 \times \cdots A_{i-1} \times \{e\} \times A_{i+1} \times \cdots \times A_n.$$

$\tilde{A}_i'$ is also normal, $\tilde{A}_i \cap \tilde{A}_i' = \{e\}$, and $\tilde{A}_i\tilde{A}_i' = P$. Thus P is the (internal) direct product of $\tilde{A}_i$ and $\tilde{A}_i'$.

It follows that $xy = yx$ if $x \in \tilde{A}_i$ and $y \in \tilde{A}_i'$. We can also check this directly.

Note that if $i \neq j$, then $\tilde{A}_j \subseteq \tilde{A}_i'$. Consequently, $\tilde{A}_i \cap \tilde{A}_j = \{e\}$ and $xy = yx$ if $x \in \tilde{A}_i$ and $y \in \tilde{A}_j$. Again, we can check these assertions directly.

Since the $\tilde{A}_i$ are mutually commuting subgroups, the subgroup generated by any collection of them is their product:

$$\langle \tilde{A}_{i_1}, \tilde{A}_{i_2} \dots, \tilde{A}_{i_s} \rangle = \tilde{A}_{i_1}\tilde{A}_{i_2} \cdots \tilde{A}_{i_s}.$$

In fact, the product is

$$\{(x_1, x_2, \dots, x_n) \in P : x_j = e \text{ if } j \notin \{i_1, \dots, i_s\}\}.$$

In particular, $\tilde{A}_1\tilde{A}_2 \cdots \tilde{A}_n = P$.

Example 3.1.9. Suppose the local animal shelter houses several collections of animals of different types: four African aardvarks, five Brazilian bears, seven Canadian canaries, three Dalmatian dogs, and two Ethiopian elephants, each collection in a different room of the shelter.

Let A denote the group of permutations of the aardvarks, $A \cong S_4$.[1] Likewise, let B denote the group of permutations of the bears, $B \cong S_5$; let C denote the group of permutations of the canaries, $C \cong S_7$; let D denote the group of permutations of the dogs, $D \cong S_4$;

[1]Imagine, if you like, that the aardvarks are kept in separate cages in the aardvark room, and after the aardvark exercise period, they may be mixed up when they are returned to their cages, since the shelter staff can't tell one aardvark from another.

and finally, let E denote the group of permutations of the elephants, $E \cong S_2 \cong \mathbb{Z}_2$. The symmetry group of the entire zoo is $P = A \times B \times C \times D \times E$.

In this situation, it is slightly artificial to distinguish between A and $\tilde{A}$! The group of permutations of aardvarks, forgetting that any other animals exist, is A. On the other hand, $\tilde{A} = A \times \{e\} \times \{e\} \times \{e\}$ is the group of those permutations of the entire zoo that leave the bears, canaries, dogs, and elephants in place and permute only aardvarks.

The subgroup $\tilde{A}'$ of P "complementary" to $\tilde{A}$ is the group of permutations of the shelter population that leave the aardvarks in place and permute only bears, canaries, dogs, and elephants. The product $\tilde{A}\tilde{C}$ is the group of permutations of the zoo that leave bears, dogs, and elephants in place and permute only aardvarks and canaries.

Example 3.1.10. $\mathbb{Z}_{30} \cong \mathbb{Z}_5 \times \mathbb{Z}_3 \times \mathbb{Z}_2$ as groups. This can be obtained using Example 3.1.3 repeatedly, $\mathbb{Z}_{30} \cong \mathbb{Z}_5 \times \mathbb{Z}_6 \cong \mathbb{Z}_5 \times \mathbb{Z}_3 \times \mathbb{Z}_2$. In general, if $n = p_1^{\alpha_1} \cdots p_k^{\alpha_k}$ is the prime decomposition of n, then

$$\mathbb{Z}_n \cong \mathbb{Z}_{p_1^{\alpha_1}} \times \cdots \mathbb{Z}_{p_k^{\alpha_k}}$$

as groups. This follows from Example 3.1.3 and induction on k.

The following proposition gives conditions for a group to be isomorphic to a direct product of several groups:

Proposition 3.1.11. *Suppose $N_1, N_2, \dots, N_r$ are normal subgroups of a group G such that for all i,*

$$N_i \cap (N_1 \dots N_{i-1} N_{i+1} \dots N_r) = \{e\}.$$

Then $N_1 N_2 \dots N_r$ is a subgroup of G and $(n_1, n_2, \dots, n_r) \mapsto n_1 n_2 \dots n_r$ is an isomorphism of $P = N_1 \times N_2 \times \cdots \times N_r$ onto $N_1 N_2 \dots N_r$. In particular, if $N_1 N_2 \dots N_r = G$, then $G \cong N_1 \times N_2 \times \cdots \times N_r$.

Proof. Because $N_i \cap N_j = \{e\}$ for $i \neq j$, it follows from Proposition 3.1.4 that $xy = yx$ for $x \in N_i$ and $y \in N_j$. Using this, it is straightforward to show that the map $(n_1, n_2, \dots, n_r) \mapsto n_1 n_2 \dots n_r$ is a homomorphism of P onto $N_1 N_2 \dots N_r$. It remains to check that the map is injective. If $(n_1, n_2, \dots, n_r)$ is in the kernel, then $n_1 n_2 \dots n_r = e$, so for each i, we have

$$n_i = (n_{i-1}^{-1} \cdots n_2^{-1} n_1^{-1})(n_r^{-1} \cdots n_{i+1}^{-1}) = n_1^{-1} \cdots n_{i-1}^{-1} n_{i+1}^{-1} \cdots n_r^{-1}$$
$$\in N_i \cap (N_1 \dots N_{i-1} N_{i+1} \dots N_r) = \{e\}.$$

Thus $n_i = e$ for all i. ■

Corollary 3.1.12. *Let* $N_1, N_2, \dots, N_r$ *be normal subgroups of a group* G *such that* $N_1N_2\cdots N_r = G$. *Then* G *is the internal direct product of* $N_1, N_2, \dots, N_r$ *if, and only if, whenever* $x_i \in N_i$ *for* $1 \le i \le r$ *and* $x_1x_2\cdots x_r = e$, *then* $x_1 = x_2 = \cdots = x_r = e$.

Proof. The condition is equivalent to

$$N_i \cap (N_1 \dots N_{i-1}N_{i+1} \dots N_r) = \{e\}$$

for all i (Exercise 3.1.14). ■

Corollary 3.1.13. *Let* G *be an abelian group, with group operation* $+$. *Suppose* $N_1, N_2, \dots, N_r$ *are subgroups with* $N_1+N_2+\cdots+N_r = G$. *Then* G *is the internal direct product of* $N_1, N_2, \dots, N_r$ *if, and only if, whenever* $x_i \in N_i$ *for* $1 \le i \le r$ *and* $\sum_i x_i = 0$, *then* $x_1 = x_2 = \cdots = x_r = 0$.

Remark 3.1.14. *Caution:* When $r > 2$, it does not suffice that $N_i \cap N_j = \{e\}$ for $i \ne j$ and $N_1N_2 \dots N_r = G$ in order for G to be isomorphic to $N_1 \times N_2 \times \cdots \times N_r$. For example, take G to be $\mathbb{Z}_2 \times \mathbb{Z}_2$. G has *three* normal subgroups of order 2; the intersection of any two is $\{e\}$ and the product of any two is G. G is not isomorphic to the direct product of the three normal subgroups (i.e., G is not isomorphic to $\mathbb{Z}_2 \times \mathbb{Z}_2 \times \mathbb{Z}_2$.)

Exercises 3.1

3.1.1. Show that $A \times B$ is a group with identity (e_A, e_B) and with the inverse of an element (a, b) equal to (a^{-1}, b^{-1}).

3.1.2. Verify that the two subgroups $A \times \{e_B\}$ and $\{e_a\} \times B$ are normal in $A \times B$, and that the two subgroups have trivial intersection.

3.1.3. Verify that the map $[x]_{ab} \mapsto ([x]_a, [x]_b)$ from $\mathbb{Z}_{ab}$ to $\mathbb{Z}_a \times \mathbb{Z}_b$ is a group isomorphism.

3.1.4. Verify that $\pi_1 : (a, b) \mapsto a$ is a surjective homomorphism of $A \times B$ onto A with kernel $\{e_A\} \times B$. Likewise, $\pi_2 : (a, b) \mapsto b$ is a surjective homomorphism of $A \times B$ onto B with kernel $A \times \{e_B\}$.

3.1.5. Suppose G, A, and B are groups and $\psi_1 : G \to A$ and $\psi_2 : G \to B$ are surjective homomorphisms. Suppose, moreover that $\ker(\psi_1) \cap \ker(\psi_2) = \{e\}$.

(a) Show that $\psi : G \to A \times B$, defined by $\psi(g) = (\psi_1(g), \psi_2(g)$, is an injective homomorphism of G into $A \times B$. Moreover, $\pi_i \circ \psi = \psi_i$ for $i = 1, 2$, where π_i are as in the previous exercise.

(b) We cannot expect ψ to be surjective in general, even though ψ_1 and ψ_2 are surjective. For example, take $A = B$, take G to be the diagonal subgroup, $G = \{(a, a) : a \in A\} \subseteq A \times A$, and define $\psi_i : G \to A$ by $\psi_i((a, a)) = a$ for $i = 1, 2$. Show that $\psi : G \to A \times B$ is just the inclusion of G into $A \times B$, which is not surjective.

3.1.6. What sort of conditions on the maps ψ_1, ψ_2 in the previous exercise will ensure that $\psi : G \to A \times B$ is an isomorphism? Show that a necessary and sufficient condition for ψ to be an isomorphism is that there exist maps $\theta_1 : A \to G$ and $\theta_2 : B \to G$ satisfying the following conditions:

(a) $\psi_1 \circ \theta_1(a) = a$, while $\psi_2 \circ \theta_1(a) = e_B$ for all $a \in A$; and

(b) $\psi_2 \circ \theta_2(b) = b$, while $\psi_1 \circ \theta_2(b) = e_A$ for all $b \in B$.

3.1.7. Suppose that M and N are normal subgroups of a group G and that $M \cap N = \{e\}$. Show that $MN = \{mn : m \in M \text{ and } n \in M\}$ is a subgroup, and that $(m, n) \mapsto mn$ is a homomorphism of $M \times N$ onto MN.

3.1.8. Verify that the coordinate–by–coordinate multiplication on the Cartesian product $P = A_1 \times \cdots \times A_n$ of groups makes P into a group.

3.1.9. Verify that the $\tilde{A}_i = \{e\} \times \dots \{e\} \times A_i \times \{e\} \times \dots \{e\}$ is a normal subgroup of $P = A_1 \times \cdots \times A_n$, and $A_i \cong \tilde{A}_i$.

3.1.10. Verify that

$$\tilde{A}_i' = A_1 \times \cdots A_{i-1} \times \{e\} \times A_{i+1} \times \cdots \times A_n$$

is a normal subgroup of $P = A_1 \times \cdots \times A_n$, that $\tilde{A}_i \cap \tilde{A}_i' = \{e\}$, and $\tilde{A}_i \tilde{A}_i' = P$.

3.1.11. Consider a direct product of groups $P = A_1 \times \cdots \times A_n$. For each i, define $\pi_i : P \to A_i$ by $\pi_i : (a_1, a_2, \dots, a_n) \mapsto a_i$. Show that π_i surjective homomorphism of P onto A_i with kernel equal to $\tilde{A}_i'$. Show that

$$\ker(\pi_1) \cap \cdots \cap \ker(\pi_n) = \{e\},$$

and, for all i, the restriction of π_i to $\cap_{j \neq i} \ker(\pi_j)$ maps this subgroup onto A_i.

3.1.12. Show that the direct product has an associativity property:

$$A \times (B \times C) \cong (A \times B) \times C \cong A \times B \times C.$$

3.1.13.

(a) Suppose G is a group and $\varphi_i : G \longrightarrow A_i$ is a homomorphism for $i = 1, 2, \dots, r$. Suppose $\ker(\varphi_1) \cap \dots \ker(\varphi_r) = \{e\}$. Show that $\varphi : x \mapsto (\varphi_1(x), \dots, \varphi_r(x))$ is an injective homomorphism into $A_1 \times \cdots \times A_r$.

(b) Explore conditions for φ to be surjective.

3.1.14. Let $N_1, N_2, \dots, N_r$ be normal subgroups of a group G. Show that

$$N_i \cap (N_1 \dots N_{i-1} N_{i+1} \dots N_r) = \{e\}$$

for all i if, and only if, whenever $x_i \in N_i$ for $1 \leq i \leq r$ and $x_1 x_2 \cdots x_r = e$, then $x_1 = x_2 = \cdots = x_r = e$.

3.2. Semidirect Products

We now consider a slightly more complicated way in which two groups can be fit together to form a larger group.

Example 3.2.1. Consider the dihedral group D_n of order $2n$, the rotation group of the regular n-gon. D_n has a normal subgroup N of index 2 consisting of rotations about the axis through the centroid of the faces of the n-gon. N is cyclic of order n, generated by the rotation through an angle of $2\pi/n$. D_n also has a subgroup A of order 2, generated by a rotation j through an angle π about an axis through the centers of opposite edges (if n is even), or through a vertex and the opposite edge (if n is odd). We have $D_n = NA$, $N \cap A = \{e\}$, and $A \cong D_n/N$, just as in a direct product. But A is not normal, and the nonidentity element j of A does not commute with N. Instead, we have a commutation relation $jr = r^{-1}j$ for $r \in N$.

Example 3.2.2. Recall the affine or "$Ax + b$" group $\mathrm{Aff}(n)$ from Exercises 2.4.20 and 2.7.4 and Examples 2.7.5 and 2.7.9. It has a normal subgroup N consisting of transformations $T_{\mathbf{b}} : \mathbf{x} \mapsto \mathbf{x} + \mathbf{b}$ for $\mathbf{b} \in \mathbb{R}^n$. And it has the subgroup $GL(n, \mathbb{R})$, which is not normal. We have $N GL(n, \mathbb{R}) = \mathrm{Aff}(n)$, $N \cap GL(n, \mathbb{R}) = \{E\}$, and $\mathrm{Aff}(n)/N \cong GL(n, \mathbb{R})$. $GL(n, \mathbb{R})$ does not commute with N; instead, we have the commutation relation $AT_{\mathbf{b}} = T_{A\mathbf{b}}A$ for $A \in GL(n, \mathbb{R})$ and $\mathbf{b} \in \mathbb{R}^n$.

Both of the last examples are instances of the following situation: A group G has a normal subgroup N and another subgroup A, which is not normal. We have $G = NA = AN$, $A \cap N = \{e\}$, and $A \cong G/N$. Since N is normal, for each $a \in A$, the inner automorphism c_a of G restricts to an automorphism of N, and we have the commutation relation $an = c_a(n)a$ for $a \in A$ and $n \in N$. [Recall that the inner automorphism c_a is defined by $c_a(x) = axa^{-1}$.]

Now, if we have groups N and A, and we have a homomorphism $\alpha : a \mapsto \alpha_a$ from A into the automorphism group $\text{Aut}(N)$ of N, we can build from these data a new group $N \underset{\alpha}{\rtimes} A$, called the semidirect product of A and N. The semidirect product $N \underset{\alpha}{\rtimes} A$ has the following features: It contains (isomorphic copies of) A and N as subgroups, with N normal; the intersection of these subgroups is the identity, and the product of these subgroups is $N \underset{\alpha}{\rtimes} A$; and we have the commutation relation $an = \alpha_a(n)a$ for $a \in A$ and $n \in N$.

The construction is straightforward. As a set, $N \underset{\alpha}{\rtimes} A$ is $N \times A$, but now the product is defined by $(n, a)(n', a') = (n\alpha_a(n'), aa')$.

Proposition 3.2.3. *Let* N *and* A *be groups, and* $\alpha : A \to \text{Aut}(N)$ *a homomorphism of* A *into the automorphism group of* N. *The Cartesian product* $N \times A$ *is a group under the multiplication* $(n, a)(n', a') = (n\alpha_a(n'), aa')$. *This group is denoted* $N \underset{\alpha}{\rtimes} A$. $\tilde{N} = \{(n, e) : n \in N\}$ *and* $\tilde{A} = \{(e, a) : a \in A\}$ *are subgroups of* $N \underset{\alpha}{\rtimes} A$, *with* $\tilde{N} \cong N$ *and* $\tilde{A} \cong A$, *and* $\tilde{N}$ *is normal in* $N \underset{\alpha}{\rtimes} A$. *We have* $(e, a)(n, e) = (\alpha_a(n), e)(e, a) = (\alpha_a(n), a)$ *for all* $n \in N$ *and* $a \in A$.

Proof. We first have to check the associativity of the product on $N \underset{\alpha}{\rtimes} A$. Let (n, a), (n', a'), and (n'', a'') be elements of $N \underset{\alpha}{\rtimes} A$. We compute

$$\begin{aligned}((n, a)(n', a'))(n'', a'') &= (n\alpha_a(n'), aa')(n'', a'') \\ &= (n\alpha_a(n')\alpha_{aa'}(n''), aa'a'').\end{aligned}$$

On the other hand,

$$\begin{aligned}(n, a)((n', a')(n'', a'')) &= (n, a)(n'\alpha_{a'}(n''), a'a'') \\ &= (n\alpha_a(n'\alpha_{a'}(n'')), aa'a'').\end{aligned}$$

We have

$$n\alpha_a(n'\alpha_{a'}(n'')) = n\alpha_a(n')\alpha_a(\alpha_{a'}(n'')) = n\alpha_a(n')\alpha_{aa'}(n''),$$

where the first equality uses that α_a is an automorphism of N, and the second uses that α is a homomorphism of A into $\text{Aut}(N)$. This proves associativity of the product.

It is easy to check that (e, e) serves as the identity of $N \underset{\alpha}{\rtimes} A$.

Finally, we have to find the inverse of an element of $N \underset{\alpha}{\rtimes} A$. If (n', a') is to be the inverse of (n, a), we must have $aa' = e$ and $n\alpha_a(n') = e$. Thus $a' = a^{-1}$ and $n' = \alpha_a^{-1}(n^{-1}) = \alpha_{a^{-1}}(n^{-1})$. Thus our candidate for $(n, a)^{-1}$ is $(\alpha_{a^{-1}}(n^{-1}), a^{-1})$. Now we have to check this candidate by multiplying by (n, a) on either side. This is left as an exercise. ■

Remark 3.2.4. The direct product is a special case of the semidirect product, with the homomorphism α trivial, $\alpha(a) = \mathrm{id}_N$ for all $a \in A$.

Corollary 3.2.5. *Suppose* G *is a group,* N *and* A *are subgroups with* N *normal,* $G = NA = AN$*, and* $A \cap N = \{e\}$*. Then there is a homomorphism* $\alpha : A \to \mathrm{Aut}(N)$ *such that* G *is isomorphic to the semidirect product* $N \underset{\alpha}{\rtimes} A$.

Proof. We have a homomorphism α from A into $\mathrm{Aut}(N)$ given by $\alpha(a)(n) = ana^{-1}$. Since $G = NA$ and $N \cap A = \{e\}$, every element $g \in G$ can be written in exactly one way as a product $g = na$, with $n \in N$ and $a \in A$. Furthermore, $(n_1a_1)(n_2a_2) = [n_1(a_1n_2a_1^{-1})][a_1a_2]$ $= [n_1\alpha(a_1)(n_2)][a_1a_2]$. Therefore, the map $(n, a) \mapsto na$ is an isomorphism from $N \underset{\alpha}{\rtimes} A$ to G. ■

Example 3.2.6. $\mathbb{Z}_7$ has an automorphism φ of order 3, $\varphi([x]) = [2x]$; this gives a homomorphism $\alpha : \mathbb{Z}_3 \to \mathrm{Aut}(\mathbb{Z}_7)$, defined by $\alpha([k]) = \varphi^k$. The semidirect product $\mathbb{Z}_7 \underset{\alpha}{\rtimes} \mathbb{Z}_3$ is a nonabelian group of order 21. This group is generated by two elements a and b satisfying the relations $a^7 = b^3 = e$, and $bab^{-1} = a^2$.

Exercises 3.2

3.2.1. Complete the proof that $N \underset{\alpha}{\rtimes} A$ (as defined in the text) is a group by verifying that $(\alpha_{a^{-1}}(n^{-1}), a^{-1})$ is the inverse of (a, n).

3.2.2. Show that D_n is isomorphic to a semidirect product of $\mathbb{Z}_2$ and $\mathbb{Z}_n$.

3.2.3. Show that the affine group $\mathrm{Aff}(n)$ is isomorphic to a semidirect product of $\mathrm{GL}(n, \mathbb{R})$ and the additive group $\mathbb{R}^n$.

3.2.4. Show that the permutation group S_n is a semidirect product of $\mathbb{Z}_2$ and the group of even permutations A_n.

3.2.5. Consider the set G of n-by-n matrices with entries in $\{0, \pm 1\}$ that have exactly one nonzero entry in each row and column. These are called signed permutation matrices. Show that G is a group, and that G is a semidirect product of S_n and the group of diagonal matrices with entries in $\{\pm 1\}$. S_n acts on the group of diagonal matrices by permutation of the diagonal entries.

One final example shows that direct products and semidirect products do not exhaust the ways in which a normal subgroup N and the quotient group G/N can be fit together to form a group G:

3.2.6. $\mathbb{Z}_4$ has a subgroup isomorphic to $\mathbb{Z}_2$, namely the subgroup generated by $[2]$. The quotient $\mathbb{Z}_4/Z_2$ is also isomorphic to $\mathbb{Z}_2$. Nevertheless, $\mathbb{Z}_4$ is not a direct or semidirect product of two copies of $\mathbb{Z}_2$.

3.3. Finite Abelian Groups

In this section, we will obtain a definitive structure theorem and classification of finite abelian groups. The theorem states that any finite abelian group is a direct product of cyclic groups each of order a power of a prime; furthermore, the number of the cyclic subgroups appearing in the direct product decomposition, and their orders, are unique.

Two finite abelian groups are isomorphic if, and only if, they have the same decomposition into a direct product of cyclic groups of prime power order.

All the groups in this section will be abelian, and, following a common convention, we will use additive notation for the group operation. In particular, the sth power of an element x will be written as sx, and the order of an element x is the smallest natural number s such that $sx = 0$. All subgroups are normal, and the subgroup generated by a family $A_1, \dots A_s$ of subgroups is $A_1 + \cdots + A_s = \{a_1 + \cdots + a_s : a_i \in A_i \text{ for all } i\}$.

We have the following elementary results on direct products of abelian groups, which may remind you of a result from linear algebra.

Proposition 3.3.1. *Let* G *be an abelian group with subgroups* $A_1, \dots A_s$ *such that* $G = A_1 + \cdots + A_s$. *Then the following conditions are equivalent:*

(a) $(a_1, \dots, a_s) \mapsto a_1 + \cdots + a_s$ *is an isomorphism of* $A_1 \times \cdots \times A_s$ *onto* G.

(b) *Each element* $g \in G$ *can be expressed as a sum* $x = a_1 + \cdots + a_s$, *with* $a_i \in A_i$ *for all* i, *in exactly one way.*

(c) *If* $0 = a_1 + \cdots + a_s$, *with* $a_i \in A_i$ *for all* i, *then* $a_i = 0$ *for all* i.

Proof. This can be obtained from results about direct products for general (not necessarily abelian) groups, but the proof for abelian groups is very direct. The map in part (a) is a homomorphism, because the groups are abelian. (Check this.) By hypothesis the homomorphism is surjective, so (a) is equivalent to the injectivity of the map. But (b) also states that the map is injective, and (c) states that the kernel of the map is trivial. So all three assertions are equivalent. ■

Let G be a finite abelian group. For each prime number p define

$$G[p] = \{g \in G : o(g) \text{ is a power of } p\}.$$

It is straightforward to check that $G[p]$ is a subgroup of G. Note that $x \in G[p] \Leftrightarrow p^j x = 0$ for some $j \Leftrightarrow p^r x = 0$, where p^r is the largest power of p dividing the order of the group. $G[p] = \{0\}$ if p does not divide the order of G.

The first step in our analysis of finite abelian groups is to show that G is the (internal) direct product of the subgroups $G[p]$ for p dividing the order of G.

Theorem 3.3.2. *Let* G *be a finite abelian group and let* $p_1, \dots, p_k$ *be the primes dividing* $|G|$. *Then* $G \cong G[p_1] \times \cdots \times G[p_k]$.

Proof. Write $n = |G|$, and let $p_1^{k_1} \cdots p_s^{k_s}$ be the prime decomposition of n.

For each index i let $r_i = n/p_i^{k_i}$; that is, r_i is the largest divisor of n that is relatively prime to p_i. For all $x \in G$, we have $r_i x \in G[p_i]$, because $p_i^{k_i}(r_i x) = nx = 0$. Furthermore, if $x \in G[p_j]$ for some $j \neq i$, then $r_i x = 0$, because $p_j^{k_j}$ divides r_i.

The greatest common divisor of $\{r_1, \dots, r_s\}$ is 1. Therefore, there exist $t_1, \dots, t_s$ such that $t_1 r_1 + \cdots + t_s r_s = 1$. Hence for any $x \in G$, $x = 1x = t_1 r_1 x + \cdots + t_s r_s x \in G[p_1] + G[p_2] + \cdots + G[p_s]$. Thus $G = G[p_1] + \cdots + G[p_s]$.

Suppose that $x_j \in G[p_j]$ for $1 \le j \le s$ and $\sum_j x_j = 0$. Fix an index i. Since $r_i x_j = 0$ for $j \neq i$, we have

$$0 = r_i(\sum_j x_j) = \sum_j r_i x_j = r_i x_i.$$

Because r_i is relatively prime to the order of each nonzero element of $G[p_i]$, it follows that $x_i = 0$. Thus by Proposition 3.3.1 $G = G[p_1] \times \cdots \times G[p_s]$. ■

The following lemma is the most subtle item in this section. You might prefer to skip the proof on the first reading.

Lemma 3.3.3. *Let G be a finite abelian group. Let m denote the maximum of the orders of elements of G, and let $a \in G$ be an element order m. Let $\pi : G \longrightarrow G/\langle a \rangle$ be the quotient map. For every $\bar{b} \in G/\langle a \rangle$, there is an element $b \in G$ such that $\pi(b) = \bar{b}$, and $o(b) = o(\bar{b})$.*

Proof. Denote the order of $\bar{b}$ in $G/\langle a \rangle$ by r. Let b_1 be any element of B such that $\pi(b_1) = \bar{b}$; this is possible because π is surjective. Then

$$o(b_1)\bar{b} = o(b_1)\pi(b_1) = \pi(o(b_1)b_1) = \pi(0) = 0;$$

therefore, r divides the order of b_1. On the other hand, $o(b_1) \le m$ by hypothesis.

Since $0 = r\bar{b} = \pi(rb_1)$, we have $rb_1 \in \langle a \rangle$, say $rb_1 = na$ for some integer n with $0 \le n \le m - 1$.

I claim that n is divisible by r, say $n = qr$. Assuming this for the moment, we have $rb_1 = rqa$, or $r(b_1 - qa) = 0$. So, putting $b = b_1 - qa$, we have $\pi(b) = \pi(b_1) = \bar{b}$, and $rb = 0$; therefore, $o(b)|r$. On the other hand, because $\pi(b) = \bar{b}$, we have $r|o(b)$ (just as for b_1). It follows that $o(b) = r$, as desired.

It remains to show that n is divisible by r. Write $n = qr + s$ with $0 \le s < r$. Then $rb_1 = qra + sa$, or $r(b_1 - qa) = sa$. We have to show that $s = 0$. Assume $s > 0$, in order to reach a contradiction.

Again, set $b = b_1 - qa$, so $\pi(b) = \pi(b_1) = \bar{b}$. We know that $o(sa) = m/\alpha$, where $\alpha = \text{g.c.d.}(s, m)$, by Proposition 2.2.33. Since $o(b)$ is divisible by r,

$$o(b) = r\,o(rb) = r\,o(sa) = r\,(m/\alpha) \ge rm/s > m.$$

Since $o(b) \le m$, this is a contradiction. ■

Proposition 3.3.4. *Let G be a finite abelian group. Then G is a direct product of cyclic groups.*

Proof. We prove this by induction on the order of the abelian group, there being nothing to prove if $|G| = 1$. So assume $|G| > 1$ and that the assertion holds for all finite abelian groups of size strictly less than $|G|$. Let a_1 be an element of G of maximum order $r_1 > 1$ and put $A_1 = \langle a_1 \rangle$.

Applying the induction hypothesis to G/A_1, suppose that G/A_1 is a direct product of subgroups $\bar{A}_2, \bar{A}_3, \dots, \bar{A}_k$, where $\bar{A}_i = \langle \bar{a}_i \rangle$ is cyclic of order r_i for $2 \le i \le k$. Let $\pi : G \longrightarrow G/A_1$ be the quotient map. Using Lemma 3.3.3, let a_i be a pre-image of $\bar{a}_i$ of order r_i for each i, $2 \le i \le k$, and let $A_i = \langle a_i \rangle$. We will show that $G = A_1 \times A_2 \times \cdots A_k$.

For each $g \in G$, there exist n_i for $2 \le i \le k$ such that $\pi(g) = \sum_{i \ge 2} n_i \bar{a}_i$. Thus, $g - \sum_{i \ge 2} n_i a_i \in A_1 = \langle a_i \rangle$, so there is an n_1 such that $g = \sum_{i \ge 1} n_i a_i$. This shows that $G = A_1 + A_2 + \cdots + A_k$.

Since each a_i has order r_i, an arbitrary element of A_i can be written as $n_i a_i$, with $0 \le n_i < r_i$. Suppose a sum of such elements is zero: $\sum_{i \ge 1} n_i a_i = 0$. Applying π gives $\sum_{i \ge 2} n_i \bar{a}_i = 0$. Since G/A_1 is the internal direct product of the subgroups $\bar{A}_i = \langle \bar{a}_i \rangle$ for $2 \le i \le k$, it follows from Proposition 3.3.1 that $n_i \bar{a}_i = 0$ for $i \ge 2$. Because the order of $\bar{a}_i$ is r_i and $0 \le n_i < r_i$, we have $n_i = 0$ for $i \ge 2$. Hence also $n_i a_i = 0$ for $i \ge 2$, and, therefore, $n_1 a_1 = 0$ as well.

By Proposition 3.3.1, G is the direct product of the subgroups A_i, $1 \le i \le k$. ■

Corollary 3.3.5. *If G is a finite abelian group and p is a prime dividing the order of G, then G has an element of order p.*

Proof. Since G is a direct product of cyclic groups, p divides the order of some cyclic subgroup C of G, and C has an element of order p by Proposition 2.2.32. ■

Corollary 3.3.6. *If G is a finite abelian group, then for each prime p, the order of $G[p]$ is the largest power of p dividing $|G|$.*

Proof. $G[p]$ has, by definition, no element of order q, where q is a prime different from p. Therefore, by the previous corollary, the order of $G[p]$ is a power of p. Since $|G| = \prod_p |G[p]|$, it follows that $|G[p]|$ is the largest power of p dividing $|G|$. ■

Corollary 3.3.7. *Any finite abelian group is a direct product of cyclic groups, each of which has order a power of a prime.*

Proof. A finite abelian group G is a direct product of its subgroups G[p]. Each G[p] is in turn a direct product of cyclic groups by Proposition 3.3.4, which must each be of order a power of p by Corollary 3.3.6 and Lagrange's theorem. ■

We can now obtain a complete structure theorem for abelian groups whose order is a power of a prime.

Theorem 3.3.8.

(a) *Let G be an abelian group of order p^n, where p is a prime. There exist natural numbers $n_1 \geq n_2 \geq \cdots \geq n_s$ such that $\sum_i n_i = n$, and $G \cong \mathbb{Z}_{p^{n_1}} \times \cdots \times \mathbb{Z}_{p^{n_s}}$.*

(b) *The sequence of exponents in part (b) is unique. That is, if $m_1 \geq m_2 \geq \cdots \geq m_r$, $\sum_j m_j = n$, and $G \cong \mathbb{Z}_{p^{m_1}} \times \cdots \times \mathbb{Z}_{p^{m_r}}$, then $s = r$ and $n_i = m_i$ for all i.*

Proof. By Proposition 3.3.4, G is a direct product of cyclic groups, each of which must have order a power of p by Lagrange's theorem. This gives (a).

We prove the uniqueness statement (b) by induction on n, the case $n = 1$ being trivial. So suppose the uniqueness statement holds for all abelian groups of order $p^{n'}$ where $n' < n$. Consider the homomorphism $\varphi(x) = x^p$ of G into itself. Suppose

$$(n_1, \dots, n_s) = (n_1, \dots, n_{s'}, 1, 1, \dots, 1)$$

and

$$(m_1, \dots, m_r) = (m_1, \dots, m_{r'}, 1, 1, \dots, 1),$$

where $n_{s'} > 1$ and $n_{r'} > 1$. Then the isomorphism

$$G \cong \mathbb{Z}_{p^{n_1}} \times \cdots \times \mathbb{Z}_{p^{n_s}}$$

gives

$$\varphi(G) \cong \mathbb{Z}_{p^{n_1 - 1}} \times \cdots \times \mathbb{Z}_{p^{n_{s'}-1}},$$

and the isomorphism

$$G \cong \mathbb{Z}_{p^{m_1}} \times \cdots \times \mathbb{Z}_{p^{m_r}}$$

gives

$$\varphi(G) \cong \mathbb{Z}_{p^{m_1 - 1}} \times \cdots \times \mathbb{Z}_{p^{m_{r'}-1}}.$$

The first isomorphism yields

$$\log_p |\varphi(G)| = \sum_{i=1}^{s'} (n_i - 1) = \sum_{i=1}^{s} (n_i - 1) = n - s,$$

and similarly the second isomorphism yields $\log_p |\varphi(G)| = n - r$. In particular, $r = s$. Now applying the induction hypothesis to $\varphi(G)$ gives also $s' = r'$ and $n_i - 1 = m_i - 1$ for $1 \le i \le s'$. This implies $n_i = m_i$ for $1 \le i \le s$. ■

Example 3.3.9. Every abelian groups of order 32 is isomorphic to one of the following: $\mathbb{Z}_{32}$, $\mathbb{Z}_{16} \times \mathbb{Z}_2$, $\mathbb{Z}_8 \times \mathbb{Z}_4$, $\mathbb{Z}_8 \times \mathbb{Z}_2 \times \mathbb{Z}_2$, $\mathbb{Z}_4 \times \mathbb{Z}_4 \times \mathbb{Z}_2$, $\mathbb{Z}_4 \times \mathbb{Z}_2 \times \mathbb{Z}_2 \times \mathbb{Z}_2$, $\mathbb{Z}_2 \times \mathbb{Z}_2 \times \mathbb{Z}_2 \times \mathbb{Z}_2 \times \mathbb{Z}_2$.

Let G be an abelian group of order p^n. There exist uniquely determined natural numbers $n_1 \ge n_2 \ge \cdots \ge n_s$ such that $\sum_i n_i = n$, and $G \cong \mathbb{Z}_{p^{n_1}} \times \cdots \times \mathbb{Z}_{p^{n_s}}$. The sequence $(n_1, \ldots, n_s)$ is called the *type* of G. The type of an abelian group of prime power order determines the group up to isomorphism; two groups each of which have order a power of p are isomorphic if, and only if, they have the same type.

Example 3.3.10. A *partition* of a natural number n is a sequence of natural numbers $n_1 \ge n_2 \ge \cdots \ge n_s$ such that $\sum_i n_i = n$. The type of an abelian group of order p^n is a partition of n, and the number of different isomorphism classes of abelian groups of order p^n is the number of partitions of n. (The number does not depend on p.) For example, the distinct partitions of 7 are (7), (6, 1), (5, 2), (5, 1, 1), (4, 3), (4, 2, 1), (4, 1, 1, 1), (3, 3, 1), (3, 2, 2), (3, 2, 1, 1), (3, 1, 1, 1, 1), (2, 2, 2, 1), (2, 2, 1, 1, 1), (2, 1, 1, 1, 1, 1), and (1, 1, 1, 1, 1, 1, 1). So there are 15 different isomorphism classes of abelian groups of order p^7 for any prime p.

Lemma 3.3.11. *Suppose a finite abelian group G is an internal direct product of a collection $\{C_i\}$ of cyclic subgroups. Then for each prime p, the sum of those C_i whose order is a power of p is equal to $G[p]$.*

Proof. Denote by $A[p]$ the sum of those C_i whose order is a power of p. Then $A[p] \subseteq G[p]$ and G is the internal direct product of the subgroups $A[p]$. Since G is also the internal direct product of the subgroups $G[p]$, it follows that $A[p] = G[p]$ for all p. ■

Example 3.3.12. Consider $G = \mathbb{Z}_{30} \times \mathbb{Z}_{50} \times \mathbb{Z}_{28}$. Then

$$\begin{aligned} G &\cong (\mathbb{Z}_3 \times \mathbb{Z}_2 \times \mathbb{Z}_5) \times (\mathbb{Z}_{25} \times \mathbb{Z}_2) \times (\mathbb{Z}_4 \times \mathbb{Z}_7) \\ &\cong (\mathbb{Z}_4 \times \mathbb{Z}_2 \times \mathbb{Z}_2) \times \mathbb{Z}_3 \times (\mathbb{Z}_{25} \times \mathbb{Z}_5) \times \mathbb{Z}_7. \end{aligned}$$

Thus $G[2] \cong \mathbb{Z}_4 \times \mathbb{Z}_2 \times \mathbb{Z}_2$, $G[3] \cong \mathbb{Z}_3$, $G[5] \cong \mathbb{Z}_{25} \times \mathbb{Z}_5$, and $G[7] \cong \mathbb{Z}_7$. $G[p] = 0$ for all other primes p.

Theorem 3.3.13. *(Fundamental theorem of finite abelian groups). Every finite abelian group is isomorphic to a direct product of cyclic groups of prime power order. The number of cyclic groups of each order appearing in such a direct product decomposition is uniquely determined, and, moreover, determines the group up to isomorphism.*

Proof. Let G be a finite abelian group. By Corollary 3.3.7, G is isomorphic to a direct product of cyclic groups of prime power order. Suppose $\{C_i : 1 \le i \le N\}$ and $\{D_j : 1 \le j \le M\}$ are two families of cyclic subgroups of G of prime power order such that

$$G = C_1 \times \cdots \times C_N = D_1 \times \cdots \times D_M.$$

Group each family of cyclic subgroups according to the primes dividing $|G|$,

$$\{C_i\} = \bigcup_p \{C_i^p : 1 \le i \le N(p)\}, \quad \text{and}$$

$$\{D_j\} = \bigcup_p \{D_j^p : 1 \le j \le M(p)\},$$

where each C_i^p and D_j^p has order a power of p. According to the previous lemma, $\sum_{i=1}^{N(p)} C_i^p = \sum_{j=1}^{M(p)} D_j^p = G[p]$ for each prime p dividing $|G|$. It follows from Theorem 3.3.8 and Corollary 3.3.6 that $N(p) = M(p)$ and

$$\{|C_i^p| : 1 \le i \le N(p)\} = \{|D_j^p| : 1 \le j \le N(p)\}.$$

It follows that $M = N$ and $\{|C_i| : 1 \le i \le N\} = \{|D_j| : 1 \le j \le N\}$. ■

Example 3.3.14. Consider the example $G = \mathbb{Z}_{30} \times \mathbb{Z}_{50} \times \mathbb{Z}_{28}$ again. The unique decomposition of G into cyclic groups of prime power order is

$$G \cong (\mathbb{Z}_4 \times \mathbb{Z}_2 \times \mathbb{Z}_2) \times \mathbb{Z}_3 \times (\mathbb{Z}_{25} \times \mathbb{Z}_5) \times \mathbb{Z}_7.$$

Another canonical direct product decomposition is obtained by regrouping the factors as follows:

$$\begin{aligned} G &\cong (\mathbb{Z}_4 \times \mathbb{Z}_3 \times \mathbb{Z}_{25} \times \mathbb{Z}_7) \times (\mathbb{Z}_2 \times \mathbb{Z}_5) \times \mathbb{Z}_2 \\ &\cong \mathbb{Z}_{4\cdot 3\cdot 25\cdot 7} \times \mathbb{Z}_{2\cdot 5} \times \mathbb{Z}_2 \\ &\cong \mathbb{Z}_{2100} \times \mathbb{Z}_{10} \times \mathbb{Z}_2. \end{aligned}$$

Corollary 3.3.15. *If* G *is a finite abelian group, there exist unique natural numbers* $m_1, m_2, \dots, m_r$ *such that* $m_i \geq 2$*,* m_{i+1} *divides* m_i *for all* i *and* $G \cong \mathbb{Z}_{m_1} \times \cdots \times \mathbb{Z}_{m_r}$.

Proof. Exercise 3.3.6. ■

Corollary 3.3.16. *If* G *is a finite abelian group and* m *is the maximum of orders of elements of* G*, then the order of any element of* G *divides* m.

Proof. Exercise 3.3.8. ■

Corollary 3.3.17. *Let* K *be a finite field of order* n*. Then the multiplicative group of units of* K *is cyclic of order* $n-1$*. In particular, for* p *a prime number, the multiplicative group* $\Phi(p)$ *of units of* $\mathbb{Z}_p$ *is cyclic of order* $p-1$.

Proof. Let K^* denote the multiplicative group of nonzero elements of K. Then K^* is abelian of order $n-1$. Let m denote the maximum of the orders of elements of K^*. We want to show that $m = n-1$. On the one hand, $m \leq n-1$, since every element has order dividing the order of the group. On the other hand, according to Corollary 3.3.16, $x^m = 1$ for all elements $x \in K^*$, so that the equation $x^m - 1 = 0$ has $n-1$ distinct solutions in the field K. But the number of distinct roots of a polynomial in a field is never more than the degree of the polynomial (Corollary 1.8.25), so $n-1 \leq m$. ■

Remark 3.3.18. Note that while the proof insures that the group of units of K^* is cyclic, it does not provide a means of actually *finding* a generator! In particular, it is not obvious how to find a nonzero element of $\mathbb{Z}_p$ of multiplicative order $p-1$.

The rest of this section is devoted to working out the structure of the group $\Phi(N)$ of units in $\mathbb{Z}_N$. Recall that $\Phi(N)$ has order $\varphi(N)$, where φ is the Euler φ function. We have just determined that for a prime p, $\Phi(p)$ is cyclic of order $p-1$.

Proposition 3.3.19.

(a) *If* N *has prime decomposition* $N = p_1^{k_1} p_2^{k_2} \cdots p_s^{k_s}$, *then*

$$\Phi(N) \cong \Phi(p_1^{k_1}) \times \Phi(p_2^{k_2}) \times \cdots \times \Phi(p_s^{k_s}).$$

(b) $\Phi(2)$ *and* $\Phi(4)$ *are cyclic.* $\Phi(2^n) \cong \mathbb{Z}_2 \times \mathbb{Z}_{2^{n-2}}$ *if* $n \geq 3$.

(c) *If* p *is an odd prime, then for all* n, $\Phi(p^n) \cong \mathbb{Z}_{p^{n-1}(p-1)} \cong \mathbb{Z}_{p^{n-1}} \times \mathbb{Z}_{p-1}$.

Proof. You are asked to prove part (a) in Exercise 3.3.9.

The groups $\Phi(2)$ and $\Phi(4)$ are of orders 1 and 2, respectively, so they are necessarily cyclic. For $n \geq 3$, we have already seen in Example 2.2.34 that $\Phi(2^n)$ is not cyclic and that $\Phi(2^n)$ contains three distinct elements of order 2, and in Exercise 2.2.30 that [3] has order 2^{n-1} in $\Phi(2^n)$. The cyclic subgroup $\langle[3]\rangle$ contains exactly one of the three elements of order 2. If a is an element of order 2 *not* contained in $\langle[3]\rangle$, then $\langle a\rangle \cap \langle[3]\rangle = [1]$, so the subgroup generated by $\langle a\rangle$ and $\langle[3]\rangle$ is a direct product, isomorphic to $\mathbb{Z}_2 \times \mathbb{Z}_{2^{n-1}}$. Because $|\Phi(2^n)| = |\mathbb{Z}_2 \times \mathbb{Z}_{2^{n-1}}| = 2^{n-1}$, we have $\Phi(2^n) = \langle a\rangle \times \langle[3]\rangle \cong \mathbb{Z}_2 \times \mathbb{Z}_{2^{n-1}}$. This completes the proof of part (b).

Now let p be an odd prime, and let $n \geq 1$. Lemma 3.3.17 already shows that $\Phi(p^n)$ is cyclic when $n = 1$, so we can assume $n \geq 2$.

Using Lemma 3.3.17, we obtain a natural number a such that $a^{p-1} \equiv 1 \pmod{p}$ and $a^\ell \not\equiv 1 \pmod{p}$ for $\ell < p-1$.

We claim that the order of $[a^{p^{n-1}}]$ in $\Phi(p^n)$ is $(p-1)$. In any case, $(a^{p^{n-1}})^{p-1} = a^{p^{n-1}(p-1)} \equiv 1 \pmod{p^n}$ so the order ℓ of $[a^{p^{n-1}}]$ divides $p-1$. But we have $a^p \equiv a \pmod{p}$, so $a^{p^{n-1}} \equiv a \pmod{p}$, and $a^{p^{n-1}\ell} \equiv a^\ell \pmod{p}$. If $\ell < p-1$, then $a^{p^{n-1}\ell} \equiv a^\ell$ is not congruent to 1 modulo p, so it is not congruent to 1 modulo p^n. Therefore, the order of $[a^{p^{n-1}}]$ is $p-1$ as claimed.

It follows from Exercise 1.9.10 that the order of $[p+1]$ in $\Phi(p^n)$ is p^{n-1}.

We now have elements $x = [a^{p^{n-1}}]$ of order $p-1$ and $y = [p-1]$ of order p^{n-1} in $\Phi(p^n)$. Since the orders of x and y are relatively prime, the order of the product xy is the product of the orders $p^{n-1}(p-1)$. Thus $\Phi(p^n)$ is cyclic.

Another way to finish the proof of part (c) is to observe that $\langle x\rangle \cap \langle y\rangle = \{1\}$, since the orders of these cyclic subgroups are relatively prime. It follows then that the subgroup generated by x and y is the direct product $\langle x\rangle \times \langle y\rangle \cong \mathbb{Z}_{p-1} \times \mathbb{Z}_{p^{n-1}} \cong \mathbb{Z}_{p^{n-1}(p-1)}$. ■

Remark 3.3.20. All of the isomorphisms here are explicit, as long as we are able to find a generator for $\Phi(p)$ for all primes p appearing in the decompositions.

Exercises 3.3

3.3.1. Find all abelian groups of order 108.

3.3.2. Find all abelian groups of order 144.

3.3.3. How many abelian groups are there of order 128, up to isomorphism?

3.3.4. How many abelian groups are there of order p^5q^4, where p and q are distinct primes?

3.3.5. Show that $\mathbb{Z}_a \times \mathbb{Z}_b$ is not cyclic if $\text{g.c.d.}(a, b) \geq 2$.

3.3.6. Prove Corollary 3.3.15. Hint: You need to work out how the m_i are related to the orders of the cyclic groups of prime power order appearing in the fundamental theorem. The uniqueness follows because the m_i and the orders of the cyclic groups of prime power order determine one another.

3.3.7. Determine the decomposition $G \cong \mathbb{Z}_{m_1} \times \cdots \times \mathbb{Z}_{m_r}$ given in the last exercise for finite abelian groups G or order 108 and 144.

3.3.8. Prove Corollary 3.3.16.

3.3.9. Recall that if a and b are relatively prime natural numbers, then $\mathbb{Z}_{ab} \cong \mathbb{Z}_a \times \mathbb{Z}_b$ as rings.

(a) If a, b are relatively prime natural numbers, show that the ring isomorphism $\mathbb{Z}_{ab} \cong \mathbb{Z}_a \times \mathbb{Z}_b$ implies that $\Phi(ab) \cong \Phi(a) \times \Phi(b)$.

(b) Show that if $N = p_1^{k_1} \cdots p_s^{k_s}$ is the prime decomposition of N, then

$$\Phi(N) \cong \Phi(p_1^{k_1}) \times \cdots \times \Phi(p_s^{k_s}).$$

(c) Since these group isomorphisms are obtained independently of our earlier computations of $\varphi(N)$, show that we can recover the multiplicativity of the Euler φ function from the group theory results. Namely, conclude from from parts (a) - (c) that $\varphi(ab) = \varphi(a)\varphi(b)$ if a, b are relatively prime, and that if $N = p_1^{k_1} \cdots p_s^{k_s}$, then $\varphi(N) = \prod_i \varphi(p_i^{k_i})$.

3.3.10. Find the structure of the group $\Phi(n)$ for $n \leq 20$.

3.4. Vector Spaces

You can use your experience with group theory to gain a new appreciation of linear algebra. In this section K denotes one of the fields $\mathbb{Q}$, $\mathbb{R}$, $\mathbb{C}$, or $\mathbb{Z}_p$, or any other favorite field of yours.

Definition 3.4.1. A *vector space* V over a field K is a abelian group with a product $K \times V \to V$, $(s, v) \mapsto sv$ satisfying the following conditions:

(a) $1v = v$ for all $v \in V$.
(b) $(st)v = s(tv)$ for all $s, t \in K$, $v \in V$.
(c) $s(v + w) = sv + sw$ for all $s \in K$ and $v, w \in V$.
(d) $(s + t)v = sv + tv$ for all $s, t \in K$ and $v \in V$.

Compare this definition with that contained in your linear algebra text; notice that we were able to state the definition more concisely by referring to the notion of an abelian group. It follows from the definition that $0\,v = \alpha 0 = 0$, and $-1(v) = -v$, for all $\alpha \in K$ and all $v \in V$; here 0 is begin used to denote both the additive identity in K and is the additive identity element in V, and $-v$ is the additive inverse of v.

A vector space over K is also called a K–vector space. A vector space over $\mathbb{R}$ is also called a real vector space and a vector space over $\mathbb{C}$ a complex vector space.

Example 3.4.2.

(a) K^n is a vector space over K, and any vector subspace of K^n is a vector space over K.
(b) The set of K–valued functions on a set X is a vector space over K, with pointwise addition of functions and the usual multiplication of functions by scalars.
(c) The set of *continuous* real–valued functions on $[0, 1]$ (or, in fact, on any other metric or topological space) is a vector space over $\mathbb{R}$ with pointwise addition of functions and the usual multiplication of functions by scalars.
(d) The set of polynomials $K[x]$ is a vector space over K, as is the set of polynomials of degree $\le n$, for any natural number n.

We define linear independence, span, and basis for abstract vector spaces, and linear transformations between abstract vector spaces exactly as for vector subspaces of K^n. (Vector subspaces of K^n are treated in Appendix E.) We give the definitions here for the sake of completeness.

Definition 3.4.3. A *linear combination* of a subset S of a vector space V is any element of V of the form $\alpha_1 v_1 + \alpha_2 v_2 + \cdots + \alpha_s v_s$, where for all i, $\alpha_i \in K$ and $v_i \in S$. The *span* of S is the set of all linear combinations of S. We denote the span of S by $\text{span}(S)$.

The span of the empty set is the set containing only the zero vector $\{0\}$.

Definition 3.4.4. A subset S of vector space V is *linearly independent* if for all natural numbers s, for all $\boldsymbol{\alpha} = \begin{bmatrix} \alpha_1 \\ \vdots \\ \alpha_s \end{bmatrix} \in K^s$, and for all sequences $(v_1, \ldots v_s)$ of *distinct* vectors in S, if $\alpha_1 v_1 + \alpha_2 v_2 + \cdots + \alpha_s v_s = 0$, then $\boldsymbol{\alpha} = 0$. Otherwise, S is *linearly dependent*.

Note that a linear independent set cannot contain the zero vector. The empty set is linearly independent, since there are no sequences of its elements!

Example 3.4.5. Define $e_n(x) = e^{inx}$ for n an integer and $x \in \mathbb{R}$. Then $\{e_n : n \in \mathbb{Z}\}$ is a linearly independent subset of the (complex) vector space of $\mathbb{C}$–valued functions on $\mathbb{R}$. To show this , we have to prove that for all natural numbers s, any set consisting of s of the functions e_n is linearly independent. We prove this statement by induction on s. For $s = 1$, suppose $\alpha \in \mathbb{C}$, $n_1 \in \mathbb{Z}$, and $\alpha e_{n_1} = 0$. Evaluating at $x = 0$ gives $0 = \alpha e^{i n_1 0} = \alpha$. This shows that $\{e_{n_1}\}$ is linearly independent. Now fix $s > 1$ and suppose that any set consisting of fewer than s of the functions e_n is linearly independent. Let $n_1, \ldots, n_s$ be distinct integers, $\alpha_1, \ldots, \alpha_s \in \mathbb{C}$, and suppose that

$$\alpha_1 e_{n_1} + \cdots + \alpha_s e_{n_s} = 0.$$

Notice that $e_n e_m = e_{n+m}$ and $e_0 = 1$. Also, the e_n are differentiable, with $(e_n)' = i n e_n$. Multiplying our equation by e_{-n_1} and rearranging gives

$$-\alpha_1 = \alpha_2 e_{n_2 - n_1} + \cdots + \alpha_s e_{n_s - n_1}. \tag{3.4.1}$$

Now we can differentiate to get

$$0 = i(n_2 - n_1)\alpha_2 e_{n_2 - n_1} + \cdots + i(n_s - n_1)\alpha_s e_{n_s - n_1}.$$

The integers $n_j - n_1$ for $2 \le j \le s$ are all nonzero and distinct, so the induction hypothesis entails $\alpha_2 = \cdots = \alpha_s = 0$. But then Equation (3.4.1) gives $\alpha_1 = 0$ as well.

Definition 3.4.6. Let V be a vector space over K. A subset of V is called a *basis* of V if the set is linearly independent and has span equal to V.

Example 3.4.7.

(a) The set $\{1, x, x^2, \ldots, x^n\}$ is a basis of this vector space (over K) of polynomials in $K[x]$ of degree $\leq n$.

(b) The set $\{1, x, x^2, \ldots\}$ is a basis of $K[x]$.

Definition 3.4.8. Let V and W be vector spaces over K. A map $T : V \to W$ is called a *linear transformation* or *linear map* if $T(x+y) = T(x) + T(y)$ for all $x, y \in V$ and $T(\alpha x) = \alpha T(x)$ for all $\alpha \in K$ and $x \in V$.

Example 3.4.9.

(a) Fix a polynomial $f(x) \in K[x]$. The map $g(x) \mapsto f(x)g(x)$ is a linear transformation from $K[x]$ into $K[x]$.

(b) The formal derivative $\sum_k \alpha_k x^k \mapsto \sum_k k\alpha_k x^{k-1}$ is a linear transformation from $K[x]$ into $K[x]$.

(c) Let V denote the complex vector space of $\mathbb{C}$–valued continuous functions on the interval $[0, 1]$. The map $f \mapsto f(1/2)$ is a linear transformation from V to $\mathbb{C}$.

(d) Let V denote the complex vector space of $\mathbb{C}$–valued continuous functions on the interval $[0, 1]$ and let $g \in V$. The map $f \mapsto \int_0^1 f(t)g(t)\,dt$ is a linear transformation from V to $\mathbb{C}$.

Linear transformations are the homomorphisms in the theory of vector spaces; in fact, a linear transformation $T : V \to W$ between vector spaces is a homomorphism of abelian groups that additionally satisfies $T(\alpha v) = \alpha T(v)$ for all $\alpha \in K$ and $v \in V$. A linear *isomorphism* between vector spaces is a bijective linear transformation between them.

Definition 3.4.10. A *subspace* of a vector space V is a (nonempty) subset that is a vector space with the operations inherited from V.

As with groups, we have a criterion for a subset of a vector space to be a subspace, in terms of closure under the vector space operations:

Proposition 3.4.11. *For a nonempty subset of a vector space to be a subspace, it suffices that the subset be closed under addition and under scalar multiplication.*

Proof. Exercise 3.4.3. ■

Again as with groups, the kernel of a vector space homomorphism (linear transformation) is a subspace of the domain, and the range of a vector space homomorphism is a subspace of the codomain.

Proposition 3.4.12. *Let* $T : V \to W$ *be a linear map between vector spaces. Then the range of* T *is a subspace of* W *and the kernel of* T *is a subspace of* V.

Proof. Exercise 3.4.5. ■

If V is a vector space over K and W is a subspace, then in particular W is a subgroup of the abelian group V, so we can form the quotient *group* V/W, whose elements are cosets $v+W$ of W in V. The additive group operation in V/W is $(x+W)+(y+W) = (x+y)+W$. Now attempt to define a multiplication by scalars on V/W in the obvious way: $\alpha(v + W) = (\alpha v + W)$. We have to check that this this is well–defined. But this follows from the closure of W under scalar multiplication; namely, if $v + W = v' + W$ and, then $\alpha v - \alpha v' = \alpha(v - v') \in \alpha W \subseteq W$. Thus $\alpha v + W = \alpha v' + W$, and the scalar multiplication on V/W is well-defined.

Theorem 3.4.13. *If* W *is subspace of a vector space* V *over* K*, then* V/W *has the structure of a vector space, and the quotient map* $v \mapsto v + W$ *is a surjective linear map from* V *to* V/W *with kernel equal to* W.

Proof. Once we have checked that the scalar multiplication in V/W is well defined, it is straightforward to check the vector space axioms for V/W. That the quotient map is linear follows directly from the definition of the operations on V/W. ■

V/W is called a *quotient vector space* and $v \mapsto v + W$ the *quotient map*. We have a homomorphism theorem for vector spaces that is analogous to, and in fact follows from, the homomorphism theorem for groups.

Theorem 3.4.14. *(Homomorphism theorem for vector spaces). Let* $T : V \longrightarrow W$ *be a surjective linear map between vector spaces with kernel* N. *Let* $\pi : V \longrightarrow V/N$ *be the quotient map. There is linear isomorphism* $\tilde{T} : V/N \longrightarrow W$ *satisfying* $\tilde{T} \circ \pi = T$. *(See the following diagram.)*

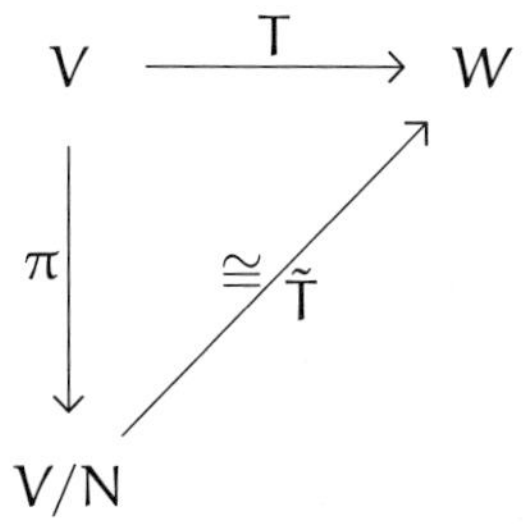

Proof. The homomorphism theorem for groups (Theorem 2.7.6) gives us an isomorphism of abelian groups $\tilde{T}$ satisfying $\tilde{T} \circ \pi = T$. We have only to verify that $\tilde{T}$ also respects multiplication by scalars. But this follows at once from the definitions: $\tilde{T}(\alpha(x + N)) = \tilde{T}(\alpha x + N) = T(\alpha x) = \alpha T(x) = \alpha \tilde{T}(x + N)$. ■

The next three propositions are analogues for vector spaces and linear transformations of results that we have established for groups and group homomorphisms. Each is proved by adapting the proof from the group situation. I leave this for you to do.

Proposition 3.4.15. *Let $T : V \to W$ be a surjective linear transformation, with kernel N. There is a one to one correspondence between subspaces of W and subspaces of V containing N.*

Proof. Exercise 3.4.6. ■

Proposition 3.4.16. *Let $T : V \to W$ be a surjective linear transformation. Let $\overline{M}$ be a subspace of W and let $M = T^{-1}(\overline{M})$. Then $x + M \mapsto T(x) + \overline{M}$ defines an isomorphism of V/M to $W/\overline{M}$.*

Proof. Exercise 3.4.7. ■

Proposition 3.4.17. *Let A and N be subspaces of a vector space V. Let π denote the quotient map $\pi : V \to V/N$. Then $\pi^{-1}(\pi(A)) = A + N$ is a subspace of V containing both A and N. Furthermore, $(A + N)/N \cong \pi(A) \cong A/(A \cap N)$.*

Proof. Exercise 3.4.8. ■

We now consider bases and dimension for abstract vector spaces. This rests on the theory developed for subspaces of K^n in Appendix E.

Definition 3.4.18. A vector space is said to be *finite–dimensional* if it has a finite basis. Otherwise, V is said to be *infinite–dimensional*.

If V has a basis with $\{v_1, \dots, v_n\}$ with n elements, then we have a linear isomorphism $K^n \to V$ defined by $\begin{bmatrix} \alpha_1 \\ \vdots \\ \alpha_n \end{bmatrix} \mapsto \sum \alpha_i v_i$. Since a linear isomorphism carries linearly independent sets to linearly independent sets, spanning sets to spanning sets, and bases to bases, it follows (from the corresponding results for K^n) that

1. Any linearly independent set in V has no more than n elements.
2. Any spanning set in V has at least n elements.
3. Any basis in V has exactly n elements.
4. Any subspace of V is finite–dimensional.
5. Any linearly independent subset of V is contained in a basis.
6. Any spanning subset of V contains a basis as a subset.

See Appendix E for the results for K^n. The unique cardinality of a basis of a finite–dimensional vector space V is called the *dimension* of V and denoted $\dim(V)$. If V is infinite–dimensional, we write $\dim(V) = \infty$.

Proposition 3.4.19. *Any two n–dimensional vector spaces over K are linearly isomorphic.*

Proof. The case $n = 0$ is left to the reader.

For $n \geq 1$, any two n–dimensional vector spaces over K are each isomorphic to K^n, and hence isomorphic to each other. ■

This proposition reveals that (finite–dimensional) vector spaces are not very interesting, as they are completely classified by their dimension. That is why the actual subject of finite–dimensional linear algebra is not vector spaces but rather linear maps, which have more interesting structure than vector spaces themselves.

Proposition 3.4.20. *Let V be a vector space over K and let S be a basis of V. Then any function $f : S \to W$ from S into a vector space W extends uniquely to a linear map $T : V \to W$.*

Proof. We will assume that $S = \{v_1, \dots, v_n\}$ is finite, in order to simplify the notation, although the result is equally valid if S is infinite.

Let $f : S \to W$ be a function. Any element $v \in V$ has a unique expression as a linear combination of elements of S, $v = \sum_i \alpha_i v_i$. There is only one possible way to define $T(v)$, namely $T(v) = \sum_i \alpha_i f(v_i)$. It is then straightforward to check that T is linear. ■

An *ordered basis* of a finite–dimensional vector space is a finite sequence whose entries are the elements of a basis listed without repetition; that is, an ordered basis is just a basis endowed with a particular linear order. Corresponding to an ordered basis $B = (v_1, \dots, v_n)$ of an n–dimensional vector space V over K, we have a linear isomorphism $S_B : V \to K^n$ given by $\sum_i \alpha_i v_i \mapsto \sum_i \alpha_i \hat{\mathbf{e}}_i$, where $(\hat{\mathbf{e}}_1, \dots, \hat{\mathbf{e}}_n)$ is the standard ordered basis of K^n.

Now let V be an n–dimensional vector space over K with an ordered basis B and W an m–dimensional vector space over K with ordered basis C. Furthermore, let $T : V \to W$ be a linear map. The *matrix of* T *with respect to the ordered bases* B *and* C is defined to be the standard matrix of $S_C T S_B^{-1} : K^n \to K^m$. (The standard matrix is discussed in Appendix E.) We denote the matrix of T with respect to B and C by $T_{C,B}$. Properties of the assignment $T \mapsto T_{C,B}$ are explored in the Exercises.

The *direct sum* of two vector spaces V and W is the Cartesian product endowed with component-by-component operations:

$$(v, w) + (v', w') = (v + v', w + w'),$$

and

$$\alpha(v, w) = (\alpha v, \alpha w).$$

We generally denote the direct sum by $V \oplus W$.

Now suppose that M, N are subspaces of a vector space W such that $M \cap N = \{0\}$ and $M + N = V$. Define a map from the direct sum $M \oplus N$ to V by $(m, n) \mapsto m + n$. This map is evidently linear, it is surjective by the hypothesis $M + N = V$, and it is injective because if (m, n) is in the kernel, then $m = -n \in M \cap N = \{0\}$. In this case, we say that V is the *internal direct sum* of M and N, and we write $V = M \oplus N$.

Let M be a subspace of a vector space V. A subspace N of V is said to be a *complement* of M if $V = M \oplus N$. The fact is that subspaces of finite–dimensional vector spaces *always* have a complement, as we shall now explain.

Proposition 3.4.21. *Let* $T : V \to W$ *be a surjective linear map of a finite–dimensional vector space* V *onto a vector space* W. *Then* T *admits a right inverse; that is, there exists a linear map* $S : W \to V$ *such that* $TS = \mathrm{id}_W$.

Proof. First, let's check that W is finite–dimensional, with dimension no greater than $\dim(V)$. If $\{v_1, \dots, v_n\}$ is a basis of V, then $\{T(v_1), \dots, T(v_n)\}$ is a spanning subset of W, so contains a basis of W as a subset.

Now let $\{w_1, \dots, w_s\}$ be a basis of W. For each basis element w_i, let x_i be a preimage of w_i in V (i.e., choose x_i such that $T(x_i) = w_i$). The map $w_i \mapsto x_i$ extends uniquely to a linear map $S : W \to V$, defined by $S(\sum_i \alpha_i w_i) = \sum_i \alpha_i x_i$. We have $TS(\sum_i \alpha_i w_i) = T(\sum_i \alpha_i x_i) = \sum_i \alpha_i T(x_i) = \sum_i \alpha_i w_i$. Thus $TS = \mathrm{id}_W$. ■

In the situation of the previous proposition, let W' denote the image of S. I claim that

$$V = \ker(T) \oplus W' \cong \ker(T) \oplus W.$$

Suppose $v \in \ker(T) \cap W'$. Since $v \in W'$, there is a $w \in W$ such that $v = S(w)$. But then $0 = T(v) = T(S(w)) = w$, and, therefore, $v = S(w) = S(0) = 0$. This shows that $\ker(T) \cap W' = \{0\}$. For any $v \in V$, we can write $v = S(T(v)) + (v - ST(v))$. The first summand is evidently in W', and the second is in the kernel of T, as $T(v) = TST(v)$. This shows that $\ker(T) + W' = V$. We have shown that $V = \ker(T) \oplus W'$. Finally, note that S is an isomorphism of W onto W', so we also have $V \cong \ker(T) \oplus W$. We have shown the following:

Proposition 3.4.22. *If* $T : V \to W$ *is a linear map and* V *is finite–dimensional, then* $V \cong \ker(T) \oplus \mathrm{range}(T)$. *In particular,* $\dim(V) = \dim(\ker(T)) + \dim(\mathrm{range}(T))$.

Now let V be a finite–dimensional vector space and let N be a subspace. The quotient map $\pi : V \to V/N$ is a a surjective linear map with kernel N. Let S be a right inverse of π, as in the proposition, and let M be the image of S. The preceding comments show that $V = N \oplus M \cong N \oplus V/N$. We have proved the following:

Proposition 3.4.23. *Let* V *be a finite–dimensional vector space and let* N *be a subspace. Then* $V \cong N \oplus V/N$. *In particular,* $\dim(V) = \dim(N) + \dim(V/N)$.

Exercises 3.4

3.4.1. Show that the intersection of an arbitrary family of linear subspaces of a vector space is a linear subspace.

3.4.2. Let S be a subset of a vector space. Show that $\mathrm{span}(S) = \mathrm{span}(\mathrm{span}(S))$. Show that $\mathrm{span}(S)$ is the unique smallest linear subspace of V containing S as a subset, and that it is the intersection of all linear subspaces of V that contain S as a subset.

3.4.3. Prove Proposition 3.4.11.

3.4.4. Show that any composition of linear transformations is linear. Show that the inverse of a linear isomorphism is linear.

3.4.5. Let $T : V \to W$ be a linear map between vector spaces. Show that the range of T is a subspace of W and the kernel of T is a subspace of V.

3.4.6. Prove Proposition 3.4.15.

3.4.7. Prove Proposition 3.4.16.

3.4.8. Prove Proposition 3.4.17.

3.4.9. Let A and B be finite–dimensional subspaces of a not necessarily finite–dimensional vector space V. Show that $A+B$ is finite–dimensional and that $\dim(A + B) + \dim(A \cap B) = \dim(A) + \dim(B)$.

3.4.10. Show that the following conditions are equivalent for a vector space V:

(a) V is finite dimensional.
(b) V has a finite spanning set.
(c) Every linearly independent subset of V is finite.

[To prove that (c) implies (a), you need the fact that every vector space has a maximal linearly independent subset; this does not follow from the theory we have presented, but requires Zorn's lemma or the Hausdorff maximal principal.]

3.4.11. Show that the following conditions are equivalent for a vector space V:

(a) V is infinite–dimensional.
(b) V has an infinite linearly independent subset.
(c) For every $n \in \mathbb{N}$, V has a linearly independent subset with n elements.

3.4.12. Let V and W be vector spaces over K. Let $\mathrm{Hom}(V, W)$ denote the set of all linear transformations from V to W. Show that $\mathrm{Hom}(V, W)$ is a vector space over K.

3.4.13. Let V be an n–dimensional vector space over K with ordered basis $B = (v_1, \dots, v_n)$, and let W be an m–dimensional vector space over K with ordered basis $C = (w_1, \dots, w_m)$. Let $T : V \to W$ be a linear map. Show that the jth column of $T_{C,B}$ is $\begin{bmatrix} \alpha_1 \\ \vdots \\ \alpha_m \end{bmatrix}$, where the α_i are the coefficients in the expansion of $T(v_j)$ with respect to C, namely $T(v_j) = \sum_i \alpha_i w_i$.

3.4.14. Consider the $\mathbb{R}$-vector space $\mathcal{P}_n$ of polynomials of degree $\le n$ with $\mathbb{R}$-coefficients, with the ordered basis $(1, x, x^2, \dots, x^n)$.

(a) Find the matrix of differentiation $\frac{d}{dx} : \mathcal{P}_7 \to \mathcal{P}_7$.

(b) Find the matrix of integration $\int : \mathcal{P}_6 \to \mathcal{P}_7$.

(c) Observe that multiplication by $1 + 3x + 2x^2$ is linear from $\mathcal{P}_5$ to $\mathcal{P}_7$, and find the matrix of this linear map.

3.4.15. Let V be an n–dimensional vector space over K with ordered basis B, and let W be an m–dimensional vector space over K with ordered basis C. Show that the map $T \mapsto T_{C,B}$ is a linear isomorphism of $\mathrm{Hom}(V, W)$ onto the set $\mathrm{Mat}_{m,n}(K)$ of m-by-n matrices over K. (You have to show that the map is linear, one to one, and onto.)

3.4.16. Let V, W, X be finite–dimensional vector spaces over K with ordered bases B, C, and D. Let $T \in \mathrm{Hom}(V, W)$ and $S \in \mathrm{Hom}(W, X)$. Show that $(S \circ T)_{D,B} = S_{D,C} T_{C,B}$.

3.4.17. Let V, W be finite–dimensional vector spaces over K. Let B, B' be two ordered bases of V, and let C, C' be two ordered bases of W. Let $T \in \mathrm{Hom}(V, W)$.

(a) Show that $T_{C',B'} = \mathrm{id}_{C',C}\, T_{C,B}\, \mathrm{id}_{B,B'}$, where $\mathrm{id}_{C',C}$ denotes the matrix of the identity transformation of W with respect to C and C', and $\mathrm{id}_{B,B'}$ denotes the matrix of the identity transformation of V with respect to B and B'.

(b) Show that $\mathrm{id}_{B,B'}$ is an invertible matrix with inverse $\mathrm{id}_{B',B}$.

(c) Let T be the linear transformation of $\mathbb{R}^3$ with standard matrix $\begin{bmatrix} 1 & 5 & 2 \\ 2 & 1 & 3 \\ 1 & 1 & 4 \end{bmatrix}$. Find the matrix of $T_{B,B}$ of T with respect to the ordered basis $B = \left(\begin{bmatrix} 1 \\ 1 \\ 1 \end{bmatrix}, \begin{bmatrix} 1 \\ 1 \\ 0 \end{bmatrix}, \begin{bmatrix} 1 \\ 0 \\ 0 \end{bmatrix} \right)$.

3.4.18. Let V be a vector space over K. Show that $\mathrm{Hom}(V, V)$ is a ring. [We generally write $\mathrm{End}(V)$ for $\mathrm{Hom}(V, V)$.]

3.4.19. Let V be an n–dimensional vector space over K with ordered basis B. Show that $T \mapsto T_{B,B}$ is a ring isomorphism from $\mathrm{End}(V)$ to $\mathrm{Mat}_n(K)$.

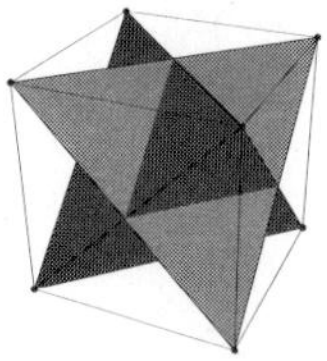

CHAPTER 4

Symmetries of Polyhedra

4.1. Rotations of Regular Polyhedra

There are five regular polyhedra (Figure 4.1.1): the tetrahedron, the cube, the octahedron, the dodecahedron (12 faces), and the icosahedron (20 faces).

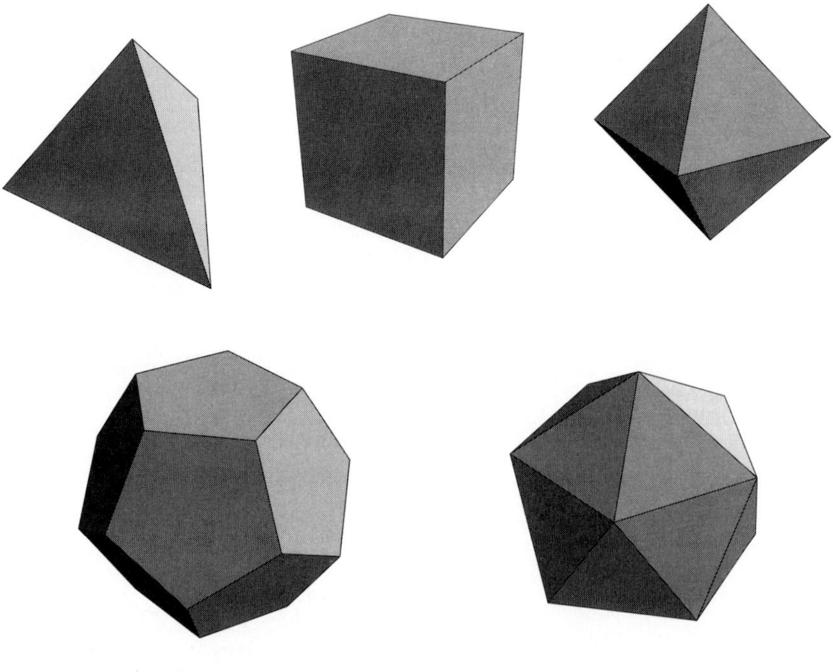

Figure 4.1.1. The regular polyhedra.

In this section, we will work out the rotational symmetry groups of the tetrahedron, cube, and octahedron, and in the following section we will treat the dodecahedron and icosahedron. In later sections, we will work out the full symmetry groups, with reflections allowed as well as rotations.

It is convenient to have physical models of the regular polyhedra that you can handle while studying their properties. In case you can't get any ready-made models, I have provided you (at the end of the book, Appendix E) with patterns for making paper model. I urge you to obtain or construct models of the regular polyhedra before continuing with your reading.

Now that you have your models of the regular polyhedra, we can proceed to obtain their rotation groups. We start with the tetrahedron.

Definition 4.1.1. A line is an n*–fold axis of symmetry* for a geometric figure if the rotation by $2\pi/n$ about this line is a symmetry.

For each n–fold axis of symmetry, there are $n-1$ nonidentity symmetries of the figure, namely the rotations by $2k\pi/n$ for $1 \leq k \leq n-1$.

The tetrahedron has four 3–fold axes of rotation, each of which passes through a vertex and the centroid of the opposite face. These give eight nonidentity group elements, each of order 3. See Figure 4.1.2.

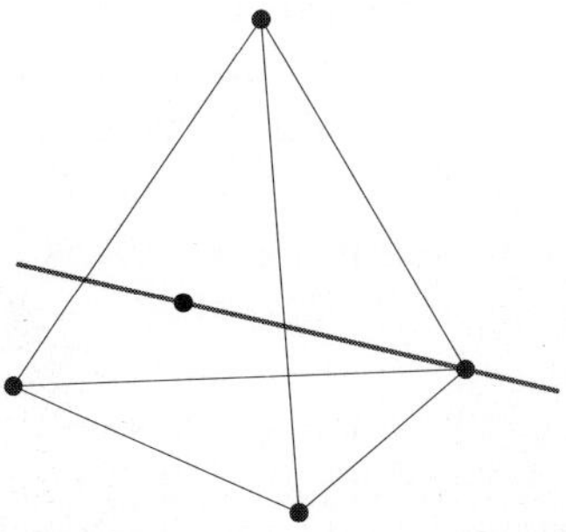

Figure 4.1.2. Three-fold axis of the tetrahedron.

The tetrahedron also has three 2–fold axes of symmetry, each of which passes through the centers of a pair of opposite edges. These contribute three nonidentity group elements, each of order 2. See Figure 4.1.3.

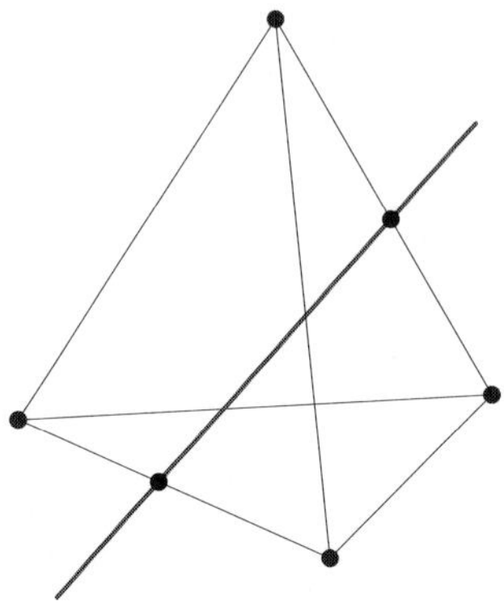

Figure 4.1.3. Two–fold axis of the tetrahedron.

Including the identity, we have 12 rotations. Are these all of the rotational symmetries of the tetrahedron? According to Proposition 1.4.1, every symmetry of the tetrahedron is realized by a linear isometry of $\mathbb{R}^3$; the rotational symmetries of the tetrahedron are realized by rotations of $\mathbb{R}^3$. Furthermore, every symmetry permutes the vertices of the tetrahedron, by Exercise 1.4.7, and fixes the center of mass of the tetrahedron, which is the average of the vertices. It follows that a symmetry also carries edges to edges and faces to faces. Using these facts, we can show that we have accounted for all of the rotational symmetries. See Exercise 4.1.7.

The rotation group acts faithfully as permutations of the four vertices; this means there is an injective homomorphism of the rotation group into S_4. Under this homomorphism the eight rotations of order 3 are mapped to the eight 3–cycles in S_4. The three rotations of order 2 are mapped to the three elements $(12)(34)$, $(13)(24)$, and $(14)(23)$. Thus the image in S_4 is precisely the group of even permutations A_4.

Proposition 4.1.2. *The rotation group of the tetrahedron is isomorphic to the group A_4 of even permutations of four objects.*

Let us also work out the matrices that implement the rotations of the tetrahedron. First we need to figure out how to write the matrix for a rotation through an angle θ about the axis determined by a unit vector $\hat{\mathbf{v}}$. Of course, there are two possible such rotations, which are inverses of each other; let's agree to find the one determined by the "right–hand rule," as in Figure 4.1.4.

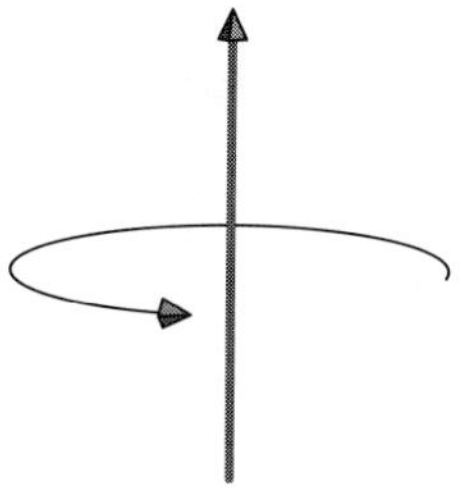

Figure 4.1.4. Right–hand rule.

If $\hat{\mathbf{v}}$ is the first standard coordinate vector

$$\hat{\mathbf{e}}_1 = \begin{bmatrix} 1 \\ 0 \\ 0 \end{bmatrix},$$

then the rotation matrix is

$$R_\theta = \begin{bmatrix} 1 & 0 & 0 \\ 0 & \cos\theta & \sin\theta \\ 0 & -\sin\theta & \cos\theta \end{bmatrix}.$$

To compute the matrix for the rotation about the vector $\hat{\mathbf{v}}$, first we need to find additional vectors $\hat{\mathbf{v}}_2$ and $\hat{\mathbf{v}}_3$ such that $\{\hat{\mathbf{v}}_1 = \hat{\mathbf{v}}, \hat{\mathbf{v}}_2, \hat{\mathbf{v}}_3\}$ form a right–handed orthonormal basis of $\mathbb{R}^3$; that is, the three vectors are of unit length, mutually orthogonal, and the determinant of the matrix $V = [\hat{\mathbf{v}}_1, \hat{\mathbf{v}}_2, \hat{\mathbf{v}}_3]$ with columns $\hat{\mathbf{v}}_i$ is 1, or equivalently, $\hat{\mathbf{v}}_3$ is the vector cross-product $\hat{\mathbf{v}}_1 \times \hat{\mathbf{v}}_2$. V is the matrix that rotates the standard right–handed orthonormal basis $\{\hat{\mathbf{e}}_1, \hat{\mathbf{e}}_2, \hat{\mathbf{e}}_3\}$ onto the basis $\{\hat{\mathbf{v}}_1, \hat{\mathbf{v}}_2, \hat{\mathbf{v}}_3\}$. The inverse of V is the transposed matrix V^t, because the matrix entries of V^tV are the inner products $\langle \hat{\mathbf{v}}_i, \hat{\mathbf{v}}_j \rangle = \delta_{ij}$. The matrix we are looking for is $VR_\theta V^t$, because the matrix first rotates the orthonormal basis$\{\hat{\mathbf{v}}_i\}$ onto the standard orthonormal basis$\{\hat{\mathbf{e}}_i\}$, then rotates through an angle θ about $\hat{\mathbf{e}}_1$, and then rotates the standard basis $\{\hat{\mathbf{e}}_i\}$ back to the basis $\{\hat{\mathbf{v}}_i\}$.

Consider the points

$$\begin{bmatrix} 1 \\ 1 \\ 1 \end{bmatrix}, \quad \begin{bmatrix} 1 \\ -1 \\ -1 \end{bmatrix}, \quad \begin{bmatrix} -1 \\ -1 \\ 1 \end{bmatrix}, \quad \text{and} \quad \begin{bmatrix} -1 \\ 1 \\ -1 \end{bmatrix}.$$

They are equidistant from each other, and the sum of the four is $\mathbf{0}$, so the four points are the vertices of a tetrahedron whose center of mass is at the origin.

One 3–fold axis of the tetrahedron passes through the origin and the point $\begin{bmatrix}1\\1\\1\end{bmatrix}$. Thus the right-handed rotation through the angle $2\pi/3$ about the unit vector $(1/\sqrt{3})\begin{bmatrix}1\\1\\1\end{bmatrix}$ is one symmetry of the tetrahedron. In Exercise 4.1.1, you are asked to compute the matrix of this rotation. The result is the permutation matrix

$$R = \begin{bmatrix}0 & 0 & 1\\1 & 0 & 0\\0 & 1 & 0\end{bmatrix}.$$

In Exercise 4.1.2, you are asked to show that the matrices for rotations of order 2 are the diagonal matrices with two entries of -1 and one entry of 1. These matrices generate the group of order 4 consisting of diagonal matrices with diagonal entries of ± 1 and determinant equal to 1.

Finally, you can show (Exercise 4.1.3) that these diagonal matrices and the permutation matrix R generate the group of rotation matrices for the tetrahedron. Consequently, we have the following result:

Proposition 4.1.3. *The group of rotational symmetries of the tetrahedron is isomorphic to the group of signed permutation matrices that can be written in the form* DR^k, *where* D *is a diagonal signed permutation matrix with determinant 1,* $R = \begin{bmatrix}0 & 0 & 1\\1 & 0 & 0\\0 & 1 & 0\end{bmatrix}$, *and* $0 \le k \le 2$.

Now we consider the cube. The cube has four 3–fold axes through pairs of opposite vertices, giving eight rotations of order 3. See Figure 4.1.5.

There are also three 4–fold axes through the centroids of opposite faces, giving nine nonidentity group elements, three of order 2 and six of order 4. See Figure 4.1.6.

Finally, the cube has six 2–fold axes through centers of opposite edges, giving six order 2 elements. See Figure 4.1.7. With the identity, we have 24 rotations.

As for the tetrahedron, we can show that we have accounted for all of the rotational symmetries of the cube. See Exercise 4.1.8.

Now we need to find an injective homomorphism into some permutation group induced by the action of the rotation group on

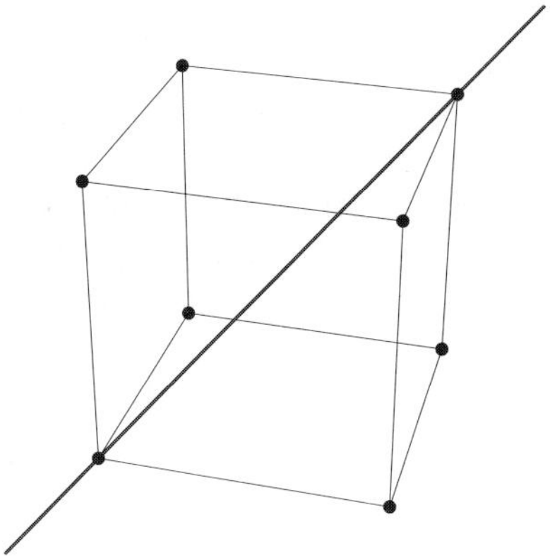

Figure 4.1.5. Three–fold axis of the cube.

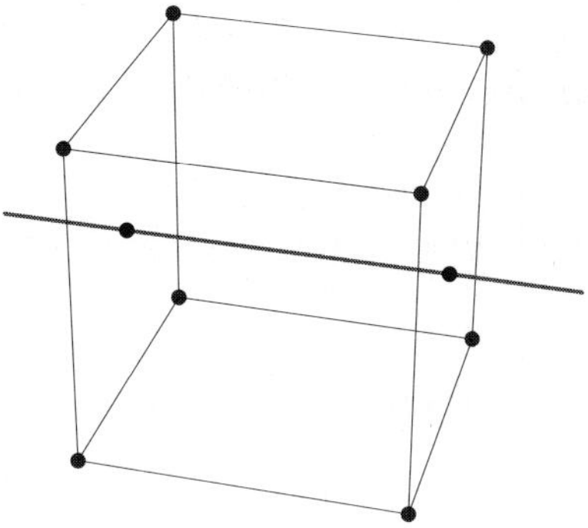

Figure 4.1.6. Four–fold axis of the cube.

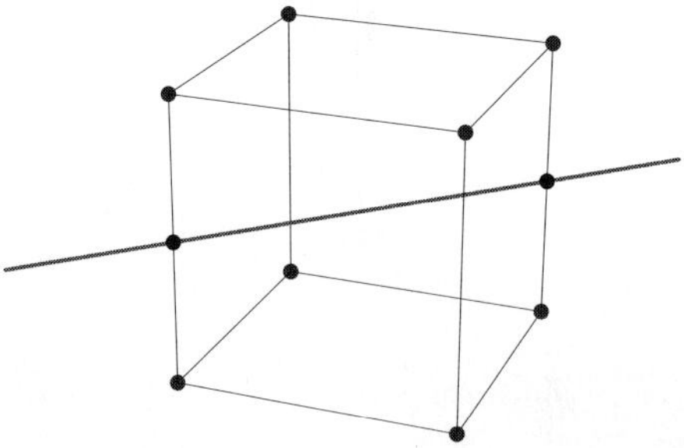

Figure 4.1.7. Two–fold axis of the cube.

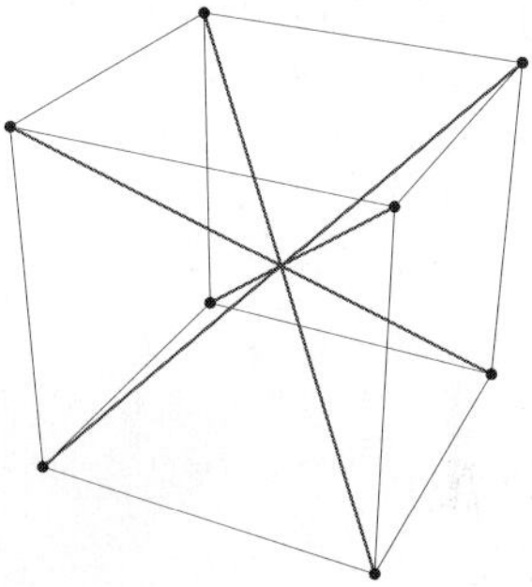

Figure 4.1.8. Diagonals of cube.

some set of geometric objects associated with the cube. If we choose vertices, or edges, or faces, we get homomorphisms into S_8, S_{12}, or S_6 respectively. None of these choices look very promising, since our group has only 24 elements.

The telling observation now is that vertices (as well as edges and faces) are really permuted in pairs. So we consider the action of the rotation group on pairs of opposite vertices, or, what amounts to the same thing, on the four diagonals of the cube. See Figure 4.1.8. This gives a homomorphism of the rotation group of the cube into S_4. Since both the rotation group and S_4 have 24 elements, to show that this is an isomorphism, it suffices to show that it is injective, that is, that no rotation leaves all four diagonals fixed. This is easy to check.

Proposition 4.1.4. *The rotation group of the cube is isomorphic to the permutation group* S_4.

The close relationship between the rotation groups of the tetrahedron and the cube suggests that there should be tetrahedra related geometrically to the cube. In fact, the choice of coordinates for the vertices of the tetrahedron in the preceding discussion shows how to embed a tetrahedron in the cube: Take the vertices of the cube at the points $\begin{bmatrix} \pm 1 \\ \pm 1 \\ \pm 1 \end{bmatrix}$. Then those four vertices that have the property that the product of their coordinates is 1 are the vertices of an embedded tetrahedron. The remaining vertices (those for which the product of the coordinates is -1) are the vertices of a complementary tetrahedron. The even permutations of the diagonals of the cube preserve the two tetrahedra; the odd permutations interchange the two tetrahedra See Figure 4.1.9.

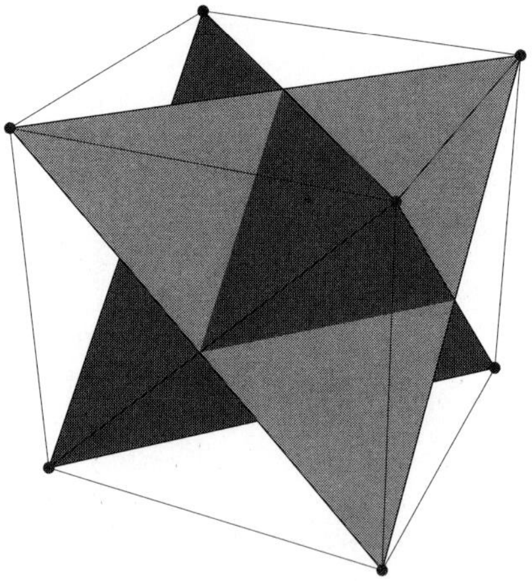

Figure 4.1.9. Cube with inscribed tetrahedra.

This observation also lets us compute the 24 matrices for the rotations of the cube very easily. Note that the 3–fold axes for the tetrahedron are also 3–fold axes for the cube, and the 2–fold axes for the tetrahedron are 4–fold axes for the cube. In particular, the symmetries of the tetrahedron form a subgroup of the symmetries of the cube of index 2. Using these considerations, we obtain the following result; see Exercise 4.1.5.

Proposition 4.1.5. *The group of rotation matrices of the cube is isomorphic to the group of 3–by–3 signed permutation matrices with determinant 1. This group is the semidirect product of the group of order 4 consisting of diagonal signed permutation matrices with determinant 1, and the group of 3–by–3 permutation matrices.*

Corollary 4.1.6. S_4 *is isomorphic to the group of 3–by–3 signed permutation matrices with determinant 1.*

Each convex polyhedron T has a *dual polyhedron* whose vertices are at the centroids of the faces of T; two vertices of the dual are joined by an edge if the corresponding faces of T are adjacent. *The dual polyhedron has the same symmetry group as does the original polyhedron.*

Proposition 4.1.7. *The octahedron is dual to the cube, so its group of rotations is also isomorphic to* S_4 *(Figure 4.1.10).*

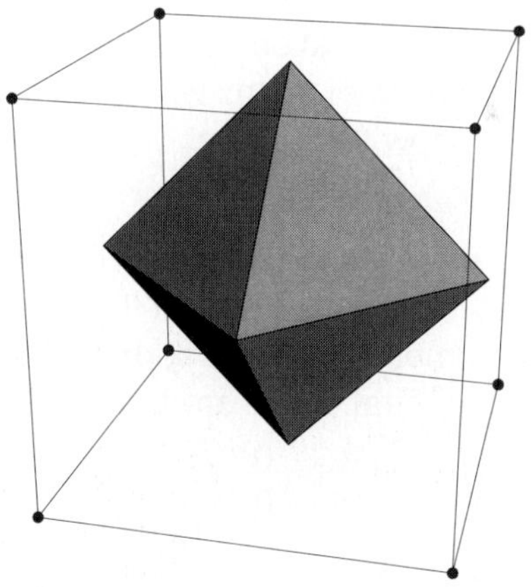

Figure 4.1.10. Cube and octahedron.

Exercises 4.1

4.1.1.

(a) Given a unit vector $\hat{\mathbf{v}}_1$, explain how to find two further unit vectors $\hat{\mathbf{v}}_2$ and $\hat{\mathbf{v}}_3$ such that the $\{\hat{\mathbf{v}}_i\}$ form a right–handed orthonormal basis.

(b) Carry out the procedure for $\hat{\mathbf{v}}_1 = (1/\sqrt{3})\begin{bmatrix}1\\1\\1\end{bmatrix}$.

(c) Compute the matrix of rotation through $2\pi/3$ about the vector $\begin{bmatrix}1\\1\\1\end{bmatrix}$, and explain why the answer has such a remarkably simple form.

4.1.2. Show that the midpoints of the edges of the tetrahedron are at the six points

$$\pm\hat{\mathbf{e}}_1, \pm\hat{\mathbf{e}}_2, \pm\hat{\mathbf{e}}_3.$$

Show that the matrix of the rotation by π about any of the 2–fold axes of the tetrahedron is a diagonal matrix with two entries equal to -1 and one entry equal to 1. Show that the set of these matrices generates a group of order 4.

4.1.3. Show that the matrices computed in Exercises 4.1.1 and 4.1.2 generate the group of rotation matrices of the tetrahedron; the remaining matrices can be computed by matrix multiplication. Show that the rotation matrices for the tetrahedron are the matrices

$$\begin{bmatrix}\pm1 & 0 & 0\\ 0 & \pm1 & 0\\ 0 & 0 & \pm1\end{bmatrix}, \quad \begin{bmatrix}0 & 0 & \pm1\\ \pm1 & 0 & 0\\ 0 & \pm1 & 0\end{bmatrix}, \quad \begin{bmatrix}0 & \pm1 & 0\\ 0 & 0 & \pm1\\ \pm1 & 0 & 0\end{bmatrix},$$

where the product of the entries is 1. That is, there are no -1's or else two -1's.

4.1.4. Show that the group of rotational matrices of the tetrahedron is the semidirect product of the group $\mathcal{V}$ consisting of diagonal permutation matrices with determinant 1 (which is a group of order 4) and the cyclic group of order 3 generated by the permutation matrix $R = \begin{bmatrix}0 & 0 & 1\\ 1 & 0 & 0\\ 0 & 1 & 0\end{bmatrix}$.

4.1.5.

(a) If $\mathcal{T}$ denotes the group of rotation matrices for the tetrahedron and S is any rotation matrix for the cube that is not contained in $\mathcal{T}$, show that the group of rotation matrices for the cube is $\mathcal{T} \cup \mathcal{T}s$.

(b) Show that the group of rotation matrices for the cube consists of signed permutation matrices with determinant 1, that is, matrices with entries in $\{0, \pm 1\}$ with exactly one nonzero entry in each row and in each column, and with determinant 1.

(c) Show that the group of rotation matrices for the cube is the semidirect product of the group $\mathcal{V}$ consisting of diagonal permutation matrices with determinant 1 (which is a group of order 4) and the group of 3–by–3 permutation matrices.

4.1.6. Let $\mathcal{R}$ be a convex polyhedron.

(a) Show that if a rotational symmetry τ of $\mathcal{R}$ maps a certain face to itself, then it fixes the centroid of the face. Conclude that τ is a rotation about the line containing the centroid of $\mathcal{R}$ and the centroid of the face.

(b) Similarly, if a rotational symmetry τ of $\mathcal{R}$ maps a certain edge onto itself, then τ is a rotation about the line containing the midpoint of the edge and the centroid of $\mathcal{R}$.

4.1.7. Show that we have accounted for all of the rotational symmetries of the tetrahedron. Hint: Let τ be a symmetry and ν a vertex. Show that the orbit of ν under τ, namely the set $\{\tau^n(\nu) : n \in \mathbb{Z}\}$, consists of one, two, or three vertices.

4.1.8. Show that we have accounted for all of the rotational symmetries of the cube.

4.2. Rotations of the Dodecahedron and Icosahedron

The dodecahedron and icosahedron are dual to each other and so have the same rotational symmetry group. We need only work out the rotation group of the dodecahedron. See Figure 4.2.1.

The dodecahedron has six 5–fold axes through the centroids of opposite faces, giving 24 rotations of order 5. See Figure 4.2.2.

There are ten 3–fold axes through pairs of opposite vertices, giving 20 elements of order 3. See Figure 4.2.3.

Finally, the dodecahedron has 15 2–fold axis through centers of opposite edges, giving 15 elements of order 2. See Figure 4.2.4.

With the identity element, the dodecahedron has 60 rotational symmetries. Now considering that the rotation groups of the "smaller"

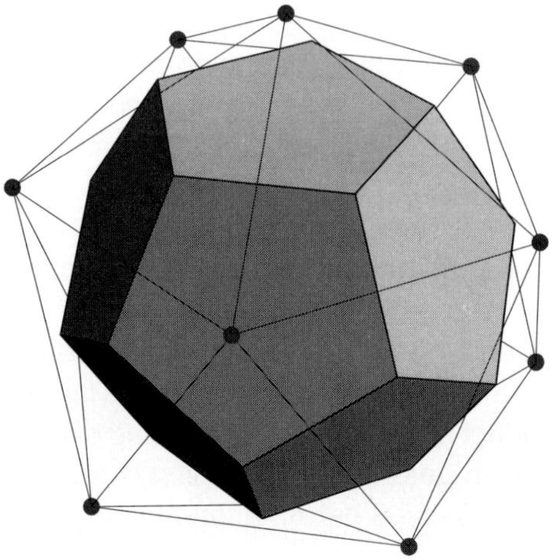

Figure 4.2.1. Dodecahedron and icosahedron.

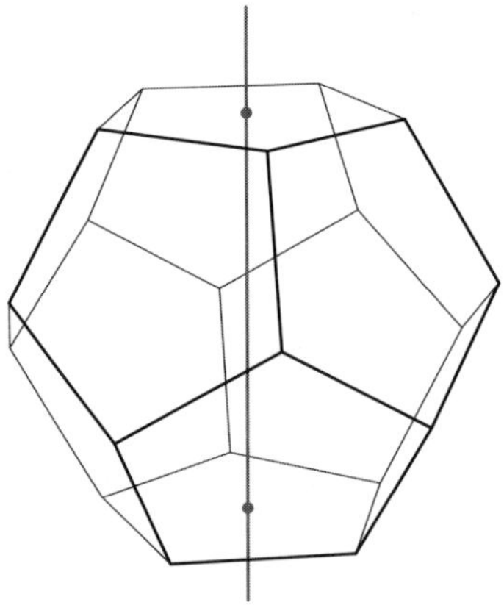

Figure 4.2.2. Five–fold axis of the dodecahedron.

regular polyhedra are A_4 and S_4, and suspecting that there ought to be a lot of regularity in this subject, we might guess that the rotation group of the dodecahedron is isomorphic to the group A_5 of even permutations of five objects. So we are led to look for five geometric objects that are permuted by this rotation group. Finding the five objects is a perhaps a more subtle task than picking out the four diagonals that are permuted by the rotations of the cube. But since each face has five edges, we might suspect that each object is some equivalence class of edges that includes one edge from each

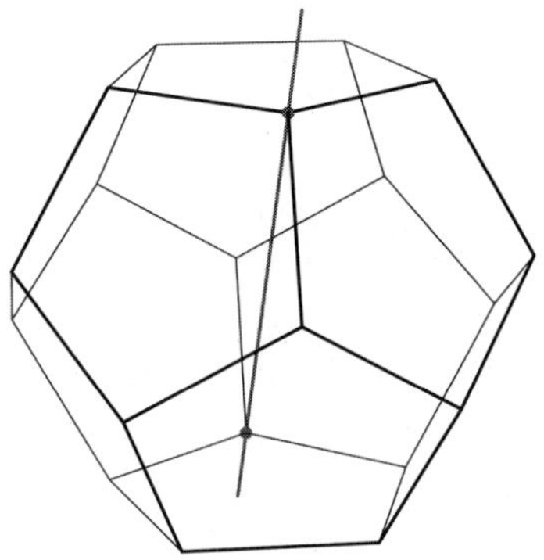

Figure 4.2.3. Three–fold axis of the dodecahedron.

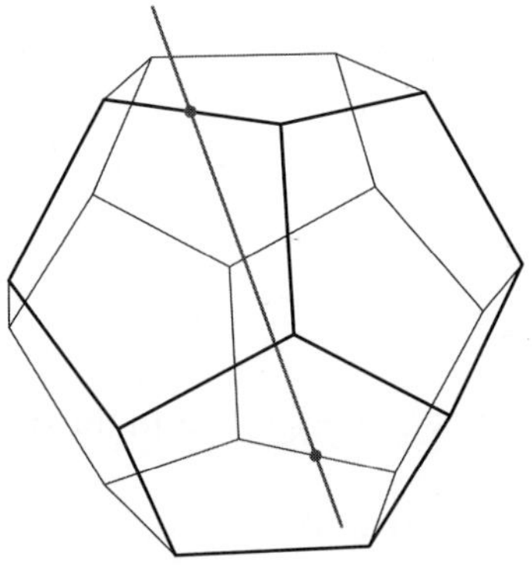

Figure 4.2.4. Two–fold axis of the dodecahedron.

face. Since there are 30 edges, each such equivalence class of edges should contain six edges.

Pursuing this idea, consider any edge of the dodecahedron and its opposite edge. Notice that the plane containing these two edges bisects another pair of edges, and the plane containing *that* pair of edges bisects a third pair of edges. The resulting family of six edges contains one edge in each face of the dodecahedron. There are five such families that are permuted by the rotations of the dodecahedron. To understand these ideas, you are well advised to look closely at your physical model of the dodecahedron.

There are several other ways to pick out five objects that are permuted by the rotation group. Consider one of our families of six edges. It contains three pairs of opposite edges. Take the three lines joining the centers of the pairs of opposite edges. These three lines are mutually orthogonal; they are the axes of a cartesian coordinate system. There are five such coordinate systems that are permuted by the rotation group.

Finally, given one such coordinate system, we can locate a cube whose faces are parallel to the coordinate planes and whose edges lie on the faces of the dodecahedron. Each edge of the cube is a diagonal of a face of the dodecahedron, and exactly one of the five diagonals of each face is an edge of the cube. There are five such cubes that are permuted by the rotation group. See Figure 4.2.5.

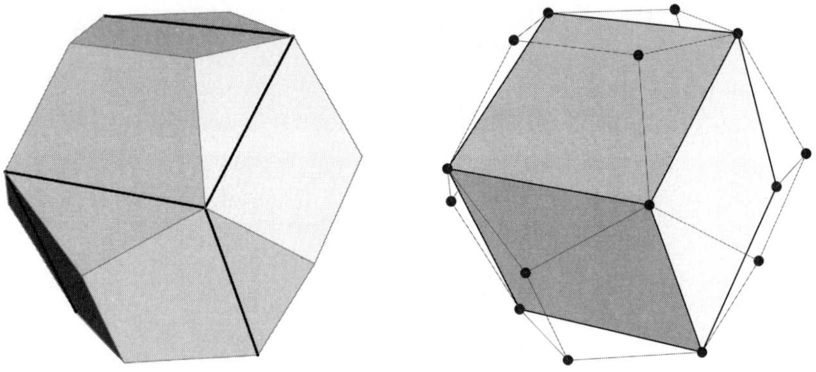

Figure 4.2.5. Cube inscribed in the dodecahedron.

You are asked to show in Exercise 4.2.1 that the action of the rotation group on the set of five inscribed cubes is faithful; that is, the homomorphism of the rotation group into S_5 is injective.

Now, it remains to show that the image of the rotation group in S_5 is the group of even permutations A_5. We could do this by explicit computation. However, by using a previous result, we can avoid doing any computation at all. We established earlier that for each n, A_n is the unique subgroup of S_n of index 2 (Exercise 2.5.16). Since the image of the rotation group has 60 elements, it follows that it must be A_5.

Proposition 4.2.1. *The rotation groups of the dodecahedron and the icosahedron are isomorphic to the group of even permutations* A_5.

Exercises 4.2

4.2.1. Show that no rotation of the dodecahedron leaves each of the five inscribed cubes fixed. Thus the action of the rotation group on the set of inscribed cubes induces an injective homomorphism of the rotation group into S_5.

4.2.2. Let

$$A = \{ \begin{bmatrix} \cos 2k\pi/5 \\ \sin 2k\pi/5 \\ 1/2 \end{bmatrix} : 1 \le k \le 5\}$$

and

$$B = \{ \begin{bmatrix} \cos (2k+1)\pi/5 \\ \sin (2k+1)\pi/5 \\ -1/2 \end{bmatrix} : 1 \le k \le 5\}.$$

Show that

$$\{ \begin{bmatrix} 0 \\ 0 \\ \pm\sqrt{5}/2 \end{bmatrix} \} \cup A \cup B$$

is the set of vertices of an icosahedron.

4.2.3. Each vertex of the icosahedron lies on a 5–fold axis, each midpoint of an edge on a 2–fold axis, and each centroid of a face on a 3–fold axis. Using the data of the previous exercise and the method of Exercises 4.1.1, 4.1.2, and 4.1.3, you can compute the matrices for rotations of the icosahedron. (I have only done this numerically and I don't know if the matrices have a nice closed form.)

4.3. What about Reflections?

When you thought about the nature of symmetry when you first began reading this text, you might have focused especially on reflection symmetry. (People are particularly attuned to reflection symmetry since human faces and bodies are important to us.)

A reflection in $\mathbb{R}^3$ through a plane P is the transformation that leaves the points of P fixed and sends a point $\mathbf{x} \notin P$ to the point on the line through $\mathbf{x}$ and perpendicular to P, which is equidistant from P with $\mathbf{x}$ and on the opposite side of P.

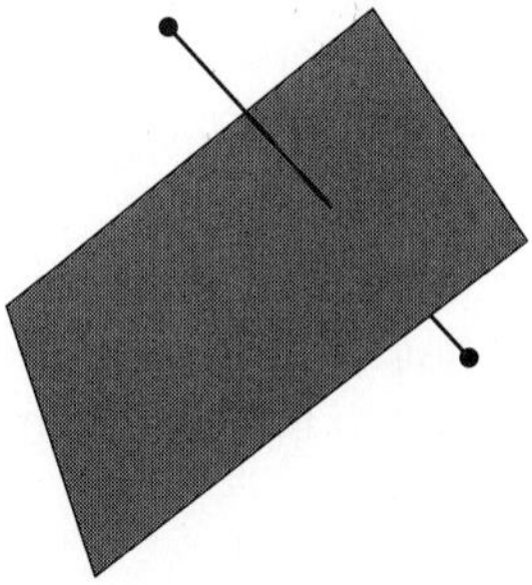

Figure 4.3.1. A reflection.

For a plane P through the origin in $\mathbb{R}^3$, the reflection through P is given by the following formula. Let $\boldsymbol{\alpha}$ be a unit vector perpendicular to P. For any $\mathbf{x} \in \mathbb{R}^3$, the reflection j_{α} of $\mathbf{x}$ through P is given by $j_{\alpha}(\mathbf{x}) = \mathbf{x} - 2\langle \mathbf{x}|\boldsymbol{\alpha}\rangle\boldsymbol{\alpha}$, where $\langle \cdot|\cdot \rangle$ denotes the inner product in $\mathbb{R}^3$.

In the Exercises, you are asked to verify this formula and to compute the matrix of a reflection, with respect to the standard basis of $\mathbb{R}^3$. You are also asked to find a formula for the reflection through a plane that does not pass through the origin.

A reflection that sends a geometric figure onto itself is a type of symmetry of the figure. It is not an actual motion that you could perform on a physical model of the figure, but it is an ideal motion.

Let's see how we can bring reflection symmetry into our account of the symmetries of some simple geometric figures. Consider a thickened version of our rectangular card: a rectangular brick. Place the brick with its faces parallel to the coordinate planes and with its centroid at the origin of coordinates. See Figure 4.3.2.

The rotational symmetries of the brick are the same as those of the rectangular card. There are four rotational symmetries: the nonmotion e, and the rotations r_1, r_2, and r_3 through an angle of π about the x–, y–, and z–axes. The same matrices E, R_1, R_2, and R_3 listed in Section 1.5 implement these rotations.

In addition, the reflections in each of the coordinate planes are symmetries; write j_i for $j_{\hat{\mathbf{e}}_i}$, the reflection in the plane orthogonal to the standard unit vector $\hat{\mathbf{e}}_i$. See Figure 4.3.2.

The symmetry j_i is implemented by the diagonal matrix J_i with -1 in the ith diagonal position and 1's in the other diagonal positions. For example,

$$J_2 = \begin{bmatrix} 1 & 0 & 0 \\ 0 & -1 & 0 \\ 0 & 0 & 1 \end{bmatrix}.$$

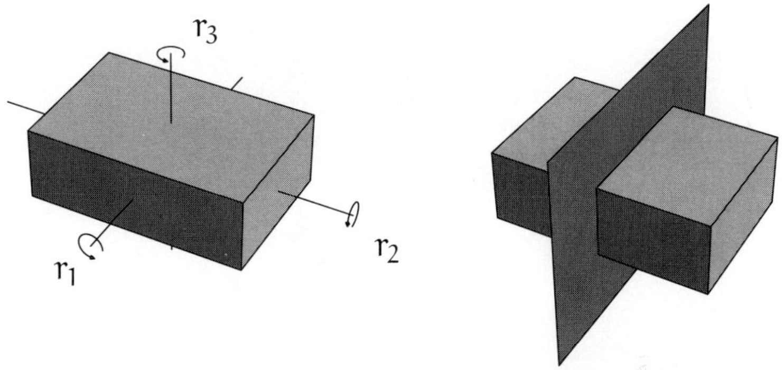

Figure 4.3.2. Rotations and reflections of a brick.

There is one more symmetry that must be considered along with these, which is neither a reflection nor a rotation but is a product of a rotation and a reflection in several different ways. This is the inversion, which sends each corner of the rectangular solid to its opposite corner. This is implemented by the matrix $-E$. Note that $-E = J_1R_1 = J_2R_2 = J_3R_3$, so the inversion is equal to j_ir_i for each i.

Having included the inversion as well as the three reflections, we again have a group. It is very easy to check closure under multiplication and inverse and to compute the multiplication table. The eight symmetries are represented by the eight 3—by—3 diagonal matrices with 1's and -1's on the diagonal; this set of matrices is clearly closed under matrix multiplication and inverse, and products of symmetries can be obtained immediately by multiplication of matrices. The product of symmetries (or of matrices) is a priori associative.

Now consider a thickened version of the square card: a square tile, which we place with its centroid at the origin of coordinates, its square faces parallel with the (x, y)–plane, and its other faces parallel with the other coordinate planes. This figure has the same rotational symmetries as does the square card, and these are implemented by the matrices given in Section 1.5. See Figure 4.3.3.

In addition, we can readily detect five reflection symmetries: For each of the five axes of symmetry, the plane perpendicular to the axis and passing through the origin is a plane of symmetry. See Figure 4.3.4

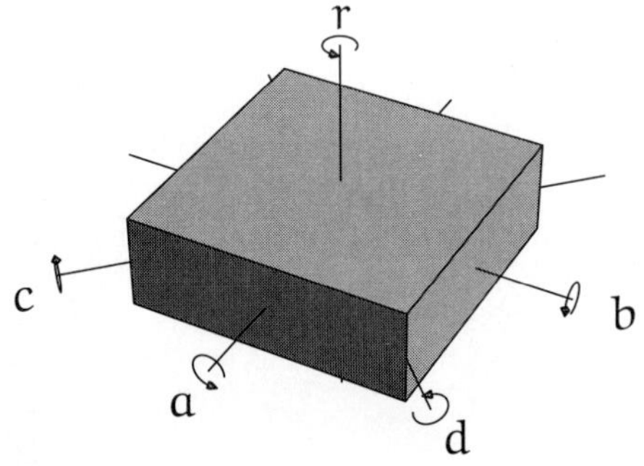

Figure 4.3.3. Rotations of the square tile.

Let us label the reflection through the plane perpendicular to the axis of the rotation a by j_a, and similarly for the other four rotation axes. The reflections j_a, j_b, j_c, j_d, and j_r are implemented by the following matrices:

$$J_a = \begin{bmatrix} -1 & 0 & 0 \\ 0 & 1 & 0 \\ 0 & 0 & 1 \end{bmatrix} \quad J_b = \begin{bmatrix} 1 & 0 & 0 \\ 0 & -1 & 0 \\ 0 & 0 & 1 \end{bmatrix} \quad J_r = \begin{bmatrix} 1 & 0 & 0 \\ 0 & 1 & 0 \\ 0 & 0 & -1 \end{bmatrix}$$

$$J_c = \begin{bmatrix} 0 & -1 & 0 \\ -1 & 0 & 0 \\ 0 & 0 & 1 \end{bmatrix} \quad J_d = \begin{bmatrix} 0 & 1 & 0 \\ 1 & 0 & 0 \\ 0 & 0 & 1 \end{bmatrix}.$$

I claim that there are three additional symmetries that we must consider along with the five reflections and eight rotations; these symmetries are neither rotations nor reflections, but are products of a rotation and a reflection. One of these we can guess from our experience with the brick, the inversion, which is obtained, for example, as the product aj_a, and which is implemented by the matrix $-E$.

If we can't find the other two by insight, we can find them by computation: If τ_1 and τ_2 are any two symmetries, then their composition product $\tau_1\tau_2$ is also a symmetry; and if the symmetries τ_1

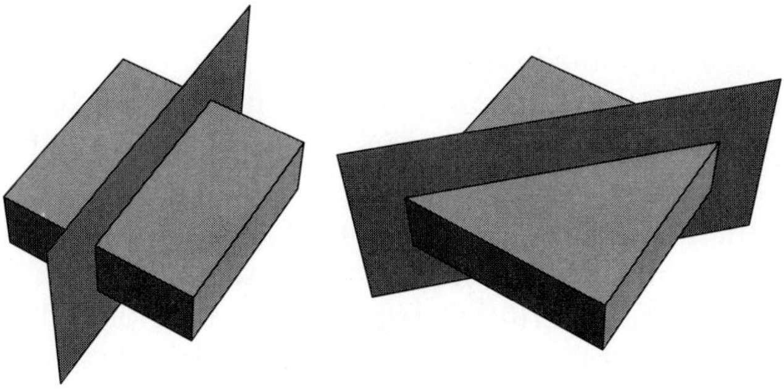

Figure 4.3.4. Reflections of the square tile.

and τ_2 are implemented by matrices F_1 and F_2, then $\tau_1\tau_2$ is implemented by the matrix product F_1F_2. So we can look for other symmetries by examining products of the matrices implementing the known symmetries.

We have 14 matrices, so we can compute a lot of products before finding something new. If you are lucky, after a bit of trial and error you will discover that the combinations to try are powers of R multiplied by the reflection matrix J_r:

$$J_rR = RJ_r = \begin{bmatrix} 0 & -1 & 0 \\ 1 & 0 & 0 \\ 0 & 0 & -1 \end{bmatrix}$$

$$J_rR^3 = R^3J_r = \begin{bmatrix} 0 & 1 & 0 \\ -1 & 0 & 0 \\ 0 & 0 & -1 \end{bmatrix}.$$

These are new matrices. What symmetries do they implement? Both are reflection-rotations, reflections in a plane followed by a rotation about an axis perpendicular to the plane.

(Here are all the combinations that, as we find by experimentation, will *not* give something new: We already know that the product of two rotations of the square tile is again a rotation. The product of two reflections appears always to be a rotation. The product of a reflection and a rotation by π about the axis perpendicular to the plane of the reflection is always the inversion.)

Now we have sixteen symmetries of the square tile, eight rotations (including the nonmotion), five reflections, one inversion, and two reflection-rotations. This set of sixteen symmetries is a group. It seems a bit daunting to work this out by computing 256 matrix products and recording the multiplication table, but we could do it in an hour or two, or we could get a computer to work out the multiplication table in no time at all.

However, there is a more thoughtful method to work this out that seriously reduces the necessary computation; this method is outlined in the Exercises.

In the next section, we will develop a more general conceptual framework in which to place this exploration, which will allow us to understand the experimental observation that for both the brick and the square tile, the total number of symmetries is twice the number of rotations. We will also see, for example, that the product of two rotations matrices is again a rotation matrix, and the product of two matrices, each of which is a reflection or a rotation–reflection, is a rotation.

Exercises 4.3

4.3.1. Verify the formula for the reflection $J_{\boldsymbol{\alpha}}$ through the plane perpendicular to $\boldsymbol{\alpha}$.

4.3.2. $J_{\boldsymbol{\alpha}}$ is linear. Find its matrix with respect to the standard basis of $\mathbb{R}^3$. (Of course, the matrix involves the coordinates of $\boldsymbol{\alpha}$.)

4.3.3. Consider a plane P that does not pass through the origin. Let $\boldsymbol{\alpha}$ be a unit normal vector to P and let $\mathbf{x}_0$ be a point on P. Find a formula (in terms of $\boldsymbol{\alpha}$ and $\mathbf{x}_0$) for the reflection of a point $\mathbf{x}$ through P. Such a reflection through a plane not passing through the origin is called an affine reflection.

4.3.4. Here is a method to determine all the products of the symmetries of the square tile. Write J for J_r, the reflection in the (x, y)–plane.

(a) The eight products αJ, where α runs through the set of eight rotation matrices of the square tile, are the eight nonrotation matrices. Which matrix corresponds to which nonrotation symmetry?

(b) Show that J commutes with the eight rotation matrices; that is, $J\alpha = \alpha J$ for all rotation matrices α.

(c) Check that the information from parts (a) and (b), together with the multiplication table for the rotational symmetries, suffices to compute all products of symmetries.

(d) Verify that the sixteen symmetries form a group.

4.3.5. Another way to work out all the products, and to understand the structure of the group of symmetries, is the following. Consider the matrix

$$S = J_c = \begin{bmatrix} 0 & 1 & 0 \\ 1 & 0 & 0 \\ 0 & 0 & 1 \end{bmatrix}.$$

(a) The set $\mathcal{D}$ of eight diagonal matrices with ± 1's on the diagonal is a subset of the matrices representing symmetries of the square tile. This set of diagonal matrices is closed under matrix multiplication. Show that every symmetry matrix for the square tile is either in $\mathcal{D}$ or is a product DS, where $D \in \mathcal{D}$.

(b) For each $D \in \mathcal{D}$, there is a $D' \in \mathcal{D}$ (which is easy to compute) that satisfies $SD = D'S$, or, equivalently, $SDS = D'$. Find the rule for determining D' from D, and use this to show how all products can be computed. Compare the results with those of the previous exercise.

4.4. Linear Isometries

The main purpose of this section is to investigate the linear isometries of three–dimensional space. However, much of the work can be done without much extra effort in n-dimensional space.

Consider Euclidean n-space, $\mathbb{R}^n$ with the usual inner product $\langle \mathbf{x}|\mathbf{y}\rangle = \sum_{i=1}^n x_i y_i$, norm $||\mathbf{x}|| = \langle \mathbf{x}, \mathbf{x}\rangle^{1/2}$, and distance function $d(\mathbf{x}, \mathbf{y}) = ||\mathbf{x} - \mathbf{y}||$. Recall that transposes of matrices and inner products are related by

$$\langle A\mathbf{x}, \mathbf{y}\rangle = \langle \mathbf{x}, A^t \mathbf{y}\rangle$$

for all n–by–n matrices A, and all $\mathbf{x}, \mathbf{y} \in \mathbb{R}^n$.

Definition 4.4.1. An *isometry* of $\mathbb{R}^n$ is a map $\tau : \mathbb{R}^n \longrightarrow \mathbb{R}^n$ that preserves distance, $d(\tau(\mathbf{a}), \tau(\mathbf{b})) = d(\mathbf{a}, \mathbf{b})$, for all $\mathbf{a}, \mathbf{b} \in \mathbb{R}^n$.

You are asked to show in the Exercises that the set of isometries is a group, and the set of isometries τ satisfying $\tau(\mathbf{0}) = \mathbf{0}$ is a subgroup.

Lemma 4.4.2. *Let τ be an isometry of $\mathbb{R}^n$ such that $\tau(\mathbf{0}) = \mathbf{0}$. Then $\langle \tau(\mathbf{a}), \tau(\mathbf{b})\rangle = \langle \mathbf{a}, \mathbf{b}\rangle$ for all $\mathbf{a}, \mathbf{b} \in \mathbb{R}^n$.*

Proof. Since $\tau(\mathbf{0}) = \mathbf{0}$, τ preserves norm as well as distance, $||\tau(\mathbf{x})|| = ||\mathbf{x}||$ for all $\mathbf{x}$. But since $\langle \mathbf{a}, \mathbf{b}\rangle = (1/2)(||\mathbf{a}||^2 + ||\mathbf{b}||^2 - d(\mathbf{a}, \mathbf{b})^2)$, it follows that τ also preserves inner products. ■

Remark 4.4.3. Recall the following property of orthonormal bases of $\mathbb{R}^n$: If $\mathbb{F} = \{\mathbf{f}_i\}$ is an orthonormal basis, then the expansion of a vector $\mathbf{x}$ with respect to $\mathbb{F}$ is $\mathbf{x} = \sum_i \langle \mathbf{x}, \mathbf{f}_i \rangle \mathbf{f}_i$. If $\mathbf{x} = \sum_i x_i \mathbf{f}_i$ and $\mathbf{y} = \sum_i y_i \mathbf{f}_i$, then $\langle \mathbf{x}, \mathbf{y} \rangle = \sum_i x_i y_i$.

Definition 4.4.4. A matrix A is said to be *orthogonal* if A^t is the inverse of A.

Exercise 4.4.3 gives another characterization of orthogonal matrices.

Lemma 4.4.5. *If A is an orthogonal matrix, then the linear map* $\mathbf{x} \mapsto A\mathbf{x}$ *is an isometry.*

Proof. For all $\mathbf{x} \in \mathbb{R}^n$, $||A\mathbf{x}||^2 = \langle A\mathbf{x}, A\mathbf{x} \rangle = \langle A^t A\mathbf{x}, \mathbf{x} \rangle = \langle \mathbf{x}, \mathbf{x} \rangle = ||\mathbf{x}||^2$. ■

Lemma 4.4.6. *Let* $\mathbb{F} = \{\mathbf{f}_i\}$ *be an orthonormal basis. A set* $\{\mathbf{v}_i\}$ *is an orthonormal basis if and only if the matrix* $A = [a_{ij}] = [\langle \mathbf{v}_i, \mathbf{f}_j \rangle]$ *is an orthogonal matrix.*

Proof. The ith row of A is the coefficient vector of $\mathbf{v}_i$ with respect to the orthonormal basis $\mathbb{F}$. According to Remark 4.4.3, the inner product of $\mathbf{v}_i$ and $\mathbf{v}_j$ is the same as the inner product of the ith and j^{th} rows of A. Hence, the $\mathbf{v}_i$'s are an orthonormal basis if and only if the rows of A are an orthonormal basis. By Exercise 4.4.3, this is true if and only if A is orthogonal. ■

Theorem 4.4.7. *Let* τ *be a linear map on* $\mathbb{R}^n$. *The following are equivalent:*

(a) τ *is an isometry.*
(b) τ *preserves inner products.*
(c) *For some orthonormal basis* $\{\mathbf{f}_1, \ldots, \mathbf{f}_n\}$ *of* $\mathbb{R}^n$, *the set* $\{\tau(\mathbf{f}_1), \ldots, \tau(\mathbf{f}_n)\}$ *is also orthonormal.*
(d) *For every orthonormal basis* $\{\mathbf{f}_1, \ldots, \mathbf{f}_n\}$ *of* $\mathbb{R}^n$, *the set* $\{\tau(\mathbf{f}_1), \ldots, \tau(\mathbf{f}_n)\}$ *is also orthonormal.*
(e) *The matrix of* τ *with respect to some orthonormal basis is orthogonal.*
(f) *The matrix of* τ *with respect to every orthonormal basis is orthogonal.*

Proof. Condition (a) implies (b) by Lemma 4.4.2. The implications (b) $\Rightarrow$ (d) $\Rightarrow$ (c) are trivial, and (c) implies (e) by Lemma 4.4.6. Now assume the matrix A of τ with respect to the orthonormal basis $\mathbb{F}$ is orthogonal. Let $U(\mathbf{x})$ be the coordinate vector of $\mathbf{x}$ with respect to $\mathbb{F}$; then by Remark 4.4.3, U is an isometry. The linear map τ is $U^{-1} \circ M_A \circ U$, where M_A means multiplication by A. By Lemma 4.4.5, M_A is an isometry, so τ is an isometry. Thus (e) $\Rightarrow$(a). Similarly, we have (a) $\Rightarrow$(b) $\Rightarrow$(d) $\Rightarrow$(f) $\Rightarrow$(a). ■

Proposition 4.4.8. *The determinant of an orthogonal matrix is* ± 1.

Proof. Let A be an orthogonal matrix. Since $\det(A) = \det(A^t)$, we have

$$1 = \det(E) = \det(A^t A) = \det(A^t)\det(A) = \det(A)^2.$$

■

Remark 4.4.9. If τ is a linear transformation of $\mathbb{R}^n$ and A and B are the matrices of τ with respect to two different bases of $\mathbb{R}^n$, then $\det(A) = \det(B)$, because A and B are related by a similarity, $A = VBV^{-1}$, where V is a change of basis matrix. Therefore, we can, without ambiguity, define the determinant of τ to be the determinant of the matrix of τ with respect to any basis.

Corollary 4.4.10. *The determinant of a linear isometry is* ± 1.

Since $\det(AB) = \det(A)\det(B)$, $\det : O(n, \mathbb{R}) \longrightarrow \{1, -1\}$ is a group homomorphism.

Definition 4.4.11. The set of orthogonal n–by–n matrices with determinant equal to 1 is called the special orthogonal group and denoted $SO(n, \mathbb{R})$.

Evidently, the special orthogonal group $SO(n, \mathbb{R})$ is a normal subgroup of the orthogonal group of index 2, since it is the kernel of $\det : O(n, \mathbb{R}) \longrightarrow \{1, -1\}$.

We next restrict our attention to three-dimensional space and explore the role of rotations and orthogonal reflections in the group of linear isometries.

Recall from Section 4.3 that for any unit vector $\boldsymbol{\alpha}$ in $\mathbb{R}^3$, the plane $P_{\boldsymbol{\alpha}}$ is $\{\mathbf{x} : \langle \mathbf{x}, \boldsymbol{\alpha} \rangle = 0\}$. The orthogonal reflection in $P_{\boldsymbol{\alpha}}$ is the linear map $j_{\boldsymbol{\alpha}} : \mathbf{x} \mapsto \mathbf{x} - 2\langle \mathbf{x}, \boldsymbol{\alpha} \rangle \boldsymbol{\alpha}$. The orthogonal reflection $j_{\boldsymbol{\alpha}}$ fixes $P_{\boldsymbol{\alpha}}$ pointwise and sends $\boldsymbol{\alpha}$ to $-\boldsymbol{\alpha}$. Let $J_{\boldsymbol{\alpha}}$ denote the matrix of $j_{\boldsymbol{\alpha}}$ with respect to the standard basis of $\mathbb{R}^3$. Call a matrix of the form $J_{\boldsymbol{\alpha}}$ a *reflection matrix*.

Let's next sort out the role of reflections and rotations in $SO(2, \mathbb{R})$. Consider an orthogonal matrix $\begin{bmatrix} \alpha & \gamma \\ \beta & \delta \end{bmatrix}$. Orthogonality implies that $\begin{bmatrix} \alpha \\ \beta \end{bmatrix}$ is a unit vector and $\begin{bmatrix} \gamma \\ \delta \end{bmatrix} = \pm \begin{bmatrix} -\beta \\ \alpha \end{bmatrix}$. The vector $\begin{bmatrix} \alpha \\ \beta \end{bmatrix}$ can be written as $\begin{bmatrix} \cos(\theta) \\ \sin(\theta) \end{bmatrix}$ for some angle θ, so the orthogonal matrix has the form $\begin{bmatrix} \cos(\theta) & -\sin(\theta) \\ \sin(\theta) & \cos(\theta) \end{bmatrix}$ or $\begin{bmatrix} \cos(\theta) & \sin(\theta) \\ \sin(\theta) & -\cos(\theta) \end{bmatrix}$. The matrix $R_\theta = \begin{bmatrix} \cos(\theta) & -\sin(\theta) \\ \sin(\theta) & \cos(\theta) \end{bmatrix}$ is the matrix of the rotation through an angle θ, and has determinant equal to 1. The matrix $\begin{bmatrix} \cos(\theta) & \sin(\theta) \\ \sin(\theta) & -\cos(\theta) \end{bmatrix}$ equals $R_\theta J$, where $J = \begin{bmatrix} 1 & 0 \\ 0 & -1 \end{bmatrix}$ is the reflection matrix $J = J_{\hat{\mathbf{e}}_2}$. The determinant of $R_\theta J$ is equal to -1.

Now consider the situation in three dimensions. Any real 3–by–3 matrix has a real eigenvalue, since the characteristic polynomial is cubic with real coefficients. A real eigenvalue of an orthogonal matrix must be ± 1 because the matrix implements an isometry.

Lemma 4.4.12. *Any element of* $SO(3, \mathbb{R})$ *has* $+1$ *as an eigenvalue.*

Proof. Let $A \in SO(3, \mathbb{R})$, let τ be the linear isometry $\mathbf{x} \mapsto A\mathbf{x}$, and let $\mathbf{v}$ be an eigenvector with eigenvalue ± 1. If the eigenvalue is $+1$, there is nothing to do. So suppose the eigenvalue is -1. The plane $P = P_{\mathbf{v}}$ orthogonal to $\mathbf{v}$ is invariant under A, because if $\mathbf{x} \in P$, then $\langle \mathbf{v}, A\mathbf{x} \rangle = -\langle A\mathbf{v}, A\mathbf{x} \rangle = -\langle \mathbf{v}, \mathbf{x} \rangle = 0$. The restriction of τ to P is also orthogonal, and since $1 = \det(\tau) = (-1)(\det(\tau_{|P})$, $\tau_{|P}$ must be a reflection. But a reflection has an eigenvalue of $+1$, so in any case A has an eigenvalue of $+1$. ■

Proposition 4.4.13. *An element* $A \in O(3, \mathbb{R})$ *has determinant* 1 *if and only if* A *implements a rotation.*

Proof. Suppose $A \in SO(3, \mathbb{R})$. Let τ denote the corresponding linear isometry $\mathbf{x} \mapsto A\mathbf{x}$. By the lemma, A has an eigenvector $\mathbf{v}$ with eigenvalue 1. The plane P orthogonal to $\mathbf{v}$ is invariant under τ, and $\det(\tau_{|P}) = \det(\tau) = 1$, so $\tau_{|P}$ is a rotation of P. Hence, τ is a rotation about the line spanned by $\mathbf{v}$. On the other hand, if τ is a rotation, then the matrix of τ with respect to an appropriate orthonormal basis has the form

$$\begin{bmatrix} 1 & 0 & 0 \\ 0 & \cos(\theta) & -\sin(\theta) \\ 0 & \sin(\theta) & \cos(\theta) \end{bmatrix},$$

so τ has determinant 1. ■

Proposition 4.4.14. *An element of* $O(3, \mathbb{R}) \setminus SO(3, \mathbb{R})$ *implements either an orthogonal reflection, or a reflection-rotation, that is, the product of a reflection* j_α *and a rotation about the line spanned by* $\boldsymbol{\alpha}$*.*

Proof. Suppose $A \in O(3, \mathbb{R}) \setminus SO(3, \mathbb{R})$. Let τ denote the corresponding linear isometry $\mathbf{x} \mapsto A\mathbf{x}$. Let $\mathbf{v}$ be an eigenvector of A with eigenvalue ± 1. If the eigenvalue is 1, then the restriction of τ to the plane P orthogonal to $\mathbf{v}$ has determinant -1, so is a reflection. Then τ itself is a reflection. If the eigenvalue is -1, then the restriction of τ to P has determinant 1, so is a rotation. In this case τ is the product of the reflection $j_{\mathbf{v}}$ and a rotation about the line spanned by $\mathbf{v}$. ■

Exercises 4.4

4.4.1.

(a) Show that j_α is isometric.
(b) Show that $\det(j_\alpha) = -1$.
(c) If τ is a linear isometry, show that $\tau j_\alpha \tau^{-1} = j_{\tau(\alpha)}$.
(d) If A is any orthogonal matrix, show that $AJ_\alpha A^{-1} = J_{A\alpha}$.
(e) Conclude that the matrix of j_α with respect to any orthonormal basis is a reflection matrix.

4.4.2. Show that the set of isometries of $\mathbb{R}^n$ is a group. Show that the set of isometries τ satisfying $\tau(\mathbf{0}) = \mathbf{0}$ is a subgroup.

4.4.3. Show that the following are equivalent for a matrix A:

(a) A is orthogonal.
(b) The columns of A are an orthonormal basis.
(c) The rows of A are an orthonormal basis.

4.4.4.

(a) Show that the matrix $R_{2\theta}J = R_\theta J R_{-\theta}$ is the matrix of the reflection in the line spanned by

$$\begin{bmatrix} \cos\theta \\ \sin\theta \end{bmatrix}.$$

Write $J_\theta = R_{2\theta} J$.

(b) The reflection matrices are precisely the elements of $O(2, \mathbb{R})$ with determinant equal to -1.

(c) Compute $J_{\theta_1} J_{\theta_2}$.

(d) Show that any rotation matrix R_θ is a product of two reflection matrices.

4.4.5. Show that an element of $SO(3, \mathbb{R})$ is a product of two reflection matrices. A matrix of a rotation–reflection is a product of three reflection matrices. Thus any element of $O(3, \mathbb{R})$ is a product of at most three reflection matrices.

4.5. The Full Symmetry Group and Chirality

All the geometric figures in three-dimensional space that we have considered — the polygonal tiles, bricks, and the regular polyhedra — admit reflection symmetries. For "reasonable" figures, every symmetry is implemented by by a linear isometry of $\mathbb{R}^3$; see Proposition 1.4.1. The rotation group of a geometric figure is the group of $g \in SO(3, \mathbb{R})$ that leave the figure invariant. The full symmetry group is the group of $g \in O(3, \mathbb{R})$ that leave the figure invariant. The existence of reflection symmetries means that the full symmetry group is strictly larger than the rotation group.

In the following discussion, I do not distinguish between linear isometries and their standard matrices.

Theorem 4.5.1. *Let S be a geometric figure with full symmetry group G and rotation group $\mathcal{R} = G \cap SO(3, \mathbb{R})$. Suppose S admits a reflection symmetry J. Then $\mathcal{R}$ is an index 2 subgroup of G and $G = \mathcal{R} \cup \mathcal{R}J$.*

Proof. Suppose A is an element of $G \setminus \mathcal{R}$. Then $\det(A) = -1$ and $\det(AJ) = 1$, so $AJ \in \mathcal{R}$. Thus $A = (AJ)J \in \mathcal{R}J$. ■

For example, the full symmetry group of the cube has 48 elements. What group is it? We could compute an injective homomorphism of the full symmetry group into S_6 using the action on the faces of the cube, or into S_8 using the action on the vertices. A more efficient method is given in Exercise 4.5.1; the result is that the full symmetry group is $S_4 \times \mathbb{Z}_2$.

Are there geometric figures with a nontrivial rotation group but with no reflection symmetries? Such a figure must exhibit chirality or "handedness"; it must come in two versions that are mirror images of each other. Consider, for example, a belt that is given n half twists ($n \geq 2$) and then fastened. There are two mirror image versions, with right–hand and left–hand twists. Either version has rotation group D_n, the rotation group of the n–gon, but no reflection symmetries. Reflections convert the right–handed version into the left–handed version. See Figure 4.5.1.

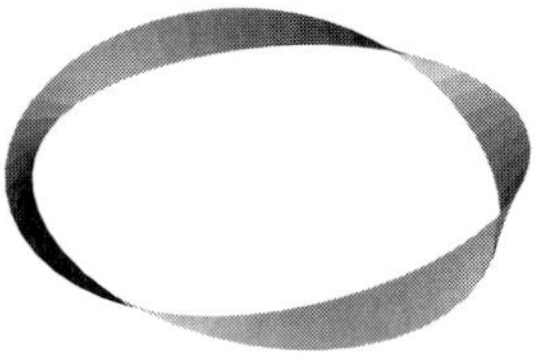

Figure 4.5.1. Twisted band with symmetry D_3.

There exist chiral convex polyhedra with two types of regular polygonal faces, for example, the "snubcube," shown in Figure 4.5.2.

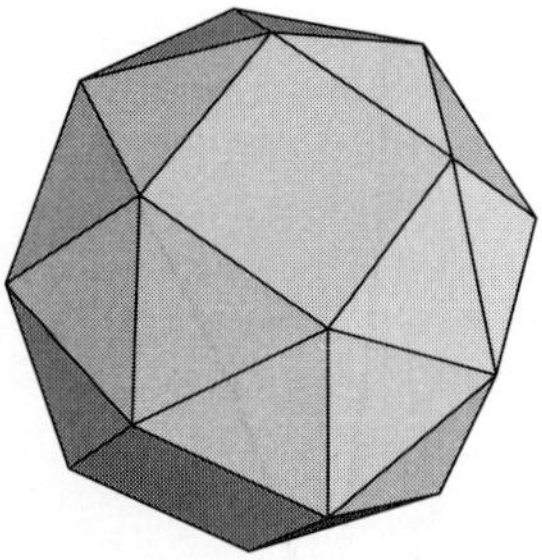

Figure 4.5.2. Snubcube.

Exercises 4.5

4.5.1. Let G denote the full symmetry group of the cube and $\mathcal{R}$ the rotation group. The inversion $i : x \mapsto -x$ with matrix $-E$ is an element of $G \setminus \mathcal{R}$, so $G = \mathcal{R} \cup \mathcal{R}i$. Observe that $i^2 = 1$, and that for any rotation r, $ir = ri$. Conclude that $G \cong S_4 \times \mathbb{Z}_2$.

4.5.2. Show that the same trick works for the dodecahedron, and that the full symmetry group is isomorphic to $A_5 \times \mathbb{Z}_2$. Show that this group is not isomorphic to S_5.

4.5.3. Show that the full symmetry group of the tetrahedron is S_4.

4.5.4. What is the full symmetry group of a brick?

4.5.5. What is the full symmetry group of a square tile?

4.5.6. What is the full symmetry group of a tile in the shape of a regular n–gon?

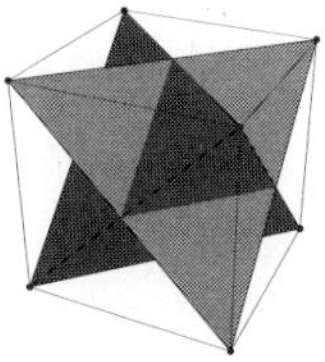

CHAPTER 5

Actions of Groups

5.1. Group Actions on Sets

We have observed that the symmetry group of the cube acts on various geometric sets associated with the cube, the set of vertices, the set of diagonals, the set of edges, and the set of faces. In this section we look more closely at the concept of a group acting on a set.

Definition 5.1.1. An action of a group G on a set X is a homomorphism from G into $\mathrm{Sym}(X)$.

Let φ be an action of G on X. For each $g \in G$, $\varphi(g)$ is a bijection of X. The homomorphism property of φ means that for $x \in X$ and $g_1, g_2 \in G$, $\varphi(g_2 g_1)(x) = \varphi(g_2)(\varphi(g_1)(x))$. Thus, if $\varphi(g_1)$ sends x to x', and $\varphi(g_2)$ sends x' to x'', then $\varphi(g_2 g_1)$ sends x to x''. When it cannot cause ambiguity, it is convenient to write gx for $\varphi(g)(x)$. With this simplified notation, the homomorphism property reads like a mixed associative law: $(g_2 g_1)x = g_2(g_1 x)$.

Lemma 5.1.2. *Given an action of* G *on* X, *define a relation on* X *by* $x \sim y$ *if there exists a* $g \in G$ *such that* $gx = y$. *This relation is an equivalence relation.*

Proof. Exercise 5.1.1. ■

Definition 5.1.3. Given an action of G on X, the equivalence classes of the equivalence relation associated to the action are called the *orbits* of the action. The orbit of x will be denoted $\mathcal{O}(x)$.

Example 5.1.4. Any group G acts on itself by left multiplication. That is, for $g \in G$ and $x \in G$, gx is just the usual product of g and x

in G. The homomorphism, or associative, property of the action is just the associative law of G. There is only one orbit. The action of G on itself by left multiplication is often called the *left regular action*.

Definition 5.1.5. An action of G on X is called *transitive* if there is only one orbit. That is, for any two elements $x, x' \in X$, there is a $g \in G$ such that $gx = x'$. A subgroup of $\mathrm{Sym}(X)$ is called *transitive* if it acts transitively on X.

Example 5.1.6. Let G be any group and H any subgroup. Then G acts on the set G/H of left cosets of H in G by left multiplication, $g(aH) = (ga)H$. The action is transitive.

Example 5.1.7. Any group G acts on itself by conjugation: For $g \in G$, define $c_g \in \mathrm{Aut}(G) \subseteq \mathrm{Sym}(G)$ by $c_g(x) = gxg^{-1}$. It was shown in Exercise 2.7.7 that the map $g \mapsto c_g$ is a homomorphism. The orbits of this action are called the *conjugacy classes* of G; two elements x and y are *conjugate* if there is a $g \in G$ such that $gxg^{-1} = y$. For example, it was shown in Exercise 2.4.14 that two elements of the symmetric group S_n are conjugate if, and only if, they have the same cycle structure.

Example 5.1.8. Let G be any group and X the set of subgroups of G. Then G acts on X by conjugation, $c_g(H) = gHg^{-1}$. Two subgroups in the same orbit are called *conjugate*. (You are asked in Exercise 5.1.5 to verify that this is an action.)

In Section 5.4, we shall pursue the idea of classifying groups of small order, up to isomorphism. Another organizational scheme for classifying small groups is to classify those that act transitively on small sets, that is, to classify transitive subgroups of S_n for small n.

Example 5.1.9. (See Exercise 5.1.9.) The transitive subgroups of S_3 are exactly S_3 and A_3.

Example 5.1.10. (See Exercise 5.1.20.) The transitive subgroups of S_4 are

(a) S_4
(b) A_4, which is normal
(c) D_4 (three conjugate copies)
(d) $\mathcal{V} = \{e, (12)(34), (13)(24), (14)(23)\}$ (which is normal)
(e) $\mathbb{Z}_4$ (three conjugate copies)

Definition 5.1.11. Let G act on X. For $x \in X$, the stabilizer of x in G is $\text{Stab}(x) = \{g \in G : gx = x\}$. If it is necessary to specify the group, we will write $\text{Stab}_G(x)$.

Lemma 5.1.12. *For any action of a group G on a set X and any $x \in X$, $\text{Stab}(x)$ is a subgroup of G.*

Proof. Exercise 5.1.6. ■

Proposition 5.1.13. *Let G act on X, and let $x \in X$. Then $\psi : a\text{Stab}(x) \mapsto ax$ defines a bijection from $G/\text{Stab}(x)$ onto $\mathcal{O}(x)$, which satisfies*

$$\psi(g(a\text{Stab}(x))) = g\psi(a\text{Stab}(x))$$

for all $g, a \in G$.

Proof. Note that $a\text{Stab}(x) = b\text{Stab}(x) \Leftrightarrow b^{-1}a \in \text{Stab}(x) \Leftrightarrow b^{-1}ax = x \Leftrightarrow bx = ax$. This calculation shows that ψ is well defined and injective. If $y \in \mathcal{O}(x)$, then there exists $a \in G$ such that $ax = y$, so $\psi(a\text{Stab}(x)) = y$; thus ψ is surjective as well. The relation

$$\psi(g(a\text{Stab}(x))) = g\psi(a\text{Stab}(x))$$

for all $g, a \in G$ is evident from the definition of ψ. ■

Corollary 5.1.14. *Suppose G is finite. Then*

$$|\mathcal{O}(x)| = [G : \text{Stab}(x)] = \frac{|G|}{|\text{Stab}(x)|}.$$

In particular, $|\mathcal{O}(x)|$ divides $|G|$.

Proof. This follows immediately from Proposition 5.1.13 and Lagrange's theorem. ■

Definition 5.1.15. Consider the action of a group G on its subgroups by conjugation. The stabilizer of a subgroup H is called the *normalizer* of H in G and denoted $N_G(H)$.

According to Corollary 5.1.14, if G is finite, then the number of distinct subgroups xHx^{-1} for $x \in G$ is

$$[G : N_G(H)] = \frac{|G|}{|N_G(H)|}.$$

Since (clearly) $N_G(H) \supseteq H$, the number of such subgroups is no more than $[G : H]$.

Definition 5.1.16. Consider the action of a group G on itself by conjugation. The stabilizer of an element $g \in G$ is called the *centralizer* of g in G and denoted $\text{Cent}(g)$, or when it is necessary to specify the group by $\text{Cent}_G(x)$.

Again, according to the corollary the size of the conjugacy class of g, that is, of the orbit of g under conjugacy, is

$$[G : \text{Cent}(g)] = \frac{|G|}{|\text{Cent}(g)|}.$$

Example 5.1.17. What is the size of each conjugacy class in the symmetric group S_4?

Recall that two elements of a symmetric group S_n are conjugate in S_n precisely if they have the same cycle structure (i.e., if when written as a product of disjoint cycles, they have the same number of cycles of each length). Cycle structures are parameterized by partitions of n.

(a) There is only one element of cycle structure 1^4, namely, the identity.

(b) There are six 2–cycles. We can compute this number by dividing the size of the group, 24, by the size of the centralizer of any particular 2-cycle. The centralizer of the 2-cycle $(1,2)(3)(4)$ is the set of permutations that leave invariant the sets $\{1,2\}$ and $\{3,4\}$, and the size of the centralizer is 4. So the number of 2–cycles is $24/4 = 6$.

(c) There are three elements with cycle structure 2^2. In fact, the centralizer of $(1,2)(3,4)$ is the group of size 8 generated by $\{(1,2),(3,4),(1,3)(2,4)\}$. Hence the number of elements with cycle structure 2^2 is $24/8 = 3$.

(d) There are eight 3–cycles. The centralizer of the 3-cycle $(1,2,3)$ is the group generated by $(1,2,3)$, and has size 3. Therefore, the number of 3–cycles is $24/3 = 8$.

(e) There are six 4–cycles. The computation is similar to that for 3–cycles.

Example 5.1.18. How large is the conjugacy class of elements with cycle structure 4^3 in S_{12}? The centralizer in S_{12} of the element

$$(1,2,3,4)(5,6,7,8)(9,10,11,12)$$

is the semidirect product of $\mathbb{Z}_4 \times \mathbb{Z}_4 \times \mathbb{Z}_4$ and S_3. Here $\mathbb{Z}_4 \times \mathbb{Z}_4 \times \mathbb{Z}_4$ is generated by the commuting 4–cycles

$$\{(1,2,3,4),\ (5,6,7,8),\ (9,10,11,12)\},$$

while S_3 acts by permutations of the set of three ordered 4-tuples

$$\{(1,2,3,4),(5,6,7,8),(9,10,11,12)\}.$$

Thus the size of the centralizer is $4^3 \times 6$, and the size of the conjugacy class is $\dfrac{12!}{4^3 6} = 1247400$.

The *kernel* of an action of a group G on a set X is the kernel of the corresponding homomorphism $\varphi : G \to \mathrm{Sym}(X)$; that is,

$$\{g \in G : gx = x \quad \text{for all} \quad x \in X\}.$$

According to the general theory of homomorphisms, the kernel is a normal subgroup of G. The kernel is evidently the intersection of the stabilizers of all $x \in X$. For example, the kernel of the action of G on itself by conjugation is the center of G.

Application: Counting Formulas.

It is possible to obtain a number of well–known counting formulas by means of the proposition and its corollary.

Example 5.1.19. The number of k-element subsets of a set with n elements is

$$\binom{n}{k} = \frac{n!}{k!(n-k)!}.$$

Proof. Let X be the family of k element subsets of $\{1, 2, \dots, n\}$. S_n acts transitively on X by $\pi\{a_1, a_2, \dots, a_k\} = \{\pi(a_1), \pi(a_2), \dots, \pi(a_k)\}$. (Verify!) The stabilizer of $x = \{1, 2, \dots, k\}$ is $S_k \times S_{n-k}$, the group of permutations that leaves invariant the sets $\{1, 2, \dots, k\}$ and $\{k + 1, \dots, n\}$. Therefore, the number of k-element subsets is the size of the orbit of x, namely,

$$\frac{|S_n|}{|S_k \times S_{n-k}|} = \frac{n!}{k!(n-k)!}.$$

■

Example 5.1.20. The number of ordered sequences of k items chosen from a set with n elements is

$$\frac{n!}{(n-k)!}.$$

Proof. The proof is similar to that for the previous example. This time let S_n act on the set of ordered sequences of k-elements from $\{1, 2, \ldots, n\}$. ■

Example 5.1.21. The number of sequences of r_1 1's, r_2 2's, and so forth, up to r_k k's, is

$$\frac{(r_1 + r_2 + \cdots + r_k)!}{r_1!r_2!\ldots r_k!}.$$

Proof. Let $n = r_1 + r_2 + \cdots + r_k$. S_n acts transitively on sequences of r_1 1's, r_2 2's, ..., and r_k k's. The stabilizer of

$$(1, \ldots, 1, 2, \ldots, 2, \ldots, k, \ldots, k)$$

with r_1 consecutive 1's, r_2 consecutive 2's, and so on is

$$S_{r_1} \times S_{r_2} \times \cdots \times S_{r_k}.$$

■

Example 5.1.22. How many distinct arrangements are there of the letters of the word MISSISSIPPI? There are 4 I's, 4 S's, 2 P's and 1 M in the word, so the number of arrangements of the letters is $\dfrac{11!}{2!\,4!\,4!} = 34650.$

Exercises 5.1

5.1.1. Let the group G act on a set X. Define a relation on X by $x \sim y$ if, and only if, there is a $g \in G$ such that $gx = y$. Show that this is an equivalence relation on X, and the orbit (equivalence class) of $x \in X$ is $Gx = \{gx : g \in G\}$.

5.1.2. Verify all the assertions made in Example 5.1.4.

5.1.3. The symmetric group S_n acts naturally on the set $\{1, 2, \ldots, n\}$. Let $\sigma \in S_n$. Show that the cycle decomposition of σ can be recovered by considering the orbits of the action of the cyclic subgroup $\langle\sigma\rangle$ on $\{1, 2, \ldots, n\}$.

5.1.4. Verify the assertions made in Example 5.1.6.

5.1.5. Verify that any group G acts on the set X of its subgroups by $c_g(H) = gHg^{-1}$. Compute the example of S_3 acting by conjugation of the set X of (six) subgroups of S_3. Verify that there are four orbits, three of which consist of a single subgroup, and one of which contains three subgroups.

5.1.6. Let G act on X, and let $x \in X$. Verify that $\mathrm{Stab}(x)$ is a subgroup of G. Verify that if x and y are in the same orbit, then the subgroups $\mathrm{Stab}(x)$ and $\mathrm{Stab}(y)$ are conjugate subgroups.

5.1.7. Let $H = \{e, (1,2)\} \subseteq S_3$. Find the orbit of H under conjugation by G, the stabilizer of H in G, and the family of left cosets of the stabilizer in G, and verify explicitly the bijection between left cosets of the stabilizer and conjugates of H.

5.1.8. Show that $N_G(aHa^{-1}) = aN_G(H)a^{-1}$ for $a \in G$.

5.1.9. Show that the transitive subgroups of S_3 are exactly S_3 and A_3.

5.1.10. Suppose A is a subgroup of $N_G(H)$. Show that AH is a subgroup of $N_G(H)$, $AH = HA$, and

$$|AH| = \frac{|A|\,|H|}{|A \cap H|}.$$

Hint: H is normal in $N_G(H)$.

5.1.11. Let A be a subgroup of $N_G(H)$. Show that there is a homomorphism $\alpha : A \longrightarrow \mathrm{Aut}(H)$ such that $ah = \alpha(a)(h)a$ for all $a \in A$ and $h \in H$. The product in HA is determined by

$$(h_1a_1)(h_2a_2) = h_1\alpha(a_1)(h_2)a_1a_2.$$

5.1.12. Count the number of ways to arrange four red beads, three blue beads, and two yellow beads on a straight wire.

5.1.13. How may elements are there in S_8 of cycle structure 4^2?

5.1.14. How may elements are there in S_8 of cycle structure 2^4?

5.1.15. How may elements are there in S_{12} of cycle structure $3^2 2^3$?

5.1.16. How may elements are there in S_{26} of cycle structure $4^3 3^2 2^4$?

5.1.17. Let $r_1 > r_2 > \ldots > r_s \geq 1$ and let $m_i \in \mathbb{N}$ for $1 \leq i \leq s$, such that $\sum_i m_i r_i = n$. How many elements does S_n have with m_1 cycles of length r_1, m_2 cycles of length r_2, and so on?

5.1.18. Verify that the formula

$$\pi\{a_1, a_2, \ldots, a_k\} = \{\pi(a_1), \pi(a_2), \ldots, \pi(a_k)\}$$

does indeed define an action of S_n on the set X of k-element subsets of $\{1, 2, \ldots, n\}$. (See Example 5.1.19.)

5.1.19. Give the details of the proof of Example 5.1.20. In particular, define an action of S_n on the set of ordered sequences of k-elements from $\{1, 2, \ldots, n\}$, and verify that it is indeed an action. Show that the action is transitive. Calculate the size of the stabilizer of a particular k-element sequence.

5.1.20. Show that the transitive subgroups of S_4 are

- S_4
- A_4, which is normal
- $D_4 = \langle(1\,2\,3\,4), (1\,2)(3\,4)\rangle$, and two conjugate subgroups
- $V = \{e, (12)(34), (13)(24), (14)(23)\}$, which is normal
- $\mathbb{Z}_4 = \langle(1\,2\,3\,4), (1\,2)(3\,4)\rangle$, and two conjugate subgroups

5.2. Group Actions—Counting Orbits

How many different necklaces can we make from four red beads, three white beads and two yellow beads? Two arrangements of beads on a circular wire must be counted as the same necklace if one can be obtained from the other by sliding the beads around the wire or turning the wire over. So what we actually need to count is orbits of the action of the dihedral group D_9 (symmetries of the nonagon) on the

$$\frac{9!}{4!3!2!}$$

arrangements of the beads.

Let's consider a simpler example that we can work out by inspection:

Example 5.2.1. Consider necklaces made of two blue and two white beads. There are six arrangements of the beads at the vertices of the square, but only two orbits under the action of the dihedral group D_4, namely, that with two blue beads adjacent and that with the two blue beads at opposite corners. One orbit contains four arrangements and the other two arrangements.

We see from this example that the orbits will have different sizes, so we cannot expect the answer to the problem simply to be some divisor of the number of arrangements of beads.

In order to count orbits for the action of a finite group G on a finite set X, consider the set $F = \{(g, x) \in G \times X : gx = x\}$. For $g \in G$, let $\text{Fix}(g) = \{x \in X : gx = x\}$, and let

$$\mathbf{1}_F(g, x) = \begin{cases} 1 & \text{if } (g, x) \in F \\ 0 & \text{otherwise.} \end{cases}$$

We can count F in two different ways:

$$|F| = \sum_{x \in X} \sum_{g \in G} \mathbf{1}_F(x, g) = \sum_{x \in X} |\text{Stab}(x)|$$

and

$$|F| = \sum_{g \in G} \sum_{x \in X} \mathbf{1}_F(x, g) = \sum_{g \in G} |\mathrm{Fix}(g)|.$$

Dividing by $|G|$, we get

$$\frac{1}{|G|} \sum_{g \in G} |\mathrm{Fix}|(g) = \sum_{x \in X} \frac{|\mathrm{Stab}(x)|}{|G|} = \sum_{x \in X} \frac{1}{|\mathcal{O}(x)|}.$$

The last sum can be decomposed into a double sum:

$$\sum_{x \in X} \frac{1}{|\mathcal{O}(x)|} = \sum_{\mathcal{O}} \sum_{x \in \mathcal{O}} \frac{1}{|\mathcal{O}|},$$

where the outer sum is over distinct orbits. But

$$\sum_{\mathcal{O}} \sum_{x \in \mathcal{O}} \frac{1}{|\mathcal{O}|} = \sum_{\mathcal{O}} \frac{1}{|\mathcal{O}|} \sum_{x \in \mathcal{O}} 1 = \sum_{\mathcal{O}} 1,$$

which is the number of orbits! Thus, we have the following result, known as Burnside's lemma.

Proposition 5.2.2. *(Burnside's lemma). Let a finite group* G *act on a finite set* X*. Then the number of orbits of the action is*

$$\frac{1}{|G|} \sum_{g \in G} |\mathrm{Fix}(g)|.$$

Example 5.2.3. Let's use this result to calculate the number of necklaces that can be made from four red beads, three white beads, and two yellow beads. X is the set of

$$\frac{9!}{4!3!2!} = 1260$$

arrangements of the beads, which we locate at the nine vertices of a nonagon. Let g be an element of D_9 and consider the orbits of $\langle g \rangle$ acting on vertices of the nonagon. An arrangement of the colored beads is fixed by g if, and only if, all vertices of each orbit of the action of $\langle g \rangle$ are of the same color. Every arrangement is fixed by e.

Let r be the rotation of $2\pi/9$ of the nonagon. For any k ($1 \le k \le 8$), r^k either has order 9, and $\langle r^k \rangle$ acts transitively on vertices, or r^k has order 3, and $\langle r^k \rangle$ has three orbits, each with three vertices. In either case, there are no fixed arrangements, since it is not possible to place beads of one color at all vertices of each orbit.

Now consider any rotation j of π about an axis through one vertex v of the nonagon and the center of the opposite edge. The subgroup $\{e, j\}$ has one orbit containing the one vertex v and four orbits containing two vertices. In any fixed arrangement, the vertex v must have a white bead. Of the remaining four orbits, two must be colored red, one white and one yellow; there are

$$\frac{4!}{2!1!1!} = 12$$

ways to do this. Thus, j has 12 fixed points in X. Since there are 9 such elements, there are

$$\frac{1}{|G|}\sum_{g\in G}|\text{Fix}(g)| = \frac{1}{18}(1260 + 9(12)) = 76$$

possible necklaces.

Example 5.2.4. How many different necklaces can be made with nine beads of three different colors, if any number of beads of each color can be used? Now the set X of arrangements of beads has 3^9 elements; namely, each of the nine vertices of the nonagon can be occupied by a bead of any of the three colors. Likewise, the number of arrangements fixed by any $g \in D_9$ is $3^{N(g)}$, where $N(g)$ is the number of orbits of $\langle g\rangle$ acting on vertices; each orbit of $\langle g\rangle$ must have beads of only one color, but any of the three colors can be used. We compute the following data:

n-fold rotation axis, n =	order of rotation	N(g)	number of such group elements
*	1	9	1
9	9	1	6
9	3	3	2
2	2	5	9

Thus, the number of necklaces is

$$\frac{1}{|G|}\sum_{g\in G}|\text{Fix}(g)| = \frac{1}{18}(3^9 + 2\times 3^3 + 6\times 3 + 9\times 3^5) = 1219.$$

Example 5.2.5. How many different ways are there to color the faces of a cube with three colors? Regard two colorings to be the same if they are related by a rotation of the cube.

It is required to count the orbits for the action of the rotation group G of the cube on the set X of 3^6 colorings of the faces of the cube. For each $g \in G$ the number of $x \in X$ that are fixed by g is

$3^{N(g)}$, where $N(g)$ is the number of orbits of $\langle g \rangle$ acting on faces of the cube. We compute the following data:

n-fold rotation axis, n =	order of rotation	N(g)	number of such group elements
*	1	6	1
2	2	3	6
3	3	2	8
4	4	3	6
4	2	4	3

Thus, the number of colorings of the faces of the cube with three colors is

$$\frac{1}{|G|}\sum_{g \in G} |\mathrm{Fix}(g)| = \frac{1}{24}(3^6 + 8 \times 3^2 + 6 \times 3^3 + 3 \times 3^4 + 6 \times 3^3) = 57.$$

Exercises 5.2

5.2.1. How many necklaces can be made with six beads of three different colors?

5.2.2. How many necklaces can be made with two red beads, two green beads, and two violet beads?

5.2.3. Count the number of ways to color the edges of a cube with four colors. Count the number of ways to color the edges of a cube with r colors; the answer is a polynomial in r.

5.2.4. Count the number of ways to color the vertices of a cube with three colors. Count the number of ways to color the vertices of a cube with r colors.

5.2.5. Count the number of ways to color the faces of a dodecahedron with three colors. Count the number of ways to color the faces of a dodecahedron with r colors.

5.3. Symmetries of Groups

A mathematical object is a set with some structure. A bijection of the set that preserves the structure is undetectable insofar as that structure is concerned. For example, a rotational symmetry of the cube moves the individual points of the cube around but preserves the structure of the cube. A structure preserving bijection of any sort of object can be regarded as a symmetry of the object, and the set of symmetries always constitutes a group.

If, for example, the structure under consideration is a group, then a structure preserving bijection is a group automorphism. In this section, we will work out a few examples of automorphism groups of groups.

Recall that the inner automorphisms of a group are those of the form $c_g(x) = gxg^{-1}$ for some g in the group. The map $g \mapsto c_g$ is a homomorphism of G into $\mathrm{Aut}(G)$ with image the subgroup $\mathrm{Int}(G)$ of inner automorphisms. The kernel of this homomorphism is the center of the group and, therefore, $\mathrm{Int}(G) \cong G/Z(G)$. Observe that the group of inner automorphisms of an abelian group is trivial, since $G = Z(G)$.

Proposition 5.3.1. $\mathrm{Int}(G)$ *is a normal subgroup of* $\mathrm{Aut}(G)$.

Proof. Compute that for any automorphism α of G, $\alpha c_g \alpha^{-1} = c_{\alpha(g)}$. ■

Remark 5.3.2. The symmetric group S_n has trivial center, so $\mathrm{Int}(S_n) \cong S_n/Z(S_n) \cong S_n$. We showed that every automorphism of S_3 is inner, so $\mathrm{Aut}(S_3) \cong S_3$. An interesting question is whether this is also true for S_n for all $n \geq 3$. The rather unexpected answer that it is true except when $n = 6$. [$S_6 \cong \mathrm{Int}(S_6)$ is an index 2 subgroup of $\mathrm{Aut}(S_6)$.] See W. R. Scott, *Group Theory*, Dover Publications, 1987, pp. 309–314 (original edition, Prentice-Hall, 1964).

Proposition 5.3.3. *The automorphism group of a cyclic group* $\mathbb{Z}$ *or* $\mathbb{Z}_n$ *is isomorphic to the group of units of the* ring $\mathbb{Z}$ *or* $\mathbb{Z}_n$. *Thus* $\mathrm{Aut}(\mathbb{Z}) \cong \{\pm 1\}$ *and* $\mathrm{Aut}(\mathbb{Z}_n) \cong \Phi(n)$.

Proof. Let α be an *endomorphism* $\alpha : \mathbb{Z} \to \mathbb{Z}$, and let $r = \alpha(1)$. By the homomorphism property, for all n, $\alpha(n) = \alpha(n \cdot 1) = n \cdot \alpha(1) = rn$. Clearly $\alpha \mapsto \alpha(1)$ is a bijection from the set of endomorphisms of $\mathbb{Z}$ to $\mathbb{Z}$.

If α and β are two endomorphisms of $\mathbb{Z}$, with $\alpha(1) = r$ and $\beta(1) = s$, then $\alpha\beta(1) = \alpha(s) = rs$. It follows that for α to be invertible, it is necessary and sufficient that $\alpha(1)$ be a unit of $\mathbb{Z}$, and $\alpha \mapsto \alpha(1)$ is a group isomorphism from $\text{Aut}(\mathbb{Z})$ to the group of units of $\mathbb{Z}$, namely, $\{\pm 1\}$.

The proof for $\mathbb{Z}_n$ is similar, and is left to the reader; see Exercise 5.3.1. ∎

Corollary 5.3.4. $\text{Aut}(\mathbb{Z}_p)$ *is cyclic of order* $p - 1$.

Proof. This follows from the previous result and Corollary 3.3.17. ∎

The automorphism groups of several other abelian groups are determined in the Exercises.

Exercises 5.3

5.3.1. Show that $\alpha \mapsto \alpha([1])$ is an isomorphism from $\text{Aut}(\mathbb{Z}_n)$ onto the group of units of $\mathbb{Z}_n$.

5.3.2. Show that a homomorphism of the additive group $\mathbb{Z}^2$ into itself is determined by a 2-by-2 matrix of integers. Show that the homomorphism is injective if, and only if, the determinant of the matrix is nonzero, and bijective if, and only if, the determinant of the matrix is ± 1. Conclude that the group of automorphisms of $\mathbb{Z}^2$ is isomorphic to the group of 2-by-2 matrices with integer coefficients and determinant equal to ± 1.

5.3.3. Generalize the previous problem to describe the automorphism group of $\mathbb{Z}^n$.

5.3.4. Show that any homomorphism of the additive group $\mathbb{Q}$ into itself has the form $x \mapsto rx$ for some $r \in \mathbb{Q}$. Show that a homomorphism is an automorphism unless $r = 0$. Conclude that the automorphism group of $\mathbb{Q}$ is isomorphic to $\mathbb{Q}^*$, namely, the multiplicative group of nonzero rational numbers.

5.3.5. Show that any group homomorphism of the additive group $\mathbb{Q}^2$ is determined by rational 2-by-2 matrix. Show that any group homomorphism is actually a linear map, and the group of automorphisms is the same as the group of invertible linear maps.

5.3.6. Think about whether the results of the last two exercises hold if $\mathbb{Q}$ is replaced by $\mathbb{R}$. What issue arises?

5.3.7. We can show that $\mathrm{Aut}(\mathbb{Z}_2 \times \mathbb{Z}_2) \cong S_3$ in two ways.

(a) One way is to show that any automorphism is determined by a 2-by-2 matrix with entries in $\mathbb{Z}_2$, that there are six such matrices, and that they form a group isomorphic to S_3. Work out the details of this approach.

(b) Another way is to recall that $\mathbb{Z}_2 \times \mathbb{Z}_2$ can be described as a group with four elements e, a, b, c, with each nonidentity element of order 2 and the product of any two nonidentity elements equal to the third. Show that any permutation of $\{a, b, c\}$ determines an automorphism and, conversely, any automorphism is given by a permutation of $\{a, b, c\}$.

5.3.8. Describe the automorphism group of $\mathbb{Z}_n \times \mathbb{Z}_n$. [The description need not be quite as explicit as that of $\mathrm{Aut}(\mathbb{Z}_2 \times \mathbb{Z}_2)$.] Can you describe the automorphism group of $(\mathbb{Z}_n)^k$?

5.4. Group Actions and Group Structure

In this section, we consider some applications of the idea of group actions to the study of the structure of groups.

Consider the action of a group G on itself by conjugation. Recall that the stabilizer of an element is called its *centralizer* and the orbit of an element is called its *conjugacy class*. The set of elements z whose conjugacy class consists of z alone is precisely the center of the group. If G is finite, the decomposition of G into disjoint conjugacy classes gives the equation

$$|G| = |Z(G)| + \sum_g \frac{|G|}{|\mathrm{Cent}(g)|},$$

where $Z(G)$ denotes the center of G, $\mathrm{Cent}(g)$ the centralizer of g, and the sum is over representatives of distinct conjugacy classes in $G \setminus Z(g)$. This is called the *class equation*.

Example 5.4.1. Let's compute the right side of the class equation for the group S_4. We saw in Example 5.1.17 that S_4 has only one element in its center, namely, the identity. Its nonsingleton conjugacy classes are of sizes 6, 3, 8, and 6. This gives $24 = 1 + 6 + 3 + 8 + 6$.

Consider a group of order p^n, where p is a prime number and n a positive integer. Every subgroup has order a power of p by Lagrange's theorem, so for $g \in G \setminus Z(G)$, the size of the conjugacy class of g, namely,

$$\frac{|G|}{|\mathrm{Cent}(g)|},$$

is a positive power of p. Since p divides $|G|$ and $|Z(G)| \geq 1$, it follows that p divides $|Z(G)|$. We have proved the following:

Proposition 5.4.2. *If $|G|$ is a power of a prime number, then the center of G contains nonidentity elements.*

We discovered quite early that any group of order 4 is either cyclic or isomorphic to $\mathbb{Z}_2 \times \mathbb{Z}_2$. We can now generalize this result to groups of order p^2 for any prime p.

Corollary 5.4.3. *Any group of order p^2, where p is a prime, is either cyclic or isomorphic to $\mathbb{Z}_p \times \mathbb{Z}_p$.*

Proof. Suppose G, of order p^2, is not cyclic. Then any nonidentity element must have order p. Using the proposition, choose a nonidentity element $g \in Z(G)$. Since $o(g) = p$, it is possible to choose $h \in G \setminus \langle g \rangle$. Then g and h are both of order p, and they commute.

I claim that $\langle g \rangle \cap \langle h \rangle = \{e\}$. In fact, $\langle g \rangle \cap \langle h \rangle$ is a subgroup of $\langle g \rangle$, so if it is not equal to $\{e\}$, then it has cardinality p; but then it is equal to $\langle g \rangle$ and to $\langle h \rangle$. In particular, $h \in \langle g \rangle$, a contradiction.

It follows from this that $\langle g \rangle \langle h \rangle$ contains p^2 distinct elements of G, hence $G = \langle g \rangle \langle h \rangle$. Therefore, G is abelian.

Now $\langle g \rangle$ and $\langle h \rangle$ are two normal subgroups with $\langle g \rangle \cap \langle h \rangle = \{e\}$ and $\langle g \rangle \langle h \rangle = G$. Hence $G \cong \langle g \rangle \times \langle h \rangle \cong \mathbb{Z}_p \times \mathbb{Z}_p$. ■

Look now at Exercise 5.4.1, in which you are asked to show that a group of order p^3 (p a prime) is either abelian or has center of size p.

Corollary 5.4.4. *Let G be a group of order p^n, $n > 1$. Then G has a normal subgroup $\{e\} \subsetneq N \subsetneq G$. Furthermore, N can be chosen so that every subgroup of N is normal in G.*

Proof. If G is nonabelian, then by the proposition, $Z(G)$ has the desired properties. If G is abelian, every subgroup is normal. If g is a nonidentity element, then g has order p^s for some $s \geq 1$. If $s < n$, then $\langle g \rangle$ is a proper subgroup. If $s = n$, then g^p is an element of order p^{n-1}, so $\langle g^p \rangle$ is a proper subgroup. ■

Corollary 5.4.5. *Suppose* $|G| = p^n$ *is a power of a prime number. Then* G *has a sequence of subgroups*

$$\{e\} = G_0 \subseteq G_1 \subseteq G_2 \subseteq \cdots \subseteq G_n = G$$

such that the order of G_k *is* p^k, *and* G_k *is normal in* G *for all* k.

Proof. We prove this by induction on n. If $n = 1$, there is nothing to do. So suppose the result holds for all groups of order $p^{n'}$, where $n' < n$. Let N be a proper normal subgroup of G, with the property that every subgroup of N is normal in G. The order of N is p^s for some s, $1 \leq s < n$. Apply the induction hypothesis to N to obtain a sequence

$$\{e\} = G_0 \subseteq G_1 \subseteq G_2 \subseteq \cdots \subseteq G_s = N$$

with $|G_k| = p^k$. Apply the induction hypothesis again to $\bar{G} = G/N$ to obtain a sequence of subgroups

$$\{e\} = \bar{G}_0 \subseteq \bar{G}_1 \subseteq \bar{G}_2 \subseteq \cdots \subseteq \bar{G}_{n-s} = \bar{G}$$

with $|\bar{G}_k| = p^k$ and $\bar{G}_k$ normal in $\bar{G}$. Then put $G_{s+k} = \pi^{-1}\bar{G}_k$, for $1 \leq k \leq n - s$, where $\pi : G \longrightarrow G/N$ is the quotient map. Then the sequence $(G_k)_{0 \leq k \leq n}$ has the desired properties. ■

We now use similar techniques to investigate the existence of subgroups of order a power of a prime. The first result in this direction is Cauchy's theorem:

Theorem 5.4.6. *(Cauchy's theorem). Suppose the prime* p *divides the order of a group* G. *Then* G *has an element of order* p.

The proof given here, due to McKay,[1] is simpler and shorter than other known proofs.

Proof. Let X be the set consisting of sequences $(a_1, a_2, \ldots, a_p)$ of elements of G such that $a_1 a_2 \ldots a_p = e$. Note that a_1 through a_{p-1} can be chosen arbitrarily, and $a_p = (a_1 a_2 \ldots a_{p-1})^{-1}$. Thus the cardinality of X is $|G|^{p-1}$. Recall that if $a, b \in G$ and $ab = e$, then also $ba = e$. Hence if $(a_1, a_2, \ldots, a_p) \in X$, then $(a_p, a_1, a_2, \ldots, a_{p-1}) \in X$ as well. Hence, the cyclic group of order p acts on X by cyclic permutations of the sequences.

Each element of X is either fixed under the action of $\mathbb{Z}_p$, or it belongs to an orbit of size p. Thus $|X| = n + kp$, where n is the

[1] J. H. McKay, Amer. Math. Monthly **66** (1959), p. 119.

number of fixed points and k is the number of orbits of size p. Note that $n \geq 1$, since $(e, e, \ldots, e)$ is a fixed point of X. But p divides $|X| - kp = n$, so X has a fixed point $(a, a, \ldots, a)$ with $a \neq e$. But then a has order p. ■

Theorem 5.4.7. *(First Sylow theorem). Suppose p is a prime, and p^n divides the order of a group G. Then G has a subgroup of order p^n.*

Proof. We prove this statement by induction on n, the case $n = 1$ being Cauchy's theorem. We assume inductively that G has a subgroup H of order p^{n-1}. Then $[G : H]$ is divisible by p.

Let H act on G/H by left multiplication. We know that $[G : H]$ is equal to the number of fixed points plus the sum of the cardinalities of nonsingleton orbits. The size of every nonsingleton orbit divides the cardinality of H, so is a power of p. Since p divides $[G : H]$, and p divides the size of each nonsingleton orbit, it follows that p also divides the number of fixed points. The number of fixed points is nonzero, since H itself is fixed.

Let's look at the condition for a coset xH to be fixed under left multiplication by H. This is so if, and only if, for each $h \in H$, $hxH = xH$. That is, for each $h \in H$, $x^{-1}hx \in H$. Thus x is in the *normalizer* of H in G (i.e., the set of $g \in G$ such that $gHg^{-1} = H$).

We conclude that the normalizer $N_G(H) \underset{\neq}{\supset} H$. More precisely, the number of fixed points for the action of H on G/H is the index $[N_G(H) : H]$, which is thus divisible by p. Of course, H is normal in $N_G(H)$, so we can consider $N_G(H)/H$, which has size divisible by p. By Cauchy's theorem, $N_G(H)/H$ has a subgroup of order p. The inverse image of this subgroup in $N_G(H)$ is a subgroup H_1 of cardinality p^n. ■

Definition 5.4.8. If p^n is the largest power of the prime p dividing the order of G, a subgroup of order p^n is called a p-*Sylow subgroup.*

The first Sylow theorem asserts, in particular, the existence of a p-Sylow subgroup for each prime p.

Theorem 5.4.9. *Let G be a finite group, p a prime number, H a subgroup of G of order p^s, and P a p-Sylow subgroup of G. Then there is a $a \in G$ such that $aHa^{-1} \subseteq P$.*

Proof. Let X be the family of conjugates of P in G. According to Exercise 5.4.3, the cardinality of X is not divisible by p. Now let H act on X by conjugation. Any nonsingleton orbit must have cardinality a power of p. Since $|X|$ is not divisible by p, it follows that X has a fixed point under the action of H. That is, for some $g \in G$, conjugation by elements of H fixes gPg^{-1}. Equivalently, $H \subseteq N_G(gPg^{-1}) = gN_G(P)g^{-1}$, or $g^{-1}Hg \subseteq N_G(P)$. Since $|g^{-1}Hg| = |H| = p^s$, it follows from Exercise 5.4.4 that $g^{-1}Hg \subseteq P$. ∎

Corollary 5.4.10. *(Second Sylow theorem). Let P and Q be two p-Sylow subgroups of a finite group G. Then P and Q are conjugate subgroups.*

Proof. According to the theorem, there is an $a \in G$ such that $aQa^{-1} \subseteq P$. Since the two groups have the same size, it follows that $aQa^{-1} = P$ ∎

Theorem 5.4.11. *(Third Sylow theorem). Let G be a finite group and let p be a prime number. The number of p-Sylow subgroups of G divides $|G|$ and is congruent to $1 \pmod{p}$. In other words, the number of p-Sylow subgroups can be written in the form $mp + 1$.*

Proof. Let P be a p-Sylow subgroup. The family X of p-Sylow subgroups is the set of conjugates of P, according to the second Sylow theorem. Let P act on X by conjugation. If Q is a p-Sylow subgroup distinct from P, then Q is not fixed under the action of P; for if Q were fixed, then $P \subseteq N_G(Q)$, and by Exercise 5.4.4, $P \subseteq Q$. Therefore, there is exactly one fixed point for the action of P on X, namely, P. All the nonsingleton orbits for the action of P on X have size a power of p, so $|X| = mp + 1$.

On the other hand, G acts transitively on X by conjugation, so

$$|X| = \frac{|G|}{|N_G(P)|}$$

divides the order of G. ∎

We can summarize the three theorems of Sylow as follows: If p^n is the largest power of a prime p dividing the order of a finite group G, then G has a subgroup of order p^n. Any two such subgroups are conjugate in G and the number of such subgroups is conjugate to $1 \pmod{p}$.

Example 5.4.12. Let p and q be primes with $p > q$. If q does not divide $p - 1$, then any group of order pq is cyclic. If q divides $p - 1$, then any group of order pq is either cyclic or a semidirect product $\mathbb{Z}_p \rtimes_\alpha \mathbb{Z}_q$. In this case, up to isomorphism, there exactly only one nonabelian group of order pq.

Proof. Let G be a group of order pq. Then G has a p-Sylow subgroup P of order p and a q-Sylow subgroup of order q; P and Q are cyclic, and since the orders of P and Q are relatively prime, $P \cap Q = \{e\}$. It follows from the third Sylow theorem that P is normal in G, since 1 is the only natural number that divides pq and is congruent to 1 (mod p). Therefore, $PQ = QP$ is a subgroup of G of order pq, so $PQ = G$.

According to Corollary 3.2.5, there is a homomorphism $\alpha : \mathbb{Z}_q \to \text{Aut}(\mathbb{Z}_p)$ such that $G \cong \mathbb{Z}_p \rtimes_\alpha \mathbb{Z}_q$. Since $\mathbb{Z}_q$ is simple, α is either trivial or injective; in the latter case, $\alpha(\mathbb{Z}_q)$ is a cyclic subgroup of $\text{Aut}(\mathbb{Z}_p)$ of order q. But, by Corollary 5.3.4, $\text{Aut}(\mathbb{Z}_p) \cong \mathbb{Z}_{p-1}$. Therefore, if q does not divide $p - 1$, then α must be trivial, so $G \cong \mathbb{Z}_p \times \mathbb{Z}_q \cong \mathbb{Z}_{pq}$.

On the other hand, if q divides $p - 1$, then $\text{Aut}(\mathbb{Z}_p) \cong \mathbb{Z}_{p-1}$ has a unique subgroup of order q, and there exists an injective homomorphism α of $\mathbb{Z}_q$ into $\text{Aut}(\mathbb{Z}_p)$. Thus there exists a nonabelian semidirect product $\mathbb{Z}_p \rtimes_\alpha \mathbb{Z}_q$.

It remains to show that if α and β are distinct injective homomorphisms of $\mathbb{Z}_q$ into $\text{Aut}(\mathbb{Z}_p)$, then $\mathbb{Z}_p \rtimes_\alpha \mathbb{Z}_q \cong \mathbb{Z}_p \rtimes_\beta \mathbb{Z}_q$. This is left as an exercise for the reader; see Exercise 5.4.5. ■

Example 5.4.13. Since 3 does not divide $(5 - 1)$, the only group of order 15 is cyclic. Since 3 divides $(7 - 1)$, there is a unique nonabelian group of order 21, as well as a unique (cyclic) abelian group of order 21.

Example 5.4.14. We know several groups of order 30, namely, $\mathbb{Z}_{30}$, D_{15}, $\mathbb{Z}_3 \times D_5$, and $\mathbb{Z}_5 \times D_3$. We can show that these groups are mutually nonisomorphic; see Exercise 5.4.6.

Are these the only possible groups of order 30? Let G be a group of order 30. Then G has (cyclic) Sylow subgroups P, Q, and R of orders 2, 3, and 5. R and Q are both normal, by the third Sylow theorem, and $N = QR \cong Q \times R$ is, therefore, a normal, cyclic subgroup of order 15. According to Corollary 3.2.5, there is a homomorphism $\alpha : \mathbb{Z}_2 \to \text{Aut}(\mathbb{Z}_{15})$, such that $G \cong \mathbb{Z}_{15} \rtimes_\alpha \mathbb{Z}_2$. To complete the classification of groups of order 30, we have to classify such homomorphisms; the nontrivial homomorphisms are determined by order 2 elements of $\text{Aut}(\mathbb{Z}_{15})$.

We have $\text{Aut}(\mathbb{Z}_{15}) \cong \text{Aut}(\mathbb{Z}_5) \times \text{Aut}(\mathbb{Z}_3) \cong \Phi(5) \times \Phi(3) \cong \mathbb{Z}_4 \times \mathbb{Z}_2$. In particular, if θ is an automorphism of $\mathbb{Z}_5 \times \mathbb{Z}_3$, then there exist unique automorphisms θ' of $\mathbb{Z}_5$ and θ'' of $\mathbb{Z}_3$ such that for all

$([a], [b]) \in \mathbb{Z}_5 \times \mathbb{Z}_3$, $\theta(([a], [b])) = (\theta'([a]), \theta''([b]))$. The reader is asked to check the details of these assertions in Exercise 5.4.8.

It is easy to locate one order 2 automorphism of $\mathbb{Z}_n$ for any n, namely, the automorphism given by $[k] \mapsto [-k]$. Since $\text{Aut}(\mathbb{Z}_5)$ and $\text{Aut}(\mathbb{Z}_3)$ are cyclic, each of these groups has exactly one element of order 2. Then $\text{Aut}(\mathbb{Z}_5) \times \text{Aut}(\mathbb{Z}_3)$ has exactly three elements of order 2, namely, $\varphi_1(([a], [b])) = ([-a], [b])$, $\varphi_2(([a], [b])) = ([a], [-b])$, and $\varphi_3(([a], [b])) = ([-a], -[b])$.

It is straightforward to check that

(a) $(\mathbb{Z}_5 \times \mathbb{Z}_3) \rtimes_{\varphi_1} \mathbb{Z}_2 \cong D_5 \times \mathbb{Z}_3$,

(b) $(\mathbb{Z}_5 \times \mathbb{Z}_3) \rtimes_{\varphi_2} \mathbb{Z}_2 \cong \mathbb{Z}_5 \times D_3$, and

(c) $(\mathbb{Z}_5 \times \mathbb{Z}_3) \rtimes_{\varphi_3} \mathbb{Z}_2 \cong D_{15}$;

see Exercise 5.4.7.

Thus these three groups are the only nonabelian groups of order 30 (and $\mathbb{Z}_{30}$ is the only abelian group of order 30).

Exercises 5.4

5.4.1. Suppose $|G| = p^3$, where p is a prime. Show that either $|Z(G)| = p$ or G is abelian.

5.4.2. Let p be the largest prime dividing the order of a finite group G, and let P be a p-Sylow subgroup of G. Show that P is normal and that P is the unique p-Sylow subgroup of G.

5.4.3. Let P be a p-Sylow subgroup of a finite group G. Consider the set of conjugate subgroups gPg^{-1} with $g \in G$. According to Corollary 5.1.14, the number of such conjugates is the index of the normalizer of P in G, $[G : N_G(P)]$. Show that the number of conjugates is not divisible by p.

5.4.4. Let P be a p-Sylow subgroup of a finite group G. Let H be a subgroup of $N_G(P)$ such that $|H| = p^s$. Show that $H \subseteq P$. Hint: Refer to Exercise 5.1.10, where it is shown that HP is a subgroup of $N_G(P)$ with

$$|HP| = \frac{|P|\,|H|}{|H \cap P|}.$$

5.4.5. Let $p > q$ be prime numbers such that q divides $p - 1$. If α and β are two distinct injective homomorphisms of $\mathbb{Z}_q$ into $\text{Aut}(\mathbb{Z}_p) \cong \mathbb{Z}_{p-1}$, show that $\mathbb{Z}_p \rtimes_{\alpha} \mathbb{Z}_q \cong \mathbb{Z}_p \rtimes_{\beta} \mathbb{Z}_q$.

5.4.6. Show that the groups $\mathbb{Z}_{30}$, D_{15}, $\mathbb{Z}_3 \times D_5$, and $\mathbb{Z}_5 \times D_3$ are mutually nonisomorphic.

5.4.7. Verify the following isomorphisms:

(a) $(\mathbb{Z}_5 \times \mathbb{Z}_3) \rtimes_{\varphi_1} \mathbb{Z}_2 \cong D_5 \times \mathbb{Z}_3$

(b) $(\mathbb{Z}_5 \times \mathbb{Z}_3) \rtimes_{\varphi_2} \mathbb{Z}_2 \cong \mathbb{Z}_5 \times D_3$

(c) $(\mathbb{Z}_5 \times \mathbb{Z}_3) \rtimes_{\varphi_3} \mathbb{Z}_2 \cong D_{15}$

5.4.8. Verify the assertion made about $\mathrm{Aut}(\mathbb{Z}_{15})$ in Example 5.4.14.

5.4.9. Show that an abelian group is the direct product of p-Sylow subgroups for primes p dividing $|G|$.

We have classified all abelian groups and all groups of orders p, p^2, and pq completely (p and q primes).

5.4.10. For which $n \leq 30$ can you say that all groups of order n are abelian? For which $n \leq 30$ can you classify the nonabelian groups?

The Sylow theorems provide good tools for investigating the possible structures of the nonabelian groups of order no more than 30.

5.4.11. Show that a group of order 28 has a normal subgroup N of order 7; the automorphism group of N is cyclic of order 6. A 2-Sylow subgroup of order 4 is either $\mathbb{Z}_4$ or $\mathbb{Z}_2 \times \mathbb{Z}_2$. Show that there is a unique nontrivial homomorphism of $\mathbb{Z}_4$ into $\mathrm{Aut}(N)$, and an essentially unique nontrivial homomorphism of $\mathbb{Z}_2 \times \mathbb{Z}_2$ into $\mathrm{Aut}(N)$. Conclude that there are, up to isomorphism, two nonabelian groups of order 28. One of these has to be the dihedral group D_{14}. Show that $D_{14} \cong D_7 \times \mathbb{Z}_2$. Can you describe the other nonabelian group of order 28?

5.4.12. Do a similar analysis of the nonabelian group(s) of order 20.

5.4.13. Try a similar analysis of the nonabelian group(s) of order 18. What new issues do you run into?

5.4.14. Try a similar analysis of the nonabelian group(s) of order 12. What new issues do you run into?

5.5. Application: Transitive Subgroups of S_5

This section can be omitted without loss of continuity. However, in the discussion of Galois groups in Chapter 9, we shall refer to the results of this section.

In this brief section, we will use the techniques of the previous sections to classify the transitive subgroups of S_5. Of course, S_5 itself and the alternating group A_5 are transitive subgroups. Also, we can readily think of $Z_5 = \langle(12345)\rangle$ (and its conjugates), and $D_5 = \langle(12345), (12)\rangle$ (and its conjugates). (We write $\langle a, b, c, \dots\rangle$ for the subgroup generated by elements $a, b, c, \dots$.)

We know that the only subgroup of index 2 in S_5 is A_5; see Exercise 2.4.15. We will need the fact, which is proved later in the text (Section 9.3), that A_5 *is the only normal subgroup of* S_5 *other than* S_5 *and* $\{e\}$.

Lemma 5.5.1. *Let* G *be a subgroup of* S_5 *that is not equal to* S_5 *or* A_5. *Then* $[S_5 : G] \geq 5$; *that is,* $|G| \leq 24$.

Proof. Write $d = [S_5 : G]$. Consider the action of S_5 on the set X of left cosets of G in S_5 by left multiplication.

I claim that this action is faithful [i.e., that the corresponding homomorphism of S_5 into $\mathrm{Sym}(X) \cong S_d$ is injective]. In fact, the kernel of the homomorphism is the intersections of the stabilizers of the points of X, and, in particular, is contained in G, which is the stabilizer of $G = eG \in X$. On the other hand, the kernel is a normal subgroup of S_5. Since G does not contain S_5 or A_5, it follows from the fact just mentioned that the kernel is $\{e\}$.

But this means that S_5 is isomorphic to a subgroup of $\mathrm{Sym}(X) \cong S_d$, and, consequently, $5! \leq d!$, or $5 \leq d$. ■

Remark 5.5.2. More, generally, A_n is the only nontrivial normal subgroup of S_n for $n \geq 5$. So the same argument shows that a subgroup of S_n that is not equal to S_n or A_n must have index at least n in S_n.

Now, suppose that G is a transitive subgroup of S_5, not equal to S_5 or A_5. We know that for any finite group acting on any set, the size of any orbit divides the order of the group. By hypothesis, G, acting on $\{1, \dots, 5\}$, has one orbit of size 5, so 5 divides $|G|$. But, by the lemma, $|G| \leq 24$. Thus,

$$|G| = 5k, \quad \text{where} \quad 1 \leq k \leq 4.$$

By Sylow theory, G has a normal subgroup of order 5, necessarily cyclic. Let $\sigma \in G$ be an element of order 5. Since $G \subseteq S_5$, the only possible cycle structure for σ is a 5-cycle. Without loss of generality, assume $\sigma = (1\,2\,3\,4\,5)$.

That $\langle\sigma\rangle$ is normal in G means that $G \subseteq N_{S_5}(\langle\sigma\rangle)$. What is $N_{S_5}(\langle a\rangle)$? We know that for any $\rho \in S_5$, $\rho\sigma\rho^{-1} = (\rho(1)\,\rho(2)\ldots\rho(5))$. For ρ to normalize $\langle\sigma\rangle$, it is necessary and sufficient that the element $(\rho(1)\,\rho(2)\ldots\rho(5))$ be a power of $(1\,2\,3\,4\,5)$. We can readily find one permutation ρ that will serve, namely, $\rho = (2\,3\,5\,4)$, which satisfies $\rho\sigma\rho^{-1} = \sigma^2$.

Observe that A_5 does *not* normalize $\langle\sigma\rangle$, so the lemma, applied to $N_{S_5}(\langle\sigma\rangle)$ gives that the cardinality of this group is no more than 20. On the other hand, $\langle\sigma,\rho\rangle$ is a subgroup of $N_{S_5}(\langle\sigma\rangle)$ isomorphic to $\mathbb{Z}_4 \ltimes \mathbb{Z}_5$ and has cardinality 20. Therefore, $N_{S_5}(\langle\sigma\rangle) = \langle\sigma,\rho\rangle$.

Now, we have $\langle\sigma\rangle \subseteq G \subseteq \langle\sigma,\rho\rangle$. The possibilities for G are $\langle\sigma\rangle \cong \mathbb{Z}_5$, $\langle\sigma,\rho^2\rangle \cong D_5$, and, finally, $\langle\sigma,\rho\rangle \cong \mathbb{Z}_4 \ltimes \mathbb{Z}_5$. The normalizer of each of these groups is $\langle\sigma,\rho\rangle$; hence the number of conjugates of each of the groups is 5.

To summarize,

Proposition 5.5.3. *The transitive subgroups of S_5 are*

(a) S_5
(b) A_5
(c) $\langle(1\,2\,3\,4\,5), (2\,3\,5\,4)\rangle \cong \mathbb{Z}_4 \ltimes \mathbb{Z}_5$ *(and its five conjugates)*
(d) $\langle(1\,2\,3\,4\,5), (2\,5)(3\,4)\rangle \cong D_5$ *(and its five conjugates)*
(e) $\langle(1\,2\,3\,4\,5)\rangle \cong \mathbb{Z}_5$ *(and its five conjugates)*

Remark 5.5.4. There are 16 conjugacy classes of transitive subgroups of S_6, and seven of S_7. (See J. D. Dixon and B. Mortimer, *Permutation Groups*, Springer–Verlag, 1996, pp. 58–64.) Transitive subgroups of S_n at least for $n \le 11$ have been classified. Consult Dixon and Mortimer for further details.

5.6. Additional Exercises for Chapter 5

5.6.1. Let G be a finite group and let H be a subgroup. Let Y denote the set of conjugates of H in G, $Y = \{gHg^{-1} : g \in G\}$. As usual, G/H denotes the set of left cosets of H in G, $G/H = \{gH : g \in G\}$.

(a) Show that $\dfrac{\#(G/H)}{\#Y} = [N_G(H) : H]$.

(b) Consider the map from G/H to Y defined by $gH \mapsto gHg^{-1}$. Show that this map is well defined and surjective.

(c) Show that the map in part (b) is one to one if, and only if, $H = N_G(H)$. Show that, in general, the map is $[N_G(H) : H]$ to one [i.e., the preimage of each element of Y has size $[N_G(H) : H]$].

Definition 5.6.1. Suppose a group G acts on sets X and Y. We say that a map $\varphi : X \to Y$ is G-*equivariant* if for all $x \in X$,

$$\varphi(g \cdot x) = g \cdot (\varphi(x)).$$

5.6.2. Let G act transitively on a set X. Fix $x_0 \in X$, let $H = \mathrm{Stab}(x_0)$, and let Y denote the set of conjugates of H in G. Show that there is a G-equivariant surjective map from X to Y given by $x \mapsto \mathrm{Stab}(x)$, and this map is $[N_G(H) : H]$ to one.

5.6.3. Let $D_4 \subseteq S_4$ be the subgroup generated by (1234) and $(14)(23)$. Show that $N_{S_4}(D_4) = D_4$. Conclude that there is an S_4-equivariant bijection from S_4/D_4 onto the set of conjugates of D_4 in S_4.

5.6.4. Let G be the rotation group of the tetrahedron, acting on the set of faces of the tetrahedron. Show that map $F \mapsto \mathrm{Stab}(F)$ is bijective, from the set of faces to the set of stabilizer subgroups of faces.

5.6.5. Let G be the rotation group of the cube, acting on the set of faces of the cube. Show that map $F \mapsto \mathrm{Stab}(F)$ is 2-to-1, from the set of faces to the set of stabilizer subgroups of faces.

5.6.6. Let $G = S_n$ and $H = \mathrm{Stab}(n) \cong S_{n-1}$. Show that H is its own normalizer, so that the cosets of H correspond 1-to-1 with conjugates of H. Describe the conjugates of H explicitly.

5.6.7. Identify the group G of rotations of the cube with S_4, via the action on the diagonals of the cube. G also acts transitively on the set of set of three 4–fold rotation axes of the cube; this gives a homomorphism of S_4 into S_3.

(a) Compute the resulting homomorphism ψ of S_4 to S_3 explicitly. (For example, compute the image of a set of generators of S_4.) Show that ψ is surjective. Find the kernel of ψ.

(b) Show that the stabilizer of each 4–fold rotation axis is conjugate to $D_4 \subseteq S_4$.

(c) Show that $L \mapsto \mathrm{Stab}(L)$ is a bijection between the set of 4–fold rotation axes and the stabilizer subgroups of these axes in G. This map is G-equivariant, where G acts on the set of stabilizer subgroups by conjugation.

5.6.8. Let H be a proper subgroup of a finite group G. Show that G contains an element that is not in any conjugate of H.

5.6.9. Find all (2- and 3-) Sylow subgroups of S_4.

5.6.10. Find all (2- and 3-) Sylow subgroups of A_4.

5.6.11. Find all (2- and 3-) Sylow subgroups of D_6.

5.6.12. Let G be a finite group, p a prime, P a p-Sylow subgroup of G, and N a normal subgroup of G. Show that PN/N is a p-Sylow subgroup of G/N and that $P \cap N$ is a p-Sylow subgroup of N.

5.6.13. Let G be a finite group, p a prime, and P a p-Sylow subgroup of G. Show that $N_G(N_G(P)) = N_G(P)$.

Definition 5.6.2. Let p be a prime. A group (not necessarily finite) is called a p-group if every element has finite order p^k for some $k \geq 0$.

5.6.14. Show that a finite group G is a p-group if, and only if, the order of G is equal to a power of p.

5.6.15. Let N be a normal subgroup of a group G (not necessarily finite). Show that G is a p-group if, and only if, both N and G/N are p-groups.

5.6.16. Let H be a subgroup of a finite p-group G, with $H \neq \{e\}$. Show that $H \cap Z(G) \neq \{e\}$.

5.6.17. Let G be a finite group, p a prime, and P a p-Sylow subgroup. Suppose H is a normal subgroup of G of order p^k for some k. Show that $H \subseteq P$.

5.6.18. Show that a group of order $2^n 5^m$, $m, n \geq 1$, has a normal 5-Sylow subgroup. Can you generalize this statement?

5.6.19. Show that a group G of order 56 has a normal Sylow subgroup. Hint: Let P be a 7-Sylow subgroup. If P is not normal, count the elements in $\bigcup_{g \in G} gPg^{-1}$.

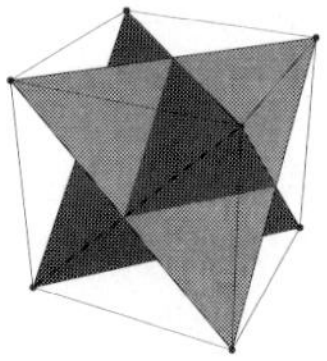

CHAPTER 6

Rings

6.1. A Recollection of Rings

We encountered the definitions of rings and fields in Section 1.11. Let us recall them here for convenience.

Definition 6.1.1. A *ring* is a nonempty set R with two operations: addition, denoted here by $+$, and multiplication, denoted by juxtaposition, satisfying the following requirements:

(a) Under addition, R is an abelian group.
(b) Multiplication is associative.
(c) Multiplication distributes over addition: $a(b+c) = ab+ac$, and $(b+c)a = ba+ca$ for all $a, b, c \in R$.

A ring is called commutative if multiplication is commutative, $ab = ba$ for all elements a, b in the ring. Recall that a multiplicative identity in a ring is an element 1 such that $1a = a1 = a$ for all elements a in the ring. An element a in a ring with multiplicative identity 1 is a *unit* or *invertible* if there exists an element b such that $ab = ba = 1$.

Some authors include the the existence of a multiplicative identity in the definition of a ring, but as this requirement excludes many natural examples, we will not follow this practice.

Definition 6.1.2. A *field* is a commutative ring with multiplicative identity element 1 (different from 0) in which *every nonzero element is a unit.*

We gave a number of examples of rings and fields in Section 1.11, which you should review now. There are (at least) four main sources of ring theory:

1. *Numbers.* The familiar number systems $\mathbb{Z}$, $\mathbb{Q}$, $\mathbb{R}$, and $\mathbb{C}$ are rings. In fact, all of them but $\mathbb{Z}$ are fields.

2. *Polynomial rings in one or several variables.* We have discussed polynomials in one variable over a field in Section 1.8. Polynomials in several variables, with coefficients in any commutative ring R, have a similar description: Let $x_1, \dots, x_n$ be variables, and let $I = (i_1, \dots, i_n)$ be a so–called *multi-index*, namely, a sequence of nonnegative integers of length n. Let x^I denote the monomial $x^I = x_1^{i_1} x_2^{i_2} \cdots x_n^{i_n}$. A polynomial in the variables $x_1, \dots, x_n$ with coefficients in R is an expression of the form $\sum_I \alpha_I x^I$, where the sum is over multi-indices, the α_I are elements of R, and $\alpha_I = 0$ for all but finitely many multi-indices I.

Example 6.1.3. $7xyz + 3x^2yz^2 + 2yz^3$ is an element of $\mathbb{Q}[x, y, z]$. The three nonzero terms correspond to the multi-indices

$$(1, 1, 1,),\ (2, 1, 2),\ \text{and}\ (0, 1, 3).$$

Polynomials in several variables are added and multiplied according to the following rules:

$$\sum_I \alpha_I x^I + \sum_I \beta_I x^I = \sum_I (\alpha_I + \beta_I) x^I,$$

and

$$(\sum_I \alpha_I x^I)(\sum_J \beta_J x^J) = \sum_I \sum_J \alpha_I \beta_J x^{I+J} = \sum_L \gamma_L x^L,$$

where $\gamma_L = \sum_{\substack{I,J \\ I+J=L}} \alpha_I \beta_J$.

Example 6.1.4. Let $p(x, y, z) = 7xyz + 3x^2yz^2 + 2yz^3$ and $q(x, y, z) = 2 + 3xz + 2xyz$. Then

$$p(x, y, z) + q(x, y, z) = 2 + 3xz + 9xyz + 3x^2yz^2 + 2yz^3,$$

and

$$\begin{aligned} p(x, y, z)q(x, y, z) = {} & 14\,x\,y\,z + 27\,x^2\,y\,z^2 + 14\,x^2\,y^2\,z^2 + 4\,y\,z^3 \\ & + 9\,x^3\,y\,z^3 + 6\,x^3\,y^2\,z^3 + 6\,x\,y\,z^4 + 4\,x\,y^2\,z^4. \end{aligned}$$

The set $R[x_1, \dots, x_n]$ of polynomials in the variables $\{x_1 \dots, x_n\}$ with coefficients in R is a commutative ring, with these operations.

3. *Rings of functions.* Let X be any set and let R be a field. Then the set of functions defined on X with values in R is a ring, with

the operations defined pointwise: $(f+g)(x) = f(x) + g(x)$, and $(fg)(x) = f(x)g(x)$.

If X is a metric space (or a topological space) and R is equal to one of the fields $\mathbb{R}$ or $\mathbb{C}$, then the set of *continuous* R–valued functions on X, with pointwise operations, is a ring. The essential point here is that the sum and product of continuous functions are continuous. (If you are not familiar with metric or topological spaces, just think of X as a subset of $\mathbb{R}$.)

If X is an open subset of $\mathbb{C}$, then the set of *holomorphic* $\mathbb{C}$–valued functions on X is a ring. (If you are not familiar with holomorphic functions, just ignore this example.)

4. *Endomorphism rings and matrix rings.* Let V be a vector space over a field K. The set $\text{End}(V) = \text{Hom}(V, V)$ of linear maps from V to V has two operations: Addition of linear maps is defined pointwise, $(S+T)(v) = S(v) + T(v)$. Multiplication of linear maps, however, is defined by composition: $ST(v) = S(T(v))$. With these operations, End(V) is a ring.

The set $\text{Mat}_n(K)$ of n-by-n matrices with entries in K is a ring, with the usual operations of addition and multiplication of matrices.

If V is n-dimensional over K, then the rings End(V) and $\text{Mat}_n(K)$ are isomorphic. In fact, for any ordered basis $B = (v_1, \dots, v_n)$ of V the map that assigns to each linear map $T : V \to V$ its matrix $T_{B,B}$ with respect to B is a ring isomorphism from End(V) to $\text{Mat}_n(K)$. (The matrix of a linear transformation with respect to an ordered basis was defined at the end of Section 3.4. That $T \mapsto T_{B,B}$ is a ring isomorphism is the content of Exercise 3.4.19.)

Let's now make the notion of subring, which was introduced informally in Section 1.11, more precise.

Definition 6.1.5. A nonempty subset S of a ring R is called a *subring* if S is a ring *with the two ring operations inherited from* R.

For S to be a subring of R, it is necessary and sufficient that

1. For all elements x and y of S, the sum and product $x+y$ and xy are elements of S.
2. For all $x \in S$, the additive opposite $-x$ is an element of S.

We gave a number of examples of subrings in Example 1.11.5. You are asked to verify these examples, and others, in the Exercises.

For any ring R and any subset $\mathcal{S} \subseteq R$ there is a smallest subring of R that contains $\mathcal{S}$, which is called the *subring generated by* $\mathcal{S}$ and denoted $\langle\langle \mathcal{S} \rangle\rangle$. When $\mathcal{S} = \{a\}$ is a singleton, the subring generated by $\mathcal{S}$ is denoted by $\langle\langle a \rangle\rangle$. We say that R *is generated by* $\mathcal{S}$ or that $\mathcal{S}$ *generates* R if $R = \langle\langle \mathcal{S} \rangle\rangle$.

A "constructive" view of $\langle\langle \mathcal{S} \rangle\rangle$ is that it consists of all possible finite sums of finite products $\pm T_1 T_2 \cdots T_n$, where $T_i \in \mathcal{S}$. In particular, the subring generated by a single element $T \in R$ is the set of all sums $\sum_{i=1}^{n} n_i T^i$. (Note there is no term for $i = 0$.) The subring generated by T and the multiplicative identity 1 (assuming R has a multiplicative identity) is the set of all sums $n_0 1 + \sum_{i=1}^{n} n_i T^i = \sum_{i=0}^{n} n_i T^i$, where we use the convention $T^0 = 1$.

Another view of $\langle\langle S \rangle\rangle$, which is sometimes useful, is that it is the intersection of the family of all subrings of R that contain S; this family is nonempty since R itself is such a subring. (As for subgroups, the intersection of an arbitrary nonempty collection of subrings is a subring.)

Example 6.1.6. Let $\mathcal{S}$ be a subset of End(V) for some vector space V. There are two subrings of End(V) associated to $\mathcal{S}$. One is the subring $\langle\langle \mathcal{S} \rangle\rangle$ generated by $\mathcal{S}$ that consists of all finite sums of products of elements of $\mathcal{S}$. Another is

$$\mathcal{S}' = \{T \in \mathrm{End}(V) : TS = ST \text{ for all } S \in \mathcal{S}\},$$

the so–called *commutant* of $\mathcal{S}$ in End(V).

Example 6.1.7. Let G be a subgroup of GL(V), the group of invertible linear transformations of V. I claim that the subring $\langle\langle G \rangle\rangle$ of End(V) generated by G is the set of finite sums $\sum_{g \in G} n_g g$, where $n_g \in \mathbb{Z}$. In fact, we can easily check that this set is closed under taking sums, additive opposites, and products.

Example 6.1.8. The previous example inspires the following construction. Let G be any finite group. Consider the set $\mathbb{Z}G$ of formal linear combinations of group elements, with coefficients in $\mathbb{Z}$, $\sum_{g \in G} a_g\, g$. (If you like, you can identify such a sum with the a function $g \mapsto a_g$ from G to $\mathbb{Z}$.) Two such expressions are added coefficient-by-coefficient,

$$\sum_{g \in G} a_g\, g + \sum_{g \in G} b_g\, g = \sum_{g \in G} (a_g + b_g)\, g,$$

and multiplied according to the rule

$$\sum_{g \in G} a_g\, g \sum_{h \in G} b_h\, h = \sum_{g \in G} \sum_{h \in G} a_g b_h\, gh = \sum_{\ell \in G} (\sum_{g \in G} a_g b_{g^{-1}\ell})\, \ell.$$

You are asked to verify that $\mathbb{Z}G$ is a ring in the Exercises. $\mathbb{Z}G$ is called the *integer group ring* of G.

Instead of taking coefficients in $\mathbb{Z}$, we can also take coefficients in $\mathbb{C}$, for example; the result is called the *complex group ring* of G.

Example 6.1.9. A formal power series in one variable with coefficients in a commutative ring R is a formal infinite sum $\sum_{i=0}^{\infty} \alpha_i x^i$. The set of formal power series is denoted $K[[x]]$. Formal power series are added coefficient-by-coefficient,

$$\sum_{i=0}^{\infty} \alpha_i x^i + \sum_{i=0}^{\infty} \beta_i x^i = \sum_{i=0}^{\infty} (\alpha_i + \beta_i) x^i.$$

The product of formal power series is defined as for polynomials:

$$(\sum_{i=0}^{\infty} \alpha_i x^i)(\sum_{i=0}^{\infty} \beta_i x^i) = \sum_{i=0}^{\infty} \gamma_i x^i,$$

where $\gamma_n = \sum_{j=0}^{n} \alpha_j \beta_{n-j}$. With these operations, the set of formal power series is a commutative ring.

Definition 6.1.10. Two rings R and S are *isomorphic* if there is a bijection between them that preserves both the additive and multiplicative structures. That is, there is a bijection $\varphi : R \to S$ satisfying $\varphi(a+b) = \varphi(a) + \varphi(b)$ and $\varphi(ab) = \varphi(a)\varphi(b)$ for all $a, b \in R$.

Example 6.1.11. Recall that the *direct sum* of two rings R and S is the Cartesian product of the rings endowed with the operations $(r, s) + (r', s') = (r + r', s + s')$ and $(r, s)(r', s') = (rr', ss')$. We usually denote the direct sum by $R \oplus S$. The direct sum of several rings $R_1, R_2, \dots, R_n$ is the Cartesian product endowed with the operations

$$(r_1, r_2, \dots, r_n) + (r_1', r_2', \dots, r_s') = (r_1 + r_1', r_2 + r_2', \dots, r_n + r_n')$$

and

$$(r_1, r_2, \dots, r_n)(r_1', r_2', \dots, r_s') = (r_1 r_1', r_2 r_2', \dots, r_n r_n').$$

The direct sum of $R_1, R_2, \dots, R_n$ is denoted $R_1 \oplus R_2 \oplus \cdots \oplus R_n$.

Example 6.1.12. According to the discussion at the end of Section 1.11 and in Exercise 1.11.10, if a and b are relatively prime natural numbers, then $\mathbb{Z}_{ab}$ and $\mathbb{Z}_a \oplus \mathbb{Z}_b$ are isomorphic rings.

Exercises 6.1

6.1.1. Show that if a ring R has a multiplicative identity, then the multiplicative identity is unique. Show that if an element $r \in R$ has a left multiplicative inverse r' and a right multiplicative inverse r'', then $r' = r''$.

6.1.2. Verify that $R[x_1, \dots, x_n]$ is a ring for any commutative ring R.

6.1.3. Consider the set of infinite-by-infinite matrices with real entries that have only finitely many nonzero entries. (Such a matrix has entries a_{ij}, where i and j are natural numbers. For each such matrix, there is a natural number n such that $a_{ij} = 0$ if $i \geq n$ or $j \geq n$.) Show that the set of such matrices is a ring without identity element.

6.1.4. Show that (a) the set of upper triangular matrices and (b) the set of upper triangular matrices with zero entries on the diagonal are both subrings of the ring of all n-by-n matrices with real coefficients. The second example is a ring without multiplicative identity.

6.1.5. Show that the set of matrices with integer entries is a subring of the ring of all n-by-n matrices with real entries. Show that the set of matrices with entries in $\mathbb{N}$ is closed under addition and multiplication but is not a subring.

6.1.6. Show that the set of symmetric polynomials in three variables is a subring of the ring of all polynomials in three variables. Recall that a polynomial is symmetric if it remains unchanged when the variables are permuted, $p(x, y, z) = p(y, x, z)$, and so on.

6.1.7. Experiment with variations on the preceding examples and exercises by changing the domain of coefficients of polynomials, values of functions, and entries of matrices: for example, polynomials with coefficients in the natural numbers, complex–valued functions, matrices with complex entries. What is allowed and what is not allowed for producing rings?

6.1.8. Show that $\mathrm{End}(V)$ is a ring, for any vector space V.

6.1.9. Suppose $\varphi : R \to S$ is a ring isomorphism. Show that R has a multiplicative identity if, and only if, S has a multiplicative identity. Show that R is commutative if, and only if, S is commutative.

6.1.10. Show that the intersection of any family of subrings of a ring is a subring. Show that the subring generated by a subset $\mathcal{S}$ of a ring R is the intersection of all subrings R' such that $\mathcal{S} \subseteq R' \subseteq R$.

6.1.11. Show that the set $\mathbb{R}(x)$ of rational functions $p(x)/q(x)$, where $p(x), q(x) \in \mathbb{R}[x]$ and $q(x) \neq 0$, is a field. (Note the use of parentheses to distinguish this ring $\mathbb{R}(x)$ of rational functions from the ring $\mathbb{R}[x]$ of polynomials.)

6.1.12. Let R be a ring and X a set. Show that the set $\mathrm{Fun}(X, R)$ of functions on X with values in R is a ring. Show that R is isomorphic to the subring of constant functions on X. Show that $\mathrm{Fun}(X, R)$ is commutative if, and only if, R is commutative. Suppose that R has an identity; show that $\mathrm{Fun}(X, R)$ has an identity and describe the units of $\mathrm{Fun}(X, R)$.

6.1.13. Let $\mathcal{S} \subseteq \mathrm{End}(V)$. Show that

$$\mathcal{S}' = \{T \in \mathrm{End}(V) : TS = ST \text{ for all } S \in \mathcal{S}\}$$

is a subring of $\mathrm{End}(V)$.

6.1.14. Let G be a subgroup of $GL(V)$. Show that the subring of $\mathrm{End}(V)$ generated by G is the set of all linear combinations $\sum_g n_g g$ of elements of G, with coefficients in $\mathbb{Z}$.

6.1.15. Verify that the "group ring" $\mathbb{Z}G$ of Example 6.1.8 is a ring.

6.1.16. Consider the group $\mathbb{Z}_2$ written as $\{e, \xi\}$, where $\xi^2 = e$. The complex group ring $\mathbb{C}\mathbb{Z}_2$ consists of formal sums $ae + b\xi$, with $a, b \in \mathbb{C}$. Show that the map $a + b\xi \mapsto (a + b, a - b)$ is a ring isomorphism from the group ring $\mathbb{C}\mathbb{Z}_2$ to the ring $\mathbb{C} \oplus \mathbb{C}$.

6.1.17. Show that the set of formal power series $R[[x]]$, with coefficients in a commutative ring R , is a commutative ring. If R has a multiplicative identity element, then so does $R[[x]]$.

6.2. Homomorphisms and Ideals

In this section, we return to the study of rings in general and develop some basic principles.

It should not be a big surprise that certain concepts that were fundamental to our study of groups are also important for the study of rings. One could expect these concepts to play a role for any reasonable algebraic structure. We have already discussed the idea of a subring, which is analogous to the idea of a subgroup. The next concept from group theory that we might expect to play a fundamental role in ring theory is the notion of a homomorphism.

Definition 6.2.1. A *homomorphism* $\varphi : R \to S$ of rings is a map satisfying $\varphi(x+y) = \varphi(x)+\varphi(y)$, and $\varphi(xy) = \varphi(x)\varphi(y)$ for all $x, y \in R$.

In particular, a ring homomorphism is a homomorphism for the abelian group structure of R and S, so we know, for example, that $\varphi(-x) = -\varphi(x)$ and $\varphi(0) = 0$. Even if R and S both have an identity element 1, it is not automatic that $\varphi(1) = 1$. If we want to specify that this is so, we will call the homomorphism a *unital* homomorphism.

Example 6.2.2. The map $\varphi : \mathbb{Z} \to \mathbb{Z}_n$ defined by $\varphi(a) = [a] = a + n\mathbb{Z}$ is a unital ring homomorphism. In fact, it follows from the definition of the operations in $\mathbb{Z}_n$ that $\varphi(a+b) = [a+b] = [a]+[b] = \varphi(a) + \varphi(b)$, and, similarly, $\varphi(ab) = [ab] = [a][b] = \varphi(a)\varphi(b)$ for integers a and b. (The actual work here was in verifying that the operations in $\mathbb{Z}_n$ were well defined.)

Example 6.2.3. Let R be any ring with multiplicative identity 1. The map $k \mapsto k \cdot 1$ is a ring homomorphism from $\mathbb{Z}$ to R. The map is just the usual *group* homomorphism from $\mathbb{Z}$ to the additive subgroup $\langle 1 \rangle$ generated by 1; see Example 2.4.7. It is necessary to check that $\langle 1 \rangle$ is closed under multiplication and that this map respects multiplication; that is, $(m \cdot 1)(n \cdot 1) = mn \cdot 1$. This follows from two observations: First, for any $a \in R$ and $n \in \mathbb{Z}$, $(n \cdot 1)a = n \cdot a$. This can be checked by induction on $|n|$. Second, $n \cdot (m \cdot a) = nm \cdot a$ is just the usual law of powers in a cyclic group. [In a group written with multiplicative notation, this law would be written as $(b^m)^n = b^{mn}$.] See Exercise 2.2.8. Putting these observations together, we have $(m \cdot 1)(n \cdot 1) = m \cdot (n \cdot 1) = mn \cdot 1$.

Warning: Such a homomorphism is not always injective. In fact, the ring homomorphism $k \mapsto [k] = k[1]$ from $\mathbb{Z}$ to $\mathbb{Z}_n$ is a homomorphism of this sort that is not injective.

Example 6.2.4. Consider the ring $C(\mathbb{R})$ of continuous real–valued functions on $\mathbb{R}$. Let S be any subset of R, for example, $S = [0, 1]$. The map $f \mapsto f_{|S}$ that associates to each function its restriction to S is a unital ring homomorphism from $C(\mathbb{R})$ to $C(S)$. Likewise, for any $t \in \mathbb{R}$ the map $f \mapsto f(t)$ is a unital ring homomorphism from $C(\mathbb{R})$ to $\mathbb{R}$.

We are used to evaluating polynomials (say with real coefficients) by substituting a number for the variable. For example, if $p(x) = x^2 + 2$, then $p(5) = 5^2 + 2 = 27$. When we do this, we are treating polynomials as functions. The following (easy) proposition justifies this practice.

Proposition 6.2.5. *(Evaluation of polynomials). Consider the ring* $R[x]$ *of polynomials over a commutative ring* R.

(a) *For any* $a \in R$, *there is a unique homomorphisms* $\mathrm{ev}_a : R[x] \to R$ *with the property that* $\mathrm{ev}_a(r) = r$ *for* $r \in R$ *and* $\mathrm{ev}_a(x) = a$. *We have*

$$\mathrm{ev}_a(\sum_i r_i x^i) = \sum_i r_i a^i.$$

We usually denote $\mathrm{ev}_a(p)$ *by* $p(a)$.

(b) *More generally, if* R *and* R' *are commutative rings, and* $\varphi : R \to R'$ *is a ring homomorphism, and* $a \in R'$ *is arbitrary, then there is a unique ring homomorphism* $\varphi_a : R[x] \to R'$ *such that* $\varphi_a(r) = \varphi(r)$ *for* $r \in R$, *and* $\varphi_a(x) = a$. *We have*

$$\varphi_a(\sum_i r_i x^i) = \sum_i \varphi(r_i) a^i.$$

Proof. It suffices to prove part (b). If φ_a is to be a homomorphism, then it must satisfy

$$\varphi_a(\sum_i r_i x^i) = \sum_i \varphi(r_i) a^i.$$

Therefore, we define φ_a by this formula. It is then straightforward to check that φ_a is a ring homomorphism. ■

Corollary 6.2.6. *If* $\psi : S \to S'$ *is a homomorphism of commutative rings, then there is a unique homomorphism* $\tilde{\psi} : S[x] \to S'[x]$ *that extends* φ. *The kernel of* $\tilde{\psi}$ *is* $I[x]$, *where* $I = \ker(\psi)$.

Proof. Apply part (b) of the proposition with the following data: Take $\varphi : S \to S'[x]$ to be the composition of $\psi : S \to S'$ with the inclusion S' into $S'[x]$, and let set $a = x$. By the proposition, there is a unique homomorphism from $S[x]$ to $S'[x]$ extending φ, and sending x to x. The extension is given by the formula

$$\tilde{\psi}(\sum_i s_i x^i) = \sum_i \psi(s_i) x^i.$$

■

Example 6.2.7. The map $\sum_i k_i x^i \mapsto \sum_i [k_i] x^i$ is a homomorphism of $\mathbb{Z}[x]$ to $\mathbb{Z}_n[x]$.

Further examples of ring homomorphisms are given in the Exercises.

The *kernel* of a ring homomorphism $\varphi : R \to S$ is the set of $x \in R$ such that $\varphi(x) = 0$. Again extrapolating from our experience with group theory, we would expect the kernel of a ring homomorphism to be a special sort of subring. Observe that a ring homomorphism is injective if, and only if, its kernel is $\{0\}$. The following definition captures the special properties of the kernel of a homomorphism.

Definition 6.2.8. An *ideal* I in a ring R is a subring of R satisfying $xr, rx \in I$ for all $x \in I$ and $r \in R$. A *left ideal* I of R is a subring of R such that $rx \in I$ whenever $r \in R$ and $x \in I$. A *right ideal* is defined similarly. Note that for commutative rings, all of these notions coincide.

Proposition 6.2.9. *If $\varphi : R \to S$ is a ring homomorphism, then* $\ker(\varphi)$ *is an ideal of of* R. *A ring homomorphism φ is injective if, and only if,* $\ker(\varphi) = \{0\}$.

Proof. Exercise 6.2.6. ■

Example 6.2.10. The kernel of the ring homomorphism $\mathbb{Z} \to \mathbb{Z}_n$ given by $k \mapsto [k]$ is $n\mathbb{Z}$. The kernel of the ring homomorphism $\mathbb{Z}[x] \to \mathbb{Z}_n[x]$ given by $\sum_i k_i x^i \mapsto \sum_i [k_i] x^i$ is the set of polynomials all of whose coefficients are divisible by n.

Example 6.2.11. The kernel of the ring homomorphism $K[x] \to K$ given by $p \mapsto p(a)$ is the set of all polynomials p having a as a root.

Example 6.2.12. The kernel of the ring homomorphism $C(\mathbb{R}) \to C(S)$ given by $f \mapsto f_{|S}$ is the set of all continuous functions whose restriction to S is zero.

Example 6.2.13. Let $\psi : S \to S'$ be a homomorphism of commutative rings, and let $\tilde{\psi} : S[x] \to S'[x]$ be the unique homomorphism extending ψ. The kernel of $\tilde{\psi}$ is $I[x]$, where $I = \ker(\psi)$. In particular, $\tilde{\psi}$ is injective if, and only if, ψ is injective.

In fact,

$$\tilde{\psi}(\sum_j \sigma_j x^j) = \sum_j \psi(a_j) x^j = 0$$

if, and only if $a_j \in I$ for all j.

Example 6.2.14. Let K be a field. Define a map φ from $K[x]$ to $\text{Fun}(K, K)$, the ring of K–valued functions on K by $\varphi(p)(a) = p(a)$. [That is, $\varphi(p)$ is the polynomial function on K corresponding to the polynomial p.] Then φ is a ring homomorphism. The homomorphism property of φ follows from the homomorphism property of $p \mapsto p(a)$ for $a \in K$. Thus $\varphi(p+q)(a) = (p+q)(a) = p(a)+q(a) = \varphi(p)(a) + \varphi(q)(a) = (\varphi(p) + \varphi(q))(a)$, and similarly for multiplication.

The kernel of φ is the set of polynomials p such that $p(a) = 0$ for all $a \in K$. If K is required to be infinite, then the kernel is $\{0\}$, since no nonzero polynomial with coefficients in a field has infinitely many roots.

If K is finite, then φ is *never* injective. That is, there always exist nonzero polynomials $p \in K[x]$ such that $p(a) = 0$ for all $a \in K$. Indeed, we need merely take $p(x) = \prod_{a \in K}(x - a)$.

The most direct way to obtain examples of ideals is the following:

Example 6.2.15. If R is any ring and $x \in R$, then $Rx = \{yx : y \in R\}$ is a left ideal. Similarly, xR is a right ideal and RxR is a (two-sided) ideal. Verification of these statements is left as an exercise.

Definition 6.2.16. xR is called the principal right ideal generated by x in R, Rx the principal left ideal generated by x in R, and RxR the principal ideal generated by x in R.

In the ring of integers, and in the ring $K[x]$ of polynomials in one variable over a field, every ideal is principal:

Proposition 6.2.17.

(a) *For a subset $S \subseteq \mathbb{Z}$, the following are equivalent:*
  (i) *S is a subgroup of $\mathbb{Z}$.*
  (ii) *S is a subring of $\mathbb{Z}$.*
  (iii) *S is an ideal of $\mathbb{Z}$.*

(b) *Every ideal in the ring of integers is principal.*

(c) *Every ideal in $K[x]$, where K is a field, is principal.*

Proof. Clearly an ideal is always a subring, and a subring is always a subgroup. If S is a nonzero subgroup of $\mathbb{Z}$, then $S = \mathbb{Z}d$, where d is the least positive element of S, according to Proposition 2.2.21. If $S = \{0\}$, then $S = \mathbb{Z}0$. In either case, S is a principal ideal of $\mathbb{Z}$. This proves (a) and (b).

The proof of (c) is similar to that of Proposition 2.2.21. The zero ideal of $K[x]$ is clearly principal. Let J be a nonzero ideal, and let $f \in J$ be a nonzero element of least degree in J. If $g \in J$, write $g = qf + r$, where $q \in K[x]$, and $\deg(r) < \deg(f)$. Then $r = g - qf \in J$. Since $\deg(r) < \deg(f)$ and f was a nonzero element of least degree in J, it follows that $r = 0$. Thus $g = qf \in K[x]f$, and $J = K[x]f$. ■

In Exercise 6.2.10, you are asked to show that the ring M of n-by-n matrices with real entries has no ideals other than $\{0\}$ and M itself. This holds equally well for matrix rings over any field. A ring R with no ideals other than $\{0\}$ and R itself is said to be *simple*.

Proposition 6.2.18.

(a) *Let $\{I_\alpha\}$ be any collection of ideals in a ring R. Then $\cap_\alpha I_\alpha$ is an ideal of R.*

(b) *Let I_n be an increasing sequence of ideals in a ring R. Then $\cup_n I_n$ is an ideal of R.*

(c) *Let I and J be two ideals in a ring R. Then*

$$IJ = \{a_1 b_1 + a_2 b_2 + \cdots + a_s b_s : s \geq 1, a_i \in I, b_i \in J\}$$

is an ideal in R, and $IJ \subseteq I \cap J$.

(d) *Let I and J be two ideals in a ring R. Then $I + J = \{a + b : a \in I$ and $b \in J\}$ is an ideal in R.*

Proof. We leave parts (a), (c), and (d) as exercises. For part (b), let $x, y \in I = \cup_n I_n$. Then there exist $k, \ell \in \mathbb{N}$ such that $x \in I_k$ and $y \in I_\ell$. If $n = \max\{k, \ell\}$, then $x \in I_k \subseteq I_n$ and $y \in I_\ell \subseteq I_n$. Therefore, $x + y \in I_n \subseteq I$. If $x \in I$ and $r \in R$, then there exists $n \in \mathbb{N}$ such that $x \in I_n$. Then $rx, xr \in I_n \subseteq I$. Thus I is an ideal. ■

If $\mathcal{S}$ is a subset of a ring R, there is a smallest ideal of R that contains $\mathcal{S}$, called the *ideal generated by* S, which is the intersection of all ideals of R containing $\mathcal{S}$. Assuming that R has an identity 1, the ideal generated by $\mathcal{S}$ consists of all sums of elements of the form $r_1 s r_2$, where $r_i \in R$ and $s \in \mathcal{S}$.

The ideal generated by $\mathcal{S}$ is in general larger than the subring generated by $\mathcal{S}$; for example, the subring generated by the identity element consists of integer multiples of the identity, but the ideal generated by the identity element is all of R.

Example 6.2.19. Consider a direct sum of rings $R = R_1 \oplus \cdots \oplus R_n$. For each i, $\tilde{R}_i = \{0\} \oplus \cdots \oplus \{0\} \oplus R_i \oplus \{0\} \oplus \cdots \oplus \{0\}$ is an ideal of R.

When is a ring R isomorphic to the direct sum of several rings $A_1, A_2, \dots, A_n$? On the one hand, according to the previous example, the component subrings must actually be ideals. On the other hand, the ring must be isomorphic to the direct product of the A_i, regarded as abelian groups.

Proposition 6.2.20. *Let R be a ring with ideals $A_1, \dots A_s$ such that $R = A_1 + \cdots + A_s$. Then the following conditions are equivalent:*

(a) $(a_1, \dots, a_s) \mapsto a_1 + \cdots + a_s$ *is a group isomorphism of* $A_1 \times \cdots \times A_s$ *onto* R.

(b) $(a_1, \dots, a_s) \mapsto a_1 + \cdots + a_s$ *is a ring isomorphism of* $A_1 \oplus \cdots \oplus A_s$ *onto* R.

(c) *Each element* $x \in R$ *can be expressed as a sum* $x = a_1 + \cdots + a_s$, *with* $a_i \in A_i$ *for all* i, *in exactly one way.*

(d) *If* $0 = a_1 + \cdots + a_s$, *with* $a_i \in A_i$ *for all* i, *then* $a_i = 0$ *for all* i.

Proof. The equivalence of (a), (c), and (d) is by Proposition 3.3.1. Clearly (b) implies (a). Let us assume (a) and show that the map

$$(a_1, \dots, a_s) \mapsto a_1 + \cdots + a_s$$

is actually a ring isomorphism. We have $A_i A_j \subseteq A_i \cap A_j = \{0\}$ if $i \neq j$ [using condition (d)]. Therefore,

$$(a_1 + \cdots + a_s)(b_1 + \cdots + b_s) = a_1 b_1 + \cdots + a_s b_s,$$

whenever $a_i, b_i \in A_i$ for all i. It follows that the map is a ring isomorphism. ■

Exercises 6.2

6.2.1. Show that $A \mapsto \begin{bmatrix} A & 0 \\ 0 & A \end{bmatrix}$ and $A \mapsto \begin{bmatrix} A & 0 \\ 0 & 0 \end{bmatrix}$ are homomorphisms of the ring of 2-by-2 matrices into the ring of 4-by-4 matrices. The former is unital, but the latter is not.

6.2.2. Define a map φ from the ring $\mathbb{R}[x]$ of polynomials with real coefficients into the ring M of 3-by-3 matrices by

$$\varphi(\sum a_i x^i) = \begin{bmatrix} a_0 & a_1 & a_2 \\ 0 & a_0 & a_1 \\ 0 & 0 & a_0 \end{bmatrix}.$$

Show that φ is a unital ring homomorphism. What is the kernel of this homomorphism?

6.2.3. If $\varphi : R \to S$ is a ring homomorphism and R has an identity element 1, show that $e = \varphi(1)$ satisfies $e^2 = e$ and $ex = xe = exe$ for all $x \in \varphi(R)$.

6.2.4. Show that if $\varphi : R \to S$ is a ring homomorphism, then $\varphi(R)$ is a subring of S.

6.2.5. Show that if $\varphi : R \to S$ and $\psi : S \to T$ are ring homomorphisms, then the composition $\psi \circ \varphi$ is a ring homomorphism.

6.2.6. Show that if $\varphi : R \to S$ is a ring homomorphism, then $\ker(\varphi)$ is an ideal of of R and that φ is injective if, and only if, the kernel is (0).

6.2.7. Let S be a subset of a set X. Let R be the ring of real–valued functions on X, and let I be the set of real–valued functions on X whose restriction to S is zero. Show that I is an ideal in R.

6.2.8. Let R be the ring of 3-by-3 upper triangular matrices and I be the set of upper triangular matrices that are zero on the diagonal. Show that I is an ideal in R.

6.2.9. If R is any ring and $x \in R$, then $Rx = \{yx : y \in R\}$ is a left ideal. Similarly, xR is a right ideal and RxR is a (two-sided) ideal.

6.2.10. Show that the ring M of n-by-n matrices over $\mathbb{R}$ has no ideals other than 0 and M. Conclude that any ring homomorphism $\varphi : M \to S$ is either identically zero or is injective. Hint: To begin with, work in the 2-by-2 or 3-by-3 case; when you have done these cases, you will understand the general case as well. Let I be a nonzero ideal, and let $x \in I$ be a nonzero element. Introduce the matrix units E_{ij}, which are matrices with a 1 in the (i, j) position and zeros elsewhere. Observe that the set of E_{ij} is a basis for the linear space of matrices. Show that $E_{ij}E_{kl} = \delta_{jk}E_{il}$. Note that the identity E matrix satisfies $E = \sum_{i=1}^{n} E_{ii}$, and write $x = ExE = \sum_{i,j} E_{ii}xE_{jj}$. Conclude that $y = E_{ii}xE_{jj} \neq 0$ for some pair (i, j). Now, since y is a matrix, it is possible to write $y = \sum_{r,s} y_{rs}E_{r,s}$. Conclude that $y = y_{i,j}E_{i,j}$, and $y_{i,j} \neq 0$, and that, therefore, $E_{ij} \in I$. Now use the multiplication rules for the matrix units to conclude that $E_{rs} \in I$ for all (r, s), and hence $I = M$.

6.2.11. An element e of a ring is called an idempotent if $e^2 = e$. What are the idempotents in the ring of real–valued functions on a set X? What are the idempotents in the ring of continuous real–valued functions on $[0, 1]$?

6.2.12. Find a nontrivial idempotent (i.e., an idempotent different from 0 or 1) in the ring of 2-by-2 matrices with real entries.

6.2.13. Let e be a nontrivial idempotent in a commutative ring R with identity. Show that $R \cong Re \oplus R(1-e)$ as rings.

6.2.14. Find a nontrivial idempotent e in the ring $\mathbb{Z}_{35}$. Show that the decomposition $\mathbb{Z}_{35} \cong \mathbb{Z}_5 \oplus \mathbb{Z}_7$ corresponds to the decomposition $\mathbb{Z}_{35} = \mathbb{Z}_{35}e \oplus \mathbb{Z}_{35}(1-e)$.

6.2.15. Show that a nonzero homomorphism of a simple ring is injective. In particular, a nonzero homomorphism of a field is injective.

6.2.16. Show that the intersection of any family of ideals in a ring is an ideal. Show that the ideal generated by a subset $\mathcal{S}$ of a ring R is the intersection of all ideals J of R such that $\mathcal{S} \subseteq J \subseteq R$.

6.2.17. Let I and J be two ideals in a ring R. Show that

$$IJ = \{a_1b_1 + a_2b_2 + \cdots + a_sb_s : s \geq 1, a_i \in I, b_i \in J\}$$

is an ideal in R, and $IJ \subseteq I \cap J$.

6.2.18. Let I and J be two ideals in a ring R. Show that

$$I + J = \{a + b : a \in I \text{ and } b \in J\}$$

is an ideal in R.

6.2.19. Let R be a ring without identity and $a \in R$. Show that the ideal generated by a in R is equal to $\mathbb{Z}a + Ra + aR + RaR$, where $\mathbb{Z}a$ is the abelian subgroup generated by a, $Ra = \{ra : r \in R\}$, and so on. Show that if R is commutative, then the ideal generated by a is $\mathbb{Z}a + Ra$.

6.2.20. Let M be an ideal in a ring R with identity, and $a \in R \setminus M$. Show that $M + RaR$ is the ideal generated by M and a. How must this statement be altered if R does not have an identity?

6.3. Quotient Rings

In Section 2.7, it was shown that given a group G and a normal subgroup N, we can construct a quotient group G/N and a natural homomorphism from G onto G/N. The program of Section 2.7 can be carried out more or less verbatim with rings and ideals in place of groups and normal subgroups:

For a ring R and an ideal I, we can form the quotient *group* R/I, whose elements are cosets $a + I$ of I in R. The additive group operation in R/I is $(a+I)+(b+I) = (a+b)+I$. Now attempt to define a multiplication in R/I in the obvious way: $(a+I)(b+I) = (ab+I)$. We have to check that this this is well defined. But this follows from the closure of I under multiplication by elements of R; namely, if $a + I = a' + I$ and $b + I = b' + I$, then

$$(ab - a'b') = a(b - b') + (a - a')b' \in aI + Ib \subseteq I.$$

Thus, $ab + I = a'b' + I$, and the multiplication in R/I is well defined.

Theorem 6.3.1. *If I is an ideal in a ring R, then R/I has the structure of a ring, and the quotient map $a \mapsto a + I$ is a surjective ring homomorphism from R to R/I with kernel equal to I. If R has a multiplicative identity, then so does R/I, and the quotient map is unital.*

Proof. Once we have checked that the multiplication in R/I is well defined, it is straightforward to check that it is associative, that the distributive law holds, and that the quotient map is a ring homomorphism. If 1 is the multiplicative identity in R, then $1 + I$ is the multiplicative identity in R/I. ■

Example 6.3.2. The ring $\mathbb{Z}_n$ is the quotient of the ring $\mathbb{Z}$ by the principal ideal $n\mathbb{Z}$. The homomorphism $a \mapsto [a] = a + n\mathbb{Z}$ is the quotient homomorphism.

Example 6.3.3. For K a field, any ideal in $K[x]$ is of the form $(f) = fK[x]$ for some polynomial f according to Proposition 6.2.17. For any $g(x) \in K[x]$, there exist polynomials q, r such that $g(x) = q(x)f(x) + r(x)$, and $\deg(r) < \deg(f)$. Thus $g(x) + (f) = r(x) + (f)$. In other words, $K[x]/(f) = \{r(x) + (f) : \deg(r) < \deg(f)\}$. The multiplication in $K[x]/(f)$ is as follows: Given polynomials $r(x)$ and $s(x)$ each of degree less than the degree of f,the product $(r(x) + (f))(s(x) + (f)) = r(x)s(x) + (f) = a(x) + (f)$, where $a(x)$ is the remainder upon division of $r(x)s(x)$ by $f(x)$.

Let's look at the particular example $K = \mathbb{R}$ and $f(x) = x^2 + 1$. Then $\mathbb{R}[x]/(f)$ consists of cosets $a + bx + (f)$ represented by linear polynomials. Furthermore, we have the computational rule

$$x^2 + (f) = x^2 + 1 - 1 + (f) = -1 + (f).$$

Thus

$$(a + bx + (f))(a' + b'x + (f)) = (aa' - bb') + (ab' + a'b)x + (f).$$

All of the homomorphism theorems for groups, which were presented in Section 2.7, have analogues for rings. The basic homomorphism theorem for rings is the following.

Theorem 6.3.4. *(Homomorphism theorem for rings). Let $\varphi : R \longrightarrow S$ be a surjective homomorphism of rings with kernel I. Let $\pi : R \longrightarrow R/I$ be the quotient homomorphism. There is a ring isomorphism $\tilde{\varphi} : R/I \longrightarrow S$ satisfying $\tilde{\varphi} \circ \pi = \varphi$. (See the following diagram.)*

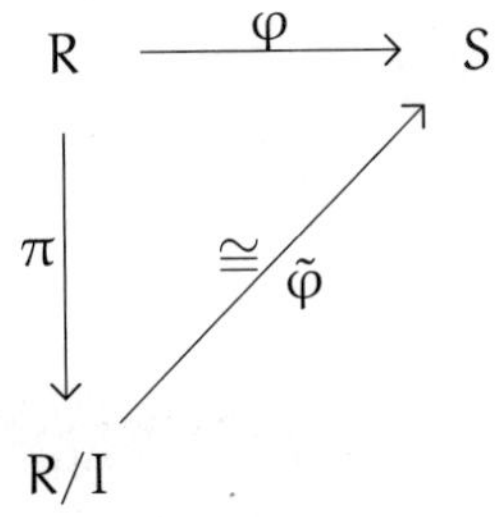

Proof. The homomorphism theorem for groups (Theorem 2.7.6) gives us an isomorphism of abelian groups $\tilde{\varphi} : R/I \to S$ satisfying $\tilde{\varphi} \circ \pi = \varphi$. We have only to verify that $\tilde{\varphi}$ also respects multiplication. But this follows at once from the definition of the product on R/I:

$$\begin{aligned}\tilde{\varphi}(a + I)(b + I) &= \tilde{\varphi}(ab + I) \\ &= \varphi(ab) = \varphi(a)\varphi(b) = \tilde{\varphi}(a + I)\tilde{\varphi}(b + I).\end{aligned}$$

■

Example 6.3.5. Define a homomorphism $\varphi : \mathbb{R}[x] \to \mathbb{C}$ by evaluation of polynomials at $i \in \mathbb{C}$, $\varphi(g(x)) = g(i)$. For example, $\varphi(x^3 - 1) = i^3 - 1 = -i - 1$. This homomorphism is surjective because $\varphi(a + bx) = a + bi$. The kernel of φ consists of all polynomials g such that $g(i) = 0$. The kernel contains at least the ideal $(x^2 + 1) = (x^2 + 1)\mathbb{R}[x]$ because $i^2 + 1 = 0$. On the other hand, if $g \in \ker(\varphi)$, write $g(x) = (x^2 + 1)q(x) + (a + bx)$; evaluating at i, we get $0 = a + bi$, which is possible only if $a = b = 0$. Thus g is a multiple of $x^2 + 1$. That is $\ker(\varphi) = (x^2 + 1)$. By the homomorphism theorem for rings, $\mathbb{R}[x]/(x^2 + 1) \cong \mathbb{C}$ as rings. In particular, since $\mathbb{C}$ is a field, $\mathbb{R}[x]/(x^2 + 1)$ is a field. Note that we have already calculated explicitly in Example 6.3.3 that multiplication in $\mathbb{R}[x]/(x^2 + 1)$ satisfies the same rule as multiplication in $\mathbb{C}$.

Example 6.3.6. Let R be a ring with identity containing ideals B_1, $\dots$, B_s. Let $B = \cap_i B_i$. Suppose that $B_i + B_j = R$ for all $i \neq j$. Then $R/B \cong R/B_1 \oplus \cdots \oplus R/B_s$. In fact, $\varphi : r \mapsto (r + B_1, \dots, r + B_s)$ is a homomorphism of R *into* $R/B_1 \oplus \cdots \oplus R/B_s$ with kernel B, so $R/B \cong \varphi(R)$. The problem is to show that φ is surjective. Fix i and for each $j \neq i$ find $r_j' \in B_i$ and $r_j \in B_j$ such that $r_j' + r_j = 1$. Consider the product of all the $(r_j' + r_j)$ (in any order). When the product is expanded, all the summands except for one contain at least one factor r_j' in the ideal B_i, so all of these summands are in B_i. The remaining summand is the product of all of the $r_j \in B_j$, so it lies in $\cap_{j \neq i} B_j$. Thus we get $1 = a_i + b_i$, where $b_i \in B_i$ and $a_i \in \cap_{j \neq i} B_j$. The image of a_i in R/B_j is zero for $j \neq i$, but $a_i + B_i = 1 + B_i$. Now if $(r_1, \ldots, r_s)$ is an arbitrary sequence of elements of R, then $\varphi(r_1 a_1 + r_2 a_2 + \cdots + r_s a_s) = (r_1 + B_1, r_2 + B_2, \ldots, r_s + B_s)$, so φ is surjective.

Proposition 6.3.7. *Let $\varphi : R \longrightarrow S$ be a ring homomorphism of R onto S, and let I denote its kernel.*

(a) *If B is a subring of S, then $\varphi^{-1}(B)$ is a subring of R containing I.*
(b) *$B \mapsto \varphi^{-1}(B)$ is a bijection between the subrings of S and the subrings of R containing I.*
(c) *$J \subseteq S$ is an ideal of S if, and only if, $\varphi^{-1}(J)$ is an ideal of R.*

Proof. By Proposition 2.7.12, $B \mapsto \varphi^{-1}(B)$ is a bijection between the subgroups of S and the subgroups of R containing I. It remains to show that this bijection carries subrings to subrings and ideals to ideals. We leave this as an exercise. ■

Proposition 6.3.8. *Let $\varphi : R \longrightarrow S$ be a surjective ring homomorphism. Let $\bar{I}$ be an ideal of S and let $I = \varphi^{-1}(\bar{I})$. Then $x + I \mapsto \varphi(x) + \bar{I}$ is an isomorphism of R/I onto $S/\bar{I}$.*

Proof. Exercise 6.3.8. ■

Proposition 6.3.9. *Let $\varphi : R \longrightarrow S$ be a surjective homomorphism of rings with kernel I. Let A be a subring of R. Then $\varphi^{-1}(\varphi(A)) = A + I = \{a + r : a \in A \text{ and } r \in I\}$ is a subring of R containing I. Furthermore,*

$$(A + I)/I \cong \varphi(A) \cong A/(A \cap I).$$

Proof. Exercise 6.3.5. ■

An ideal M in a ring R is called *proper* if $M \neq R$. An ideal is said to be *nontrivial* if $M \neq \{0\}$.

Definition 6.3.10. An ideal M in a ring R is called *maximal* if $M \neq R$ there are no ideals strictly between M and R; that is, the only ideals containing M are M and R.

Recall that a ring is called simple if it has no ideals other that the trivial ideal $\{0\}$ and the whole ring; so a nonzero ring is simple precisely when $\{0\}$ is a maximal ideal.

Proposition 6.3.11. *A proper ideal M in R is maximal if, and only if, R/M is simple.*

Proof. Exercise 6.3.6. ■

Proposition 6.3.12. *A (nonzero) commutative ring R with multiplicative identity is a field if, and only if, R is simple.*

Proof. Suppose R is simple and $x \in R$ is a nonzero element. The ideal Rx is nonzero since $x = 1x \in Rx$; because R is simple, $R = Rx$. Hence there is a $y \in R$ such that $1 = yx$. Conversely, suppose R is a field and M is a nonzero ideal. Since M contains a nonzero element x, it also contains $r = rx^{-1}x$ for any $r \in R$; that is, $M = R$. ■

Corollary 6.3.13. *If M is a proper ideal in a commutative ring R with 1, then R/M is a field if, and only if, M is maximal.*

Proof. This follows from Propositions 6.3.11 and 6.3.12. ■

Exercises 6.3

6.3.1. Work out the rule of computation in the ring $\mathbb{R}[x]/(f)$, where $f(x) = x^2 - 1$. Note that the quotient ring consists of elements $a + bx + (f)$. Compare Example 6.3.3.

6.3.2. Work out the rule of computation in the ring $\mathbb{R}[x]/(f)$, where $f(x) = x^3 - 1$. Note that the quotient ring consists of elements $a + bx + cx^2 + (f)$. Compare Example 6.3.3.

6.3.3. Complete the proof of Proposition 6.3.7.

6.3.4. Prove Proposition 6.3.8, following the pattern of the proof of Proposition 2.7.17.

6.3.5. Prove Proposition 6.3.9, following the pattern of the proof of Proposition 2.7.18.

6.3.6. Prove that an ideal M in R is maximal if, and only if, R/M is simple.

6.3.7.

(a) Show that $n\mathbb{Z}$ is maximal ideal in $\mathbb{Z}$ if, and only if, $\pm n$ is a prime.
(b) Show that $(f) = fK[x]$ is a maximal ideal in $K[x]$ if, and only if, f is irreducible.
(c) Conclude that $\mathbb{Z}_n = \mathbb{Z}/n\mathbb{Z}$ is a field if, and only if, $\pm n$ is prime, and that $K[x]/(f)$ is a field if, and only if, f is irreducible.

6.3.8. If J is an ideal of the ring R, show that $J[x]$ is an ideal in $R[x]$ and furthermore $R[x]/J[x] \cong (R/J)[x]$. Hint: Find a natural homomorphism from $R[x]$ onto $(R/J)[x]$ with kernel $J[x]$.

6.3.9. For any ring R, and any natural number n, we can define the matrix ring $\mathrm{Mat}_n(R)$ consisting of n-by-n matrices with entries in R. If J is an ideal of R, show that $\mathrm{Mat}_n(J)$ is an ideal in $\mathrm{Mat}_n(R)$ and furthermore $\mathrm{Mat}_n(R)/\mathrm{Mat}_n(J) \cong \mathrm{Mat}_n(R/J)$. Hint: Find a natural homomorphism from $\mathrm{Mat}_n(R)$ onto $\mathrm{Mat}_n(R/J)$ with kernel $\mathrm{Mat}_n(J)$.

6.3.10. Let R be a commutative ring. Show that $R[x]/xR[x] \cong R$.

6.3.11. This exercise gives a version of the Chinese remainder theorem.

(a) Let R be a ring, P and Q ideals in R, and suppose that $P \cap Q = \{0\}$, and $P + Q = R$. Show that the map $x \mapsto (x + P, x + Q)$ is an isomorphism of R onto $R/P \oplus R/Q$. Hint: Injectivity is clear. For surjectivity, show that for each $a, b \in R$, there exist $x \in R$, $p \in P$, and $q \in Q$ such that $x + p = a$, and $x + q = b$.
(b) More generally, if $P+Q = R$, show that $R/(P \cap Q) \cong R/P \oplus R/Q$.

6.3.12.

(a) Show that integers m and n are relatively prime if, and only if, $m\mathbb{Z} + n\mathbb{Z} = \mathbb{Z}$ if, and only if, $m\mathbb{Z} \cap n\mathbb{Z} = mn\mathbb{Z}$. Conclude that if m and n are relatively prime, then $\mathbb{Z}_{mn} \cong \mathbb{Z}_m \oplus \mathbb{Z}_n$ as rings.
(b) State and prove a generalization of this result for the ring of polynomials $K[x]$ over a field K.

6.4. Integral Domains

Definition 6.4.1. An integral domain is a commutative ring with identity element 1 in which the product of any two nonzero elements is nonzero.

You may think at first that the product of nonzero elements in a ring is always nonzero, but you already know of examples where this is not the case! Let R be the ring of real–valued functions on a set X and let A be a proper subset of X. Let f be the characteristic function of A, that is, the function satisfying $f(a) = 1$ if $a \in A$ and $f(x) = 0$ if $x \in X \setminus A$. Then f and $1 - f$ are nonzero elements of R whose product is zero.

For another example, let x be the 2-by-2 matrix $\begin{bmatrix} 0 & 1 \\ 0 & 0 \end{bmatrix}$. Compute that $x \neq 0$ but $x^2 = 0$. (An element n in a ring is said to be *nilpotent* if $n^k = 0$ for some k.)

Example 6.4.2.

(a) The ring of integers $\mathbb{Z}$ is an integral domain.
(b) Any field is an integral domain.
(c) If R is an integral domain, then $R[x]$ is an integral domain. In particular, $K[x]$ is an integral domain for any field K.
(d) If R is an integral domain and R' is a subring containing the identity, then R' is an integral domain.
(e) The ring of formal power series $R[[x]]$ with coefficients in an integral domain is an integral domain.

You are asked to verify these examples in the Exercises.

There are two common constructions of fields from integral domains. One construction is treated in Corollary 6.3.13 and Exercise 6.3.7. Namely, if R is a commutative ring with 1 and M is a maximal ideal, then the quotient ring R/M is a field.

Another construction is that of the *field of fractions*[1] of an integral domain. This construction is known to you from the formation of the rational numbers as fractions of integers. Given an integral domain R, we wish to construct a field from symbols of the form a/b, where $a, b \in R$ and $b \neq 0$. You know that in the rational numbers, $2/3 = 8/12$; the rule is $a/b = a'/b'$ if $ab' = a'b$. It seems prudent to adopt the same rule for any integral domain R.

[1] Do not confuse the terms *field of fractions* and *quotient ring* or *quotient field*.

Lemma 6.4.3. *Let R be an integral domain. Let S be the set of symbols of the form a/b, where $a, b \in R$ and $b \neq 0$. The relation $a/b \sim a'/b'$ if $ab' = a'b$ is an equivalence relation on S.*

Proof. Exercise 6.4.9. ■

Let us denote the quotient of S by the equivalence relation $\sim$ by $Q(R)$. For $a/b \in S$, write $[a/b]$ for the equivalence class of a/b, an element of $Q(R)$.

Now, we attempt to define operations on $Q(R)$, following the guiding example of the rational numbers. The sum $[a/b] + [c/d]$ will have to be defined as $[(ad+bc)/bd]$ and the product $[a/b][c/d]$ as $[ac/bd]$. *As always, when we define operations on equivalence classes in terms of representatives of those classes, the next thing that has to be done is to check that the operations are well defined.* You are asked to check that addition and multiplication are well defined in Exercise 6.4.10.

It is now straightforward, if slightly tedious, to check that $Q(R)$ is a field, and that the map $a \mapsto [a/1]$ is an injective ring homomorphism of R into $Q(R)$; thus R can be considered as a subring of $Q(R)$. Moreover, any injective homomorphism of R into a field F extends to an injective homomorphism of $Q(R)$ into F:

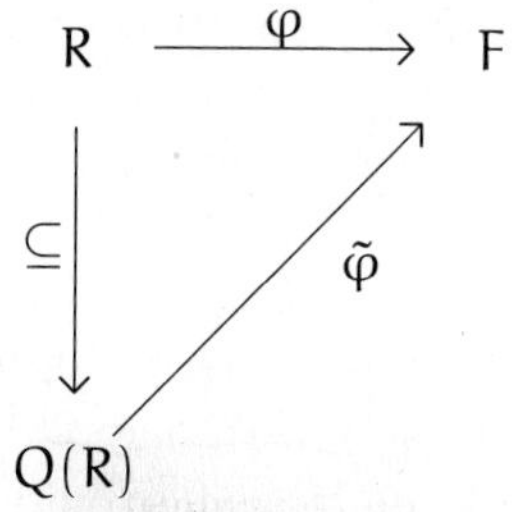

The main steps in establishing these facts are the following:

1. $Q(R)$ is a ring with zero element $0 = [0/1]$ and multiplicative identity $1 = [1/1]$. In $Q(R)$, $0 \neq 1$.
2. $[a/b] = 0$ if, and only if, $a = 0$. If $[a/b] \neq 0$, then $[a/b][b/a] = 1$. Thus $Q(R)$ is a field.
3. $a \mapsto [a/1]$ is an injective homomorphism of R into $Q(R)$.
4. If $\varphi : R \to F$ is an injective homomorphism of R into a field F, then $\tilde{\varphi} : [a/b] \mapsto \varphi(a)/\varphi(b)$ defines an injective homomorphism of $Q(R)$ into F, which extends φ.

See Exercises 6.4.11 and 6.4.12.

Proposition 6.4.4. *If* R *is an integral domain, then* $Q(R)$ *is a field containing* R *as a subring. Moreover, any injective homomorphism of* R *into a field* F *extends to an injective homomorphism of* $Q(R)$ *into* F.

Example 6.4.5. $Q(\mathbb{Z}) = \mathbb{Q}$.

Example 6.4.6. $Q(K[x])$ is the field of rational functions in one variable. This field is denoted $K(x)$.

Example 6.4.7. $Q(K[x_1, \dots, x_n])$ is the field of rational functions in n variables. This field is denoted $K(x_1, \dots, x_n)$.

Example 6.4.8. If R is the ring of symmetric polynomials in n variables, then $Q(R)$ is the ring of symmetric rational functions in n variables.

We have observed in Example 6.2.3 that in a ring R with multiplicative identity 1, the additive subgroup $\langle 1 \rangle$ generated by 1 is a subring and the map $k \mapsto k \cdot 1$ is a ring homomorphism from $\mathbb{Z}$ onto $\langle 1 \rangle \subseteq R$. If this ring homomorphism is injective, then $\langle 1 \rangle \cong \mathbb{Z}$ as rings. Otherwise, the kernel of the homomorphism is $n\mathbb{Z}$ for some $n \in \mathbb{N}$ and $\langle 1 \rangle \cong \mathbb{Z}/n\mathbb{Z} = \mathbb{Z}_n$ as rings.

If R is an integral domain, then any subring containing the identity is also an integral domain, so, in particular, $\langle 1 \rangle$ is an integral domain. If $\langle 1 \rangle$ is finite, so ring isomorphic to $\mathbb{Z}_n$ for some n, then n must be prime. ($\mathbb{Z}_n$ is an integral domain if, and only if, n is prime.)

Definition 6.4.9. If the subring $\langle 1 \rangle$ of an integral domain R generated by the identity is isomorphic to $\mathbb{Z}$, the integral domain is said to have *characteristic* 0. If the subring C is isomorphic to $\mathbb{Z}_p$ for a prime p, then R is said to have characteristic p.

A quotient of an integral domain need not be an integral domain; for example, $\mathbb{Z}_{12}$ is a quotient of $\mathbb{Z}$. On the other hand, a quotient of a ring with nontrivial zero divisors can be an integral domain; $\mathbb{Z}_3$ is a quotient of $\mathbb{Z}_{12}$. Those ideals J in a ring R such that R/J has no nontrivial zero divisors can be easily characterized. An ideal J is said to be *prime* if for all $a, b \in R$, if $ab \in J$, then $a \in J$ or $b \in J$.

Proposition 6.4.10. *Let* J *be an ideal in a ring* R. J *is prime if, and only if,* R/J *has no nontrivial zero divisors. In particular, if* R *is commutative with identity, then* J *is prime if, and only if,* R/J *is an integral domain.*

Proof. Exercise 6.4.14. ∎

Example 6.4.11. Every ideal in $\mathbb{Z}$ has the form $d\mathbb{Z}$ for some $d > 0$. The ideal $d\mathbb{Z}$ is prime if, and only if, d is prime. The proof of this assertion is left as an exercise.

Exercises 6.4

6.4.1. Let R be a commutative ring.

(a) Show that the set N of nilpotent elements of R is an ideal in R.
(b) Show that R/N has no nonzero nilpotent elements.
(c) Show that if S is an integral domain and $\varphi : R \to S$ is a homomorphism, then $N \subseteq \ker(\varphi)$.

6.4.2. If x is a nilpotent element in a ring with identity, show that $1 - x$ is invertible. Hint: Think of the power series expansion for $\dfrac{1}{1-t}$, when t is a real variable.

6.4.3. An element e in a ring is called an idempotent if $e^2 = e$. Show that the only idempotents in an integral domain are 1 and 0. (We say that an idempotent is nontrivial if it is different from 0 or 1. So the result is that an integral domain has no nontrivial idempotents.)

6.4.4. Show that if R is an integral domain, then the ring of polynomials $R[x]$ with coefficients in R is an integral domain.

6.4.5. Generalize the results of the previous problem to rings of polynomials in several variables.

6.4.6. Show that if R is an integral domain, then the ring of formal power series $R[[x]]$ with coefficients in R is an integral domain. What are the units in $R[[x]]$?

6.4.7.

(a) Show that a subring R' of an integral domain R is an integral domain, if $1 \in R'$.
(b) The Gaussian integers are the complex numbers whose real and imaginary parts are integers. Show that the set of Gaussian integers is an integral domain.
(c) Show that the ring of symmetric polynomials in n variables is an integral domain.

6.4.8. Show that a quotient of an integral domain need not be an integral domain.

6.4.9. Prove Lemma 6.4.3.

6.4.10. Show that addition and multiplication on $Q(R)$ is well defined. This amounts to the following: Suppose $a/b \sim a'/b'$ and $c/d \sim c'/d'$ and show that $(ad+bc)/bd \sim (a'd'+c'b')/b'd'$ and $ac/bd \sim a'c'/b'd'$.

6.4.11.

(a) Show that $Q(R)$ is a ring with identity element $[1/1]$ and $0 = [0/1]$.

(b) Show that $[a/b] = 0$ if, and only if, $a = 0$.

(c) Show that if $[a/b] \neq 0$, then $[b/a]$ is the multiplicative inverse of $[a/b]$. Thus $Q(R)$ is a field.

6.4.12.

(a) Show that $a \mapsto [a/1]$ is an injective unital ring homomorphism of R into $Q(R)$. In this sense $Q(R)$ is a field containing R.

(b) If F is a field such that $F \supseteq R$, show that there is an injective unital homomorphism $\varphi : Q(R) \to F$ such that $\varphi([a/1]) = a$ for $a \in R$.

6.4.13. If R is the ring of Gaussian integers, show that $Q(R)$ is isomorphic to the subfield of $\mathbb{C}$ consisting of complex numbers with rational real and imaginary parts.

6.4.14. Let J be an ideal in a ring R. Show that J is prime if, and only if, R/J has no nontrivial zero divisors. (In particular, if R is commutative with identity, then J is prime if, and only if, R/J is an integral domain.)

6.4.15. Every ideal in $\mathbb{Z}$ has the form $d\mathbb{Z}$ for some $d > 0$. Show that the ideal $d\mathbb{Z}$ is prime if, and only if, d is prime.

6.4.16. Show that a maximal ideal in a commutative ring with identity is prime.

6.5. Euclidean Domains, Principal Ideal Domains, and Unique Factorization

We have seen two examples of integral domains with a good theory of factorization, the ring of integers $\mathbb{Z}$ and the ring of polynomials $K[x]$ over a field K. In both of these rings R, every nonzero, noninvertible element has an essentially unique factorization into irreducible factors.

The common feature of these rings, which was used to establish unique factorization is a *Euclidean function* $d : R \setminus \{0\} \to \mathbb{N}$ with the property that $d(fg) \geq \max\{d(f), d(g)\}$ and for each $f, g \in R \setminus \{0\}$ there exist $q, r \in R$ such that $f = qg + r$ and $r = 0$ or $d(r) < d(g)$.

For the integers, the Euclidean function is $d(n) = |n|$. For the polynomials, the function is $d(f) = \deg(f)$.

Definition 6.5.1. Call an integral domain R a *Euclidean domain* if it admits a Euclidean function.

Let us consider one more example of a Euclidean domain, in order to justify having made the definition.

Example 6.5.2. Let $\mathbb{Z}[i]$ be the ring of Gaussian integers, namely, the complex numbers with integer real and imaginary parts. For the "degree" map d take $d(z) = |z|^2$. We have $d(zw) = d(z)d(w)$. The trick to establishing the Euclidean property is to work temporarily in the field of fractions, the set of complex numbers with rational coefficients. Let z, w be nonzero elements in $\mathbb{Z}[i]$. There is at least one point $q \in \mathbb{Z}[i]$ whose distance to the complex number z/w is minimal; q satisfies $|\Re(q-z/w)| \leq 1/2$ and $|\Im(q-z/w)| \leq 1/2$. Write $z = qw + r$. Since z and $qw \in \mathbb{Z}[i]$, it follows that $r \in \mathbb{Z}[i]$. But $r = (z - qw) = (z/w - q)w$, so $d(r) = |z/w - q|^2 d(w) \leq (1/2)d(w)$. This completes the proof that the ring of Gaussian integers is Euclidean.

Let us introduce several definitions related to divisibility before we continue the discussion:

Definition 6.5.3. Let a be a nonzero, nonunit element of a commutative ring with identity. A *proper factorization* of a is an equality $a = bc$, where neither b nor c is a unit. The elements b and c are said to be *proper factors* of a. An element is said to be *irreducible* if it has no proper factorizations.

Definition 6.5.4. Two elements in an integral domain are said to be *associates* if each divides the other. In this case, each of the elements is equal to a unit times the other.

Definition 6.5.5. A greatest common divisor (gcd) of two elements $a_i, \ldots, a_s$ in an integral domain R is an element c such that

(a) c divides each a_i, and

(b) For all d, if d divides each a_i, then d divides c.

Elements $a_i, \ldots, a_s$ in an integral domain are said to be *relatively prime* if 1 is a gcd of $\{a_i\}$.

Given any Euclidean domain R, we can follow the proofs given for the integers and the polynomials over a field to establish the following results:

Theorem 6.5.6. *Let* R *be a Euclidean domain.*

(a) *Two nonzero elements* f *and* g ∈ R *have a greatest common divisor which is contained in the ideal* $Rf + Rg$. *The greatest common divisor is unique up to multiplication by a unit.*

(b) *Two elements* f *and* g ∈ R *are relatively prime if, and only if,* $1 \in Rf + Rg$.

(c) *Every ideal in* R *is principal.*

(d) *If an irreducible* p *divides the product of two nonzero elements, then it divides one or the other of them.*

(e) *Every nonzero, nonunit element has a factorization into irreducibles, which is unique (up to units and up to order of the factors).*

Proof. Exercises 6.5.1 through 6.5.3. ■

This result suggests making the following definitions:

Definition 6.5.7. An integral domain R is a *principal ideal domain* (PID) if every ideal of R is principal.

Definition 6.5.8. An integral domain is a *unique factorization domain* (UFD) if every nonzero element has a factorization by irreducibles that is unique up to order and multiplication by units.

According to Theorem 6.5.6, every Euclidean domain is both a principal ideal domain and a unique factorization domain. The rest of this section will be devoted to a further study of principal ideal domains and unique factorization domains; the main result will be that every principal ideal domain is a unique factorization domain. Thus we have the implications:

$$\text{Euclidean Domain} \implies \text{Principal Ideal Domain}$$
$$\implies \text{Unique Factorization Domain} \implies \text{Integral Domain}$$

It is natural to ask whether any of these implications can be reversed. The answer is no.

Example 6.5.9. $\mathbb{Z}[(1 + i\sqrt{19})/2]$ *is a principal ideal domain that is not Euclidean.* The proof of this is somewhat intricate and will not be presented here. See D. S. Dummit and R. M. Foote, *Abstract Algebra*, 2nd ed., Prentice Hall, 1999, pp. 278 and 283.

Example 6.5.10. $\mathbb{Z}[x]$ *is a unique factorization domain that is not a principal ideal domain.* We will show later that if R is a unique factorization domain, then the polynomial ring R[x] is also a unique factorization domain; see Theorem 6.6.1. This implies that $\mathbb{Z}[x]$ is a UFD. Let's check that the ideal $3\mathbb{Z}[x] + x\mathbb{Z}[x]$ is not principal; this ideal is equal to the set of polynomials in $\mathbb{Z}[x]$ whose constant coefficient is a multiple of 3. A divisor of 3 in $\mathbb{Z}[x]$ must be of degree 0 (i.e., an element of $\mathbb{Z} \subseteq \mathbb{Z}[x]$, and thus a divisor of 3 in $\mathbb{Z}$). The only possibilities are $\pm 1, \pm 3$. But 3 does not divide x in $\mathbb{Z}[x]$. Therefore, the only common divisors of 3 and x in $\mathbb{Z}[x]$ are the units ± 1. It follows that $3\mathbb{Z}[x] + x\mathbb{Z}[x]$ is not a principal ideal in $\mathbb{Z}[x]$.

Example 6.5.11. $\mathbb{Z}[\sqrt{-5}]$ *is an integral domain that is not a unique factorization domain.* $\mathbb{Z}[\sqrt{-5}]$ denotes the subring of $\mathbb{C}$ generated by $\mathbb{Z}$ and $\sqrt{-5} = i\sqrt{5}$; $\mathbb{Z}[\sqrt{-5}] = \{x + iy\sqrt{5} : x, y \in \mathbb{Z}\}$. It is useful to consider the function $N(z) = |z|^2$ on $\mathbb{Z}[\sqrt{-5}]$. N is multiplicative, $N(zw) = N(z)N(w)$. Furthermore, $N(x+iy\sqrt{5}) = x^2+5y^2 \geq 6$ if $x \neq 0$ and $y \neq 0$. Using this, it is easy to see that the only units in $\mathbb{Z}[\sqrt{-5}]$ are ± 1, and that $2, 3$ and $1 \pm i\sqrt{5}$ are all irreducible elements, no two of which are associates. We have $6 = 2 \cdot 3 = (1+i\sqrt{5})(1-i\sqrt{5})$. These are two essentially different factorizations of 6 by irreducibles.

We can use the function $N(z)$ to show that every nonzero, nonunit element of $\mathbb{Z}[\sqrt{-5}]$ has at least one factorization by irreducibles; $\mathbb{Z}[\sqrt{-5}]$ is not a unique factorization domain because some elements admit two essentially different irreducible factorizations. The next example is of an integral domain that fails to be a unique factorization domain because it has elements with no irreducible factorization at all.

Example 6.5.12. *The ring* $R = \mathbb{Z} + x\mathbb{Q}[x]$ *has elements admitting no factorization by irreducibles.* R consists of all polynomials with rational coefficients whose constant term is an integer. R is an integral domain, as it is a subring of $\mathbb{Q}[x]$. The units in R are ± 1. No rational multiple of x is irreducible, because $x = 2(\frac{1}{2}x)$ is a proper factorization. If x is written as a product of several elements in R, exactly one of these elements must be a rational multiple of x, while the remaining factors must be integers. Therefore, x has no factorization by irreducibles.

Factorization in Principal Ideal Domains

We are going to show that a principal ideal domain is a unique factorization domain. The proof has two parts: First we show that every nonzero, nonunit element of a PID has at least one factorization by irreducibles. Then we show that an element cannot have two essentially different factorizations by irreducibles.

Lemma 6.5.13. *Suppose that R is a principal ideal domain. If $a_1R \subseteq a_2R \subseteq a_3R \cdots$ is an increasing sequence of ideals in R, then there exists an $n \in \mathbb{N}$ such that $\cup_{m \geq 1} a_m R = a_n R$.*

Proof. Let $I = \cup_n a_n R$. Then I is an ideal of R, by Proposition 6.2.18. Since R is a PID, there exists an element $b \in I$ such that $I = bR$. Since $b \in I$, there exists an n such that $b \in a_n R$. It follows that $bR \subseteq a_n R \subseteq I = bR$; so $I = a_n R$. ■

Lemma 6.5.14. *Let R be a commutative ring with multiplicative identity element, and let $a, b \in R$. Then $a|b \Leftrightarrow bR \subseteq aR$. Moreover, a is a proper factor of $b \Leftrightarrow bR \underset{\neq}{\subset} aR$.*

Proof. Exercise 6.5.13. ■

Lemma 6.5.15. *Let R be a principal ideal domain. Then every nonzero element of R that is not a unit has at least one factorization by irreducible elements.*

Proof. Let R be a principal ideal domain. Suppose that R has a nonzero element a that is not a unit and has no factorization by irreducible elements. Then a itself cannot be irreducible, so a admits a proper factorization $a = bc$. At least one of b and c does not admit a factorization by irreducibles; suppose without loss of generality that b has this property. Now we have $aR \underset{\neq}{\subset} bR$, where a and b are nonunits that do not admit a factorization by irreducibles.

An induction based on the previous paragraph gives a sequence $a_1, a_2, \dots$ in R such that for all n, a_n is a nonunit that does not admit a factorization by irreducibles, and

$$a_1R \underset{\neq}{\subset} a_2R \underset{\neq}{\subset} \cdots \underset{\neq}{\subset} a_nR \underset{\neq}{\subset} a_{n+1}R \underset{\neq}{\subset} \cdots .$$

But the existence of such a sequence contradicts Lemma 6.5.13. ■

In general integral domains, we have to distinguish between irreducible elements and prime elements.

Definition 6.5.16. Let R be an integral domain. A nonunit element $a \in R$ is said to be *prime* if whenever a divides a product bc, then a divides one of the elements b or c.

An easy induction shows that whenever a prime element a divides a product of several elements $b_1 b_2 \cdots b_s$, then a divides one of the elements b_i.

In any integral domain, every prime element is irreducible (Exercise 6.5.19). According to Proposition 6.5.6(d), in a Euclidean domain, every irreducible element is prime. In the next lemma, we show that principal ideal domains also have the property that every irreducible element is prime.

Lemma 6.5.17. *Let R be a principal ideal domain. Then every irreducible element of R is prime.*

Proof. Suppose $a \in R$ is irreducible and a divides bc. We have to show that a divides at least one of b and c; to do this, we assume that a does not divide b and show that a must divide c.

Since a does not divide b, b is not an element of the ideal aR, and, therefore, the ideal $aR + bR$ properly contains aR. Since R is a principal ideal domain, there is an element d such that $dR = aR + bR$; in particular, d is a common divisor of a and b and $d = as + bt$ for some $s, t \in R$. Write $a = dq$; since a is irreducible, one of d or q must be a unit.

If q were a unit, then $d = aq^{-1} \in aR$. Hence $b \in dR \subseteq aR$. But this would contradict the assumption that a does not divide b. So we conclude that q is not a unit and d is a unit.

Then we have $1 = dd^{-1} = (as + bt)d^{-1}$, and thus $c = (as + bt)d^{-1}c = asd^{-1}c + btd^{-1}c$. Since a divides bc, a divides both summands on the right–hand side, and, therefore, a divides c. ■

Theorem 6.5.18. *Every principal ideal domain is a unique factorization domain.*

Proof. Let R be a principal ideal domain, and let $a \in R$ be a nonzero, nonunit element. According to Lemma 6.5.15, a has at least one factorization by irreducibles. We have to show that in any two such factorizations, the factors are the same, up to order and multiplication by units.

So suppose that $0 \le r \le s$, and that $p_1, p_2, \ldots, p_r, q_1, q_2, \ldots, q_s$ are irreducibles, and

$$p_1 p_2 \cdots p_r = q_1 q_2 \cdots q_s. \tag{6.5.1}$$

I claim that $r = s$, and (possibly after permuting the q_i) there exist units $c_1, \ldots, c_r$ such that $q_i = c_i p_i$ for all i.

If $r = 0$, Equation (6.5.1) reads $1 = q_1 \cdots q_s$. It follows that $s = 0$, since irreducibles are not invertible. If $r = 1$, Equation (6.5.1) reads $p_1 = q_1 \cdots q_s$; it follows that $s = 1$ and $p_1 = q_1$, since otherwise p_1 has a proper factorization.

Suppose $r \ge 2$. We can assume inductively that the assertion holds when r and s are replaced with r' and s' with $r' < r$ and $s' < s$.

Since p_1 is prime, according to Lemma 6.5.17, p_1 must divide one of the q_i; we can suppose without loss of generality that p_1 divides q_1. But since q_1 is irreducible, and p_1 is not a unit, $q_1 = p_1 c$, where c is a unit. Thus $p_1 p_2 \cdots p_r = p_1 (c q_2) q_3 \cdots q_s$. Because we are working in an integral domain, we can cancel p_1, obtaining $p_2 \cdots p_r = (c q_2) q_3 \cdots q_s$. By the induction assumption, $r = s$ and there exist units $c_2, \ldots, c_r$ such that (possibly after permuting the q_i) $q_i = c_i p_i$ for $2 \le i \le r$. ■

Exercises 6.5

The first three exercises are devoted to the proof of Theorem 6.5.6; for each part, you should check that the proofs given for the ring of integers and the ring of polynomials over a field go through essentially without change for any Euclidean domain.

6.5.1. Let R be a Euclidean domain. Show that

(a) Two nonzero elements f and $g \in R$ have a greatest common divisor that is contained in the ideal $Rf + Rg$. The greatest common divisor is unique up to multiplication by a unit.

(b) Two elements f and $g \in R$ are relatively prime if, and only if, $1 \in Rf + Rg$.

6.5.2. Let R be a Euclidean domain. Show that

(a) Every ideal in R is principal.

(b) If an irreducible p divides the product of two nonzero elements, then it divides one or the other of them.

6.5.3. Let R be a Euclidean domain. Show that every nonzero, nonunit element has a factorization into irreducibles, which is unique (up to units and up to order of the factors).

6.5.4. Let $\mathbb{Z}[\sqrt{-2}]$ be the subring of $\mathbb{R}$ generated by $\mathbb{Z}$ and $\sqrt{-2}$. Show that $\mathbb{Z}[\sqrt{-2}] = \{a+b\sqrt{-2} : a, b \in \mathbb{Z}\}$. Show that $\mathbb{Z}[\sqrt{-2}]$ is a Euclidean domain. Hint: Try $N(z) = |z|^2$ for the Euclidean function.

6.5.5. Let $\omega = \exp(2\pi i/3)$. Then ω satisfies $\omega^2 + \omega + 1 = 0$. Let $\mathbb{Z}[\omega]$ be the subring of $\mathbb{C}$ generated by $\mathbb{Z}$ and ω. Show that $\mathbb{Z}[\omega] = \{a + b\omega : a, b \in \mathbb{Z}\}$. Show that $\mathbb{Z}[\omega]$ is a Euclidean domain. Hint: Try $N(z) = |z|^2$ for the Euclidean function.

6.5.6. Compute a greatest common divisor in $\mathbb{Z}[i]$ of $14+2i$ and $21+26i$.

6.5.7. Compute a greatest common divisor in $\mathbb{Z}[i]$ of $33 + 19i$ and $18 - 16i$.

6.5.8. Show that for two elements a, b in an integral domain R, the following are equivalent:

(a) a divides b and b divides a.
(b) There exists a unit u such that $a = ub$.

6.5.9. Let R be an integral domain. Show that "a is an associate of b" is an equivalence relation on R.

6.5.10. Show that every nonzero, nonunit element of $\mathbb{Z}[\sqrt{-5}]$ has at least one factorization by irreducibles.

6.5.11. The ring $\mathbb{Z}[\sqrt{-5}]$ cannot be a principal ideal domain, since it is not a unique factorization domain. Find an ideal of $\mathbb{Z}[\sqrt{-5}]$ that is not principal.

6.5.12. The ring $\mathbb{Z} + x\mathbb{Q}[x]$ cannot be a principal ideal domain, since it is not a unique factorization domain. Find an ideal of $\mathbb{Z} + x\mathbb{Q}[x]$ that is not principal.

6.5.13. Let R be a commutative ring with multiplicative identity element, and let $a, b \in R$. Show that $a|b \Leftrightarrow bR \subseteq aR$ and, moreover, that a is a proper factor of $b \Leftrightarrow bR \underset{\neq}{\subset} aR$.

6.5.14. Let a, b be elements of an integral domain R, and let d be a greatest common divisor of a and b. Show that the set of greatest common divisors of a and b is precisely the set of associates of d. If a and b are associates, show that each is a greatest common divisor of a and b.

6.5.15. Let R be a principal ideal domain. Show that any pair of nonzero elements $a, b \in R$ have a greatest common divisor and that for any greatest common divisor d, we have $d \in aR + bR$. Show that a and b are relatively prime if, and only if, $1 \in aR + bR$.

6.5.16. Let a be an irreducible element of a principal ideal domain R. If $b \in R$ and a does not divide b, show that a and b are relatively prime.

6.5.17. Suppose a and b are relatively prime elements in a principal ideal domain R. Show that $aR \cap bR = abR$ and $aR + bR = R$. Show that $R/abR \cong R/aR \oplus R/bR$. This is a generalization of the Chinese remainder theorem, Exercise 6.3.11.

6.5.18. Consider the ring of polynomials in two variables over any field $R = K[x, y]$.

(a) Show that the elements x and y are relatively prime.
(b) Show that it is not possible to write $1 = p(x, y)x + q(x, y)y$, with $p, q \in R$.
(c) Show that R is not a principal ideal domain, hence not a Euclidean domain.

6.5.19. Show that a prime element in any integral domain is irreducible.

6.5.20. Let p be a prime element in an integral domain and suppose that p divides a product $b_1 b_2 \cdots b_s$. Show that p divides one of the b_i.

6.5.21. Let aR be an ideal in a principal ideal domain R. Show that R/aR is a ring with only finitely many ideals.

6.5.22.

(a) Let a be an element of an integral domain R. Show that R/aR is an integral domain if, and only if, a is prime.
(b) Let a be an element of a principal ideal domain R. Show that R/aR is a a field if, and only if, a is irreducible.
(c) Show that a quotient R/I of a principal ideal domain R by a nonzero proper ideal I is an integral domain if, and only if, it is a field.
(d) Show that an ideal in a principal ideal domain is maximal if, and only if, it is prime.

6.5.23. Consider the ring of formal power series $K[[x]]$ with coefficients in a field K.

(a) Show that the units of $K[[x]]$ are the power series with nonzero constant term (i.e., elements of the form $\alpha_0 + xf$, where $\alpha_0 \neq 0$, and $f \in K[[x]]$).
(b) Show that $K[[x]]$ is a principal ideal domain. Hint: Let J be an ideal of $K[[x]]$. Let n be the least integer such that J has an element of the form $\alpha_n x^n + \sum_{j>n} \alpha_j x^j$. Show that $J = x^n K[[x]]$.

(c) Show that $K[[x]]$ has a unique maximal ideal M, and $K[[x]]/M \cong K$.

6.5.24. Fix a prime number p and consider the set $\mathbb{Q}_p$ of rational numbers a/b, where b is not divisible by p. (The notation $\mathbb{Q}_p$ is not standard.) Show that $\mathbb{Q}_p$ is a principal ideal domain with a unique maximal ideal M. Show that $\mathbb{Q}_p/M \cong \mathbb{Z}_p$.

6.6. Unique Factorization Domains

The main result of this section is the following theorem:

Theorem 6.6.1. *If R is a unique factorization domain, then $R[x]$ is a unique factorization domain.*

It follows from this result and induction on the number of variables that polynomial rings $K[x_1, \cdots, x_n]$ over a field K have unique factorization; see Exercise 6.6.2. Likewise, $\mathbb{Z}[x_1, \cdots, x_n]$ is a unique factorization domain, since $\mathbb{Z}$ is.

Let R be a unique factorization domain and let F denote the field of fractions of R. The key to showing that $R[x]$ is a unique factorization domain is to compare factorizations in $R[x]$ with factorizations in the Euclidean domain $F[x]$.

Call an element of $R[x]$ *primitive* if its coefficients are relatively prime. Any element $g(x) \in R[x]$ can be written as $g(x) = df(x)$, where $d \in R$ is a greatest common divisor of the coefficients of $g(x)$ and $f(x)$ is primitive. This decomposition is unique up to units of R. Furthermore, any element $\varphi(x) \in F[x]$ can be written as $\varphi(x) = (1/b)g(x)$, where b is a nonzero element of R and $g(x) \in R[x]$. For example, just take b to be the product of the denominators of the coefficients of $\varphi(x)$ (in $F[x]$). Combining these two observations, we can write

$$\varphi(x) = (d/b)f(x), \tag{6.6.1}$$

where $f(x)$ is primitive in $R[x]$. This decomposition is unique up to units in R.

Example 6.6.2. Take $R = \mathbb{Z}$. $7/10 + 14/5x + 21/20x^3 = (7/20)(2 + 8x + 3x^3)$.

Lemma 6.6.3. *(Gauss's lemma). Let* R *be a unique factorization domain with field of fractions* F*.*

(a) *The product of two primitive elements of* $R[x]$ *is primitive.*

(b) *Suppose* $f(x) \in R[x]$*. Then* $f(x)$ *has a factorization* $f(x) = \varphi(x)\psi(x)$ *in* $F[x]$ *with* $\deg(\varphi), \deg(\psi) \geq 1$ *if, and only if,* $f(x)$ *has such a factorization in* $R[x]$*.*

Proof. Suppose that $f(x) = \sum a_i x^i$ and $g(x) = \sum b_j x^j$ are primitive in $R[x]$. Suppose p is an irreducible in R. There is a first index r such that p does not divide a_r and a first index s such that p does not divide b_s. The coefficient of x^{r+s} in $fg(x)$ is $a_r b_s + \sum_{i<r} a_i b_{r+s-i} + \sum_{j<s} a_{r+s-j} b_j$. By assumption, all the summands are divisible by p, except for $a_r b_s$, which is not. So the coefficient of x^{r+s} in $fg(x)$ is not divisible by p. It follows that $fg(x)$ is also primitive. This proves part (a).

Suppose that $f(x)$ has the factorization $f(x) = \varphi(x)\psi(x)$ in $F[x]$ with $\deg(\varphi), \deg(\psi) \geq 1$. Write $f(x) = e\tilde{f}(x)$, $\varphi(x) = (a/b)\tilde{\varphi}(x)$ and $\psi(x) = (c/d)\tilde{\psi}(x)$, where $\tilde{f}(x)$, $\tilde{\varphi}(x)$, and $\tilde{\psi}(x)$ are primitive in $R[x]$. Then $f(x) = e\tilde{f}(x) = (ac/bd)\tilde{\varphi}(x)\tilde{\psi}(x)$. By part (a), the product $\tilde{\varphi}(x)\tilde{\psi}(x)$ is primitive in $R[x]$. By the uniqueness of such decompositions, it follows that $eu = (ac/bd)$, where u is a unit in R, so $f(x)$ factors as $f(x) = ue\tilde{\varphi}(x)\tilde{\psi}(x)$ in $R[x]$. ■

Corollary 6.6.4. *If a polynomial in* $\mathbb{Z}[x]$ *has a nontrivial factorization in* $\mathbb{Q}[x]$*, then it has a nontrivial factorization in* $\mathbb{Z}[x]$*.*

Proof of Theorem 6.6.1. Let $g(x)$ be an element of $R[x]$. First, $g(x)$ can be written as $af(x)$, where $f(x)$ is primitive and $a \in R$; furthermore, this decomposition is unique up to units in R. The element a has a unique factorization in R, by assumption, so it remains to show that $f(x)$ has a unique factorization into irreducibles in $R[x]$. But using the factorization of $f(x)$ in $F[x]$ and the technique of the lemma, we can write

$$f(x) = p_1(x)p_2(x)\cdots p_s(x),$$

where the $p_i(x)$ are primitive elements of $R[x]$ and irreducible in $F[x]$. The uniqueness of this factorization follows from the uniqueness of irreducible factorization in $F[x]$ together with the uniqueness of the factorization in Equation (6.6.1). ■

Some Properties of UFDs

The remainder of this section can be skipped without loss of continuity.

We show that all unique factorization domains share some of the familiar properties of principal ideal domains such as $\mathbb{Z}$ or $K[x]$. In particular, pairs of elements have a greatest common divisor, irreducible elements are prime, and each principal ideal is contained in only finitely many principal ideals. The last two properties characterize unique factorization domains.

Lemma 6.6.5. *Let R be a unique factorization domain, and let $a \in R$ have irreducible factorization $a = f_1 \cdots f_n$. If b is a proper factor of a, then there exist a nonempty proper subset S of $\{1, 2, \dots, n\}$ and a unit u such that $b = u \prod_{i \in S} f_i$.*

Proof. Suppose b is a proper factor of a, so $a = bc$ with neither b nor c a unit. If $b = g_1 \cdots g_\ell$ and $c = h_1 \cdots h_m$ are irreducible factorizations of b and c, respectively, then $g_1 \cdots g_\ell h_1 \cdots h_m$ is an irreducible factorization of a. By uniqueness of irreducible factorization, $g_1 \cdots g_\ell$ is equal to a product of a unit and ℓ of the f_i. ■

Proposition 6.6.6. *In a unique factorization domain, any finite set of nonzero elements has a greatest common divisor, which is unique up to multiplication by units.*

Proof. Let $a_1, \dots, a_s$ be nonzero elements in a unique factorization domain R. Let $f_1, \dots, f_N$ be a maximal collection of pairwise nonassociate (that is, relatively prime) irreducible elements, each of which divides at least one a_i. It follows from Lemma 6.6.5 that any divisor of any of the a_i has the form $u f_1^{k_1} \cdots f_N^{k_N}$ for some $k_j \geq 0$. For each j, let $m(j)$ be the maximum nonnegative integer such that $f_j^{m(j)}$ divides a_i for all i. Then $f_1^{m(1)} \cdots f_N^{m(N)}$ is a greatest common divisor of $a_1, \dots, a_s$. ■

Remark 6.6.7. In a principal ideal domain, a greatest common divisor of two elements a and b is always an element of the ideal $aR + bR$. But in an arbitrary unique factorization domain R, a greatest common divisor of two elements a and b is not necessarily contained in the ideal $aR + bR$. For example, in $\mathbb{Z}[x]$, 1 is a greatest common divisor of 2 and x, but $1 \neq 2\mathbb{Z}[x] + x\mathbb{Z}[x]$.

We are going to characterize unique factorization domains by two properties. One property, the so–called ascending chain condition for principal ideals, implies the existence of irreducible factorizations. The other property, that irreducible elements are prime, ensures the essential uniqueness of irreducible factorizations.

Definition 6.6.8. We say that a ring R satisfies the *ascending chain condition for principal ideals* if, whenever $a_1R \subseteq a_2R \subseteq \cdots$ is an infinite increasing sequence of principal ideals, then there exists an n such that $a_mR = a_nR$ for all $m \geq n$.

Equivalently, any strictly increasing sequence of principal ideals is of finite length.

Lemma 6.6.9. *A unique factorization domain satisfies the ascending chain condition for principal ideals.*

Proof. For any nonzero, nonunit element $a \in R$, let $m(a)$ denote the number of irreducible factors appearing in any irreducible factorization of a. If b is a proper factor of a, then $m(b) < m(a)$, by Lemma 6.6.5. Now if $a_1R \underset{\neq}{\subset} a_2R \underset{\neq}{\subset} \cdots$ is a strictly increasing sequence of principal ideals, then for each i, a_{i+1} is a proper factor of a_i, and, therefore, $m(a_{i+1}) < m(a_i)$. If follows that the sequence is finite. ■

Lemma 6.6.10. *If an integral domain R satisfies the ascending chain condition for principal ideals, then every nonzero, nonunit element of R has at least one factorization by irreducibles.*

Proof. This is exactly what is shown in the proof of Lemma 6.5.15. ■

Lemma 6.6.11. *In a unique factorization domain, every irreducible is prime.*

Proof. Suppose an irreducible f in the unique factorization R divides a product ab. If $g_1 \cdots g_\ell$ and $h_1 \cdots h_m$ are irreducible factorizations of a and b, respectively, then $g_1 \cdots g_\ell h_1 \cdots h_m$ is an irreducible factorization of ab. Since f is a proper factor of ab, by Lemma 6.6.5 f is an associate of one of the g_i or of one of the h_j. Thus f divides a or b. ■

Lemma 6.6.12. *If every irreducible element in an integral domain* R *is prime, then an element of* R *can have at most one factorization by irreducibles, up to permutation of the irreducible factors, and replacing irreducible factors by associates.*

Proof. This is what was shown in the proof of Theorem 6.5.18. ■

Proposition 6.6.13. *An integral domain* R *is a unique factorization domain if, and only if,* R *has the following two properties:*

(a) R *satisfies the ascending chain condition for principal ideals.*
(b) *Every irreducible in* R *is prime.*

Proof. This follows from Lemmas 6.6.9 through 6.6.12. ■

Example 6.6.14. The integral domain $\mathbb{Z}[\sqrt{-5}]$ (see Example 6.5.11) satisfies the ascending chain condition for principal ideals. On the other hand, $\mathbb{Z}[\sqrt{-5}]$ has irreducible elements that are not prime. You are asked to verify these assertions in Exercise 6.6.4.

Example 6.6.15. The integral domain $R = \mathbb{Z} + x\mathbb{Q}[x]$ (see Example 6.5.12) does not satisfy the ascending chain condition for principal ideals. It also has irreducibles that are not prime. You are asked to verify these assertions in Exercise 6.6.5.

Exercises 6.6

6.6.1. Prove Lemma 6.6.6.

6.6.2.

(a) Let R be a commutative ring with identity 1. Show that the polynomial rings $R[x_1, \dots, x_{n-1}, x_n]$ and $(R[x_1, \dots, x_{n-1}])[x_n]$ can be identified.
(b) Assuming Theorem 6.6.1, show by induction that if K is a field, then, for all n, $K[x_1, \dots, x_{n-1}, x_n]$ is a unique factorization domain.

6.6.3. Use Gauss's lemma or the idea of its proof to show that if a polynomial $a_n x^n + \cdots + a_1 x + a_0 \in \mathbb{Z}[x]$ has a rational root r/s, where r and s are relatively prime, then s divides a_n and r divides a_0. In particular, if the polynomial is monic, then its only rational roots are integers.

6.6.4. Show that $\mathbb{Z}[\sqrt{-5}]$ satisfies the ascending chain condition for principal ideals but has irreducible elements that are not prime.

6.6.5. Show that $R = \mathbb{Z} + x\mathbb{Q}[x]$ does not satisfy the ascending chain condition for principal ideals and, moreover, has irreducibles that are not prime.

6.7. Noetherian Rings

This section can be skipped without loss of continuity.

The rings $\mathbb{Z}[x]$ and $K[x, y, z]$ are not principal ideal domains. However, we shall prove that they have the weaker property that every ideal is finitely generated—that is, for every ideal I there is a finite set S such that I is the ideal generated by S.

A condition equivalent to the finite generation property is the *ascending chain condition for ideals.*

Definition 6.7.1. A ring (not necessarily commutative, not necessarily with identity) satisfies the *ascending chain condition (ACC) for ideals* if every strictly increasing sequence of ideals has finite length.

We denote the ideal generated by a subset S of a ring R by $((S))$.

Proposition 6.7.2. *For a ring R (not necessarily commutative, not necessarily with identity), the following are equivalent:*

(a) *Every ideal of R is finitely generated.*
(b) *R satisfies the ascending chain condition for ideals.*

Proof. Suppose that every ideal of R is finitely generated. Let $I_1 \subseteq I_2 \subseteq I_3 \subseteq \cdots$ be an infinite, weakly increasing sequence of ideals of R. Then $I = \cup_n I_n$ is an ideal of R, so there exists a finite set S such that I is the ideal generated by S. Each element of S is contained in some I_n; since the I_n are increasing, there exists an N such that $S \subseteq I_N$. Since I is the smallest ideal containing S, we have $I \subseteq I_N \subseteq \cup_n I_n = I$. It follows that $I_n = I_N$ for all $n \geq N$. This shows that R satisfies the ACC for ideals.

Suppose that R has an ideal I that is not finitely generated. Then for any finite subset S of R, the ideal generated by S is properly contained in I. Hence there exists an element $f \in I \setminus ((S))$, and the ideal generated by S is properly contained in that generated by

$S \cup \{f\}$. An inductive argument gives an infinite, strictly increasing sequence of finite sets such that the corresponding ideals that they generated are also strictly increasing. Therefore, R does not satisfy the ACC for ideals. ■

Definition 6.7.3. A ring is said to be *Noetherian* if every ideal is finitely generated, or, equivalently, if it satisfies the ACC for ideals.

Noetherian rings are named after Emmy Noether. The main result of this section is the *Hilbert basis theorem*, which asserts that a polynomial ring over a Noetherian ring is Noetherian.

We prepare for the theorem with a lemma:

Lemma 6.7.4. *Suppose that* $J_0 \subseteq J$ *are ideals in a polynomial ring* $R[x]$. *If for each nonzero* $f \in J$ *there exists a* $g \in J_0$ *such that* $\deg(f-g) < \deg(f)$, *then* $J_0 = J$.

Proof. Let m denote the minimum of degrees of nonzero elements of J. Suppose $f \in J$ and $\deg(f) = m$. Choose $g \in J_0$ such that $\deg(f - g) < m$. By definition of m, $f - g = 0$, so $f = g \in J_0$. Now suppose that $f \in J$ with $\deg(f) > m$. Suppose inductively that all elements of J whose degree is strictly less than $\deg(f)$ lie in J_0. Choose $g \in J_0$ such that $\deg(f - g) < \deg(g)$. Then $f - g \in J_0$, by the induction hypothesis, so $f = g + (f - g) \in J_0$. ■

Theorem 6.7.5. *(Hilbert's basis theorem). Suppose* R *is a commutative Noetherian ring with identity element. Then* $R[x]$ *is Noetherian.*

Proof. Let J be an ideal in $R[x]$. Let m denote the minimum degree of nonzero elements of J. For each $k \geq 0$, let A_k denote

$$\{a \in R : a = 0 \text{ or there exists } f \in J \text{ with leading term } ax^k\}.$$

It is easy to check that A_k is an ideal in R and $A_k \subseteq A_{k+1}$ for all k. Let $A = \cup_k A_k$. Since every ideal in R is finitely generated, there is a natural number N such that $A = A_N$. For $m \leq k \leq N$, let $\{d_j^{(k)} : 1 \leq j \leq s(k)\}$ be a finite generating set for A_k, and for each j, let $g_j^{(k)}$ be a polynomial in J with leading term $d_j^{(k)} x^k$. Let J_0 be the ideal generated by $\{g_j^{(k)} : m \leq k \leq N,\ 1 \leq j \leq s(k)\}$. We claim that $J = J_0$.

Let $f \in J$ have degree n and leading term ax^n. If $n \geq N$, then $a \in A_n = A_N$, so there exist $r_j \in R$ such that $a = \sum_j r_j d_j^{(N)}$.

Then $\sum_j r_j x^{n-N} g_j^{(N)}$ is an element of J_0 with leading term ax^n, so $\deg(f - g) < \deg(f)$. If $m \le n < N$, then $a \in A_n$, so there exist $r_j \in R$ such that $a = \sum_j r_j d_j^{(n)}$. Then $\sum_j r_j g_j^{(n)}$ is an element of J_0 with leading term ax^n, so $\deg(f - g) < \deg(f)$. Thus for all nonzero $f \in J$, there exists $g \in J_0$ such that $\deg(f - g) < \deg(f)$. It follows from Lemma 6.7.4 that $J = J_0$. ■

Corollary 6.7.6. *If* R *is a commutative Noetherian ring, then* $R[x_1, \dots, x_n]$ *is Noetherian for all* n. *In particular,* $\mathbb{Z}[x_1, \dots, x_n]$ *and, for all fields* K, $K[x_1, \dots, x_n]$ *are Noetherian.*

Proof. This follows from the Hilbert basis theorem and induction. ■

Exercises 6.7

6.7.1. Let R be a commutative ring with identity element. Let J be an ideal in $R[x]$. Show that for each $k \ge 0$, the set

$$A_k = \{a \in R : a = 0 \text{ or there exists } f \in J \text{ with leading term } ax^k\}$$

is an ideal in R, and that $A_k \subseteq A_{k+1}$ for all $k \ge 0$.

6.7.2. Suppose that R is Noetherian and $\varphi : R \to S$ is a surjective ring homomorphism. Show that S is Noetherian.

6.7.3. We know that $\mathbb{Z}[\sqrt{-5}]$ is not a UFD and, therefore, not a PID. Show that $\mathbb{Z}[\sqrt{-5}]$ is Noetherian. Hint: Find a Noetherian ring R and a surjective homomorphism $\varphi : R \to \mathbb{Z}[\sqrt{-5}]$.

6.7.4. Show that $\mathbb{Z}[x] + x\mathbb{Q}[x]$ is not Noetherian.

6.7.5. Show that a Noetherian domain in which every irreducible element is prime is a unique factorization domain.

6.8. Irreducibility Criteria

In this section, we will consider some elementary techniques for determining whether a polynomial is irreducible.

We restrict ourselves to the problem of determining whether a polynomial in $\mathbb{Z}[x]$ is irreducible. Recall that an integer polynomial factors over the integers if, and only if, it factors over the rational numbers, according to Lemma 6.6.3 and Corollary 6.6.4.

A basic technique in testing for irreducibility is to reduce the polynomial modulo a prime. The natural homomorphism of $\mathbb{Z}$ onto $\mathbb{Z}_p$, where p is a prime, extends to a homomorphism of $\mathbb{Z}[x]$ onto $\mathbb{Z}_p[x]$, $\sum a_i x^i \mapsto \sum [a_i] x^i$.

Proposition 6.8.1. *Consider a polynomial $f(x) = \sum a_i x^i \in \mathbb{Z}[x]$ of degree n. Suppose that the leading coefficient a_n is not divisible by the prime p. Consider the polynomial $\tilde{f}(x) \in \mathbb{Z}_p[x]$ obtained by reducing all coefficients of f modulo p. If $\tilde{f}$ is irreducible in $\mathbb{Z}_p[x]$, then f is irreducible in $\mathbb{Q}[x]$.*

Proof. This follows at once, because the map $f \mapsto \tilde{f}$ is a homomorphism. If f had a factorization $f = gh$ in $\mathbb{Z}[x]$, then $\tilde{f}$ would also factor as $\tilde{f} = \tilde{g}\tilde{h}$. ■

Efficient algorithms are known for factorization in $\mathbb{Z}_p[x]$. The common computer algebra packages such as *Mathematica* and *Maple* have these algorithms built in. The *Mathematica* command **Factor[f, Modulus → p]** can be used for reducing a polynomial f modulo a prime p and factoring the reduction. Unfortunately, the condition of the proposition is merely a sufficient condition. It is quite possible (but rare) for a polynomial to be irreducible over $\mathbb{Q}$ but nevertheless for its reductions modulo every prime to be reducible.

Example 6.8.2. Let $f(x) = 83 + 82x - 99x^2 - 87x^3 - 17x^4$. The reduction of f modulo 3 is $\{2 + x + x^4\}$, which is irreducible over $\mathbb{Z}_3$. Hence, f is irreducible over $\mathbb{Q}$.

Example 6.8.3. Let $f(x) = -91 - 63x - 73x^2 + 22x^3 + 50x^4$. The reduction of f modulo 17 is $16\left(6 + 12x + 5x^2 + 12x^3 + x^4\right)$, which is irreducible over $\mathbb{Z}_{17}$. Therefore, f is irreducible over $\mathbb{Q}$.

A related sufficient condition for irreducibility is Eisenstein's criterion:

Proposition 6.8.4. *(Eisenstein's criterion). Let $f(x) \in \mathbb{Z}[x]$,*

$$f(x) = x^n + a_{n-1}x^{n-1} + \cdots + a_1 x + a_0.$$

Suppose p is a prime that divides all the coefficients a_i and such that p^2 does not divide a_0. Then $f(x)$ is irreducible over $\mathbb{Q}$.

Proof. If f has a nontrivial factorization over $\mathbb{Q}$, then it also has a nontrivial factorization over $\mathbb{Z}$, with both factors monic polynomials. Write $f(x) = a(x)b(x)$, where $a(x) = \sum_{i=0}^{r} \alpha_i x^i$ and $b(x) = \sum_{j=0}^{s} \beta_i x^j$. Since $a_0 = \alpha_0\beta_0$, exactly one of α_0 and β_0 is divisible by p; suppose without loss of generality that p divides β_0, and p does not divide α_0. Considering the equations

$$\begin{aligned} a_1 &= \beta_1\alpha_0 + \beta_0\alpha_1 \\ &\dots \\ a_{s-1} &= \beta_{s-1}\alpha_0 + \dots + \beta_0\alpha_{s-1} \\ a_s &= \alpha_0 + \beta_{s-1}\alpha_1 + \dots + \beta_0\alpha_s, \end{aligned}$$

we obtain by induction that β_j is divisible by p for all j ($0 \le j \le s-1$). Finally, the last equation yields that α_0 is divisible by p, a contradiction. ■

Example 6.8.5. $x^3+14x+7$ is irreducible by the Eisenstein criterion.

Example 6.8.6. Sometimes the Eisenstein criterion can be applied after a linear change of variables. For example, for the so–called cyclotomic polynomial

$$f(x) = x^{p-1} + x^{p-2} + \dots + x^2 + x + 1,$$

where p is a prime, we have

$$f(x+1) = \sum_{s=0}^{p-1} \binom{p}{s+1} x^s.$$

This is irreducible by Eisenstein's criterion, so f is irreducible as well. You are asked to provide the details for this example in Exercise 6.8.3.

There is a simple criterion for a polynomial in $\mathbb{Z}[x]$ to have (or not to have) a linear factor, the so–called *rational root test.*

Proposition 6.8.7. *(Rational root test). Let* $f(x) = a_n x^n + a_{n-1}x^n - 1 + \cdots a_1 x + a_0 \in \mathbb{Z}[x]$. *If* r/s *is a rational root of* f, *where* r *and* s *are relatively prime, then* s *divides* a_n *and* r *divides* a_0.

Proof. Exercise 6.8.1. ■

A quadratic or cubic polynomial is irreducible if, and only if, it has no linear factors, so the rational root test is a definitive test for irreducibility for such polynomials. The rational root test can sometimes be used as an adjunct to prove irreducibility of higher degree polynomials: If an integer polynomial of degree n has no rational root, but for some prime p its reduction mod p has irreducible factors of degrees 1 and $n-1$, then then f is irreducible (Exercise 6.8.2).

Exercises 6.8

6.8.1. Prove the rational root test.

6.8.2. Show that if a polynomial $f(x) \in \mathbb{Z}[x]$ of degree n has no rational root, but for some prime p its reduction mod p has irreducible factors of degrees 1 and $n-1$, then then f is irreducible.

6.8.3. Provide the details for Example 6.8.6.

6.8.4. Show that for each natural number n

$$(x-1)(x-2)(x-3)\cdots(x-n)-1$$

is irreducible over the rationals.

6.8.5. Show that for each natural number $n \neq 4$

$$(x-1)(x-2)(x-3)\cdots(x-n)+1$$

is irreducible over the rationals.

6.8.6. Determine whether the following polynomials are irreducible over the rationals. You may wish to do computer computations of factorizations modulo primes.

(a) $8 - 60\,x - 54\,x^2 + 89\,x^3 - 55\,x^4$
(b) $42 - 55\,x - 66\,x^2 + 44\,x^3$
(c) $42 - 55\,x - 66\,x^2 + 44\,x^3 + x^4$
(d) $5 + 49\,x + 15\,x^2 - 27\,x^3$
(e) $-96 + 53\,x - 26\,x^2 + 21\,x^3 - 75\,x^4$

CHAPTER 7

Field Extensions – First Look

7.1. A Brief History

The most traditional concern of algebra is the solution of polynomial equations and related matters such as computation with radicals. Methods of solving linear and quadratic equations were known to the ancients, and Arabic scholars preserved and augmented this knowledge during the Middle Ages.

You learned the quadratic formula for the roots of a quadratic equation in school, but more fundamental is the algorithm of completing the square, which justifies the formula. Similar procedures for equations of the third and fourth degree were discovered by several mathematicians in sixteenth century Italy, who also introduced complex numbers (with some misgivings). And there the matter stood, more or less, for another 250 years.

At the end of the eighteenth century, no general method or formula, of the sort that had worked for equations of lower degree, was known for equations of degree 5, nor was it known that no such method was possible. The sort of method sought was one of "solution by radicals," that is, by algebraic operations and by introduction of nth roots.

In 1798 C. F. Gauss showed that every polynomial equation with real or complex coefficients has a complete set of solutions in the complex numbers. Gauss published several proofs of this theorem, known as the *fundamental theorem of algebra*. The easiest proofs known involve some complex analysis, and you can find a proof in any text on that subject.

A number of mathematicians worked on the problem of solution of polynomial equations during the period from 1770 to 1820, among them J-L. Lagrange, A. Cauchy, and P. Ruffini. Their insights included a certain appreciation for the role of symmetry of the roots. Finally, N. H. Abel, between 1824 and 1829, succeeded in showing that the general fifth–degree equation could not be solved by radicals.

It was E. Galois, however, who, in the years 1829 to 1832, provided the most satisfactory solution to the problem of solution of equations by radicals and, in doing so, radically changed the nature of algebra. Galois associated with a polynomial $p(x)$ over a field K a canonical smallest field L containing K in which the polynomial has a complete set of roots *and, moreover, a canonical group of symmetries of* L, *which acts on the roots of* $p(x)$. Galois's brilliant idea was to study the polynomial equation $p(x) = 0$ by means of this symmetry group. In particular, he showed that solvability of the equation by radicals corresponded to a certain property of the group, which also became known as *solvability*. The *Galois group* associated to a polynomial equation of degree n is always a subgroup of the permutation group S_n. It turns out that subgroups of S_n for $n \leq 4$ are solvable, but S_5 is not solvable. In Galois's theory, the nonsolvability of S_5 implies the impossibility of an analogue of the quadratic formula for equations of degree 5.

Neither Abel nor Galois had much time to enjoy his success. Abel died at the age of 26 in 1829, and Galois died in a duel in 1832 at the age of 20. Galois's memoir was first published by Liouville in 1846, 14 years after Galois's death. Galois had submitted two manuscripts on his theory in 1829 to the Académie des Sciences de Paris, which apparently were lost by Cauchy. A second version of his manuscript was submitted in 1830, but Fourier, who received it, died before reading it, and that manuscript was also lost. A third version was submitted to the Academy in 1831 and referred to Poisson, who reported that he was unable to understand it and recommended revisions. But Galois was killed before he could prepare another manuscript.

In this chapter, we will take a first look at polynomial equations, field extensions, and symmetry. Chapters 9 and 10 contain a more systematic treatment of Galois theory.

7.2. Solving the Cubic Equation

Consider a cubic polynomial equation,

$$x^3 + ax^2 + bx + c = 0,$$

where the coefficients lie in some field K, which for simplicity we assume to be contained in the field of complex numbers $\mathbb{C}$. Necessarily, K contains the rational field $\mathbb{Q}$.

If $\alpha_1, \alpha_2, \alpha_3$ are the roots of the equation in $\mathbb{C}$, then

$$x^3 + ax^2 + bx + c = (x - \alpha_1)(x - \alpha_2)(x - \alpha_3),$$

from which it follows that

$$\begin{aligned} \alpha_1 + \alpha_2 + \alpha_3 &= -a \\ \alpha_1\alpha_2 + \alpha_1\alpha_3 + \alpha_2\alpha_3 &= b \\ \alpha_1\alpha_2\alpha_3 &= -c. \end{aligned}$$

It is simpler to deal with a polynomial with zero quadratic term, and this can be accomplished by a linear change of variables $y = x + a/3$. We compute that the equation is transformed into

$$y^3 + (-\frac{a^2}{3} + b)\, y + (\frac{2a^3}{27} - \frac{ab}{3} + c).$$

Changing notation, we can suppose without loss of generality that we have at the outset a polynomial equation without quadratic term

$$f(x) = x^3 + px + q = 0,$$

with roots $\alpha_1, \alpha_2, \alpha_3$ satisfying

$$\begin{aligned} \alpha_1 + \alpha_3 + \alpha_3 &= 0 \\ \alpha_1\alpha_2 + \alpha_1\alpha_3 + \alpha_2\alpha_3 &= p \\ \alpha_1\alpha_2\alpha_3 &= -q. \end{aligned} \tag{7.2.1}$$

If we experiment with changes of variables in the hope of somehow simplifying the equation, we might eventually come upon the idea of expressing the variable x as a difference of two variables $x = v - u$. The result is

$$(v^3 - u^3) + q - 3uv(v - u) + p(v - u) = 0.$$

A sufficient condition for a solution is

$$\begin{aligned} (v^3 - u^3) + q &= 0 \\ 3uv &= p. \end{aligned}$$

Now, using the second equation to eliminate u from the first gives a quadratic equation for v^3:

$$v^3 - \frac{p^3}{27v^3} + q = 0.$$

The solutions to this are

$$\nu^3 = -\frac{q}{2} \pm \sqrt{\frac{q^2}{4} + \frac{p^3}{27}}.$$

It turns out that we get the same solutions to our original equation regardless of the choice of the square root. Let ω denote the primitive third root of unity $\omega = e^{2\pi i/3}$, and let A denote one cube root of

$$-\frac{q}{2} + \sqrt{\frac{q^2}{4} + \frac{p^3}{27}}.$$

Then the solutions for ν are A, ωA, and $\omega^2 A$, and the solutions for $x = \nu - \dfrac{p}{3\nu}$ are

$$\alpha_1 = A - \frac{p}{3A}, \quad a_2 = \omega A - \omega^2 \frac{p}{3A}, \quad \alpha_3 = \omega^2 A - \omega \frac{p}{3A}.$$

These are generally known as *Cardano's formulas*, but Cardano credits Scipione del Ferro and N. Tartaglia and for their discovery (prior to 1535).

What we will be concerned with here is the structure of the *field extension* $K \subseteq K(\alpha_1, \alpha_2, \alpha_3)$ and the symmetry of the roots.

Here is some general terminology and notation: If $F \subseteq L$ are fields, we say that F is a *subfield* of L or that L is a *field extension* of F. If $F \subseteq L$ is a field extension and $S \subseteq L$ is any subset, then $F(S)$ denotes the smallest subfield of L that contains F and S. If $F \subseteq L$ is a field extension and $g(x) \in F[x]$ has a complete set of roots in L (i.e., $g(x)$ factors into linear factors in $L[x]$), then the smallest subfield of L containing F and the roots of $g(x)$ in L is called a *splitting field* of $g(x)$ over F.

Returning to our more particular situation, $K(\alpha_1, \alpha_2, \alpha_3)$ is the splitting field (in $\mathbb{C}$) of the cubic polynomial $f(x) \in K[x]$.

One noticeable feature of this situation is that the roots of $f(x)$ are obtained by rational operations and by extraction of cube and square roots (in $\mathbb{C}$); in fact, A is obtained by first taking a square root and then taking a cube root. And ω also involves a square root, namely, $\omega = 1/2 + \sqrt{-3}/2$. So it would seem that it might be necessary to obtain $K(\alpha_1, \alpha_2, \alpha_3)$ by three stages, $K \subseteq K_1 \subseteq K_2 \subseteq K_3 = K(\alpha_1, \alpha_2, \alpha_3)$, where at each stage we enlarge the field by adjoining a new cube or square root. In fact, we will see that at most two stages are necessary.

An important element for understanding the splitting field is

$$\delta = (\alpha_1 - \alpha_2)(\alpha_1 - \alpha_3)(\alpha_2 - \alpha_3) = \det \begin{bmatrix} 1 & 1 & 1 \\ \alpha_1 & \alpha_2 & \alpha_3 \\ \alpha_1^2 & \alpha_2^2 & \alpha_3^2 \end{bmatrix}.$$

The square δ^2 of this element is invariant under permutations of the α_i; a general result, which we will discuss later, says that δ^2 is, therefore, a polynomial expression in the coefficients p, q of the polynomial and, in particular, $\delta^2 \in K$. We can compute δ^2 explicitly in terms of p and q; the result is $\delta^2 = -4p^3 - 27q^2$. You are asked to verify this in Exercise 7.2.7. The element δ^2 is called the *discriminant* of the polynomial f.

Now, it turns out that the nature of the field extension $K \subseteq K(\alpha_1, \alpha_2, \alpha_3)$ depends on whether δ is in the ground field K. Before discussing this further, it will be convenient to introduce some remarks of a general nature about algebraic elements of a field extension. We shall do this in the following section, and complete the discussion of the cubic equation in Section 7.4.

Exercises 7.2

7.2.1. Show that a cubic polynomial in $K[x]$ either has a root in K or is irreducible over K.

7.2.2. Verify the reduction of the monic cubic polynomial to a polynomial with no quadratic term.

7.2.3. How can you deal with a nonmonic cubic polynomial?

7.2.4. Verify in detail the derivation of Cardano's formulas.

7.2.5. Consider the polynomial $p(x) = x^3 + 2x^2 + 2x - 3$. Show that p is irreducible over $\mathbb{Q}$. Hint: Show that if p has a rational root, then it must have an integer root, and the integer root must be a divisor of 3. Carry out the reduction of p to a polynomial f without quadratic term. Use the method described in this section to find the roots of f. Use this information to find the roots of p.

7.2.6. Repeat the previous exercise with various cubic polynomials of your choice.

7.2.7. Let V denote the Vandermonde matrix $\begin{bmatrix} 1 & 1 & 1 \\ \alpha_1 & \alpha_2 & \alpha_3 \\ \alpha_1^2 & \alpha_2^2 & \alpha_3^2 \end{bmatrix}$.

(a) Show that $\delta^2 = \det(VV^t) = \det \begin{bmatrix} 3 & 0 & \sum \alpha_i^2 \\ 0 & \sum \alpha_i^2 & \sum \alpha_i^3 \\ \sum \alpha_i^2 & \sum \alpha_i^3 & \sum \alpha_i^4 \end{bmatrix}$.

(b) Use equations (7.2.1) as well as the fact that the α_i are roots of $f(x) = 0$ to compute that $\sum \alpha_i^2 = -2p$, $\sum_i a_i^3 = -3q$, and $\sum_i a_i^4 = 2p^2$.

(c) Compute that $\delta^2 = -4p^3 - 27q^2$.

7.2.8.

(a) Show that $x^3 - 2$ is irreducible in $\mathbb{Q}[x]$. Compute its roots and also δ^2 and δ.

(b) $\mathbb{Q}(\sqrt[3]{2}) \subseteq \mathbb{R}$, so is not equal to the splitting field E of $x^3 - 2$. The splitting field is $\mathbb{Q}(\sqrt[3]{2}, \omega)$, where $\omega = e^{2\pi i/3}$. Show that ω satisfies a quadratic polynomial over $\mathbb{Q}$.

(c) Show that $x^3 - 3x + 1$ is irreducible in $\mathbb{Q}[x]$. Compute its roots, and also δ^2 and δ. Hint: The quantity A is a root of unity, and the roots are twice the real part of certain roots of unity.

7.3. Adjoining Algebraic Elements to a Field

The fields in this section are general, not necessarily subfields of the complex numbers.

A field extension L of a field K is, in particular, a *vector space* over K. You are asked to check this in the Exercises. It will be helpful to review the definition of a vector space in Section 3.4

Since a field extension L of a field K is a K-vector space, in particular, L has a dimension over K, possibly infinite. The dimension of L as a K-vector space is denoted $\dim_K(L)$ (or, sometimes, $[L : K]$.) Dimensions of field extensions have the following multiplicative property:

Proposition 7.3.1. *If $K \subseteq L \subseteq M$ are fields, then*

$$\dim_K(M) = \dim_K(L)\dim_L(M).$$

Proof. Suppose that $\{\lambda_1, \dots, \lambda_r\}$ is a subset of L that is linearly independent over K, and that $\{\mu_1, \dots, \mu_s\}$ is a subset of M that is linearly independent over L. I claim that $\{\lambda_i\mu_j : 1 \le i \le r, 1 \le j \le s\}$ is linearly independent over K. In fact, if $0 = \sum_{i,j} k_{ij}\lambda_i\mu_j = \sum_j(\sum_i k_{ij}\lambda_i)\mu_j$, with $k_{ij} \in K$, then linear independence of $\{\mu_j\}$ over L implies that $\sum_i k_{ij}\lambda_i = 0$ for all j, and then linear independence of $\{\lambda_i\}$ over K implies that $k_{ij} = 0$ for all i, j, which proves the claim.

In particular, if either $\dim_K(L)$ or $\dim_L(M)$ is infinite, then there are arbitrarily large subsets of M that are linearly independent over K, so $\dim_K(M)$ is also infinite.

Suppose now that $\dim_K(L)$ and $\dim_L(M)$ are finite, that the set $\{\lambda_1, \dots, \lambda_r\}$ is a basis of L over K, and that the set $\{\mu_1, \dots, \mu_s\}$ is a basis of M over L. The fact that $\{\mu_j\}$ spans M over L and that $\{\lambda_i\}$ spans L over K implies that the set of products $\{\lambda_i\mu_j\}$ spans M over K (exercise). Hence, $\{\lambda_i\mu_j\}$ is a basis of M over K. ■

Definition 7.3.2. A field extension $K \subseteq L$ is called *finite* if L is a finite–dimensional vector space over K.

Now, consider a field extension $K \subseteq L$. According to Proposition 6.2.5, for any element $\alpha \in L$ there is a ring homomorphism ("evaluation" at α) from $K[x]$ into L given by $\varphi_\alpha(f(x)) = f(\alpha)$. That is,

$$\varphi_\alpha(k_0 + k_1 x + \cdots + k_n x^n) = k_0 + k_1\alpha + \cdots + k_n\alpha^n. \tag{7.3.1}$$

Definition 7.3.3. An element α in a field extension L of K is said to be *algebraic over* K if there is some polynomial $p(x) \in K[x]$ such that $p(\alpha) = 0$; equivalently, there are elements $k_0, k_1, \dots k_n \in K$ such that $k_0 + k_1\alpha + \cdots + k_n\alpha^n = 0$. Any element that is not algebraic is called *transcendental*. The field extension is called *algebraic* if every element of L is algebraic over K.

In particular, the set of complex numbers that are algebraic over $\mathbb{Q}$ are called *algebraic numbers* and those that are transcendental over $\mathbb{Q}$ are called *transcendental numbers*. It is not difficult to see that there are (only) countably many algebraic numbers, and that, therefore, there are uncountably many transcendental numbers. Here is the argument: The rational numbers are countable, so for each natural number n, there are only countably many distinct polynomials with rational coefficients with degree no more than n. Consequently, there are only countably many polynomials with rational coefficients altogether, and each of these has only finitely many roots in the complex numbers. Therefore, the set of algebraic numbers is a countable union of finite sets, so countable. On the other hand, the complex numbers are uncountable, so there are uncountably many transcendental numbers.

We show that any finite field extension is algebraic:

Proposition 7.3.4. *If* $K \subseteq L$ *is a finite field extension, then*

(a) L *is algebraic over* K, *and*

(b) *there are finitely many (algebraic) elements* $a_1, \dots, a_n \in L$ *such that* $L = K(a_1, \dots, a_n)$.

Proof. Consider any element $\alpha \in L$. Since $\dim_K(L)$ is finite, the powers $1, \alpha, \alpha^2, \alpha^3, \dots$ of α cannot be linearly independent over K. Therefore, there exists a natural number N and there exist $\lambda_0, \lambda_1, \dots \lambda_N \in K$, not all zero, such that $\lambda_0 + \lambda_1\alpha + \cdots + \lambda_n\alpha^N = 0$. That is, α is algebraic over K. Since α was an arbitrary element of L, this means that L is algebraic over K.

The second statement is proved by induction on $\dim_K(L)$. If $\dim_K(L) = 1$, then $K = L$, and there is nothing to prove. So suppose that $\dim_K(L) > 1$, and suppose that whenever $K' \subseteq L'$ are fields and $\dim_{K'}(L') < \dim_K(L)$, then there exists a natural number n and there exist finitely many elements $a_2, \dots, a_n$, such that $L' = K'(a_2, \dots, a_n)$.

Since $K \neq L$, there exists an element $a_1 \in L \setminus K$. Then $K \subseteq K(a_1) \subseteq L$, $\dim_K(K(a_1)) > 1$, and

$$\dim_{K(a_1)}(L) = \dim_K(L)/\dim_K(K(a_1)) < \dim_K(L).$$

By the induction hypothesis applied to the pair $K(a_1) \subseteq L$, there exists a natural number n and there exist finitely many elements $a_2, \dots, a_n$, such that $L = K(a_1)(a_2, \dots, a_n) = K(a_1, a_2, \dots, a_n)$. This completes the proof. ■

The set I_α of polynomials $p(x) \in K[x]$ satisfying $p(\alpha) = 0$ is the kernel of the homomorphism $\varphi_\alpha : K[x] \to L$, so is an ideal in $K[x]$. We have $I_\alpha = f(x)K[x]$, where $f(x)$ is an element of minimum degree in I_α, according to Proposition 6.2.17. The polynomial $f(x)$ is necessarily irreducible; if it factored as $f(x) = f_1(x)f_2(x)$, where $\deg(f) > \deg(f_i) > 0$, then $0 = f(\alpha) = f_1(\alpha)f_2(\alpha)$. But then one of the f_i would have to be in I_α, while $\deg(f_i) < \deg(f)$, a contradiction. The generator $f(x)$ of I_α is unique up to multiplication by a nonzero element of K, so there is a unique monic polynomial $f(x)$ such that $I_\alpha = f(x)K[x]$, called the *minimal polynomial for α over K.*

We have proved the following proposition:

Proposition 7.3.5. *If $K \subseteq L$ are fields and $\alpha \in L$ is algebraic over K, then there is a unique monic irreducible polynomial $f(x) \in K[x]$ such that the set of polynomials $p(x) \in K[x]$ satisfying $p(\alpha) = 0$ is $f(x)K[x]$.*

Fix the algebraic element $\alpha \in L$, and consider the set of elements of L of the form

$$k_0 + k_1\alpha + \cdots + k_n\alpha^n, \tag{7.3.2}$$

with $n \in \mathbb{N}$ and $k_0, k_1, \dots, k_n \in K$. By Equation (7.3.1), this is the range of the homomorphism φ_α, so, by the homomorphism theorem for rings, it is isomorphic to $K[x]/I_\alpha = K[x]/(f(x))$, where we have written $(f(x))$ for $f(x)K[x]$. But since $f(x)$ is irreducible, the ideal $(f(x))$ is maximal, and, therefore, the quotient ring $K[x]/(f(x))$ is a field (Exercise 6.3.7).

Proposition 7.3.6. *Suppose* $K \subseteq L$ *are fields,* $\alpha \in L$ *is algebraic over* K*, and* $f(x) \in K[x]$ *is the minimal polynomial for* α *over* K.

(a) $K(\alpha)$*, the subfield of* L *generated by* K *and* α*, is isomorphic to the quotient field* $K[x]/(f(x))$.

(b) $K(\alpha)$ *is the set of elements of the form*

$$k_0 + k_1\alpha + \cdots + k_{d-1}\alpha^{d-1},$$

where d *is the degree of* f.

(c) $\dim_K(K(\alpha)) = \deg(f)$.

Proof. We have shown that the ring of polynomials in α with coefficients in K is a field, isomorphic to $K[x]/(f(x))$. Therefore, $K(\alpha) = K[\alpha] \cong K[x]/(f(x))$. This shows part (a). For any $p \in K[x]$, write $p = qf + r$, where $r = 0$ or $\deg(r) < \deg(f)$. Then $p(\alpha) = r(\alpha)$, since $f(\alpha) = 0$. This means that $\{1, \alpha, \ldots, \alpha^{d-1}\}$ spans $K(\alpha)$ over K. But this set is also linearly independent over K, because α is not a solution to any equation of degree less than d. This shows parts (b) and (c). ■

Example 7.3.7. Consider $f(x) = x^2 - 2 \in \mathbb{Q}(x)$, which is irreducible by Eisenstein's criterion (Proposition 6.8.4). The element $\sqrt{2} \in \mathbb{R}$ is a root of $f(x)$. (The existence of this root in $\mathbb{R}$ is a fact of *analysis*.) The field $\mathbb{Q}(\sqrt{2}) \cong \mathbb{Q}[x]/(x^2 - 2)$ consists of elements of the form $a + b\sqrt{2}$, with $a, b \in \mathbb{Q}$. The rule for addition in $\mathbb{Q}(\sqrt{2})$ is

$$(a + b\sqrt{2}) + (a' + b'\sqrt{2}) = (a + a') + (b + b')\sqrt{2},$$

and the rule for multiplication is

$$(a + b\sqrt{2})(a' + b'\sqrt{2}) = (aa' + 2bb') + (ab' + ba')\sqrt{2}.$$

The inverse of $a + b\sqrt{2}$ is

$$(a + b\sqrt{2})^{-1} = \frac{a}{a^2 - 2b^2} - \frac{b}{a^2 - 2b^2}\sqrt{2}.$$

Example 7.3.8. The polynomial $f(x) = x^3 - 2x + 2$ is irreducible over $\mathbb{Q}$ by Eisenstein's criterion. The polynomial has a real root θ by application of the intermediate value theorem of analysis. The field $\mathbb{Q}(\theta) \cong \mathbb{Q}[x]/(f)$ consists in elements of the form $a + b\theta + c\theta^2$, where $a, b, c \in \mathbb{Q}$. Multiplication is performed by using the distributive law and then reducing using the rule $\theta^3 = 2\theta - 2$ (whence $\theta^4 = 2\theta^2 - 2\theta$). To find the inverse of an element of $\mathbb{Q}(\theta)$, it is convenient to compute in $\mathbb{Q}[x]$. Given $g(\theta) = a + b\theta + c\theta^2$, there exist elements $r(x), s(x) \in \mathbb{Q}[x]$ such that $g(x)r(x) + f(x)s(x) = 1$, since g and f

are relatively prime. Furthermore, r and s can be computed by the algorithm implicit in the proof of Theorem 1.8.16. It then follows that $g(\theta)r(\theta) = 1$, so $r(\theta)$ is the desired inverse. Let us compute the inverse of $2 + 3\theta - \theta^2$ in this way. Put $g(x) = -x^2 + 3x + 2$. Then we can compute that

$$\frac{1}{118}(24 + 8x - 9x^2)g + \frac{1}{118}(35 - 9x)f = 1,$$

and, therefore,

$$(2 + 3\theta - \theta^2)^{-1} = \frac{1}{118}(24 + 8\theta - 9\theta^2).$$

The next proposition is a converse to Proposition 7.3.4(b).

Proposition 7.3.9. *Let* $L = K(a_1, \dots, a_n)$, *where the* a_i *are algebraic over* K.

(a) *Then* L *is a finite extension of* K, *and, therefore, algebraic.*
(b) $L = K[a_1, \dots, a_n]$, *the set of* polynomials *in the* a_i *with coefficients in* K.

Proof. Let $K_0 = K$, and $K_i = K(a_1, \dots, a_i)$, for $1 \le i \le n$. Consider the tower of extensions:

$$K \subseteq K_1 \subseteq \cdots \subseteq K_{n-1} \subseteq L.$$

We have $K_{i+1} = K_i(a_{i+1})$, where a_{i+1} is algebraic over K, hence algebraic over K_i. By Proposition 7.3.6, $\dim_{K_i}(K_{i+1})$ is the degree of the minimal polynomial for a_{i+1} over K_i, and moreover $K_{i+1} = K_i[a_{i+1}]$.

It follows by induction that $K_i = K[a_1, \dots, a_i]$ for all i, and, in particular, $L = K[a_1, \dots, a_n]$. Moreover,

$$\dim_K(L) = \dim_K(K_1)\dim_{K_1}(K_2)\cdots\dim_{K_{n-1}}(L) < \infty.$$

By Proposition 7.3.4, L is algebraic over K. ■

Combining Propositions 7.3.4 and 7.3.9, we have the following proposition.

Proposition 7.3.10. *A field extension* $K \subseteq L$ *is finite if, and only if, it is generated by finitely many algebraic elements. Furthermore, if* $L = K(a_1, \dots, a_n)$, *where the* a_i *are algebraic over* K, *then* L *consists of polynomials in the* a_i *with coefficients in* K.

Corollary 7.3.11. *Let $K \subseteq L$ be a field extension. The set of elements of L that are algebraic over K form a subfield of L. In particular, the set of algebraic numbers (complex numbers that are algebraic over $\mathbb{Q}$) is a countable field.*

Proof. Let A denote the set of elements of L that are algebraic over K. It suffices to show that A is closed under the field operations; that is, for all $a, b \in A$, the elements $a+b$, ab, $-a$, and b^{-1} (when $b \neq 0$) also are elements of A. For this, it certainly suffices that $K(a, b) \subseteq A$. But this follows from Proposition 7.3.9.

We have already observed that the set of algebraic numbers is countable, so this set is a countable field. ■

Exercises 7.3

7.3.1. Show that if $K \subseteq L$ are fields, then the identity of K is also the identity of L. Conclude that L is a vector space over K.

7.3.2. Fill in the details of the proof of 7.3.1 to show that $\{\lambda_i \mu_j\}$ spans M over K.

7.3.3. If $K \subseteq L$ is a finite field extension, then there exist finitely many elements $a_1, \dots, a_n \in L$ such that $L = K(a_1, \dots, a_n)$. Give a different proof of this assertion as follows: If the assertion is false, show that there exists an infinite sequence $a_1, a_2, \dots$ of elements of L such that

$$K \underset{\neq}{\subset} K(a_1) \underset{\neq}{\subset} K(a_1, a_2) \underset{\neq}{\subset} K(a_1, a_2, a_3) \underset{\neq}{\subset} \cdots.$$

Show that this contradicts the finiteness of $\dim_K(L)$.

7.3.4. Suppose $\dim_K(L) < \infty$. Show that there exists a natural number n such that

$$n \leq \log_2(\dim_K(L)) + 1$$

and there exist $a_1, \dots, a_n \in L$ such that $L = K(a_1, \dots, a_n)$.

7.3.5. Show that $\mathbb{R}$ is not a finite extension of $\mathbb{Q}$.

7.3.6.

(a) Show that the polynomial

$$p(x) = \frac{x^5 - 1}{x - 1} = x^4 + x^3 + x^2 + x + 1$$

is irreducible over $\mathbb{Q}$.

(b) According to Proposition 7.3.6 , $\mathbb{Q}[\zeta] = \{a\zeta^3 + b\zeta^2 + c\zeta^2 + d : a, b, c, d \in \mathbb{Q}\} \cong \mathbb{Q}[x]/(p(x))$, and $\mathbb{Q}[\zeta]$ is a field. Compute the inverse of $\zeta^2 + 1$ as a polynomial in ζ. Hint: Obtain a system of linear equations for the coefficients of the polynomial.

7.3.7. Let $\zeta = e^{2\pi i/5}$.

(a) Find the minimal polynomials for $\cos(2\pi/5)$ and $\sin(2\pi/5)$ over $\mathbb{Q}$.

(b) Find the minimal polynomial for ζ over $\mathbb{Q}(\cos(2\pi/5))$.

7.3.8. Show that the splitting field E of $x^3 - 2$ has dimension 6 over $\mathbb{Q}$. Refer to Exercise 7.2.8.

7.3.9. If α is a fifth root of 2 and β is a seventh root of 3, what is the dimension of $\mathbb{Q}(\alpha, \beta)$ over $\mathbb{Q}$?

7.3.10. Find $\dim_{\mathbb{Q}} \mathbb{Q}(\alpha, \beta)$, where

(a) $\alpha^3 = 2$ and $\beta^2 = 2$

(b) $\alpha^3 = 2$ and $\beta^2 = 3$

7.3.11. Show that $f(x) = x^4 + x^3 + x^2 + 1$ is irreducible in $\mathbb{Q}[x]$. For $p(x) \in Q[x]$, let $\overline{p(x)}$ denote the image of $p(x)$ in the field $Q[x]/(f(x))$. Compute the inverse of $\overline{(x^2 + 1)}$.

7.3.12. Show that $f(x) = x^3 + 6x^2 - 12x + 3$ is irreducible over $\mathbb{Q}$. Let θ be a real root of $f(x)$, which exists due to the intermediate value theorem. $\mathbb{Q}(\theta)$ consists of elements of the form $a_0 + a_1\theta + a_2\theta^2$. Explain how to compute the product in this field and find the product $(7+2\theta+\theta^2)(1+\theta^2)$. Find the inverse of $(7 + 2\theta + \theta^2)$.

7.3.13. Show that $f(x) = x^5 + 4x^2 - 2x + 2$ is irreducible over $\mathbb{Q}$. Let θ be a real root of $f(x)$, which exists due to the intermediate value theorem. $\mathbb{Q}(\theta)$ consists of elements of the form $a_0 + a_1\theta + a_2\theta^2 + a_3\theta^3 + a_4\theta^4$. Explain how to compute the product in this field and find the product $(7 + 2\theta + \theta^3)(1 + \theta^4)$. Find the inverse of $(7 + 2\theta + \theta^3)$.

7.3.14. Let $K \subseteq L$ be a field extension. Suppose a_1, a_2, and a_3 are elements of L that are algebraic over K. Show that $a_1a_2a_3$ and $a_1a_2 + a_2a_3$ are algebraic over K.

7.4. Splitting Field of a Cubic Polynomial

In Section 7.2, we considered a field K contained in the complex numbers $\mathbb{C}$ and a cubic polynomial $f(x) = x^3+px+q \in K[x]$. We obtained explicit expressions involving extraction of square and cube roots for the three roots $\alpha_1, \alpha_2, \alpha_3$ of $f(x)$ in $\mathbb{C}$, and we were beginning to study the splitting field extension $E = K(\alpha_1, a_2, a_3)$.

If $f(x)$ factors in $K[x]$, then either all the roots are in K or exactly one of them (say α_3) is in K and the other two are roots of an irreducible quadratic polynomial in $K[x]$. In this case, $E = K(\alpha_1)$ is a field extension of dimension 2 over K.

Henceforth, we assume that $f(x)$ is irreducible in $K[x]$. Let us first notice that the roots of $f(x)$ are necessarily distinct (Exercise 7.4.1).

If α_1 denotes one of the roots, we know that $K(\alpha_1) \cong K[x]/(f(x))$ is a field extension of dimension $3 = \deg(f)$ over K. Since we have $K \subseteq K(\alpha_1) \subseteq E$, it follows from the multiplicativity of dimension (Proposition 7.3.1) that 3 divides the dimension of E over K.

Recall the element $\delta = (\alpha_1 - \alpha_2)(\alpha_1 - \alpha_3)(\alpha_2 - \alpha_3) \in E$; since $\delta^2 = -4p^3 - 27q^2 \in K$, either $\delta \in K$ or $K(\delta)$ is an extension field of dimension 2 over K. In the latter case, since $K \subseteq K(\delta) \subseteq E$, it follows that 2 also divides $\dim_K(E)$.

We will show that there are only two possibilities:

1. $\delta \in K$ and $\dim_K(E) = 3$, or
2. $\delta \notin K$ and $\dim_K(E) = 6$.

We're going to do some algebraic tricks to solve for α_2 in terms of α_1 and δ. The identity $\sum_i \alpha_i = 0$ gives:

$$\alpha_3 = -\alpha_1 - \alpha_2. \tag{7.4.1}$$

Eliminating α_3 in $\sum_{i<j} \alpha_i\alpha_j = p$ gives

$$\alpha_2^2 = -\alpha_1^2 - \alpha_1\alpha_2 - p. \tag{7.4.2}$$

Since $f(\alpha_i) = 0$, we have

$$\alpha_i^3 = -p\alpha_i - q \quad (i = 1, 2). \tag{7.4.3}$$

In the Exercises, you are asked to show that

$$\alpha_2 = \frac{\delta + 2\,\alpha_1\,p + 3\,q}{2\,(3\,\alpha_1{}^2 + p)}. \tag{7.4.4}$$

Proposition 7.4.1. *Let K be a subfield of $\mathbb{C}$, let $f(x) = x^3+px+q \in K[x]$ an irreducible cubic polynomial, and let E denote the splitting field of $f(x)$ in $\mathbb{C}$. Let $\delta = (\alpha_1 - \alpha_2)(\alpha_1 - \alpha_3)(\alpha_2 - \alpha_3)$, where α_i are the roots of $f(x)$. If $\delta \in K$, then* $\dim_K(E) = 3$. *Otherwise,* $\dim_K(E) = 6$.

Proof. Suppose that $\delta \in K$. Then Equation (7.4.4) shows that α_2 and, therefore, also α_3 is contained in $K(\alpha_1)$. Thus $E = K(\alpha_1, \alpha_2, \alpha_3) = K(\alpha_1)$, and E has dimension 3 over K.

On the other hand, if $\delta \notin K$, then we have seen that $\dim_K(E)$ is divisible by both 2 and 3, so $\dim_K(E) \geq 6$. Consider the field extension $K(\delta) \subseteq E$. If $f(x)$ were not irreducible in $K(\delta)[x]$, then we would have $\dim_{K(\delta)}(E) \leq 2$, so

$$\dim_K(E) = \dim_K(K(\delta)) \dim_{K(\delta)}(E) \leq 4,$$

a contradiction. So $f(x)$ must remain irreducible in $K(\delta)[x]$. But then it follows from the previous paragraph [replacing K by $K(\delta)$] that $\dim_{K(\delta)}(E) = 3$. Therefore, $\dim_K(E) = 6$. ■

The structure of the splitting field can be better understood if we consider the possible intermediate fields between K and E, and introduce symmetry into the picture as well.

Proposition 7.4.2. *Let K be a field, let $f(x) \in K[x]$ be irreducible, and suppose α and β are two roots of $f(x)$ in some extension field L. Then there is an isomorphism of fields $\sigma : K(\alpha) \to K(\beta)$ such that $\sigma(k) = k$ for all $k \in K$ and $\sigma(\alpha) = \beta$.*

Proof. According to Proposition 7.3.6, there is an isomorphism

$$\sigma_\alpha : K[x]/(f(x)) \to K(\alpha)$$

that takes $[x]$ to α and fixes each element of K. So the desired isomorphism $K(\alpha) \cong K(\beta)$ is $\sigma_\beta \circ \sigma_\alpha^{-1}$. ■

Applying this result to the cubic equation, we obtain the following: *For any two roots α_i and α_j of the irreducible cubic polynomial $f(x)$, there is an isomorphism $K(\alpha_i) \cong K(\alpha_j)$ that fixes each element of K and takes α_i to α_j.*

Now, suppose that $\delta \in K$, so $\dim_K(E) = 3$. Then $E = K(\alpha_i)$ for each i, so for any two roots α_i and α_j of $f(x)$ there is an automorphism of E (i.e., an isomorphism of E onto itself) that fixes each element of K and takes α_i to α_j. Let us consider an automorphism σ of E that fixes K pointwise and maps α_1 to α_2. What is $\sigma(\alpha_2)$? The following general observation shows that $\sigma(\alpha_2)$ is also a root of $f(x)$. Surely, $\sigma(\alpha_2) \neq \alpha_2$, so $\sigma(\alpha_2) \in \{\alpha_1, \alpha_3\}$. We are going to show that necessarily $\sigma(\alpha_2) = \alpha_3$ and $\sigma(\alpha_3) = \alpha_1$.

Proposition 7.4.3. *Suppose $K \subseteq L$ is any field extension, $f(x) \in K[x]$, and β is a root of $f(x)$ in L. If σ is an automorphism of L that leaves F fixed pointwise, then $\sigma(\beta)$ is also a root of $f(x)$.*

Proof. If $f(x) = \sum f_i x^i$, then $\sum f_i \sigma(\beta)^i = \sigma(\sum f_i \beta^i) = \sigma(0) = 0$. ∎

The set of all automorphisms of a field L, denoted $\mathrm{Aut}(L)$, is a group. If $F \subseteq L$ is a subfield, an automorphism of L that leaves F fixed pointwise is called a F–*automorphism of* L. The set of F-automorphisms of L, denoted $\mathrm{Aut}_F(L)$, is a subgroup of $\mathrm{Aut}(L)$ (Exercise 7.4.4).

Return to the irreducible cubic $f(x) \in K[x]$, and suppose that $\delta \in K$ so that the splitting field E has dimension 3 over K. We have seen that the group $\mathrm{Aut}_K(E)$ acts as permutations of the roots of $f(x)$, and the action is transitive; that is, for any two roots α_i and α_j, there is a $\sigma \in \mathrm{Aut}_K(E)$ such that $\sigma(\alpha_i) = \alpha_j$. However, not every permutation of the roots can arise as the restriction of a K–automorphism of E! In fact, any odd permutation of the roots would map $\delta = (\alpha_1 - \alpha_2)(\alpha_1 - \alpha_3)(\alpha_2 - \alpha_3)$ to $-\delta$, so cannot arise from a K-automorphism of E. The group of permutations of the roots induced by $\mathrm{Aut}_K(E)$ is a transitive subgroup of even permutations, so must coincide with A_3. We have proved the following:

Proposition 7.4.4. *If* K *is a subfield of* $\mathbb{C}$, $f(x) \in K[x]$ *is an irreducible cubic polynomial,* E *is the splitting field of* $f(x)$ *in* $\mathbb{C}$, *and* $\dim_K(E) = 3$, *then* $\mathrm{Aut}_K(E) \cong A_3 \cong \mathbb{Z}_3$.

In particular, we have the answer to the question posed previously: In the case that $\delta \in K$, if σ is a K–automorphism of the splitting field E such that $\sigma(\alpha_1) = \alpha_2$, then necessarily, $\sigma(\alpha_2) = \alpha_3$ and $\sigma(\alpha_3) = \alpha_1$.

Now, consider the case that $\delta \notin K$. Consider any of the roots of $f(x)$, say α_1. In $K(\alpha_1)[x]$, $f(x)$ factors as $(x - \alpha_1)(x^2 + \alpha_1 x - q/\alpha_1)$, and $p(x) = x^2 + \alpha_1 x - q/\alpha_1$ is irreducible in $K(\alpha_1)[x]$. Now, E is obtained by adjoining a root of $p(x)$ to $K(\alpha_1)$, $E = K(\alpha_1)(\alpha_2)$, so by Proposition 7.4.2, there is an automorphism of E that fixes $K(\alpha_1)$ pointwise and that interchanges α_2 and α_3. Similarly, for any j, $\mathrm{Aut}_K(E)$ contains an automorphism that fixes α_j and interchanges the other two roots. It follows that $\mathrm{Aut}_K(E) \cong S_3$.

Proposition 7.4.5. *If* K *is a subfield of* $\mathbb{C}$, $f(x) \in K[x]$ *is an irreducible cubic polynomial,* E *is the splitting field of* $f(x)$ *in* $\mathbb{C}$, *and* $\dim_K(E) = 6$, *then* $\mathrm{Aut}_K(E) \cong S_3$. *Every permutation of the three roots of* $f(x)$ *is the restriction of a* K*–automorphism of* E.

The rest of this section will be devoted to working out the *Galois correspondence* between subgroups of $\mathrm{Aut}_K(E)$ and intermediate fields $K \subseteq M \subseteq E$. In the general theory of field extensions, it is crucial to obtain bounds relating the size of the groups $\mathrm{Aut}_M(E)$ and the dimensions $\dim_M(E)$. But for the cubic polynomial, these bounds can be bypassed by ad hoc arguments. I have done this in order to reconstruct something that Galois would have known before he constructed a general theory; Galois would certainly have been intimately familiar with the cubic polynomial as well as with the quartic polynomial before coming to terms with the general case.

Let us assume that $E \supseteq K$ is the splitting field of an irreducible cubic polynomial in $K[x]$. For each subgroup H of $\mathrm{Aut}_K(E)$, consider $\mathrm{Fix}(H) = \{a \in E : \sigma(a) = a \text{ for all } \sigma \in H\}$.

Proposition 7.4.6. *Let $E \supseteq K$ be the splitting field of an irreducible cubic polynomial in $K[x]$.*

(a) *For each subgroup H of $\mathrm{Aut}_K(E)$, $K \subseteq \mathrm{Fix}(H) \subseteq E$ is a field.*
(b) $\mathrm{Fix}(\mathrm{Aut}_K(E)) = K$.

Proof. We leave part (a) as an exercise.

For part (b), let $\overline{K} = \mathrm{Fix}(\mathrm{Aut}_K(E))$. We have $\overline{K} \underset{\neq}{\subset} E$, since E admits nontrivial K–automorphisms. In case $\dim_K(E) = 3$, it follows that $K = \overline{K}$, as there are no fields strictly intermediate between K and E.

Suppose now that $\dim_K(E) = 6$. Let $f(x) = x^3 + px + q \in K[x]$ be an irreducible cubic polynomial with splitting field E. Let $\delta = (\alpha_1 - \alpha_2)(\alpha_1 - \alpha_3)(\alpha_2 - \alpha_3)$, where α_i are the roots of $f(x)$. By Proposition 7.4.5, $\mathrm{Aut}_K(E) \cong S_3$, acting as permutations of the three roots of f in E; in particular, neither δ nor any of the α_i is fixed by $\mathrm{Aut}_K(E)$ (i.e., $\delta \neq \overline{K}$ and $\alpha_i \neq \overline{K}$).

Consider f as an element of $\overline{K}[x]$. If f is reducible in $\overline{K}[x]$, then it has a linear factor, which means that one of the roots of f in E is an element of $\overline{K}$, contradicting the observation at the end of the previous paragraph. Therefore, f remains irreducible as an element of $\overline{K}[x]$. Now by Proposition 7.4.1, with $\overline{K}$ in place of K, $\dim_{\overline{K}}(E) = 6$ since $\delta \neq \overline{K}$. It follows that $\overline{K} = K$.

■

In case $\dim_K(E) = 3$, there are no proper intermediate fields between K and E because of multiplicativity of dimensions, and there are no nontrivial proper subgroups of $\mathrm{Aut}_K(E) \cong \mathbb{Z}_3$. We have $\mathrm{Fix}(\mathrm{Aut}_E(E)) = \mathrm{Fix}(\{e\}) = E$, and $\mathrm{Fix}(\mathrm{Aut}_K(E)) = K$, by the previous proposition. Hence for all fields M such that $K \subseteq M \subseteq E$, we have $\mathrm{Fix}(\mathrm{Aut}_M(E)) = M$.

Let us now consider the case $\dim_K(E) = 6$. We want to show in this case as well that for all fields M such that $K \subseteq M \subseteq E$, we have $\mathrm{Fix}(\mathrm{Aut}_M(E)) = M$.

Recall that the subgroups of $\mathrm{Aut}_K(E) \cong S_3$ are

- S_3 itself,
- the three copies of S_2, $H_i = \{\sigma \in S_3 : \sigma(i) = i\}$,
- the cyclic group A_3, and
- the trivial subgroup $\{e\}$.

Now, suppose $K \subseteq M \subseteq E$ is some intermediate field. Let

$$H = \mathrm{Aut}_M(E) \subseteq \mathrm{Aut}_K(E) \cong S_3.$$

Then H must be one of the subgroups just listed. Put $\overline{M} = \mathrm{Fix}(H)$; then $M \subseteq \overline{M}$. I want to show that $M = \overline{M}$ (and that M must be one of the "known" intermediate fields.) The first step is the following proposition:

Proposition 7.4.7. *If* $K \subseteq M \subsetneq E$, *then* $\mathrm{Aut}_M(E) \neq \{e\}$.

Proof. It is no loss of generality to assume $K \neq M$. Because $\dim_M(E)$ divides $\dim_K(E) = 6$, it follows that $\dim_M(E)$ is either 2 or 3. Let $a \in E \setminus M$. Then $E = M(a)$; there are no more intermediate fields because of the multiplicativity of dimension.

Consider the polynomial

$$g(x) = \prod_{\sigma \in \mathrm{Aut}_K(E)} (x - \sigma(a)).$$

This polynomial has coefficients in E that are invariant under $\mathrm{Aut}_K(E)$; because $\mathrm{Fix}(\mathrm{Aut}_K(E)) = K$, *the coefficients are in* K. Now we can regard $g(x)$ as an element of $M[x]$; since $g(a) = 0$, $g(x)$ has an irreducible factor $h(x) \in M[x]$ such that $h(a) = 0$. Because $g(x)$ splits into linear factors in $E[x]$, so does $h(x)$. Thus E is a splitting field for $h(x)$. Since $\deg(h) = \dim_M(E) \geq 2$, and since the roots of $h(x)$ are distinct by Exercise 7.4.1, $h(x)$ has at least one root $b \in E$ other than a and, by Proposition 7.4.2, there is an M–automorphism of E that takes a to b. Therefore, $\mathrm{Aut}_M(E) \neq \{e\}$. ■

Proposition 7.4.8. *Let* $K \subseteq M \subseteq E$ *be an intermediate field. Let* $H = \mathrm{Aut}_M(E)$ *and let* $\overline{M} = \mathrm{Fix}(H)$.

(a) *If* H *is one of the* H_i, *then* $M = \overline{M} = K(\alpha_i)$.
(b) *If* $H = A_3$, *then* $M = \overline{M} = K(\delta)$.
(c) *If* $H = \mathrm{Aut}_K(E)$, *then* $M = \overline{M} = K$.
(d) *If* $H = \{e\}$, *then* $M = \overline{M} = E$.

Proof. If $H = \mathrm{Aut}_K(E) \cong S_3$, then $M = \overline{M} = K$, by Proposition 7.4.6. If $H = \{e\}$, then $M = \overline{M} = E$, by Proposition 7.4.7. To complete the proof, it suffices to consider the case that $K \underset{\neq}{\subset} M \underset{\neq}{\subset} E$ and $\{e\} \underset{\neq}{\subset} H \underset{\neq}{\subset} S_3$. Then we have $\dim_E(M) \in \{2, 3\}$, and $M \subseteq \overline{M} \subseteq E$, so either $M = \overline{M}$, or $\overline{M} = E$.

In Exercise 7.4.7, the fixed point subfield is computed for each subgroup of $\mathrm{Aut}_K(E) \cong S_3$. The result is $\mathrm{Fix}(H_i) = K(\alpha_i)$, $\mathrm{Fix}(A_3) = K(\delta)$ [and, of course, $\mathrm{Fix}(S_3) = K$ and $\mathrm{Fix}(\{e\}) = E$].

Consequently, if $H = H_i$, then $\overline{M} = K(\alpha_i) \neq E$, so $M = \overline{M} = K(\alpha_i)$. Similarly, if $H = A_3$, then $\overline{M} = K(\delta) \neq E$, so $M = \overline{M} = K(\delta)$. ■

We have proved the following:

Theorem 7.4.9. *Let* K *be a subfield of* $\mathbb{C}$, *let* $f(x) \in K[x]$ *be an irreducible cubic polynomial, and let* E *be the splitting field of* $f(x)$ *in* $\mathbb{C}$. *Then there is a bijection between subgroups of* $\mathrm{Aut}_K(E)$ *and intermediate fields* $K \subseteq M \subseteq E$. *Under the bijection, a subgroup* H *corresponds to the intermediate field* $\mathrm{Fix}(H)$, *and an intermediate field* M *corresponds to the subgroup* $\mathrm{Aut}_M(E)$.

This is a remarkable result, even for the cubic polynomial. The splitting field E is an infinite set. For any subset $S \subseteq E$, we can form the intermediate field $K(S)$. It is certainly not evident that there are at most six possibilities for $K(S)$ and, without introducing symmetries, this fact would remain obscure.

Example 7.4.10. Consider $f(x) = x^3 - 2$, which is irreducible over $\mathbb{Q}$. The three roots of f in $\mathbb{C}$ are $\sqrt[3]{2}$, $\omega\sqrt[3]{2}$, and $\omega^2\sqrt[3]{2}$, where $\omega = -1/2+\sqrt{-3}/2$ is a primitive cube root of 1. Let E denote the splitting field of f in $\mathbb{C}$. The discriminant of f is $\delta^2 = -108$, which has square root $\delta = 6\sqrt{-3}$. It follows that the Galois group $G = \mathrm{Aut}_{\mathbb{Q}}(E)$ is of order 6, and $\dim_{\mathbb{Q}}(E) = 6$. Each of the fields $\mathbb{Q}(\alpha)$, where α is one of the roots of f, is a cubic extension of $\mathbb{Q}$, and is the fixed field of

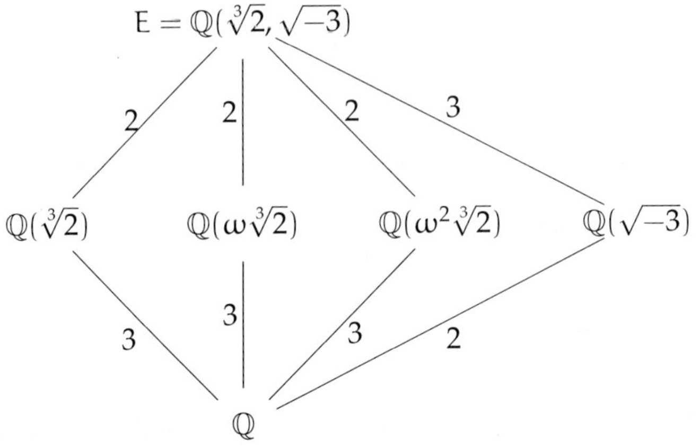

Figure 7.4.1. Intermediate fields for the splitting field of $f(x) = x^3 - 2$ over $\mathbb{Q}$.

the order 2 subgroup of G that exchanges the other two roots. The only other intermediate field between $\mathbb{Q}$ and E is $\mathbb{Q}(\delta) = \mathbb{Q}(\omega) = \mathbb{Q}(\sqrt{-3})$, which is the fixed field of the alternating group $A_3 \cong \mathbb{Z}_3$. A diagram of intermediate fields, with the dimensions of the field extensions indicated is shown as Figure 7.4.1.

Exercises 7.4

7.4.1. Let $K \subseteq \mathbb{C}$ be a field.

(a) Suppose $f(x) \in K[x]$ is an irreducible quadratic polynomial. Show that the two roots of $f(x)$ in $\mathbb{C}$ are distinct.

(b) Suppose $f(x) \in K[x]$ is an irreducible cubic polynomial. Show that the three roots of $f(x)$ in $\mathbb{C}$ are distinct. Hint: We can assume without loss of generality that $f(x) = x^3 + px + q$ has no quadratic term. If f has a double root $\alpha_1 = \alpha_2 = \alpha$, then the third root is $\alpha_3 = -2\alpha$. Now, observe that the relation

$$\sum_{i<j} \alpha_i \alpha_j = p$$

shows that α satisfies a quadratic polynomial over K.

7.4.2. Expand the expression $\delta = (\alpha_1 - \alpha_2)(\alpha_1 - \alpha_3)(\alpha_2 - \alpha_3)$, and reduce the result using Equations (7.4.1) through (7.4.3) to eliminate α_3 and to reduce higher powers of α_1 and α_2. Show that $\delta = 6\alpha_1^2\alpha_2 -$

$2\alpha_1 p + 2\alpha_2 p - 3q$. Solve for α_2 to get

$$\alpha_2 = \frac{\delta + 2\,\alpha_1\,p + 3\,q}{2\,(3\,\alpha_1{}^2 + p)}.$$

Explain why the denominator in this expression is not zero.

7.4.3. (a) Confirm the details of Example 7.4.10.

(b) Consider the irreducible polynomial $x^3 - 3x + 1$ in $\mathbb{Q}[x]$. Show that the dimension over $\mathbb{Q}$ of the splitting field is 3. Conclude that there are no fields intermediate between $\mathbb{Q}$ and the splitting field.

7.4.4. If $F \subseteq L$ is a field extension, show that $\mathrm{Aut}_F(L)$ is a subgroup of $\mathrm{Aut}(L)$.

7.4.5. Suppose $F \subseteq L$ is any field extension, $f(x) \in F[x]$, and $\beta_1, \dots, \beta_r$ are the distinct roots of $f(x)$ in L. Prove the following statements.

(a) If σ is an automorphism of L that leaves F fixed pointwise, then $\sigma_{|\{\beta_1,\dots,\beta_r\}}$ is a permutation of $\{\beta_1, \dots, \beta_r\}$.

(b) $\sigma \mapsto \sigma_{|\{\beta_1,\dots,\beta_r\}}$ is a homomorphism of $\mathrm{Aut}_F(L)$ into the group of permutations $\mathrm{Sym}(\{\beta_1, \dots, \beta_r\})$.

(c) If L is a splitting field of $f(x)$, $L = K(\beta_1, \dots, \beta_r)$, then the homomorphism $\sigma \mapsto \sigma_{|\{\beta_1,\dots,\beta_r\}}$ is injective.

7.4.6. Let $f(x) \in K[x]$ be an irreducible cubic polynomial, with splitting field E. Let H be a subgroup of $\mathrm{Aut}_K(E)$. Show that $\mathrm{Fix}(H)$ is a field intermediate between K and E.

In the following two exercises, suppose $K \subseteq E$ be the splitting field of an irreducible cubic polynomial in $K[x]$ and that $\dim_K(E) = 6$.

7.4.7. This exercise determines $\mathrm{Fix}(H)$ for each subgroup H of $\mathrm{Aut}_K(E) \cong S_3$. Evidently, $\mathrm{Fix}(\{e\}) = E$, and $\mathrm{Fix}(S_3) = K$ by Proposition 7.4.6.

(a) Observe that $E \underset{\neq}{\supset} \mathrm{Fix}(H_i) \supseteq K(\alpha_i)$. Conclude $\mathrm{Fix}(H_i) = K(\alpha_i)$.

(b) Show similarly that $\mathrm{Fix}(A_3) = K(\delta)$.

7.4.8. This exercise determines $\mathrm{Aut}_M(E)$ for each "known" intermediate field $K \subseteq M \subseteq E$. It is trivial that $\mathrm{Aut}_E(E) = \{e\}$, and $\mathrm{Aut}_K(E) \cong S_3$ is known.

(a) Show that if M is any of the fields $K(\alpha_i)$, then $\mathrm{Aut}_M(E) = H_i$.

(b) Show that if M is $K(\delta)$, then $\mathrm{Aut}_M(E) = A_3$. Hint: Replace K by $K(\delta)$, and use the case $\delta \in K$, which is already finished.

(c) Conclude from this exercise and the previous one that the equality $\mathrm{Fix}(\mathrm{Aut}_M(E)) = M$ holds for all the "known" intermediate fields $K \subseteq M \subseteq E$.

7.4.9. Let $f(x)$ be an irreducible cubic polynomial over a subfield K of $\mathbb{C}$. Let α_1, α_2, α_3 denote the three roots of f in $\mathbb{C}$ and let E denote the splitting field $E = K(\alpha_1, \alpha_2, \alpha_3)$. Let δ denote the square root of the discriminant of f, and suppose that $\delta \notin K$. Show that the lattice of intermediate fields between K and E is as shown in Figure 7.4.2.

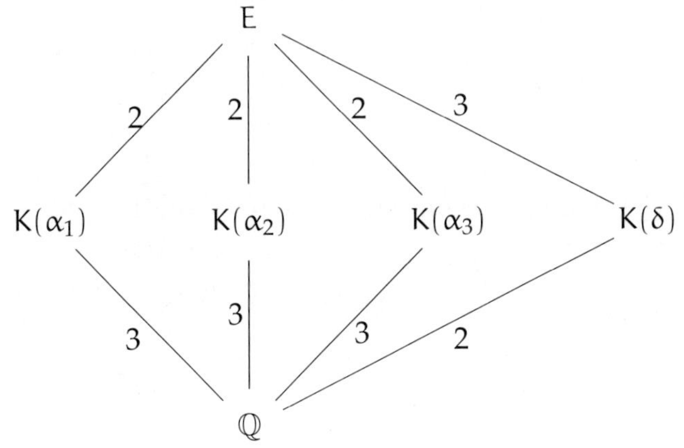

Figure 7.4.2. Intermediate fields for a splitting field with Galois group S_3.

7.4.10. For each of the following polynomials over $\mathbb{Q}$, find the splitting field E and all fields intermediate between $\mathbb{Q}$ and E.

(a) $x^3 + x^2 + x + 1$
(b) $x^3 - 3x^2 + 3$
(c) $x^3 - 3$

7.5. Splitting Fields of Polynomials in $\mathbb{C}[x]$

Our goal in this section will be to state a generalization of Theorem 7.4.9 for arbitrary polynomials in $K[x]$ for K a subfield of $\mathbb{C}$ and to sketch some ideas involved in the proof. This material will be treated systematically and in a more general context in the next chapter.

Let K be any subfield of $\mathbb{C}$, and let $f(x) \in K[x]$. According to Gauss's fundamental theorem of algebra, $f(x)$ factors into linear factors in $\mathbb{C}[x]$. The smallest subfield of $\mathbb{C}$ that contains all the roots of $f(x)$ in $\mathbb{C}$ is called the *splitting field* of $f(x)$. As for the cubic polynomial, in order to understand the structure of the splitting field, it is useful to introduce its symmetries over K. An automorphism of E is said to be a K-*automorphism* if it fixes every point of K. The set of K–automorphisms forms a group (Exercise 7.4.4).

Here is the statement of the main theorem concerning subfields of a splitting field and the symmetries of a splitting field:

Theorem 7.5.1. *Suppose* K *is a subfield of* $\mathbb{C}$, $f(x) \in K[x]$, *and* E *is the splitting field of* $f(x)$ *in* $\mathbb{C}$.

(a) $\dim_K(E) = |\mathrm{Aut}_K(E)|$.

(b) *The map* $M \mapsto \mathrm{Aut}_M(E)$ *is a bijection between the set of intermediate fields* $K \subseteq M \subseteq E$ *and the set of subgroups of* $\mathrm{Aut}_K(E)$. *The inverse map is* $H \mapsto \mathrm{Fix}(H)$. *In particular, there are only finitely many intermediate fields between* K *and* E.

This theorem asserts, in particular, that for any intermediate field $K \subseteq M \subseteq E$, there are sufficiently many M–automorphisms of E so that $\mathrm{Fix}(\mathrm{Aut}_M(E)) = M$. So the first task in proving the theorem is to see that there is an abundance of M–automorphisms of E. We will maintain the notation: K is a subfield of $\mathbb{C}$, $f(x) \in K[x]$, and E is the splitting field of $f(x)$ in $\mathbb{C}$.

Proposition 7.5.2. *Suppose* $p(x) \in K[x]$ *is an irreducible factor of* $f(x)$ *and* α *and* α' *are two roots of* $p(x)$ *in* E. *Then there is a* $\tau \in \mathrm{Aut}_K(E)$ *such that* $\tau(\alpha) = \alpha'$.

Sketch of proof. By Proposition 7.4.2, there is an isomorphism $\sigma : K(\alpha) \to K(\alpha')$ that fixes K pointwise and sends α to α'. Using an inductive argument based on the fact that E is a splitting field and a variation on the theme of 7.4.2, we show that the isomorphism σ can be extended to an automorphism of E. This gives an automorphism of E taking α to α' and fixing K pointwise. ■

Corollary 7.5.3. $\mathrm{Aut}_K(E)$ *acts faithfully by permutations on the roots of* $f(x)$ *in* E. *The action is transitive on the roots of each irreducible factor of* $f(x)$.

Proof. By Exercise 7.4.5, $\mathrm{Aut}_K(E)$ acts faithfully by permutations on the roots of $f(x)$ and, by the previous corollary, this action is transitive on the roots of each irreducible factor. ■

Theorem 7.5.4. *Suppose that* $K \subseteq \mathbb{C}$ *is a field,* $f(x) \in K[x]$, *and* E *is the splitting field of* $f(x)$ *in* $\mathbb{C}$. *Then* $\mathrm{Fix}(\mathrm{Aut}_K(E)) = K$.

Sketch of proof. We have *a priori* that $K \subseteq \mathrm{Fix}(\mathrm{Aut}_K(E))$. We must show that if $a \in L \setminus K$, then there is an automorphism of E that leaves K fixed pointwise but does not fix a. ■

Corollary 7.5.5. *If* $K \subseteq M \subseteq E$ *is any intermediate field, then* $\mathrm{Fix}(\mathrm{Aut}_M(E)) = M$.

Proof. E is also the splitting field of f over M, so the previous result applies with K replaced by M. ■

Now, we consider a converse:

Proposition 7.5.6. *Suppose* $K \subseteq E \subseteq \mathbb{C}$ *are fields,* $\dim_K(E)$ *is finite, and* $\mathrm{Fix}(\mathrm{Aut}_K(E)) = K$.

(a) *For any* $\beta \in E$, β *is algebraic over* K, *and the minimal polynomial for* β *over* K *splits in* $E[x]$.

(b) *For* $\beta \in E$, *let* $\beta = \beta_1, \dots, \beta_r$ *be a list of the distinct elements of* $\{\sigma(\beta) : \sigma \in \mathrm{Aut}_K(E)\}$. *Then* $(x-\beta_1)(\dots)(x-\beta_r)$ *is the minimal polynomial for* β *over* K.

(c) E *is the splitting field of a polynomial in* $K[x]$.

Proof. Since $\dim_K(E)$ is finite, E is algebraic over K.

Let $\beta \in E$, and let $p(x)$ denote the minimal polynomial of β over K. Let $\beta = \beta_1, \dots, \beta_r$ be the distinct elements of $\{\sigma(\beta) : \sigma \in \mathrm{Aut}_K(E)\}$. Define $g(x) = (x - \beta_1)(\dots)(x - \beta_r) \in E[x]$. Every $\sigma \in \mathrm{Aut}_K(E)$ leaves $g(x)$ invariant, so the coefficients of $g(x)$ lie in $\mathrm{Fix}(\mathrm{Aut}_K(E)) = K$. Since β is a root of $g(x)$, it follows that $p(x)$ divides $g(x)$. On the other hand, every root of $g(x)$ is of the form $\sigma(\beta)$ for $\sigma \in \mathrm{Aut}_K(E)$ and, therefore, is also a root of $p(x)$. Since the roots of $g(x)$ are simple [i.e., each root α occurs only once in the factorization $g(x) = \prod(x-\alpha)$], it follows that $g(x)$ divides $p(x)$. Hence $p(x) = g(x)$. In particular, $p(x)$ splits into linear factors over E. This proves parts (a) and (b).

Since E is finite–dimensional over K, it is generated over K by finitely many algebraic elements $\alpha_1, \dots, \alpha_s$. It follows from part (a) that E is the splitting field of $f = f_1 f_2 \cdots f_s$, where f_i is the minimal polynomial of α_i over K. ■

Definition 7.5.7. A finite–dimensional field extension $K \subseteq E \subseteq \mathbb{C}$ is said to be *Galois* if $\mathrm{Fix}(\mathrm{Aut}_K(E)) = K$.

With this terminology, the previous results say the following:

Theorem 7.5.8. *For fields* $K \subseteq E \subseteq \mathbb{C}$, *with* $\dim_K(E)$ *finite, the following are equivalent:*

(a) *The extension* E *is Galois over* K.
(b) *For all* $\alpha \in E$, *the minimal polynomial of* α *over* K *splits into linear factors over* E.
(c) E *is the splitting field of a polynomial in* $K[x]$.

Corollary 7.5.9. *If* $K \subseteq E \subseteq \mathbb{C}$ *and* E *is Galois over* K, *then* E *is Galois over every intermediate field* $K \subseteq M \subseteq E$.

Thus far we have sketched "half" of Theorem 7.5.1, namely, the map from intermediate fields M to subgroups of $\mathrm{Aut}_K(E)$, $M \mapsto \mathrm{Aut}_M(E)$, is injective, since $M = \mathrm{Fix}(\mathrm{Aut}_M(E))$. It remains to show that this map is surjective. The key to this result is the equality of the order of subgroup with the dimension of E over its fixed field: If H is a subgroup of $\mathrm{Aut}_K(E)$ and $F = \mathrm{Fix}(H)$, then

$$\dim_F(E) = |H|. \tag{7.5.1}$$

The details of the proof of the equality (7.5.1) will be given in Section 8.5, in a more general setting.

Now, consider a subgroup H of $\mathrm{Aut}_K(E)$, let F be its fixed field, and let $\overline{H}$ be $\mathrm{Aut}_F(E)$. Then we have $H \subseteq \overline{H}$, and

$$\mathrm{Fix}(\overline{H}) = \mathrm{Fix}(\mathrm{Aut}_F(E)) = F = \mathrm{Fix}(H).$$

By the equality (7.5.1) $|\overline{H}| = \dim_F(E) = |H|$, so $H = \overline{H}$. This shows that the map $M \mapsto \mathrm{Aut}_M(E)$ has as its range all subgroups of $\mathrm{Aut}_K(E)$. This completes the sketch of the proof of Theorem 7.5.1, which is known as the *fundamental theorem of Galois theory*.

Example 7.5.10. The field $E = \mathbb{Q}(\sqrt{2}, \sqrt{3})$ is the splitting field of the polynomial $f(x) = (x^2 - 2)(x^2 - 3)$, whose roots are $\pm\sqrt{2}, \pm\sqrt{3}$. The Galois group $G = \mathrm{Aut}_{\mathbb{Q}}(E)$ is isomorphic to $\mathbb{Z}_2 \times \mathbb{Z}_2$; $G = \{e, \alpha, \beta, \alpha\beta\}$, where α sends $\sqrt{2}$ to its opposite and fixes $\sqrt{3}$, and β sends $\sqrt{3}$ to its opposite and fixes $\sqrt{2}$. The subgroups of G are three copies of $\mathbb{Z}_2$ generated by α, β, and $\alpha\beta$. Therefore, there are also exactly three intermediate fields between $\mathbb{Q}$ and E, which are the fixed fields of α, β, and $\alpha\beta$. The fixed field of α is $\mathbb{Q}(\sqrt{3})$, the fixed field of β is $\mathbb{Q}(\sqrt{2})$, and the fixed field of $\alpha\beta$ is $Q(\sqrt{6})$. Figure 7.5.1 shows the lattice of intermediate fields, with the dimensions of the extensions indicated.

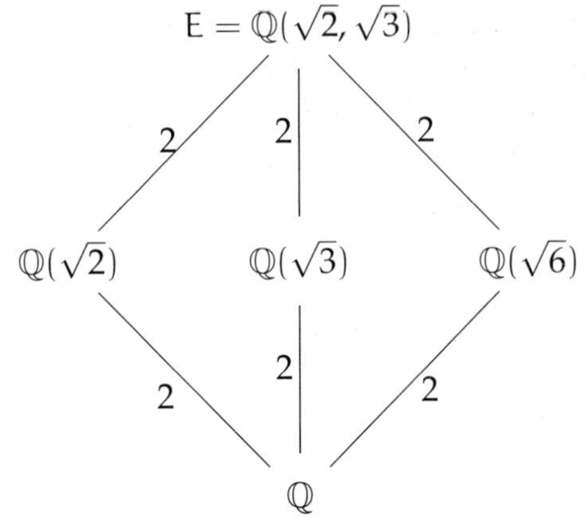

Figure 7.5.1. Lattice of intermediate fields for $\mathbb{Q} \subseteq \mathbb{Q}(\sqrt{2}, \sqrt{3})$.

There is a second part of the fundamental theorem, which describes the special role of *normal* subgroups of $\text{Aut}_K(E)$.

Theorem 7.5.11. *Suppose* K *is a subfield of* $\mathbb{C}$, $f(x) \in K[x]$, *and* E *is the splitting field of* $f(x)$ *in* $\mathbb{C}$. *A subgroup* N *of* $\text{Aut}_K(E)$ *is normal if, and only if, its fixed field* $\text{Fix}(N)$ *is Galois over* K. *In this case,* $\text{Aut}_K(\text{Fix}(N)) \cong \text{Aut}_K(E)/N$.

This result is also proved in a more general setting in Section 8.5.

Example 7.5.12. In Example 7.5.10, all the field extensions are Galois, since the Galois group is abelian.

Example 7.5.13. In Example 7.4.10, the subfield $\mathbb{Q}(\delta)$ is Galois; it is the splitting field of a quadratic polynomial, and the fixed field of the normal subgroup A_3 of the Galois group S_3.

Example 7.5.14. Consider $f(x) = x^4 - 2$, which is irreducible over $\mathbb{Q}$ by the Eisenstein criterion. The roots of f in $\mathbb{C}$ are $\pm\sqrt[4]{2}, \pm i\sqrt[4]{2}$. Let E denote the splitting field of f in $\mathbb{C}$; evidently, $E = \mathbb{Q}(\sqrt[4]{2}, i)$.

The intermediate field $\mathbb{Q}(\sqrt[4]{2})$ is of degree 4 over $\mathbb{Q}$ and $E = \mathbb{Q}(\sqrt[4]{2}, i)$ is of degree 2 over $\mathbb{Q}(\sqrt[4]{2})$, so E is of degree 8 over $\mathbb{Q}$, using Proposition 7.3.1. Therefore, it follows from the equality of dimensions in the Galois correspondence that the Galois group $G = \text{Aut}_{\mathbb{Q}}(E)$ is of order 8.

Since E is generated as a field over $\mathbb{Q}$ by $\sqrt[4]{2}$ and i, a $\mathbb{Q}$–automorphism of E is determined by its action on these two elements. Furthermore, for any automorphism σ of E over $\mathbb{Q}$, $\sigma(\sqrt[4]{2})$ must be one of the roots of f, and $\sigma(i)$ must be one of the roots of $x^2 + 1$, namely, $\pm i$. There are exactly eight possibilities for the images of $\sqrt[4]{2}$ and i, namely,

$$\sqrt[4]{2} \mapsto i^r \sqrt[4]{2} \quad (0 \le r \le 3), \quad i \mapsto \pm i.$$

As the size of the Galois group is also 8, each of these assignments must determine an element of the Galois group.

In particular, we single out two $\mathbb{Q}$–automorphisms of E:

$$\sigma : \sqrt[4]{2} \mapsto i\sqrt[4]{2}, \quad i \mapsto i,$$

and

$$\tau : \sqrt[4]{2} \mapsto \sqrt[4]{2}, \quad i \mapsto -i.$$

The automorphism τ is complex conjugation restricted to E.

Evidently, σ is of order 4, and τ is of order 2. Furthermore, we can compute that $\tau\sigma\tau = \sigma^{-1}$. It follows that the Galois group is generated by σ and τ and is isomorphic to the dihedral group D_4. You are asked to check this in the Exercises.

We identify the Galois group D_4 as a subgroup of S_4, acting on the roots $\alpha_r = i^{r-1}\sqrt[4]{2}$, $1 \le r \le 4$. With this identification, $\sigma = (1234)$ and $\tau = (24)$.

D_4 has 10 subgroups (including D_4 and $\{e\}$). The lattice of subgroups is show in Figure 7.5.2; all of the inclusions in this diagram are of index 2.

Here $\mathcal{V}$ denotes the group

$$\mathcal{V} = \{e, (12)(34), (13)(24), (14)(23)\},$$

and $\mathcal{V}'$ the group

$$\mathcal{V}' = \{e, (24), (13), (13)(24) = \sigma^2\}.$$

By the Galois correspondence, there are ten intermediate fields between $\mathbb{Q}$ and E, including the the ground field and the splitting field. Each intermediate field is the fixed field of a subgroup of G. The three subgroups of D_4 of index 2 ($\mathcal{V}$, $\mathcal{V}'$, and $\mathbb{Z}_4$) are normal, so the corresponding intermediate fields are Galois over the ground field $\mathbb{Q}$.

It is possible to determine each fixed field rather explicitly. In order to do so, we can find one or more elements that are fixed by the subgroup and that generate a field of the proper dimension. For example, the fixed field of $\mathcal{V}$ is $\mathbb{Q}(\sqrt{-2})$. You are asked in the Exercises to identify all of the intermediate fields as explicitly as possible.

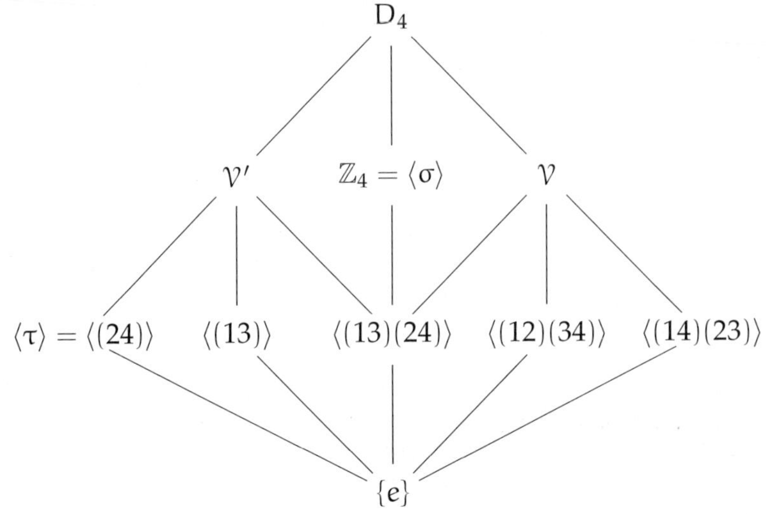

Figure 7.5.2. Lattice of subgroups of D_4.

Finally, I want to mention one more useful result, whose proof is also deferred to Section 8.5.

Theorem 7.5.15. *If $K \subseteq F \subseteq \mathbb{C}$ are fields and $\dim_K(F)$ is finite, then there is an $\alpha \in F$ such that $F = K(\alpha)$.*

It is rather easy to show that a finite–dimensional field extension is algebraic; that is, every element in the extension field is algebraic over the ground field K. Hence the extension field is certainly obtained by adjoining a finite number of algebraic elements. The tricky part is to show that if a field is obtained by adjoining two algebraic elements, then it can also be obtained by adjoining one algebraic element, $K(\alpha, \beta) = K(\gamma)$ for some γ. Then it follows by induction that a field obtained by adjoining finitely many algebraic elements can also be obtained by adjoining a single algebraic element.

Exercises 7.5

7.5.1. Check the details of Example 7.5.10.

7.5.2. In Example 7.5.14, check that the $\mathbb{Q}$–automorphisms σ and τ generate the Galois group, and that the Galois group is isomorphic to D_4.

7.5.3. Verify that the diagram of subgroups of D_4 in Example 7.5.14 is correct.

7.5.4. Identify the fixed fields of the subgroups of the Galois group in Example 7.5.14. According to the Galois correspondence, these fixed fields are all the intermediate fields between $\mathbb{Q}$ and $\mathbb{Q}(\sqrt[4]{2}, i)$.

7.5.5. Let p be a prime. Show that the splitting field of $x^p - 1$ over $\mathbb{Q}$ is $E = \mathbb{Q}(\zeta)$ where ζ is a primitive p^{th} root of unity in $\mathbb{C}$. Show that $\dim_{\mathbb{Q}}(E) = p - 1$ and that the Galois group is cyclic of order $p - 1$. For $p = 7$, analyze the fields intermediate between $\mathbb{Q}$ and E.

7.5.6. Verify that the following field extensions of $\mathbb{Q}$ are Galois. Find a polynomial for which the field is the splitting field. Find the Galois group, the lattice of subgroups, and the lattice of intermediate fields.

(a) $\mathbb{Q}(\sqrt{2}, \sqrt{7})$
(b) $\mathbb{Q}(i, \sqrt{7})$
(c) $\mathbb{Q}(\sqrt{2} + \sqrt{7})$
(d) $\mathbb{Q}(e^{2\pi i/7})$
(e) $\mathbb{Q}(e^{2\pi i/6})$

7.5.7. The Galois group G and splitting field E of $f(x) = x^8 - 2$ can be analyzed in a manner quite similar to Example 7.5.14. The Galois group G has order 16 and has a (normal) cyclic subgroup of order 8; however, G is not the dihedral group D_8. Describe the group by generators and relations (two generators, one relation). Find the lattice of subgroups. G has a normal subgroup isomorphic to D_4; in fact the splitting field E contains the splitting field of $x^4 - 2$. The subgroups of D_4 already account for a large number of the subgroups of G. Describe as explicitly as possible the lattice of intermediate fields between $\mathbb{Q}$ and the splitting field.

7.5.8. Suppose E is a Galois extension of a field K (both fields assumed, for now, to be contained in $\mathbb{C}$). Suppose that the Galois group $\text{Aut}_K(E)$ is abelian, and that an irreducible polynomial $f(x) \in K[x]$ has one root $\alpha \in E$. Show that $K(\alpha)$ is then the splitting field of f. Show by examples that this is not true if the Galois group is not abelian or if the polynomial f is not irreducible.

PART II: TOPICS

CHAPTER 8

Field Extensions – Second Look

This chapter contains a systematic introduction to Galois's theory of field extensions and symmetry. The fields in this chapter are general, not necessarily subfields of the complex numbers nor even of characteristic 0. We will restrict ours attention, however, to so called *separable algebraic* extensions.

8.1. Finite and Algebraic Extensions

In this section, we continue the exploration of finite and algebraic field extensions, which was begun in Section 7.3. Recall that a field extension $K \subseteq L$ is said to be *finite* if $\dim_K(L)$ is finite and is called *algebraic* in case each element of L satisfies a polynomial equation with coefficients in K.

Proposition 8.1.1. *Suppose that $K \subseteq M \subseteq L$ are field extensions,*

(a) *If M is algebraic over K and $b \in L$ is algebraic over M, then b is algebraic over K.*

(b) *If M is algebraic over K, and L is algebraic over M, then L is algebraic over K.*

Proof. For part (a), since b is algebraic over L, there is a polynomial

$$p(x) = a_0 + a_1x + \cdots + a_nx^n$$

with coefficients in M such that $p(b) = 0$. But this implies that b is algebraic over $K(a_0, \dots, a_n)$, and, therefore,

$$K(a_0, \dots, a_n) \subseteq K(a_0, \dots, a_n)(b) = K(a_0, \dots, a_n, b)$$

is a finite field extension, by Proposition 7.3.6. Since M is algebraic over K, the a_i are algebraic over K, and, therefore, $K \subseteq K(a_0, \dots, a_n)$ is a finite field extension, by Proposition 7.3.9. Proposition 7.3.1 implies that $K \subseteq K(a_0, \dots, a_n, b)$ is a finite field extension. It then follows from Proposition 7.3.4 that $K(a_0, \dots, a_n, b)$ is algebraic over K, so, in particular, b is algebraic over K.

Part (b) is a consequence of part (a). ■

Definition 8.1.2. Suppose that $K \subseteq L$ is a field extension and that E and F are intermediate fields, $K \subseteq E \subseteq L$ and $K \subseteq F \subseteq L$. The *composite* $E \cdot F$ of E and F is the smallest subfield of L containing E and F.

$$\begin{array}{ccccc} F & \subseteq & E \cdot F & \subseteq & L \\ \cup| & & \cup| & & \\ K & \subseteq & E & & \end{array}$$

Proposition 8.1.3. *Suppose that $K \subseteq L$ is a field extension and that E and F are intermediate fields, $K \subseteq E \subseteq L$, and $K \subseteq F \subseteq L$.*

(a) *If E is algebraic over K and F is arbitrary, then $E \cdot F$ is algebraic over F.*
(b) *If E and F are both algebraic over K, then $E \cdot F$ is algebraic over K.*
(c) $\dim_F(E \cdot F) \leq \dim_K(E)$.

Proof. Exercises 8.1.1 through 8.1.3. ■

Exercises 8.1

8.1.1. Prove Proposition 8.1.3 (a). Hint: Let $a \in E \cdot F$. Then there exist $\alpha_1, \dots, \alpha_n \in E$ such that $a \in F(\alpha_1, \dots, \alpha_n)$.

8.1.2. Prove Proposition 8.1.3 (b).

8.1.3. Prove Proposition 8.1.3 (c). Hint: In case $\dim_K(E)$ is infinite, there is nothing to be done. So assume the dimension is finite and let $\alpha_1, \dots, \alpha_n$ be a basis of E over K. Conclude successively that $E \cdot F = F(\alpha_1, \dots, \alpha_n)$, then that $E \cdot F = F[\alpha_1, \dots, \alpha_n]$, and finally that $E \cdot F = \text{span}_F\{\alpha_1, \dots, \alpha_n\}$.

8.1.4. What is the dimension of $\mathbb{Q}(\sqrt{2}+\sqrt{3})$ over $\mathbb{Q}$?

8.2. Splitting Fields

Now, we turn our point of view around regarding algebraic extensions. Given a polynomial $f(x) \in K[x]$, we *produce* extension fields $K \subseteq L$ in which f has a root, or in which f has a complete set of roots.

Proposition 7.3.6 tells us that if $f(x)$ is irreducible in $K[x]$ and α is a root of $f(x)$ in some extension field, then the field generated by K and the root α is isomorphic to $K[x]/(f(x))$; but $K[x]/(f(x))$ is a field, so we might as well choose our extension field to be $K[x]/(f(x))$! What should α be then? It will have to be the image in $K[x]/(f(x))$ of x, namely, $[x] = x + (f(x))$. (We are using $[g(x)]$ to denote the image of $g(x)$ in $K[x]/(f(x))$.) Indeed, $[x]$ is a root of $f(x)$ in $K[x]/(f(x))$, since $f([x]) = [f(x)] = 0$ in $K[x]/(f(x))$.

Proposition 8.2.1. *Let K be a field and let $f(x)$ be a monic irreducible element of $K[x]$. Then*

(a) *There is an extension field L and an element $\alpha \in L$ such that $f(\alpha) = 0$.*

(b) *For any such extension field and any such α, $f(x)$ is the minimal polynomial for α.*

(c) *If L and L' are extension fields of K containing elements α and α' satisfying $f(\alpha) = 0$ and $f(\alpha') = 0$, then there is an isomorphism $\psi : K(\alpha) \to K(\alpha')$ such that $\psi(k) = k$ for all $k \in K$ and $\psi(\alpha) = \alpha'$.*

Proof. The existence of the field extension containing a root of $f(x)$ was already shown. The minimal polynomial for α divides f and, therefore, equals f, since f is monic irreducible. For the third part, we have $K(\alpha) \cong K[x]/(f(x)) \cong K(\alpha')$, by isomorphisms that leave K pointwise fixed. ■

We will need a technical variation of part (c) of the proposition. Recall from Corollary 6.2.6 that if M and M' are fields and $\sigma : M \to M'$ is a field isomorphism, then σ extends to an isomorphism of rings $M[x] \to M'[x]$ by $\sigma(\sum m_i x^i) = \sum \sigma(m_i) x^i$.

Corollary 8.2.2. *Let K and K' be fields, let $\sigma : K \to K'$ be a field isomorphism, and let $f(x)$ be an irreducible element of $K[x]$. Suppose that α is a root of $f(x)$ in an extension $L \supseteq K$ and that α' is a root of $\sigma(f(x))$ in an extension field $L' \supseteq K'$. Then there is an isomorphism $\psi : K(\alpha) \to K'(\alpha')$ such that $\psi(k) = \sigma(k)$ for all $k \in K$ and $\psi(\alpha) = \alpha'$.*

Proof. The ring isomorphism $\sigma : K[x] \to K'[x]$ induces a field isomorphism $\tilde{\sigma} : K[x]/(f) \to K'[x]/(\sigma(f))$, satisfying $\tilde{\sigma}(k+(f)) = \sigma(k) + (\sigma(f))$ for $k \in K$. Now, use $K(\alpha) \cong K[x]/(f) \cong K'[x]/(\sigma(f)) \cong K'(\alpha')$. ■

Now, consider a monic polynomial $f(x) \in K[x]$ of degree $d > 1$, not necessarily irreducible. Factor $f(x)$ into irreducible factors. If any of these factors have degree greater than 1, then choose such a factor and adjoin a root α_1 of this factor to K, as previously. Now, regard $f(x)$ as a polynomial over the field $K(\alpha_1)$, and write it as a product of irreducible factors in $K(\alpha_1)[x]$. If any of these factors has degree greater than 1, then choose such a factor and adjoin a root α_2 of this factor to $K(\alpha_1)$. After repeating this procedure at most d times, we obtain a field in which $f(x)$ factors into linear factors. Of course, a proper proof goes by induction; see Exercise 8.2.1.

Definition 8.2.3. A *splitting field* for a polynomial $f(x) \in K[x]$ is an extension field L such that $f(x)$ factors into linear factors over L, and L is generated by K and the roots of $f(x)$ in L.

If a polynomial $p(x) \in K[x]$ factors into linear factors over a field $M \supseteq K$, and if $\{\alpha_1, \dots \alpha_r\}$ are the distinct roots of $p(x)$ in M, then $K(\alpha_1, \dots, \alpha_r)$, the subfield of M generated by K and $\{\alpha_1, \dots \alpha_r\}$, is a splitting field for $p(x)$. For polynomials over $\mathbb{Q}$, for example, it is unnecessary to refer to Exercise 8.2.1 for the existence of splitting fields; a rational polynomial splits into linear factors over $\mathbb{C}$, so a splitting field is obtained by adjoining the complex roots of the polynomial to $\mathbb{Q}$.

One consequence of the existence of splitting fields is the existence of many finite fields. See Exercise 8.2.3.

The following result says that a splitting field is unique up to isomorphism.

Proposition 8.2.4. *Let $K \subseteq L$ and $\tilde{K} \subseteq \tilde{L}$ be field extensions, let $\sigma : K \to \tilde{K}$ be a field isomorphism, and let $p(x) \in K[x]$ and $\tilde{p}(x) \in \tilde{K}[x]$ be polynomials with $\tilde{p}(x) = \sigma(p(x))$. Suppose L is a splitting field for $p(x)$ and $\tilde{L}$ is a splitting field for $\tilde{p}(x)$. Then there is a field isomorphism $\tau : L \to \tilde{L}$ such that $\tau(k) = \sigma(k)$ for all $k \in K$.*

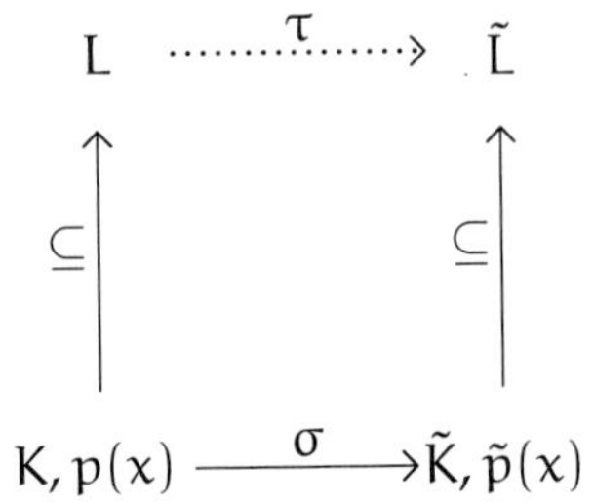

Proof. The idea is to use Proposition 8.2.2 and induction on $\dim_K(L)$. If $\dim_K(L) = 1$, there is nothing to do: The polynomial $p(x)$ factors into linear factors over K, so also $\tilde{p}(x)$ factors into linear factors over $\tilde{K}$, $K = L$, $\tilde{K} = \tilde{L}$, and σ is the required isomorphism.

We make the following induction assumption: Suppose $K \subseteq M \subseteq L$ and $\tilde{K} \subseteq \tilde{M} \subseteq \tilde{L}$ are intermediate field extensions, $\tilde{\sigma} : M \to \tilde{M}$ is a field isomorphism extending σ, and $\dim_M L < n = \dim_K(L)$. Then there is a field isomorphism $\tau : L \to \tilde{L}$ such that $\tau(m) = \tilde{\sigma}(m)$ for all $m \in M$.

Now, since $\dim_K(L) = n > 1$, at least one of the irreducible factors of $p(x)$ in $K[x]$ has degree greater than 1. Choose such an irreducible factor $p_1(x)$, and observe that $\sigma(p_1(x))$ is an irreducible factor of $\tilde{p}(x)$. Let $\alpha \in L$ and $\tilde{\alpha} \in \tilde{L}$ be roots of $p_1(x)$ and $\sigma(p_1(x))$, respectively. By Proposition 8.2.2, there is an isomorphism $\tilde{\sigma} : K(\alpha) \to \tilde{K}(\tilde{\alpha})$ taking α to $\tilde{\alpha}$ and extending σ.

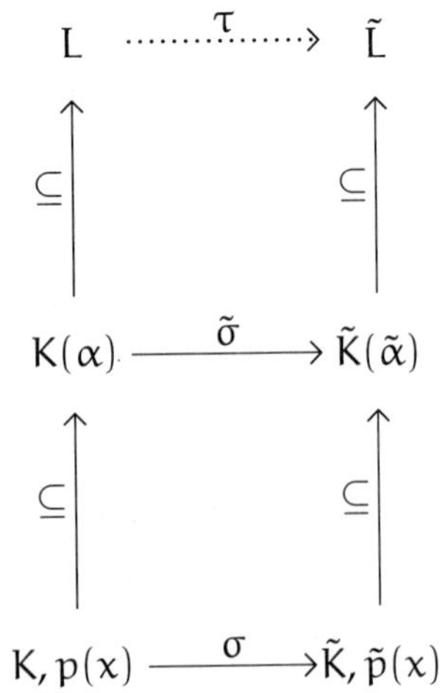

Figure 8.2.1. Stepwise extension of isomorphism.

Now, the result follows by applying the induction hypothesis with $M = K(\alpha)$ and $\tilde{M} = \tilde{K}(\tilde{\alpha})$. See Figure 8.2.1 ■

Corollary 8.2.5. *Let* $p(x) \in K[x]$, *and suppose* L *and* $\tilde{L}$ *are two splitting fields for* $p(x)$. *Then there is an isomorphism* $\tau : L \to \tilde{L}$ *such that* $\tau(k) = k$ *for all* $k \in K$.

Proof. Take $K = \tilde{K}$ and $\sigma = \text{id}$ in the proposition. ■

Exercises 8.2

8.2.1. For any polynomial $f(x) \in K[x]$ there is an extension field L of K such that $f(x)$ factors into linear factors in $L[x]$. Give a proof by induction on the degree of f.

8.2.2. Verify the following statements: The rational polynomial $f(x) = x^6 - 3$ is irreducible over $\mathbb{Q}$. It factors over $\mathbb{Q}(3^{1/6})$ as

$$(x - 3^{1/6})(x + 3^{1/6})(x^2 - 3^{1/6}x + 3^{1/3})(x^2 + 3^{1/6}x + 3^{1/3}).$$

If $\omega = e^{\pi i/3}$, then the irreducible factorization of $f(x)$ over $\mathbb{Q}(3^{1/6}, \omega)$ is

$$(x - 3^{1/6})(x + 3^{1/6})(x - \omega 3^{1/6})(x - \omega^2 3^{1/6})(x - \omega^4 3^{1/6})(x - \omega^5 3^{1/6}).$$

8.2.3.

(a) Show that if K is a finite field, then $K[x]$ has irreducible elements of arbitrarily large degree. Hint: Use the existence of infinitely many irreducibles in $K[x]$, Proposition 1.8.9.

(b) Show that if K is a finite field, then K admits field extensions of arbitrarily large finite degree.

(c) Show that a finite–dimensional extension of a finite field is a finite field.

8.3. The Derivative and Multiple Roots

In this section, we examine, by means of exercises, multiple roots of polynomials and their relation to the formal derivative.

We say that a polynomial $f(x) \in K[x]$ has a root a *with multiplicity* m in an extension field L if $f(x) = (x - a)^m g(x)$ in $L[x]$, and $g(a) \neq 0$. A root of multiplicity greater than 1 is called a *multiple root.* A root of multiplicity 1 is called a *simple root.*

The formal derivative in $K[x]$ is defined by the usual rule from calculus: We define $D(x^n) = nx^{n-1}$ and extend linearly. Thus,

$D(\sum k_n x^n) = \sum n k_n x^{n-1}$. The formal derivative satisfies the usual rules for differentiation:

8.3.1. Show that $D(f(x) + g(x)) = Df(x) + Dg(x)$ and $D(f(x)g(x)) = D(f(x))g(x) + f(x)D(g(x))$.

8.3.2.

(a) Suppose that the field K is of characteristic zero. Show that $Df(x) = 0$ if, and only if, $f(x)$ is a constant polynomial.

(b) Suppose that the field has characteristic p. Show that $Df(x) = 0$ if, and only if, there is a polynomial $g(x)$ such that $f(x) = g(x^p)$.

8.3.3. Suppose $f(x) \in K[x]$, L is an extension field of K, and $f(x)$ factors as $f(x) = (x - a)g(x)$ in $L[x]$. Show that the following are equivalent:

(a) a is a multiple root of $f(x)$.

(b) $g(a) = 0$.

(c) $Df(a) = 0$.

8.3.4. Let $K \subseteq L$ be a field extension and let $f(x), g(x) \in K[x]$. Show that the greatest common divisor of $f(x)$ and $g(x)$ in $L[x]$ is the same as the greatest common divisor in $K[x]$. Hint: Review the algorithm for computing the g.c.d., using division with remainder.

8.3.5. Suppose $f(x) \in K[x]$, L is an extension field of K, and a is a multiple root of $f(x)$ in L. Show that if $Df(x)$ is not identically zero, then $f(x)$ and $Df(x)$ have a common factor of positive degree in $L[x]$ and, therefore, by the previous exercise, also in $K[x]$.

8.3.6. Suppose that K is a field and $f(x) \in K[x]$ is irreducible.

(a) Show that if f has a multiple root in some field extension, then $Df(x) = 0$.

(b) Show that if $\mathrm{Char}(K) = 0$, then $f(x)$ has only simple roots in any field extension.

The preceding exercises establish the following theorem:

Theorem 8.3.1. *If the characteristic of a field K is zero, then any irreducible polynomial in $K[x]$ has only simple roots in any field extension.*

8.3.7. If K is a field of characteristic p and $a \in K$, then $(x+a)^p = x^p + a^p$. Hint: The binomial coefficient $\binom{p}{k}$ is divisible by p if $0 < k < p$.

Now, suppose K is a field of characteristic p and that $f(x)$ is an irreducible polynomial in $K[x]$. If $f(x)$ has a multiple root in some extension field, then $Df(x)$ is identically zero, by Exercise 8.3.6. Therefore, there is a $g(x) \in K[x]$ such that $f(x) = g(x^p) = a_0 + a_1 x^p + \dots a_r x^{rp}$. Suppose that for each a_i there is a $b_i \in K$ such that $b_i^p = a_i$. Then $f(x) = (b_0 + b_1 x + \cdots b_r x^r)^p$, which contradicts the irreducibility of $f(x)$. This proves the following theorem:

Theorem 8.3.2. *Suppose K is a field of characteristic p in which each element has a pth root. Then any irreducible polynomial in $K[x]$ has only simple roots in any field extension.*

Proposition 8.3.3. *Suppose K is a field of characteristic p. The map $a \mapsto a^p$ is a field isomorphism of K into itself. If K is a finite field, then $a \mapsto a^p$ is an automorphism of K.*

Proof. Clearly, $(ab)^p = a^p b^p$ for $a, b \in K$. But also $(a+b)^p = a^p + b^p$ by Exercise 8.3.7. Therefore, the map is a homomorphism. The homomorphism is not identically zero, since $1^p = 1$; since K is simple, the homomorphism must, therefore, be injective. If K is finite, an injective map is bijective. ■

Corollary 8.3.4. *Suppose K is a finite field. Then any irreducible polynomial in $K[x]$ has only simple roots in any field extension.*

Proof. K must have some prime characteristic p. By Proposition 8.3.3, any element of K has a pth root in K and, therefore, the result follows from Theorem 8.3.2. ■

8.4. Splitting Fields and Automorphisms

Recall that an *automorphism* of a field L is a field isomorphism of L onto L, and that the set of all automorphisms of L forms a group denoted by $\mathrm{Aut}(L)$. If $K \subseteq L$ is a field extension, we denote the set of automorphisms of L that leave each element of K fixed by $\mathrm{Aut}_K(L)$; we call such automorphisms *K-automorphisms of* L. Recall from Exercise 7.4.4 that $\mathrm{Aut}_K(L)$ is a subgroup of $\mathrm{Aut}(L)$.

Proposition 8.4.1. *Let $f(x) \in K[x]$, let L be a splitting field for $f(x)$, let $p(x)$ be an irreducible factor of $f(x)$, and finally let α and β be two roots of $p(x)$ in L. Then there is an automorphism $\sigma \in \mathrm{Aut}_K(L)$ such that $\sigma(\alpha) = \beta$.*

Proof. Using Proposition 8.2.1, we get an isomorphism from $K(\alpha)$ onto $K(\beta)$ that sends α to β and fixes K pointwise. Now, applying Corollary 8.2.5 to $K(\alpha) \subseteq L$ and $K(\beta) \subseteq L$ gives the result. ■

Proposition 8.4.2. *Let L be a splitting field for $p(x) \in K[x]$, let M, M' be intermediate fields, $K \subseteq M \subseteq L$, $K \subseteq M' \subseteq L$, and let σ be an isomorphism of M onto M' that leaves K pointwise fixed. Then σ extends to a K-automorphism of L.*

Proof. This follows from applying Proposition 8.2.4 to the situation specified in the following diagram:

$$\begin{array}{ccc} L & \overset{\tau}{\dashrightarrow} & L \\ \subseteq\uparrow & & \uparrow\subseteq \\ M, p(x) & \xrightarrow{\sigma} & M', p(x) \end{array}$$

■

Corollary 8.4.3. *Let L be a splitting field for $p(x) \in K[x]$, and let M be an intermediate field, $K \subseteq M \subseteq L$. Write $\mathrm{Iso}_K(M, L)$ for the set of field isomorphisms of M into L that leave K fixed pointwise.*

(a) *There is a bijection from the set of left cosets of $\mathrm{Aut}_M(L)$ in $\mathrm{Aut}_K(L)$ onto $\mathrm{Iso}_K(M, L)$.*

(b) $|\mathrm{Iso}_K(M, L)| = [\mathrm{Aut}_K(L) : \mathrm{Aut}_M(L)]$.

Proof. According to Proposition 8.4.2, the map $\tau \mapsto \tau_{|M}$ is a surjection of $\mathrm{Aut}_K(L)$ onto $\mathrm{Iso}_K(M, L)$. Check that $(\tau_1)_{|M} = (\tau_2)_{|M}$ if, and only if, τ_1 and τ_2 are in the same left coset of $\mathrm{Aut}_M(L)$ in $\mathrm{Aut}_K(L)$. This proves part (a), and part (b) follows. ■

Proposition 8.4.4. *Let $K \subseteq L$ be a field extension and let $f(x) \in K[x]$.*

(a) *If $\sigma \in \mathrm{Aut}_K(L)$, then σ permutes the roots of $f(x)$ in L.*

(b) *If L is a splitting field of $f(x)$, then $\mathrm{Aut}_K(L)$ acts faithfully on the roots of f in L. Furthermore, the action is transitive on the roots of each irreducible factor of $f(x)$ in $K[x]$.*

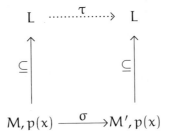

Proof. Suppose $\sigma \in \mathrm{Aut}_K(L)$,

$$f(x) = k_0 + k_1 x + \cdots + k_n x^n \in K[x],$$

and α is a root of $f(x)$ in L. Then

$$\begin{aligned} f(\sigma(\alpha)) &= k_0 + k_1\sigma(\alpha) + \cdots + k_n\sigma(\alpha^n)) \\ &= \sigma(k_0 + k_1\alpha + \cdots + k_n\alpha^n) = 0. \end{aligned}$$

Thus, $\sigma(\alpha)$ is also a root of $f(x)$. If A is the set of distinct roots of $f(x)$ in L, then $\sigma \mapsto \sigma_{|A}$ is an action of $\mathrm{Aut}_K(L)$ on A. If L is a splitting field for $f(x)$, then, in particular, $L = K(A)$, so the action of $\mathrm{Aut}_K(L)$ on A is faithful. Proposition 8.4.1 says that if L is a splitting field for $f(x)$, then the action is transitive on the roots of each irreducible factor of $f(x)$. ■

Definition 8.4.5. If $f \in K[x]$, and L is a splitting field of $f(x)$, then $\mathrm{Aut}_K(L)$ is called the *Galois group of* f, or the *Galois group of the field extension* $K \subseteq L$.

We have seen that the Galois group of an irreducible polynomial f is isomorphic to a transitive subgroup of the group of permutations the roots of f in L. At least for small n, it is possible to classify the transitive subgroups of S_n, and thus to list the possible isomorphism classes for the Galois groups of irreducible polynomials of degree n. For $n = 3, 4$, and 5, we have found all transitive subgroups of S_n, in Exercises 5.1.9 and 5.1.20 and Section 5.5.

Let us quickly recall our investigation of splitting fields of irreducible cubic polynomials in Chapter 8, where we found the properties of the Galois group corresponded to properties of the splitting field. The only possibilities for the Galois group are $A_3 = \mathbb{Z}_3$ and S_3. The Galois group is A_3 if, and only if, the field extension $K \subseteq L$ is of dimension 3, and this occurs if, and only if, the element δ defined in Chapter 8 belongs to the ground field K; in this case there are no intermediate fields between K and L. The Galois group is S_3 if, and only if, the field extension $K \subseteq L$ is of dimension 6. In this case subgroups of the Galois group correspond one to one with fields intermediate between K and L.

We are aiming at obtaining similar results in general.

Definition 8.4.6. Let H be a subgroup of $\mathrm{Aut}(L)$. Then the fixed field of H is $\mathrm{Fix}(H) = \{a \in L : \sigma(a) = a \text{ for all } \sigma \in H\}$.

Proposition 8.4.7. *Let* L *be a field,* H *a subgroup of* $\mathrm{Aut}(L)$ *and* $K \subset L$ *a subfield. Then*

(a) $\mathrm{Fix}(H)$ *is a subfield of* L.
(b) $\mathrm{Aut}_{\mathrm{Fix}(H)}(L) \supseteq H$.
(c) $\mathrm{Fix}(\mathrm{Aut}_K(L)) \supseteq K$.

Proof. Exercise 8.4.1. ■

Proposition 8.4.8. *Let* L *be a field,* H *a subgroup of* $\mathrm{Aut}(L)$, *and* $K \subset L$ *a subfield. Introduce the notation* $H^\circ = \mathrm{Fix}(H)$ *and* $K' = \mathrm{Aut}_K(L)$. *The previous exercise showed that* $H^{\circ\prime} \supseteq H$ *and* $L'^{\circ} \supseteq L$.

(a) *If* $H_1 \subseteq H_2 \subseteq \mathrm{Aut}(L)$ *are subgroups, then* $H_1^\circ \supseteq H_2^\circ$.
(b) *If* $K_1 \subseteq K_2 \subseteq L$ *are fields, then* $K_1' \supseteq K_2'$.

Proof. Exercise 8.4.2. ■

Proposition 8.4.9. *Let* L *be a field,* H *a subgroup of* $\mathrm{Aut}(L)$, *and* $K \subset L$ *a subfield.*

(a) $(H^\circ)'^\circ = H^\circ$.
(b) $(K')^{\circ\prime} = K'$.

Proof. Exercise 8.4.3. ■

Definition 8.4.10. A polynomial in $K[x]$ is said to be *separable* if each of its irreducible factors has only simple roots in some (hence any) splitting field. An algebraic element a in a field extension of K is said to be *separable over* K if its minimal polynomial is separable. An algebraic field extension L of K is said to be *separable over* K if each of its elements is separable over K.

Remark 8.4.11. Separability is automatic if the characteristic of K is zero or if K is finite, by Theorems 8.3.1 and 8.3.4.

Theorem 8.4.12. *Suppose* L *is a splitting field for a separable polynomial* $f(x) \in K[x]$. *Then* $\mathrm{Fix}(\mathrm{Aut}_K(L)) = K$.

Proof. Let $\beta_1, \dots, \beta_r$ be the distinct roots of $f(x)$ in L. Consider the tower of fields:

$$M_0 = K \subseteq \cdots \subseteq M_j = K(\beta_1, \dots, \beta_j)$$
$$\subseteq \cdots \subseteq M_r = K(\beta_1, \dots, \beta_r) = L.$$

A priori, $\text{Fix}(\text{Aut}_K(L)) \supseteq K$. We have to show that if $a \in L$ is fixed by all elements of $\text{Aut}_K(L)$, then $a \in K$. I claim that if $a \in M_j$ for some $j \geq 1$, then $a \in M_{j-1}$. It will follow from this claim that $a \in M_0 = K$.

Suppose that $a \in M_j$. If $M_{j-1} = M_j$, there is nothing to show. Otherwise, let $\ell > 1$ denote the degree of the minimal polynomial $p(x)$ for β_j in $M_{j-1}[x]$. Then $\{1, \beta_j, \cdots, \beta_j^{\ell-1}\}$ is a basis for M_j over M_{j-1}. In particular,

$$a = m_0 + m_1\beta_j + \cdots + m_{\ell-1}\beta_j^{\ell-1} \tag{8.4.1}$$

for certain $m_i \in M_{j-1}$.

Since $p(x)$ is a factor of $f(x)$ in $M_{j-1}[x]$, p is separable, and the ℓ distinct roots $\{\alpha_1 = \beta_j, \alpha_2, \dots, \alpha_\ell\}$ of $p(x)$ lie in L. According to Proposition 8.4.1, for each s, there is a $\sigma_s \in \text{Aut}_{M_{j-1}}(L) \subseteq \text{Aut}_K(L)$ such that $\sigma_s(\alpha_1) = \alpha_s$. Applying σ_s to the expression for a and taking into account that a and the m_i are fixed by σ_s, we get

$$a = m_0 + m_1\alpha_s + \cdots + m_{\ell-1}\alpha_s^{\ell-1} \tag{8.4.2}$$

for $1 \leq s \leq \ell$. Thus, the polynomial $(m_0 - a) + m_1 x + \cdots + m_{\ell-1}x^{l-1}$ of degree no more than $\ell - 1$ has at least ℓ distinct roots in L, and, therefore, the coefficients are identically zero. In particular, $a = m_0 \in M_{j-1}$. ■

The following is the converse to the previous proposition:

Proposition 8.4.13. *Suppose* $K \subseteq L$ *is a field extension,* $\dim_K(L)$ *is finite, and* $\text{Fix}(\text{Aut}_K(L)) = K$.

(a) *For any* $\beta \in L$, β *is algebraic and separable over* K, *and the minimal polynomial for* β *over* K *splits in* $L[x]$.

(b) *For* $\beta \in L$, *let* $\beta = \beta_1, \dots, \beta_n$ *be a list of the distinct elements of* $\{\sigma(\beta) : \sigma \in \text{Aut}_K(L)\}$. *Then* $(x-\beta_1)(\dots)(x-\beta_n)$ *is the minimal polynomial for* β *over* K.

(c) L *is the splitting field of a separable polynomial in* $K[x]$.

Proof. Since $\dim_K(L)$ is finite, L is algebraic over K.

Let $\beta \in L$, and let $\beta = \beta_1, \dots, \beta_r$ be the distinct elements of $\{\sigma(\beta) : \sigma \in \text{Aut}_K(L)\}$. Define $g(x) = (x - \beta_1)(\dots)(x - \beta_r) \in L[x]$. Every $\sigma \in \text{Aut}_K(L)$ leaves $g(x)$ invariant, so the coefficients of $g(x)$ lie in $\text{Fix}(\text{Aut}_K(L) = K$.

Let $p(x)$ denote the minimal polynomial of β over K. Since β is a root of $g(x)$, it follows that $p(x)$ divides $g(x)$. On the other hand, every root of $g(x)$ is of the form $\sigma(\beta)$ for $\sigma \in \text{Aut}_K(L)$ and, therefore, is also a root of $p(x)$. Since the roots of $g(x)$ are simple, it follows that $g(x)$ divides $p(x)$. Hence $p(x) = g(x)$, as both are monic. In particular, $p(x)$ splits into linear factors over L, and the roots of $p(x)$ are simple. This proves parts (a) and (b).

Since L is finite–dimensional over K, it is generated over K by finitely many algebraic elements $\alpha_1, \dots, \alpha_s$. It follows from part (a) that L is the splitting field of $f = f_1 f_2 \cdots f_s$, where f_i is the minimal polynomial of α_i over K. ■

Recall that a finite–dimensional field extension $K \subseteq L$ is said to be *Galois* if $\text{Fix}(\text{Aut}_K(L)) = K$.

Combining the last results gives the following:

Theorem 8.4.14. *For a finite–dimensional field extension $K \subseteq L$, the following are equivalent:*

(a) *The extension is Galois.*

(b) *The extension is separable, and for all $\alpha \in L$ the minimal polynomial of α over K splits into linear factors over L.*

(c) *L is the splitting field of a separable polynomial in $K[x]$.*

Corollary 8.4.15. *If $K \subseteq L$ is a finite–dimensional Galois extension and $K \subseteq M \subseteq L$ is an intermediate field, then $M \subseteq L$ is a Galois extension.*

Proof. L is the splitting field of a separable polynomial over K, and, therefore, also over M. ■

Proposition 8.4.16. *If $K \subseteq L$ is a finite–dimensional Galois extension, then*

$$\dim_K L = |\text{Aut}_K(L)|. \tag{8.4.3}$$

Proof. The result is evident if $K = L$. Assume inductively that if $K \subseteq M \subseteq L$ is an intermediate field and $\dim_M L < \dim_K L$, then $\dim_M L = |\text{Aut}_M(L)|$. Let $\alpha \in L \setminus K$ and let $p(x) \in K[x]$ be the minimal polynomial of α over K. Since L is Galois over K, p is separable and splits over L, by Theorem 8.4.14. If $\varphi \in \text{Iso}_K(K(\alpha), L)$, then $\varphi(\alpha)$ is a root of p, and φ is determined by $\varphi(\alpha)$. Therefore,

$$\deg(p) = |\text{Iso}_K(K(\alpha), L)| = [\text{Aut}_K(L) : \text{Aut}_{K(\alpha)}(L)], \tag{8.4.4}$$

where the last equality comes from Equation (8.4.3). By the induction hypothesis applied to $K(\alpha)$, $|\text{Aut}_{K(\alpha)}(L)| = \dim_{K(\alpha)} L$ is finite.

Therefore, $\mathrm{Aut}_K(L)$ is also finite, and

$$\begin{aligned}|\mathrm{Aut}_K(L)| &= \deg(p)\, |\mathrm{Aut}_{K(\alpha)}(L)| \\ &= \dim_K(K(\alpha))\ \dim_{K(\alpha)}(L) = \dim_K L,\end{aligned}$$

where the first equality comes from Equation (8.4.4), the second from the induction hypothesis and the irreducibility of p, and the final equality from the multiplicativity of dimensions, Proposition 7.3.1. ∎

Corollary 8.4.17. *Let* $K \subseteq L$ *be a finite–dimensional Galois extension and* M *an intermediate field. Then*

$$|\mathrm{Iso}_K(M, L)| = \dim_K M. \tag{8.4.5}$$

Proof.

$$\begin{aligned}|\mathrm{Iso}_K(M, L)| &= [\mathrm{Aut}_K(L) : \mathrm{Aut}_M(L)] \\ &= \frac{\dim_K(L)}{\dim_M(L)} = \dim_K M,\end{aligned}$$

using Corollary 8.4.3, Proposition 8.4.16, and the multiplicativity of dimension, Proposition 7.3.1. ∎

Corollary 8.4.18. *Let* $K \subseteq M$ *be a finite–dimensional separable field extension. Then*

$$|\mathrm{Aut}_K(M)| \leq \dim_K M.$$

Proof. There is a field extension $K \subseteq M \subseteq L$ such that L is finite–dimensional and Galois over K. (In fact, M is obtained from K by adjoining finitely many separable algebraic elements; let L be a splitting field of the product of the minimal polynomials over K of these finitely many elements.) Now, we have $|\mathrm{Aut}_K(M)| \leq |\mathrm{Iso}_K(M, L)| = \dim_K M$. ∎

Exercises 8.4

8.4.1. Prove Proposition 8.4.7.

8.4.2. Prove Proposition 8.4.8.

8.4.3. Prove Proposition 8.4.9.

8.4.4. Suppose that $f(x) \in K[x]$ is separable, and $K \subseteq M$ is an extension field. Show that f is also separable when considered as an element in $M[x]$.

8.5. The Galois Correspondence

In this section, we establish the fundamental theorem of Galois theory, a correspondence between intermediate fields $K \subseteq M \subseteq L$ and subgroups of $\mathrm{Aut}_K(L)$, when L is a Galois field extension of K.

Proposition 8.5.1. *Suppose* $K \subseteq L$ *is a finite–dimensional separable field extension. Then there is an element* $\gamma \in L$ *such that* $L = K(\alpha)$.

Proof. If K is finite, then the finite–dimensional field extension L is also a finite field. According to Corollary 3.3.17, the multiplicative group of units of L is cyclic. Then $L = K(\alpha)$, where α is a generator of the multiplicative group of units.

Suppose now that K is infinite (which is always the case if the characteristic is zero). L is generated by finitely many separable algebraic elements over K, $L = K(\alpha_1, \dots, \alpha_s)$. It suffices to show that if $L = K(\alpha, \beta)$, where α and β are separable and algebraic, then there is a γ such that $L = K(\gamma)$, for then the general statement follows by induction on s.

Suppose then that $L = K(\alpha, \beta)$. Let $K \subseteq L \subseteq E$ be a finite–dimensional field extension such that E is Galois over K. Write $n = \dim_K L = |\mathrm{Iso}_K(L, E)|$ (Corollary 8.4.17). Let $\{\varphi_1 = \mathrm{id}, \varphi_2, \dots, \varphi_n\}$ be a listing of $\mathrm{Iso}_K(L, E)$.

I claim that there is an element $k \in K$ such that the elements $\varphi_j(k\alpha + \beta)$ are all distinct. Suppose this for the moment and put $\gamma = k\alpha + \beta$. Then $K(\gamma) \subseteq L$, but $\dim_K(K(\gamma)) = |\mathrm{Iso}_K(K(\gamma), E)| \geq n = \dim_K L$. Therefore, $K(\gamma) = L$.

Now, to prove the claim, let

$$p(x) = \prod_{1 \leq i < j \leq n} (x(\varphi_i(\alpha) - \varphi_j(\alpha) + (\varphi_i(\beta) - \varphi_j(\beta)).$$

The polynomial $p(x)$ is not identically zero since the φ_i are distinct on $K(\alpha, \beta)$, so there is an element k of the infinite field K such that $p(k) \neq 0$. But then the elements $k\varphi_i(\alpha) + \varphi_i(\beta) = \varphi_i(k\alpha + \beta)$, $1 \leq i \leq n$ are distinct. ∎

Corollary 8.5.2. *Suppose* $K \subseteq L$ *is a finite–dimensional field extension, and the characteristic of* K *is zero. Then* $L = K(\alpha)$ *for some* $\alpha \in L$.

Proof. Separability is automatic in case the characteristic is zero. ■

Proposition 8.5.3. *Let* $K \subseteq L$ *be a finite–dimensional separable field extension and let* H *be a subgroup of* $\mathrm{Aut}_K(L)$. *Put* $F = \mathrm{Fix}(H)$. *Then*

(a) L *is Galois over* F.
(b) $\dim_F(L) = |H|$.
(c) $H = \mathrm{Aut}_F(L)$.

Proof. First note that $|H| \leq |\mathrm{Aut}_K(L)| \leq \dim_K(L)$, by Corollary 8.4.18, so H is necessarily finite. Then $F = \mathrm{Fix}(\mathrm{Aut}_F(L))$, by Proposition 8.4.9, so L is Galois over F and, in particular, separable by Theorem 8.4.14.

By Proposition 8.5.1, there is a β such that $L = F(\beta)$. Let $\{\varphi_1 = \mathrm{id}, \varphi_2, \dots, \varphi_n\}$ be a listing of the elements of H, and put $\beta_i = \varphi_i(\beta)$. Then, by the argument of Proposition 8.4.13, the minimal polynomial for β over F is $g(x) = (x - \beta_1)(\dots)(x - \beta_n)$. Therefore,

$$\dim_F L = \deg(g) = |H| \leq |\mathrm{Aut}_F(L)| = \dim_F(L),$$

using Proposition 8.4.16. ■

We are now ready for the fundamental theorem of Galois theory:

Theorem 8.5.4. *Let* $K \subseteq L$ *be a Galois field extension.*

(a) *There is an order-reversing bijection between subgroups of* $\mathrm{Aut}_K(L)$ *and intermediate fields* $K \subseteq M \subseteq L$, *given by* $H \mapsto \mathrm{Fix}(H)$.
(b) *The following conditions are equivalent for an intermediate field* M:
   (i) M *is Galois over* K.
   (ii) M *is invariant under* $\mathrm{Aut}_K(L)$
   (iii) $\mathrm{Aut}_M(L)$ *is a normal subgroup of* $\mathrm{Aut}_K(L)$.

   In this case,

$$\mathrm{Aut}_K(M) \cong \mathrm{Aut}_K(L)/\mathrm{Aut}_M(L).$$

Proof. If $K \subseteq M \subseteq L$ is an intermediate field, then L is Galois over M [i.e., $M = \mathrm{Fix}(\mathrm{Aut}_M(L))$]. On the other hand, if H is a subgroup of $\mathrm{Aut}_K(L)$, then, according to Proposition 8.5.3, $H = \mathrm{Aut}_{\mathrm{Fix}(H)}(L)$. Thus, the two maps $H \mapsto \mathrm{Fix}(H)$ and $M \mapsto \mathrm{Aut}_M(L)$ are inverses, which gives part (a).

Let M be an intermediate field. There is an α such that $M = K(\alpha)$ by Proposition 8.5.1. Let $f(x)$ denote the minimal polynomial of α over K. M is Galois over K if, and only if, M is the splitting field for $f(x)$, by Theorem 8.4.14. But the roots of $f(x)$ in L are the images of α under $\mathrm{Aut}_K(L)$ by Proposition 8.4.4. Therefore, M is Galois over K if, and only if, M is invariant under $\mathrm{Aut}_K(L)$.

If $\sigma \in \mathrm{Aut}_K(L)$, then $\sigma(M)$ is an intermediate field with group

$$\mathrm{Aut}_{\sigma(M)}(L) = \sigma \mathrm{Aut}_M(L)\sigma^{-1}.$$

By part (a), $M = \sigma(M)$ if, and only if,

$$\mathrm{Aut}_M(L) = \mathrm{Aut}_{\sigma(M)}(L) = \sigma \mathrm{Aut}_M(L)\sigma^{-1}.$$

Therefore, M is invariant under $\mathrm{Aut}_K(L)$ if, and only if, $\mathrm{Aut}_M(L)$ is normal.

If M is invariant under $\mathrm{Aut}_K(L)$, then $\pi : \sigma \mapsto \sigma_{|M}$ is a homomorphism of $\mathrm{Aut}_K(L)$ into $\mathrm{Aut}_K(M)$, with kernel $\mathrm{Aut}_M(L)$. I claim that this homomorphism is surjective. In fact, an element of $\sigma \in \mathrm{Aut}_K(M)$ is determined by $\sigma(\alpha)$, which is necessarily a root of $f(x)$. But by Proposition 8.4.1, there is a $\sigma' \in \mathrm{Aut}_K(L)$ such that $\sigma'(\alpha) = \sigma(\alpha)$; therefore, $\sigma = \sigma'_{|M}$. Now, the homomorphism theorem for groups gives

$$\mathrm{Aut}_K(M) \cong \mathrm{Aut}_K(L)/\mathrm{Aut}_M(L).$$

This completes the proof of part (b). ■

We shall require the following variant of Proposition 8.5.3.

Proposition 8.5.5. *Let* L *be a field,* H *a finite subgroup of* $\mathrm{Aut}(L)$, *and* $F = \mathrm{Fix}(H)$. *Then*

(a) L *is a finite–dimensional Galois field extension of* F.

(b) $H = \mathrm{Aut}_F(L)$ *and* $\dim_F(L) = |H|$.

Proof. We cannot apply Proposition 8.5.3 because it is not given that L is finite–dimensional over F. Let β be any element of L. We can adapt the argument of Proposition 8.4.13 to show that β is algebraic and separable over F, and that the minimal polynomial for β over F splits in L. Namely, let $\beta = \beta_1, \dots, \beta_r$ be the distinct elements of $\{\sigma(\beta) : \sigma \in H\}$. Define $g(x) = (x - \beta_1)(\dots)(x - \beta_r) \in L[x]$. Every $\sigma \in H$ leaves $g(x)$ invariant, so the coefficients of $g(x)$ lie in $\mathrm{Fix}(H) = F$.

Let $p(x)$ denote the minimal polynomial of β over F. Since β is a root of $g(x)$, it follows that $p(x)$ divides $g(x)$. On the other hand, every root of $g(x)$ is of the form $\sigma(\beta)$ for $\sigma \in H \subseteq \mathrm{Aut}_F(L)$ and, therefore, is also a root of $p(x)$. Since the roots of $g(x)$ are simple, it follows that $g(x)$ divides $p(x)$. Hence $p(x) = g(x)$, as both are monic. In particular, $p(x)$ splits into linear factors over L, and the roots of $p(x)$ are simple.

Note that $\deg(p) \leq |H|$, so $\dim_F(F(\beta)) = \deg(p) \leq |H|$.

Next consider an intermediate field $F \subseteq M \subseteq L$ with $\dim_F(M)$ finite. Since M is finite–dimensional and separable over F, Proposition 8.5.1 implies that there exists $\beta \in M$ such that $M = F(\alpha)$; hence $\dim_F(M) \leq |H|$. According to Exercise 8.5.1, it follows that $\dim_F(L) \leq |H|$.

We can now apply Proposition 8.5.3, with $K = F$, to reach the conclusions. ■

Let $K \subseteq L$ be a field extension and let A, B be fields intermediate between K and L. Consider the composite $A \cdot B$, namely, the subfield of L generated by $A \cup B$. We have the following diagram of field extensions:

$$\begin{array}{ccccc} & & A & \subseteq & A \cdot B & \subseteq & L \\ & & \cup| & & \cup| & & \\ K & \subseteq & A \cap B & \subseteq & B. & & \end{array}$$

The following is an important technical result that is used in the sequel.

Proposition 8.5.6. *Let $K \subseteq L$ be a finite–dimensional field extension and let A, B be fields intermediate between K and L. Suppose that B is Galois over K. Then $A \cdot B$ is Galois over A and $\mathrm{Aut}_A(A \cdot B) \cong \mathrm{Aut}_{A \cap B}(B)$.*

Proof. Exercise 8.5.2. ■

The remainder of this section can be omitted without loss of continuity.

It is not hard to obtain the inequality of Corollary 8.4.18 without the separability assumption. It follows that the separability assumption can also be removed in Proposition 8.5.3. Although we consider only separable field extensions in this text, the argument is nevertheless worth knowing.

Proposition 8.5.7. *Let L be a field. Any collection of distinct automorphisms of L is linearly independent.*

Proof. Let $\{\sigma_1, \dots, \sigma_n\}$ be a collection of distinct automorphisms of L. We show by induction on n that the collection is linearly independent (in the vector space of functions from L to L.) If $n = 1$

there is nothing to show, since an automorphism cannot be identically zero. So assume $n > 1$ and assume any smaller collection of distinct automorphisms is linearly independent. Suppose

$$\sum_{i=1}^{n} \lambda_i \, \sigma_i = 0, \tag{8.5.1}$$

where $\lambda_i \in L$. Choose $\lambda \in L$ such that $\sigma_1(\lambda) \neq \sigma_n(\lambda)$. Then for all $\mu \in L$,

$$0 = \sum_{i=1}^{n} \lambda_i \, \sigma_i(\lambda\mu) = \sum_{i=1}^{n} \lambda_i \sigma_i(\lambda) \, \sigma_i(\mu). \tag{8.5.2}$$

In other words,

$$\sum_{i=1}^{n} \lambda_i \sigma_i(\lambda) \, \sigma_i = 0. \tag{8.5.3}$$

We can now eliminate σ_1 between Equations (8.5.1) and (8.5.3) to give

$$\sum_{i=2}^{n} \lambda_i (\sigma_1(\lambda) - \sigma_i(\lambda)) \, \sigma_i = 0. \tag{8.5.4}$$

By the inductive assumption, all the coefficients of this equation must be zero. In particular, since $(\sigma_1(\lambda) - \sigma_n(\lambda)) \neq 0$, we have $\lambda_n = 0$. Now, the inductive assumption applied to the original Equation (8.5.1) gives that all the coefficients λ_i are zero. ■

Proposition 8.5.8. *Let* $K \subseteq L$ *be a field extension with* $\dim_K(L)$ *finite. Then* $|\mathrm{Aut}_K(L)| \leq \dim_K(L)$.

Proof. Suppose that $\dim_K(L) = n$ and $\{\lambda_1, \dots, \lambda_n\}$ is a basis of L over K. Suppose also that $\{\sigma_1, \dots, \sigma_{n+1}\}$ is a subset of $\mathrm{Aut}_K(L)$. (We *do not* assume that the σ_i are all distinct!) The n-by-$n+1$ matrix

$$[\sigma_j(\lambda_i)]_{1 \leq i \leq n,\ 1 \leq j \leq n+1}$$

has a nontrivial kernel by basic linear algebra. Thus, there exist $b_1, \dots, b_{n+1}$ in L, not all zero, such that

$$\sum_j \sigma_j(\lambda_i) b_j = 0$$

for all i. Now, if $k_1, \dots, k_n$ are any elements of K,

$$0 = \sum_i k_i \Big(\sum_j \sigma_j(\lambda_i) \, b_j\Big) = \sum_j b_j \sigma_j\Big(\sum_i k_i \lambda_i\Big).$$

But $\sum_i k_i \lambda_i$ represents an arbitrary element of L, so the last equation gives $\sum_j b_j \, \sigma_j = 0$. Thus, the collection of σ_j is linearly dependent. By the previous proposition, the σ_j cannot be all distinct. That is, the cardinality of $\mathrm{Aut}_K(L)$ is no more than n. ■

Exercises 8.5

8.5.1. Suppose that $K \subseteq L$ is an algebraic field extension. Show that $L = \cup\{M : K \subseteq M \subseteq L$ and M is finite–dimensional$\}$. If there is an $N \in \mathbb{N}$ such that $\dim_K(M) \leq N$ whenever $K \subseteq M \subseteq L$ and M is finite–dimensional, then also $\dim_K(L) \leq N$.

8.5.2. This exercise gives the proof of Proposition 8.5.6. We suppose that $K \subseteq L$ is a finite–dimensional field extension, that A, B are fields intermediate between K and L, and that B is Galois over K. Let α be an element of B such that $B = K(\alpha)$ (Proposition 8.5.1). Let $p(x) \in K[x]$ be the minimal polynomial for α. Then B is a splitting field for $p(x)$ over K, and the roots of $p(x)$ are distinct, by Theorem 8.4.14.

(a) Show that $A \cdot B$ is Galois over A. Hint: $A \cdot B = A(\alpha)$; show that $A \cdot B$ is a splitting field for $p(x) \in A[x]$.

(b) Show that $\tau \mapsto \tau_{|B}$ is an injective homomorphism of $\mathrm{Aut}_A(A \cdot B)$ into $\mathrm{Aut}_{A \cap B}(B)$.

(c) Check surjectivity of $\tau \mapsto \tau_{|B}$ as follows: Let
$$G' = \{\tau_{|B} : \tau \in \mathrm{Aut}_A(A \cdot B)\}.$$
Then
$$\mathrm{Fix}(G') = \mathrm{Fix}(\mathrm{Aut}_A(A \cdot B)) \cap B = A \cap B.$$
Therefore, by the Galois correspondence, $G' = \mathrm{Aut}_{A \cap B}(B)$.

8.6. Symmetric Functions

Let K be any field, and let $x_1, \dots, x_n$ be variables. For a vector $\alpha = (\alpha_1, \dots, \alpha_n)$, with nonnegative integer entries, let $x^\alpha = x_1^{\alpha_1} \dots x_n^{\alpha_n}$. The *total degree* of the *monic monomial* x^α is $|\alpha| = \sum \alpha_i$. A polynomial is said to be *homogeneous of total degree* d if it is a linear combination of monomials x^α of total degree d.

Write $K_d[x_1, \dots, x_n]$ for the set of polynomials in n variables that are homogeneous of total degree d or identically zero. Then $K_d[x_1, \dots, x_n]$ is a vector space over K and $K[x_1, \dots, x_n]$ is the direct sum over over $d \geq 0$ of the subspaces $K_d[x_1, \dots, x_n]$; see Exercise 8.6.1.

The symmetric group S_n acts on polynomials and rational functions in n variables over K by $\sigma(f)(x_1, \dots, x_n) = f(x_{\sigma(1)}, \dots, x_{\sigma(n)})$. For $\sigma \in S_n$, $\sigma(x^\alpha) = x_{\sigma(1)}^{\alpha_1} \cdots x_{\sigma(n)}^{\alpha_n}$. A polynomial or rational function is called *symmetric* if it is fixed by the S_n action. The set of symmetric polynomials is denoted $K^S[x_1, \dots, x_n]$, and the set of symmetric rational functions is denoted $K^S(x_1, \dots, x_n)$.

Note that for each d, $K_d[x_1, \dots, x_n]$ is invariant under the action of S_n, and $K^S[x_1, \dots, x_n]$ is the direct sum of the vector subspaces $K_d^S[x_1, \dots, x_n] = K_d[x_1, \dots, x_n] \cap K^S[x_1, \dots, x_n]$ for $d \geq 0$. See Exercise 8.6.3.

Lemma 8.6.1.

(a) *The action of S_n on $K[x_1, \dots, x_n]$ is an action by ring automorphisms; the action of of S_n on $K(x_1, \dots, x_n)$ is an action by by field automorphisms.*

(b) *$K^S[x_1, \dots, x_n]$ is a subring of $K[x_1, \dots, x_n]$ and $K^S(x_1, \dots, x_n)$ is a subfield of $K(x_1, \dots, x_n)$.*

(c) *The field of symmetric rational functions is the field of fractions of the ring of symmetric polynomials in n-variables.*

Proof. Exercise 8.6.2. ■

Proposition 8.6.2. *The field $K(x_1, \dots, x_n)$ of rational functions is Galois over the field $K^S(x_1, \dots, x_n)$ of symmetric rational functions, and the Galois group $\mathrm{Aut}_{K^S(x_1,\dots,x_n)}(K(x_1, \dots, x_n))$ is S_n.*

Proof. By Exercise 8.6.1, S_n acts on $K(x_1, \dots, x_n)$ by field automorphisms and $K^S(x_1, \dots, x_n)$ is the fixed field. Therefore, by Proposition 8.5.5 the extension is Galois, with Galois group is S_n. ■

We define a distinguished family of symmetric polynomials, the *elementary symmetric functions* as follows:

$$
\begin{aligned}
\varepsilon_0(x_1, \dots, x_n) &= 1 \\
\varepsilon_1(x_1, \dots, x_n) &= x_1 + x_2 + \cdots + x_n \\
\varepsilon_2(x_1, \dots, x_n) &= \sum_{1 \leq i < j \leq n} x_i x_j \\
&\cdots
\end{aligned}
$$

$$\varepsilon_k(x_1, \ldots, x_n) = \sum_{1 \le i_1 < i_2 < \cdots < i_k \le n} x_{i_1} \cdots x_{i_k}$$
$$\cdots$$
$$\varepsilon_n(x_1, \ldots, x_n) = x_1 x_2 \cdots x_n$$

We put $\varepsilon_j(x_1, \ldots x_n) = 0$ if $j > n$.

Lemma 8.6.3.

$$\begin{aligned}&(x - x_1)(x - x_2)(\cdots)(x - x_n)\\ &= x^n - \varepsilon_1 x^{n-1} + \varepsilon_2 x^{n-2} - \cdots + (-1)^n \varepsilon_n\\ &= \sum_{k=0}^{n} (-1)^k \varepsilon_k x^{n-k},\end{aligned}$$

where ε_k *is short for* $\varepsilon_k(x_1, \ldots, x_n)$.

Proof. Exercise 8.6.4. ∎

Corollary 8.6.4.

(a) *Let* $f(x) = x^n + a_{n-1}x^{n-1} + \cdots + a_0$ *be a monic polynomial in* $K[x]$ *and let* $\alpha_1, \ldots, \alpha_n$ *be the roots of* f *in a splitting field. Then* $a_i = (-1)^{n-i}\varepsilon_{n-i}(\alpha_1, \ldots, \alpha_n)$.

(b) *Let* $f(x) = a_n x^n + a_{n-1}x^{n-1} + \cdots + a_0 \in K[x]$ *be of degree* n, *and let* $\alpha_1, \ldots, \alpha_n$ *be the roots of* f *in a splitting field. Then* $a_i/a_n = (-1)^{n-i}\varepsilon_{n-i}(\alpha_1, \ldots, \alpha_n)$.

Proof. For part (a),

$$\begin{aligned}f(x) &= (x - \alpha_1)(x - \alpha_2)\cdots(x - \alpha_n)\\ &= x^n - \varepsilon_1(\alpha_1, \ldots, \alpha_n)x^{n-1} + \varepsilon_2(\alpha_1, \ldots, \alpha_n)x^{n-2} -\\ &\qquad \cdots + (-1)^n \varepsilon_n(\alpha_1, \ldots, \alpha_n).\end{aligned}$$

For part (b), apply part (a) to

$$(x - \alpha_1)(x - \alpha_2)\cdots(x - \alpha_n) = \sum_i (a_i/a_n)x^i.$$

∎

Definition 8.6.5. Let K be a field and $\{u_1, \ldots, u_n\}$ a set of elements in an extension field. We say that $\{u_1, \ldots, u_n\}$ is *algebraically independent* over K if there is no polynomial $f \in K[x_1, \ldots, x_n]$ such that $f(u_1, \ldots, u_n) = 0$.

The following is called the fundamental theorem of symmetric functions:

Theorem 8.6.6. *The set of elementary symmetric functions* $\{\varepsilon_1, \dots, \varepsilon_n\}$ *in* $K[x_1, \dots, x_n]$ *is algebraically independent over* K, *and generates* $K^S[x_1, \dots, x_n]$ *as a ring. Consequently,* $K(\varepsilon_1, \dots, \varepsilon_n) = K^S(x_1, \dots, x_n)$.

The *algebraic independence* of the ε_i is the same as *linear independence* of the monic monomials in the ε_i. First, we establish an indexing system for the monic monomials: A *partition* is a finite decreasing sequence of nonnegative integers, $\lambda = (\lambda_1, \dots, \lambda_k)$. We can picture a partition by means of an M-by-N matrix Λ of zeroes and ones, where $M \geq k$ and $N \geq \lambda_1$; $\Lambda_{rs} = 1$ if $r \leq k$ and $s \leq \lambda_r$, and $\Lambda_{rs} = 0$ otherwise. Here is a matrix representing the partition $\lambda = (5, 4, 4, 2)$:

$$\Lambda = \begin{bmatrix} 1 & 1 & 1 & 1 & 1 & 0 \\ 1 & 1 & 1 & 1 & 0 & 0 \\ 1 & 1 & 1 & 1 & 0 & 0 \\ 1 & 1 & 0 & 0 & 0 & 0 \\ 0 & 0 & 0 & 0 & 0 & 0 \\ 0 & 0 & 0 & 0 & 0 & 0 \end{bmatrix}.$$

The size of a partition λ is $|\lambda| = \sum \lambda_i$. The nonzero entries in λ are referred to as the *parts* of λ. The number of parts is called the *length* of λ. The *conjugate partition* λ^* is that represented by the transposed matrix Λ^*. Note that $\lambda_r^* = |\{i : \lambda_i \geq r\}|$, and $(\lambda^*)^* = \lambda$; see Exercise 8.6.6. For $\lambda = (5, 4, 4, 2)$, we have $\lambda^* = (4, 4, 3, 3, 1)$, corresponding to the matrix

$$\Lambda^* = \begin{bmatrix} 1 & 1 & 1 & 1 & 0 & 0 \\ 1 & 1 & 1 & 1 & 0 & 0 \\ 1 & 1 & 1 & 0 & 0 & 0 \\ 1 & 1 & 1 & 0 & 0 & 0 \\ 1 & 0 & 0 & 0 & 0 & 0 \\ 0 & 0 & 0 & 0 & 0 & 0 \end{bmatrix}.$$

For $\lambda = (\lambda_1, \lambda_2, \dots, \lambda_s)$, define

$$\varepsilon_\lambda(x_1, \dots, x_n) = \prod_i \varepsilon_{\lambda_i}(x_1, \dots, x_n).$$

For example, $\varepsilon_{(5,4,4,2)} = (\varepsilon_5)(\varepsilon_4)^2(\varepsilon_2)$. Note that $\varepsilon_\lambda(x_1, \dots, x_n) = 0$ if $\lambda_1 > n$.

We will show that the set of ε_λ with $|\lambda| = d$ and $\lambda_1 \le n$ is a basis of $K_d^S[x_1, \dots, x_n]$. In order to do this, we first produce a more obvious basis.

For a partition $\lambda = (\lambda_1, \dots \lambda_n)$ with no more than n nonzero parts, define the *monomial symmetric function*

$$m_\lambda(x_1, \dots, x_n) = (1/f) \sum_{\sigma \in S_n} \sigma(x^\lambda),$$

where f is the size of the stabilizer of x^λ under the action of the symmetric group. (Thus, m_λ is a sum of monic monomials, each occurring exactly once.) For example,

$$m_{(5,4,4,2)}(x_1, \dots, x_4) = x_1^5 x_2^4 x_3^4 x_4^2 + x_1^5 x_2^4 x_3^2 x_4^4 + x_1^5 x_2^2 x_3^4 x_4^4 + \dots,$$

a sum of 12 monic monomials.

In the Exercises, you are asked to check that the monomial symmetric functions m_λ with $\lambda = (\lambda_1, \dots, \lambda_n)$ and $|\lambda| = d$ form a linear basis of $K_d^S[x_1, \dots, x_n]$.

Define a total order on n-tuples of nonnegative integers and, in particular, on partitions by $\alpha > \beta$ if the first nonzero difference $\alpha_i - \beta_i$ is positive. Note that $\lambda \ge \sigma(\lambda)$ for any permutation σ. This total order on n-tuples α induces a total order on monomials x^α called *lexicographic order*.

Lemma 8.6.7.

(a) *The leading (i.e., lexicographically highest) monomial of ε_λ is x^{λ^*}.*

(b)

$$\varepsilon_\lambda(x_1, \dots, x_n) = \sum_{|\mu|=|\lambda|} T_{\lambda\mu} m_{\mu^*}(x_1, \dots, x_n),$$

where $T_{\lambda\mu}$ is a nonnegative integer, $T_{\lambda\lambda} = 1$, and $T_{\lambda\mu} = 0$ if $\mu^* > \lambda^*$.

(c)

$$m_\lambda = \sum_{|\mu|=|\lambda|} S_{\lambda\mu} \varepsilon_{\mu^*}(x_1, \dots, x_n),$$

where $S_{\lambda\mu}$ is a nonnegative integer, $S_{\lambda\lambda} = 1$, and $S_{\lambda\mu} = 0$ if $\mu^* > \lambda^*$.

Proof.

$$\varepsilon_\lambda(x_1, \dots, x_n) = (x_1 \cdots x_{\lambda_1})(x_1 \cdots x_{\lambda_2}) \cdots (x_1 \cdots x_{\lambda_n}) + \dots$$
$$= x_1^{\mu_1} x_2^{\mu_2} \cdots x_n^{\mu_n} + \cdots,$$

where the omitted monomials are lexicographically lower, and μ_j is the number of λ_k such that $\lambda_k \ge j$; but then $\mu_j = \lambda_j^*$, and, therefore, the leading monomial of ε_λ is x^{λ^*}. Since ε_λ is symmetric, we have $\varepsilon_\lambda = m_{\lambda^*} +$ a sum of m_{μ^*}, where $|\mu| = |\lambda|$ and $\mu < \lambda^*$ in lexicographic order. This proves parts (a) and (b).

Moreover, a triangular integer matrix with 1's on the diagonal has an inverse of the same sort. Therefore, (b) implies (c). ■

Example 8.6.8. Take $n = 4$ and $\lambda = (4, 4, 3, 3, 2)$. Then $\lambda^* = (5, 4, 4, 2)$. We have

$$\begin{aligned}\varepsilon_\lambda &= (\varepsilon_4)^2(\varepsilon_3)^2\varepsilon_1\\ &= (x_1x_2x_3x_4)^2(x_1x_2x_3 + x_1x_3x_4 + x_2x_3x_4)^2(x_1 + \dots x_4)\\ &= x_1^5x_2^4x_3^4x_4^2 + \dots,\end{aligned}$$

where the remaining monomials are less than $x_1^5x_2^4x_3^4x_4^2$ in lexicographic order.

Proof of Theorem 8.6.6. Since the m_μ of a fixed degree d form a basis of the linear space $K_d^S[x_1, \dots, x_n]$, it is immediate from the previous lemma that the ε_λ of degree d also form a basis of $K_d^S[x_1, \dots, x_n]$. Therefore, the symmetric functions ε_λ of arbitrary degree are a basis for $K^S[x_1, \dots, x_n]$.

Moreover, because the m_μ can be written as *integer* linear combinations of the ε_λ, it follows that for any ring A, the ring of symmetric polynomials in $A[x_1, \dots, x_n]$ equals $A[\varepsilon_1, \dots, \varepsilon_n]$. ■

Algorithm for expansion of symmetric polynomials in the elementary symmetric polynomials. The matrix $T_{\lambda\mu}$ was convenient for showing that the ε_λ form a linear basis of the vector space of symmetric polynomials. It is neither convenient nor necessary, however, to compute the matrix and to invert it in order to expand symmetric polynomials as linear combinations of the ε_λ. This can be done by the following algorithm instead:

Let $p = \sum a_\beta x^\beta$ be a homogeneous symmetric polynomial of degree d in variables $x_1, \dots, x_n$. Let x^α be the greatest monomial (in lexicographic order) appearing in p; since p is symmetric, α is necessarily a partition. Then

$$p = a_\alpha m_\alpha + \text{ lexicographically lower terms,}$$

and

$$\varepsilon_{\alpha^*} = m_\alpha + \text{ lexicographically lower terms.}$$

Therefore, $p_1 = p - a_\alpha e_{\alpha^*}$ is a homogeneous symmetric polynomial of the same degree that contains only monomials lexicographically lower than x^α, or $p_1 = 0$. Now, iterate this procedure. The algorithm must terminate after finitely many steps because there are only finitely many monic monomials of degree d.

We can formalize the structure of this proof by introducing the notion of induction on a totally ordered set. Let $\mathcal{P}(d, n)$ denote the totally ordered set of partitions of size d with no more than n parts. Suppose we have a predicate $P(\lambda)$ depending on a partition. In order to prove that $P(\lambda)$ holds for all $\lambda \in \mathcal{P}(d, n)$, it suffices to show that

1. $P(\kappa)$, where κ is the least partition in $\mathcal{P}(d, n)$, and
2. For all $\lambda \in \mathcal{P}(d, n)$, if $P(\mu)$ holds for all $\mu \in \mathcal{P}(d, n)$ such that $\mu < \lambda$, then $P(\lambda)$ holds.

To apply this idea to our situation, take $P(\lambda)$ to be the predicate: If p is a symmetric homogeneous polynomial of degree d in n variables, in which the lexicographically highest monomial is x^λ, then p is a linear combination of ε_μ^* with $\mu \in \mathcal{P}(d, n)$ and $\mu \leq \lambda$.

Let κ be the least partition in $\mathcal{P}(d, n)$. Then

$$m_\kappa(x_1, \dots, x_n) = \varepsilon_\kappa(x_1, \dots, x_n).$$

Moreover, if p is a symmetric homogeneous polynomial of degree d in n variables whose *highest* monomial is x^κ, then p is a multiple of $m_\kappa(x_1, \dots, x_n)$. This shows that $P(\kappa)$ holds.

Now, fix a partition $\lambda \in \mathcal{P}(d, n)$, and suppose that $P(\mu)$ holds for all partitions $\mu \in \mathcal{P}(d, n)$ such that $\mu < \lambda$. (This is the *inductive hypothesis.*) Let p be a homogeneous symmetric polynomial of degree d whose leading term is $a_\lambda x^\lambda$. As observed previously, $p_1 = p - a_\lambda \varepsilon_{\lambda^*}$ is either zero or a homogeneous symmetric polynomial of degree d with leading monomial x^β for some $\beta < \lambda$. According to the inductive hypothesis, p_1 is a linear combination of ε_μ^* with $\mu \leq \beta < \lambda$. Therefore, $p = p_1 + a_\lambda \varepsilon_{\lambda^*}$ is a linear combination of ε_μ^* with $\mu \leq \lambda$.

Example 8.6.9. We illustrate the algorithm by an example. Consider polynomials in three variables. Take $p = x^3 + y^3 + z^3$. Then

$$\begin{aligned} p_1 &= p - \varepsilon_{1^3} \\ &= -3x^2y - 3xy^2 - 3x^2z - 6xyz - 3y^2z - 3xz^2 - 3yz^2, \\ p_2 &= p_1 + 3\varepsilon_{(2,1)} \\ &= 3xyz \\ &= 3\varepsilon_3. \end{aligned}$$

Thus, $p = \varepsilon_{1^3} - 3\varepsilon_{(2,1)} + 3\varepsilon_3$.

Of course, such computations can be automated. A program in *Mathematica* for expanding symmetric polynomials in elementary symmetric functions is available on my World Wide Web site.[1]

Exercises 8.6

8.6.1.

(a) Show that the set $K_d[x_1, \dots, x_n]$ of polynomials that are homogeneous of total degree d or identically zero is a finite–dimensional vector subspace of the K-vector space $K[x_1, \dots, x_n]$.
(b) Find the dimension of $K_d[x_1, \dots, x_n]$.
(c) Show that $K[x_1, \dots, x_n]$ is the direct sum of $K_d[x_1, \dots, x_n]$, where d ranges over the nonnegative integers.

8.6.2. Prove Lemma 8.6.1.

8.6.3.

(a) For each d, $K_d[x_1, \dots, x_n]$ is invariant under the action of S_n.
(b) $K_d^S[x_1, \dots, x_n] = K_d[x_1, \dots, x_n] \cap K^S[x_1, \dots, x_n]$ is a vector subspace of $K^S[x_1, \dots, x_n]$.
(c) $K^S[x_1, \dots, x_n]$ is the direct sum of the subspaces $K_d^S[x_1, \dots, x_n]$ for $d \geq 0$.

8.6.4. Prove Lemma 8.6.3.

8.6.5. Show that every monic monomial $\varepsilon_n^{m_n} \varepsilon_{n-1}^{m_{n-1}} \dots \varepsilon_1^{m_1}$ in the ε_i is an ε_λ, and relate λ to the multiplicities m_i.

8.6.6. Show that if λ is a partition, then the conjugate partition λ^* satisfies $\lambda_j^* = |\{i : \lambda_i \geq j\}|$.

8.6.7. Show that ε_λ is homogeneous of total degree $|\lambda|$.

8.6.8. Show that the monomial symmetric functions m_λ with $\lambda = (\lambda_1, \dots, \lambda_n)$ and $|\lambda| = d$ form a linear basis of $K_d^S[x_1, \dots, x_n]$.

8.6.9. Show that a symmetric function ε_λ of degree d is an integer linear combination of monomials x^α of degree d and, therefore, an integer linear combination of monomial symmetric functions m_μ with $|\mu| = d$. (Substitute $\mathbb{Z}_q$-linear combinations in case the characteristic is q.)

8.6.10. Show that an upper triangular matrix T with 1's on the diagonal and integer entries has an inverse of the same type.

---

[1] www.math.uiowa.edu/~goodman

8.6.11. Write out the monomial symmetric functions $m_{3,3,1}(x_1, x_2, x_3)$ and $m_{3,2,1}(x_1, x_2, x_3)$, and note that they have different numbers of summands.

8.6.12. Consult the Mathematica notebook **Symmetric- Functions.nb**, which is available on my World Wide Web site. Use the Mathematica function **monomialSymmetric[ ]** to compute the monomial symmetric functions m_λ in n variables for

(a) $\lambda = [2, 2, 1, 1]$, $n = 5$
(b) $\lambda = [3, 3, 2]$, $n = 5$
(c) $\lambda = [3, 1]$, $n = 5$

8.6.13. Use the algorithm described in this section to expand the following symmetric polynomials as polynomials in the elementary symmetric functions.

(a) $x_1{}^2\, x_2{}^2\, x_3 + x_1{}^2\, x_2\, x_3{}^2 + x_1\, x_2{}^2\, x_3{}^2$
(b) $x_1^3 + x_2^3 + x_3^3$

8.6.14. Consult the Mathematica notebook **Symmetric- Functions.nb**, which is available on my World Wide Web site. Use the Mathematica function **elementaryExpand[ ]** to compute the expansion of the following symmetric functions as polynomials in the elementary symmetric functions.

(a) $[(x_1 - x_2)(x_1 - x_3)(x_1 - x_4)(x_2 - x_3)(x_2 - x_4)(x_3 - x_4)]^2$
(b) m_λ, for $\lambda = [4, 3, 3, 1]$, in four variables

8.6.15. Suppose that f is an antisymmetric polynomial in n variables; that is, for each $\sigma \in S_n$, $\sigma(f) = \varepsilon(\sigma)f$, where ε denotes the parity homomorphism. Show that f has a factorization $f = \delta(x_1, \dots, x_n)g$, where $\delta(x_1, \dots, x_n) = \prod_{i<j}(x_i - x_j)$, and g is symmetric.

8.7. The General Equation of Degree n

Consider the quadratic formula or Cardano's formulas for solutions of a cubic equation, which calculate the roots of a polynomials in terms of the coefficients; in these formulas, the coefficients may be regarded as variables and the roots as functions of these variables. This observation suggests the notion of the *general polynomial of degree* n, which is defined as follows:

Let $t_1, \dots, t_n$ be variables. The *general (monic) polynomial of degree* n *over* K is

$$P_n(x) = x^n - t_1 x^{n-1} + \cdots + (-1)^{n-1} t_n \in K(t_1, \dots, t_n)(x).$$

Let $u_1, \dots, u_n$ denote the roots of this polynomial in a splitting field E. Then $(x - u_1) \cdots (x - u_n) = P_n(x)$, and $t_j = \varepsilon_j(u_1, \dots, u_n)$ for

$1 \le j \le n$, by Corollary 8.6.4. We shall now show that the Galois group of the general polynomial of degree n is the symmetric group S_n.

Theorem 8.7.1. *Let E be a splitting field of the general polynomial $P_n(x) \in K(t_1, \dots, t_n)[x]$. The Galois group $\mathrm{Aut}_{K(t_1,\dots,t_n)}(E)$ is the symmetric group S_n.*

Proof. Introduce a new set of variables $v_1, \dots, v_n$ and let

$$f_j = \varepsilon_j(v_1, \dots, v_n) \quad \text{for } 1 \le j \le n,$$

where the ε_j are the elementary symmetric functions. Consider the polynomial

$$\tilde{P}_n(x) = (x - v_1) \cdots (x - v_n) = x^n + \sum_j (-1)^j f_j x^{n-j}.$$

The coefficients lie in $K(f_1, \dots, f_n)$, which is equal to $K^S(v_1, \dots, v_n)$ by Theorem 8.6.6. According to Proposition 8.6.2, $K(v_1, \dots, v_n)$ is Galois over $K(f_1, \dots, f_n)$ with Galois group S_n. Furthermore, $K(v_1, \dots, v_n)$ is the splitting field over $K(f_1, \dots, f_n)$ of $\tilde{P}_n(x)$.

Let $u_1, \dots, u_n$ be the roots of $P_n(x)$ in E. Then $t_j = \varepsilon_j(u_1, \dots, u_n)$ for $1 \le j \le n$, so $E = K(t_1, \dots, t_n)(u_1, \dots, u_n) = K(u_1, \dots, u_n)$.

Since $\{t_i\}$ are variables, and the $\{f_i\}$ are algebraically independent over K, according to Theorem 8.6.6, there is a ring isomorphism $K[t_1, \dots, t_n] \to K[f_1, \dots, f_n]$ fixing K and taking t_i to f_i. This ring isomorphism extends to an isomorphism of fields of fractions

$$K(t_1, \dots, t_n) \cong K(f_1, \dots, f_n),$$

and to the polynomial rings

$$K(t_1, \dots, t_n)[x] \cong K(f_1, \dots, f_n)[x];$$

the isomorphism of polynomial rings carries $P_n(x)$ to $\tilde{P}_n(x)$. Therefore, by Proposition 8.2.4, there is an isomorphism of splitting fields $E \cong K(v_1, \dots, v_n)$ extending the isomorphism

$$K(t_1, \dots, t_n) \cong K(f_1, \dots, f_n).$$

It follows that the Galois groups are isomorphic:

$$\mathrm{Aut}_{K(t_1,\dots,t_n)}(E) \cong \mathrm{Aut}_{K(f_1,\dots,f_n)}(K(v_1, \dots, v_n)) \cong S_n.$$

■

We shall see in Section 9.6 that this result implies that *there can be no analogue of the quadratic and cubic formulas for equations of degree 5 or more.* (We shall work out formulas for quartic equations in Section 8.8.)

The discriminant. We now consider some symmetric polynomials that arise in the study of polynomials and Galois groups.

Write

$$\delta = \delta(x_1, \dots, x_n) = \prod_{1 \le i < j \le n} (x_i - x_j).$$

We know that δ distinguishes even and odd permutations. Every permutation σ satisfies $\sigma(\delta) = \pm\delta$ and σ is even if, and only if, $\sigma(\delta) = \delta$. The symmetric polynomial δ^2 is called the *discriminant polynomial.*

Now let $f = \sum_i a_i x^i \in K[x]$ be polynomial of degree n. Let $\alpha_1, \dots, \alpha_n$ be the roots of $f(x)$ in a splitting field E. The element

$$\delta^2(f) = a_n^{2n-2}\delta^2(\alpha_1, \dots, \alpha_n)$$

is called the *discriminant of* f.

Now suppose in addition that f is irreducible and separable. Since $\delta^2(f)$ is invariant under the Galois group of f, it follows that $\delta^2(f)$ is an element of the ground field K. Since the roots of f are distinct, the element $\delta(f) = a_n^{n-1}\delta(\alpha_1, \dots, \alpha_n)$ is nonzero, and an element $\sigma \in \mathrm{Aut}_K(E)$ induces an even permutation of the roots of f if, and only if, $\sigma(\delta(f)) = \delta(f)$. Thus we have the following result:

Proposition 8.7.2. *Let $f \in K[x]$ be an irreducible separable polynomial. Let E be a splitting field for f, and let $\alpha_1, \dots, \alpha_n$ be the roots of $f(x)$ in E. The Galois group $\mathrm{Aut}_K(E)$, regarded as a group of permutations of the set of roots, is contained in the alternating group A_n if, and only if, the discriminant $\delta^2(f)$ of f has a square root in K.*

Proof. The discriminant has a square root in K if, and only if, $\delta(f) \in K$. But $\delta(f) \in K$ if, and only if, it is fixed by the Galois group, if, and only if, the Galois group consists of even permutations. ■

We do not need to know the roots of f in order to compute the discriminant. Because the discriminant of $f = \sum_i a_i x^i$ is the discriminant of the monic polynomial $(1/a_n)f$ multiplied by a_n^{2n-1}, it suffices to give a method for computing the discriminant of a monic polynomial. Suppose, then, that f is monic.

The discriminant is a symmetric polynomial in the roots and, therefore, a polynomial in the coefficients of f, by Theorem 8.6.6 and Corollary 8.6.4. For fixed n, we can expand the discriminant polynomial $\delta^2(x_1, \dots, x_n)$ as a polynomial in the elementary symmetric functions, say $\delta^2(x_1, \dots, x_n) = d_n(\varepsilon_1, \dots, \varepsilon_n)$. Then

$$\delta^2(f) = d_n(-a_{n-1}, a_{n-2}, \dots, (-1)^n a_0).$$

This is not the most efficient method of calculating the discriminant of a polynomial, but it works well for polynomials of low degree. (A program for computing the discriminant by this method is available on my Web site.)

Example 8.7.3. (Galois group of a cubic.) The Galois group of a monic cubic irreducible polynomial f is either $A_3 = \mathbb{Z}_3$ or S_3, according to whether $\delta^2(f) \in K$. (These are the only transitive subgroups of S_3.) We can compute that

$$\delta^2(x_1, x_2, x_3) = \varepsilon_1{}^2\,\varepsilon_2{}^2 - 4\,\varepsilon_2{}^3 - 4\,\varepsilon_1{}^3\,\varepsilon_3 + 18\,\varepsilon_1\,\varepsilon_2\,\varepsilon_3 - 27\,\varepsilon_3{}^2.$$

Therefore, a cubic polynomial $f(x) = x^3 + ax^2 + bx + c$ has

$$\delta^2(f) = a^2\,b^2 - 4\,b^3 - 4\,a^3\,c + 18\,a\,b\,c - 27\,c^2.$$

In particular, a polynomial of the special form $f(x) = x^3 + px + q$ has

$$\delta^2(f) = -4\,p^3 - 27\,q^2,$$

as computed in Chapter 8.

Consider, for example, $f(x) = x^3 - 4x^2 + 2x + 13 \in \mathbb{Z}[x]$. To test a cubic for irreducibility, it suffices to show it has no root in $\mathbb{Q}$, which follows from the rational root test. The discriminant of f is $\delta^2(f) = -3075$. Since this is not a square in $\mathbb{Q}$, it follows that the Galois group of f is S_3.

Example 8.7.4. The Galois group of an irreducible quartic polynomial f must be one of the following, as these are the only transitive subgroups of S_4, according to Exercise 5.1.20:

- $A_4, \mathcal{V} \subseteq A_4$, or
- $S_4, D_4, \mathbb{Z}_4 \not\subseteq A_4$.

We can compute that the discriminant polynomial has the expansion

$$\begin{aligned}\delta^2(x_1, \dots, x_4) = &\varepsilon_1{}^2\,\varepsilon_2{}^2\,\varepsilon_3{}^2 - 4\,\varepsilon_2{}^3\,\varepsilon_3{}^2 - 4\,\varepsilon_1{}^3\,\varepsilon_3{}^3 + 18\,\varepsilon_1\,\varepsilon_2\,\varepsilon_3{}^3 - \\ &27\,\varepsilon_3{}^4 - 4\,\varepsilon_1{}^2\,\varepsilon_2{}^3\,\varepsilon_4 + 16\,\varepsilon_2{}^4\,\varepsilon_4 + 18\,\varepsilon_1{}^3\,\varepsilon_2\,\varepsilon_3\,\varepsilon_4 - \\ &80\,\varepsilon_1\,\varepsilon_2{}^2\,\varepsilon_3\,\varepsilon_4 - 6\,\varepsilon_1{}^2\,\varepsilon_3{}^2\,\varepsilon_4 + 144\,\varepsilon_2\,\varepsilon_3{}^2\,\varepsilon_4 - 27\,\varepsilon_1{}^4\,\varepsilon_4{}^2 + \\ &144\,\varepsilon_1{}^2\,\varepsilon_2\,\varepsilon_4{}^2 - 128\,\varepsilon_2{}^2\,\varepsilon_4{}^2 - 192\,\varepsilon_1\,\varepsilon_3\,\varepsilon_4{}^2 + 256\,\varepsilon_4{}^3.\end{aligned}$$

Therefore, for $f(x) = x^4 + ax^3 + bx^2 + cx + d$,

$$\begin{aligned}\delta^2(f) = &a^2\,b^2\,c^2 - 4\,b^3\,c^2 - 4\,a^3\,c^3 + 18\,a\,b\,c^3 - 27\,c^4 - 4\,a^2\,b^3\,d + \\ &16\,b^4\,d + 18\,a^3\,b\,c\,d - 80\,a\,b^2\,c\,d - 6\,a^2\,c^2\,d + 144\,b\,c^2\,d - \\ &27\,a^4\,d^2 + 144\,a^2\,b\,d^2 - 128\,b^2\,d^2 - 192\,a\,c\,d^2 + 256\,d^3.\end{aligned}$$

For example, take $f(x) = x^4 + 3x^3 + 4x^2 + 7x - 5$. The reduction of f mod 3 is $x^4 + x^2 + x + 1$, which can be shown to be irreducible over $\mathbb{Z}_3$ by an ad hoc argument. Therefore, $f(x)$ is also irreducible over $\mathbb{Q}$. We compute that $\delta^2(f) = -212836$, which is not a square in $\mathbb{Q}$, so the Galois group must be S_4, D_4, or $\mathbb{Z}_4$.

Resultants. In the remainder of this section, we will discuss the notion of the *resultant* of two polynomials. *This material can be omitted without loss of continuity.*

The resultant of polynomials $f = \sum_{i=0}^{n} a_i x^i$, $g = \sum_{i=0}^{m} b_i x^i \in K[x]$, of degrees n and m, is defined to be the product

$$R(f, g) = a_n^m b_m^n \prod_i \prod_j (\xi_i - \eta_j), \tag{8.7.1}$$

where the ξ_i and η_j are the roots of f and g, respectively, in a common splitting field.

The product

$$R_0(f, g) = \prod_i \prod_j (\xi_i - \eta_j), \tag{8.7.2}$$

is evidently symmetric in the roots of f and in the roots of g and, therefore, is a polynomial in the quantities (a_i/a_n) and (b_j/b_n). We can show that the total degree as a polynomial in the (a_i/a_n) is m and the total degree as a polynomial in the (b_j/b_m) is m. Therefore, the resultant (8.7.1) is a polynomial in the a_i and b_j, homogeneous of degree m in the a_i and of degree n in the b_j.

Furthermore, $R(f, g) = 0$ precisely when f and g have a common root, that is, if, and only if, f and g have a a nonconstant common factor in $K[x]$. We shall find an expression for $R(f, g)$ by exploiting this observation.

If the polynomials f and g have a common divisor $q(x)$ in $K[x]$, then there exist polynomials $\varphi(x)$ and $\psi(x)$ of degrees no more than $n - 1$ and $m - 1$, respectively, such that

$$\begin{aligned} f(x) &= q(x)\varphi(x), \text{ and} \\ g(x) &= q(x)\psi(x), \text{ so} \\ f(x)\psi(x) &= g(x)\varphi(x) = q(x)\varphi(x)\psi(x). \end{aligned}$$

Conversely, the existence of polynomials $\varphi(x)$ of degree no more than $n-1$ and $\psi(x)$ of degree no more than $m-1$ such that $f(x)\psi(x) = g(x)\varphi(x)$ implies that f and g have a nonconstant common divisor (Exercise 8.7.9).

Now, write $\psi(x) = \sum_{i=0}^{m-1} \alpha_i x^i$ and $\varphi(x) = \sum_{i=0}^{n-1} \beta_i x^i$, and m coefficients in the equation $f(x)\psi(x) = g(x)\varphi(x)$ to get a syst linear equations in the variables

$$(\alpha_0, \dots, \alpha_{m-1}, -\beta_0, \dots, -\beta_{n-1}).$$

Assuming without loss of generality that $n \geq m$, the matrix $\mathcal{R}(f, g)$ of the system of linear equations has the form displayed in Figure 8.7.1.

$$\begin{bmatrix}
a_0 & & & & & b_0 & & & & & & \\
a_1 & a_0 & & & & b_1 & b_0 & & & & & \\
a_2 & a_1 & a_0 & & & b_2 & b_1 & b_0 & & & & \\
\vdots & \vdots & \vdots & \ddots & & \vdots & \vdots & \vdots & \ddots & & & \\
\vdots & \vdots & \vdots & \cdots & a_0 & \vdots & \vdots & \vdots & \cdots & & & \\
\vdots & \vdots & \vdots & \cdots & a_1 & b_m & b_{m-1} & b_{m-2} & \cdots & & & \\
\vdots & \vdots & \vdots & \cdots & a_2 & & b_m & b_{m-1} & \cdots & \cdots & \ddots & \\
\vdots & \vdots & \vdots & \cdots & \vdots & & & b_m & \cdots & \cdots & \cdots & b_0 \\
a_n & a_{n-1} & & \cdots & a_{n-m+1} & & & & \ddots & \cdots & \cdots & b_1 \\
 & a_n & \vdots & \cdots & \vdots & & & & & & \cdots & \vdots \\
 & & a_n & \cdots & \vdots & & & & & & \cdots & \vdots \\
 & & & \ddots & \vdots & & & & & & \ddots & \vdots \\
 & & & & a_n & & & & & & & b_m
\end{bmatrix}$$

Figure 8.7.1. The matrix $\mathcal{R}(f, g)$.

The matrix has m columns with a_i's and n columns with b_i's. It follows from the preceding discussion that f and g have a nonconstant common divisor if, and only if, the matrix $\mathcal{R}(f, g)$ has determinant zero.

We want to show that $\det(\mathcal{R}(f, g)) = (-1)^{n+m} R(f, g)$. Since $\det(\mathcal{R}(f, g)) = a_n^m b_m^n \det(\mathcal{R}(a_n^{-1} f, b_m^{-1} g))$, and similarly, $R(f, g) = a_n^m b_m^n R(a_n^{-1} f, b_m^{-1} g)$, it suffices to consider the case where both f and g are monic.

At this point we need to shift our point of view slightly. Regard the roots ξ_i and η_j as variables, the coefficients a_i as symmetric polynomials in the ξ_i, and the coefficients and b_j as symmetric polynomials in the η_j. Then $\det(\mathcal{R}(f, g))$ is a polynomial in the ξ_i and η_j, symmetric with respect to each set of variables, which is zero when any ξ_i and η_j agree. It follows that $\det(\mathcal{R}(f, g))$ is divisible by $\prod_i \prod_j (\xi_i - \eta_j) = R(f, g)$ (Exercise 8.7.12). But as both $\det(\mathcal{R})$

and $R(f, g)$ are polynomials in the a_i and b_j of total degree $n + m$, they are equal up to a scalar factor, and it remains to show that the scalar factor is $(-1)^{n+m}$. In fact, $\det(\mathcal{R})$ has a summand a_0^m. On the other hand $R(f, g) = (-1)^{n+m} \prod_j f(\beta_j)$, according to Exercise 8.7.10, and so has a summand $(-1)^{n+m} a_0^m$.

Proposition 8.7.5. $R(f, g) = (-1)^{n+m} \det(\mathcal{R}(f, g))$.

We now observe that the discriminant of f can be computed using the resultant of f and its formal derivative f'. In fact, from Exercise 8.7.10,

$$R(f, f') = a_n^{n-1} \prod_i f'(\xi_i).$$

Using

$$f(x) = a_n \prod_i (x - \xi_i),$$

we can verify that

$$f'(\xi_i) = a_n \prod_{j \neq i} (\xi_i - \xi_j).$$

Therefore,

$$\begin{aligned} R(f, f') &= a_n^{n-1} \prod_i f'(\xi_i) = a_n^{2n-1} \prod_i \prod_{j \neq i} (\xi_i - \xi_j) \\ &= a_n^{2n-1} (-1)^{n(n-1)/2} \prod_{i<j} (\xi_i - \xi_j)^2 \\ &= a_n (-1)^{n(n-1)/2} \delta^2(f). \end{aligned}$$

Proposition 8.7.6. $\delta^2(f) = a_n^{-1} (-1)^{n(n-1)/2} R(f, f')$.

Although determinants tend to be inefficient for computations, the matrix $\mathcal{R}$ is sparse, and it appears to be more efficient to calculate the discriminant using the determinant of $\mathcal{R}(f, f')$ than to use the method described earlier in this section.

Exercises 8.7

8.7.1. Determine the Galois groups of the following cubic polynomials:

(a) $x^3 + 2x + 1$, over $\mathbb{Q}$

(b) $x^3 + 2x + 1$, over $\mathbb{Z}_3$

(c) $x^3 - 7x^2 - 7$, over $\mathbb{Q}$

8.7.2. Verify that the Galois group of $f(x) = x^3 + 7x + 7$ over $\mathbb{Q}$ is S_3. Determine, as explicitly as possible, all intermediate fields between the rationals and the splitting field of $f(x)$.

8.7.3. Show that the discriminant polynomial $\delta^2(x_1, \dots, x_n)$ is a polynomial of degree $2n - 2$ in the elementary symmetric polynomials. Hint: Show that when $\delta^2(x_1, \dots, x_n)$ is expanded as a polynomial in the elementary symmetric functions,

$$\delta^2(x_1, \dots, x_n) = d_n(\varepsilon_1, \dots, \varepsilon_n),$$

the monomial of highest total degree in d_n comes from the lexicographically highest monomial in $\delta^2(x_1, \dots, x_n)$. Identify the lexicographically highest monomial in $\delta^2(x_1, \dots, x_n)$, say x^α, and find the degree of the corresponding monomial ε_{α^*}.

8.7.4. Let $f(x) = \sum_i a_i x^i$ have degree n and roots $\alpha_1, \dots, \alpha_n$. Using the previous exercise, show that $\delta^2(f) = a_n^{2n-2}\delta^2(\alpha_1, \dots, \alpha_n)$ is a homogeneous polynomial of degree $2n - 2$ in the coefficients of f.

8.7.5. Determine $\delta^2(f)$ as a polynomial in the coefficients of f when the degree of f is $n = 2$ or $n = 3$.

8.7.6. Check that $x^4 + x^2 + x + 1$ is irreducible over $\mathbb{Z}_3$.

8.7.7. Use a computer algebra package (e.g., Maple or Mathematica) to find the discriminants of the following polynomials. You may refer to the Mathematica notebook **Discriminants-and-Resultants.nb**, available on my World Wide Web site. The Mathematica function **Discriminant[ ]**, available in that notebook, computes the discriminant.

(a) $x^4 - 3x^2 + 2x + 5$
(b) $x^4 + 10x^3 - 3x + 4$
(c) $x^3 - 14x + 10$
(d) $x^3 - 12$

8.7.8. Each of the polynomials in the previous exercise is irreducible over the rationals. The Eisenstein criterion applies to one of the polynomials, and the others can be checked by computing factorizations of the reductions modulo small primes. For each polynomial, determine which Galois groups are consistent with the computation of the discriminant.

8.7.9. Suppose that f and g are polynomials in $K[x]$ of degrees n and m respectively, and that there exist $\varphi(x)$ of degree no more than $n-1$ and $\psi(x)$ of degree no more than $m - 1$ such that $f(x)\psi(x) = g(x)\varphi(x)$. Show that f and g have a nonconstant common divisor in $K[x]$.

8.7.10. Let f and g be polynomials of degree n and m, respectively, with roots ξ_i and η_j.

(a) Show that $R(f, g) = (-1)^{n+m}R(g, f)$.

(b) Show that

$$R(f, g) = a_n^m \prod_i g(\xi_i).$$

(c) Show that

$$R(f, g) = (-1)^{n+m} b_m^n \prod_j f(\eta_j).$$

8.7.11. Show that

$$\prod_{i=1}^{n} \prod_{j=1}^{m} (x_i - y_j)$$

is a polynomial of total degree m in the elementary symmetric functions $\epsilon_i(x_1, \dots, x_n)$ and of total degree n in the elementary symmetric functions $\epsilon_j(y_1, \dots, y_m)$.

8.7.12. Verify the assertion in the text that $\det(\mathcal{R}(f, g))$ is divisible by $\prod_i \prod_j (\xi_i - \eta_j) = R(f, g)$.

8.7.13. Use a computer algebra package (e.g., Maple or Mathematica to find the resultants:

(a) $R(x^3 + 2x + 5, x^2 - 4x + 5)$

(b) $R(3x^4 + 7x^3 + 2x - 5, 12x^3 + 21x^2 + 2)$

8.8. Quartic Polynomials

In this section, we determine the Galois groups of quartic polynomials. Consider a quartic polynomial

$$f(x) = x^4 + ax^3 + bx^2 + cx + d \tag{8.8.1}$$

with coefficients in some field K. It is convenient first to eliminate the cubic term by the linear change of variables $x = y - a/4$. This yields

$$f(x) = g(y) = y^4 + px^2 + qx + r, \tag{8.8.2}$$

with

$$\begin{aligned} p &= -\frac{3a^2}{8} + b, \\ q &= \frac{a^3}{8} - \frac{ab}{2} + c, \\ r &= -\frac{3a^4}{256} + \frac{a^2 b}{16} - \frac{ac}{4} + d. \end{aligned} \tag{8.8.3}$$

We can suppose, without loss of generality, that g is irreducible, since otherwise we could analyze g by analyzing its factors. We also suppose that g is separable. Let G denote the Galois group of g over K.

Using one of the techniques for computing the discriminant from the previous section, we compute that the discriminant of g (or f) is

$$-4\,p^3\,q^2 - 27\,q^4 + 16\,p^4\,r + 144\,p\,q^2\,r - 128\,p^2\,r^2 + 256\,r^3, \quad (8.8.4)$$

or, in terms of a, b, c, and d,

$$\begin{gathered} a^2\,b^2\,c^2 - 4\,b^3\,c^2 - 4\,a^3\,c^3 + 18\,a\,b\,c^3 - 27\,c^4 - 4\,a^2\,b^3\,d + \\ 16\,b^4\,d + 18\,a^3\,b\,c\,d - 80\,a\,b^2\,c\,d - 6\,a^2\,c^2\,d + 144\,b\,c^2\,d - \\ 27\,a^4\,d^2 + 144\,a^2\,b\,d^2 - 128\,b^2\,d^2 - 192\,a\,c\,d^2 + 256\,d^3. \end{gathered} \quad (8.8.5)$$

We can distinguish two cases, according to whether $\delta(g) \in K$:

Case 1. $\delta(g) \in K$. Then the Galois group is isomorphic to A_4 or $\mathcal{V} \cong \mathbb{Z}_2 \times \mathbb{Z}_2$, since these are the transitive subgroups of A_4, according to Exercise 5.1.20.

Case 2. $\delta(g) \notin K$. Then the Galois group is isomorphic to S_4, D_4, or $\mathbb{Z}_4$, since these are the transitive subgroups of S_4 that are not contained in A_4, according to Exercise 5.1.20.

Denote the roots of g by $\alpha_1, \alpha_2, \alpha_3, \alpha_4$, and let E denote the splitting field $K(\alpha_1, \dots, \alpha_4)$. The next idea in analyzing the quartic equation is to introduce the elements

$$\begin{aligned} \theta_1 &= (\alpha_1 + \alpha_2)(\alpha_3 + \alpha_4) \\ \theta_2 &= (\alpha_1 + \alpha_3)(\alpha_2 + \alpha_4) \\ \theta_3 &= (\alpha_1 + \alpha_4)(\alpha_2 + \alpha_3), \end{aligned} \quad (8.8.6)$$

and the cubic polynomial, called the *resolvent cubic*:

$$h(y) = (y - \theta_1)(y - \theta_2)(y - \theta_3). \quad (8.8.7)$$

Expanding $h(y)$ and identifying symmetric functions in the α_i with coefficients of g gives

$$h(y) = y^3 - 2py^2 + (p^2 - 4r)y + q^2. \quad (8.8.8)$$

The discriminant $\delta^2(h)$ turns out to be identical with the discriminant $\delta^2(g)$ (Exercise 8.8.4). We distinguish cases according to whether the resolvent cubic is irreducible.

Case 1A. $\delta(g) \in K$ and h is irreducible over K. In this case,

$$[K(\theta_1, \theta_2, \theta_3) : K] = 6,$$

and 6 divides the order of G. The only possibility is $G = A_4$.

Case 2A. $\delta(g) \notin K$ and h is irreducible over K. Again, 6 divides the order of G. The only possibility is $G = S_4$.

Case 1B. $\delta(g) \in K$ and h is not irreducible over K. Then G must be $\mathcal{V}$. (In particular, 3 does not divide the order of G, so the Galois group of h must be trivial, and h factors into linear factors over K. Conversely, if h splits over K, then the θ_i are all fixed by G, which implies that $G \subseteq \mathcal{V}$. The only possibility is then that $G = \mathcal{V}$.)

Case 2B. $\delta(g) \notin K$ and h is not irreducible over K. The Galois group must be one of D_4 and $\mathbb{Z}_4$. By the remark under Case 1B, h does not split over K, so it must factor into a linear factor and an irreducible quadratic (i.e., exactly one of the θ_i lies in K).

We can assume without loss of generality that $\theta_1 \in K$. Then G is contained in the stabilizer of θ_1, which is the copy of D_4 generated by $\{(12)(34), (13)(24)\}$,

$$D_4 = \{e, (1324), (12)(34), (1423), (34), (12), (13)(24), (14)(23)\}.$$

G is either equal to D_4 or to

$$\mathbb{Z}_4 = \{e, (1324), (12)(34), (1423)\}.$$

It remains to distinguish between these cases.

Since $K(\delta) \subseteq K(\theta_1, \theta_2, \theta_3)$ and both are quadratic over K, it follows that $K(\delta) = K(\theta_1, \theta_2, \theta_3)$. $K(\delta)$ is the fixed field of $G \cap A_4$, which is either

$$D_4 \cap A_4 = \{e, (12)(34), (13)(24), (14)(23)\} \cong \mathcal{V}, \quad \text{or}$$

$$\mathbb{Z}_4 \cap A_4 = \{e, (12)(34)\} \cong \mathbb{Z}_2.$$

It follows that the degree of the splitting field E over the intermediate field $K(\delta) = K(\theta_1, \theta_2, \theta_3)$ is either 4 or 2, according to whether G is D_4 or $\mathbb{Z}_4$. So one possibility for finishing the analysis of the Galois group in Case 2B is to compute the dimension of E over $K(\delta)$. The following lemma can be used for this.[2]

Lemma 8.8.1. *Let $g(x) = y^4 + py^2 + qy + r$ be an irreducible quartic polynomial over a field K. Let $h(x)$ denote the resolvent cubic of g. Suppose that $\delta^2 = \delta^2(g)$ is not a square in K and that h is not irreducible over K. Let θ denote the one root of the resolvent cubic which lies in K, and define*

$$H(x) = (x^2 + \theta)(x^2 + (\theta - p)x + r). \tag{8.8.9}$$

Then the Galois group of g is $\mathbb{Z}_4$ if, and only if, $H(x)$ splits over $K(\delta)$.

[2] This result is taken from L. Kappe and B. Warren, "An Elementary Test for the Galois Group of a Quartic Polynomial," *The American Mathematical Monthly* **96** (1989) 133-137.

Proof. We maintain the notation from the preceding discussion: The roots of g are denoted by α_i, the splitting field of g by E, and the roots of h by θ_i. Let $L = K(\delta) = K(\theta_1, \theta_1, \theta_3)$. Assume without loss of generality that the one root of h in K is θ_1. Now consider the polynomial

$$(x-\alpha_1\alpha_2)(x - \alpha_3\alpha_4)(x - (\alpha_1 + \alpha_2))(x - (\alpha_3 + \alpha_4) \\ = (x^2 - (\alpha_1\alpha_2 + \alpha_3\alpha_4)x + r)(x^2 + \theta_1). \tag{8.8.10}$$

Compute that $p - \theta = \alpha_1\alpha_2 + \alpha_3\alpha_4$. It follows that the preceding polynomial is none other than $H(x)$.

Suppose that $H(x)$ splits over L, so $\alpha_1\alpha_2, \alpha_3\alpha_4, \alpha_1+\alpha_2, \alpha_3+\alpha_4 \in L$. It follows that α_1 satisfies a quadratic polynomial over L,

$$(x - \alpha_1)(x - \alpha_2) = x^2 - (\alpha_1 + \alpha_2)x + \alpha_1\alpha_2 \in L[x],$$

and $[L(\alpha_1) : L] = 2$. But we can check that $L(\alpha_1) = E$. Consequently, $[E : L] = 2$ and $G = \mathbb{Z}_4$, by the discussion preceding the lemma.

Conversely, suppose that the Galois group G is $\mathbb{Z}_4$. Because $\theta_1 = (\alpha_1 + \alpha_2)(\alpha_3 + \alpha_4)$ is in K and is, therefore, fixed by the generator σ of G, we have $\sigma^{\pm 1} = (1324)$. The fixed field of $\sigma^2 = (12)(34)$ is the unique intermediate field between K and E, so L equals this fixed field. Each of the roots of $H(x)$ is fixed by σ^2 and, hence, an element of L. That is, H splits over L. ■

Examples of the use of this lemma will be given shortly. We shall now explain how to solve explicitly for the roots α_i of the quartic equation in terms of the roots θ_i of the resolvent cubic.

Note that

$$(\alpha_1 + \alpha_2)(\alpha_3 + \alpha_4) = \theta_1 \quad \text{and} \quad (\alpha_1 + \alpha_2) + (\alpha_3 + \alpha_4) = 0, \tag{8.8.11}$$

which means that $(\alpha_1 + \alpha_2)$ and $(\alpha_3 + \alpha_4)$ are the two square roots of $-\theta_1$. Similarly, $(\alpha_1+\alpha_3)$ and $(\alpha_2+\alpha_4)$ are the two square roots of $-\theta_2$, and $(\alpha_1+\alpha_4)$ and $(\alpha_2+\alpha_3)$ are the two square roots of $-\theta_3$. It is possible to choose the signs of the square roots consistently, noting that

$$\sqrt{-\theta_1}\sqrt{-\theta_2}\sqrt{-\theta_2} = \sqrt{-\theta_1\theta_2\theta_3} = \sqrt{q^2} = q. \tag{8.8.12}$$

That is, it is possible to choose the square roots so that their product is q. We can check that

$$(\alpha_1 + \alpha_2)(\alpha_1 + \alpha_3)(\alpha_1 + \alpha_4) = q, \tag{8.8.13}$$

and so put

$$(\alpha_1 + \alpha_2) = \sqrt{-\theta_1} \quad (\alpha_1 + \alpha_3) = \sqrt{-\theta_2} \quad (\alpha_1 + \alpha_4) = \sqrt{-\theta_3}. \tag{8.8.14}$$

Using this together with $\alpha_1 + \alpha_2 + \alpha_3 + \alpha_4 = 0$, we get

$$2\alpha_i = \pm\sqrt{-\theta_1} \pm \sqrt{-\theta_2} \pm \sqrt{-\theta_3}, \tag{8.8.15}$$

with the four choices of signs giving the four roots α_i (as long as the characteristic of the ground field is not 2).

Example 8.8.2. Take $K = \mathbb{Q}$ and $f(x) = x^4 + 3x^3 - 3x - 2$. Applying the linear change of variables $x = y - 3/4$ gives $f(x) = g(y) = -\frac{179}{256} + \frac{3}{8}y - \frac{27}{8}y^2 + y^4$. The reduction of f modulo 5 is irreducible over $\mathbb{Z}_5$, and, therefore, f is irreducible over $\mathbb{Q}$. The discriminant of f (or g or h) is -2183, which is not a square in $\mathbb{Q}$. Therefore, the Galois group of f is not contained in the alternating group A_4. The resolvent cubic of g is $h(y) = \frac{9}{64} + \frac{227}{16}y + \frac{27}{4}y^2 + y^3$. The reduction of $64h$ modulo 7 is irreducible over $\mathbb{Z}_7$ and, hence, h is irreducible over $\mathbb{Q}$. It follows that the Galois group is S_4.

Example 8.8.3. Take $K = \mathbb{Q}$ and $f(x) = 21 + 12\,x + 6\,x^2 + 4\,x^3 + x^4$. Applying the linear change of variables $x = y - 1$ gives $f(x) = g(y) = 12 + 8\,y + y^4$. The reduction of g modulo 5 factors as

$$\bar{g}(y) = y^4 + 3y + 2 = (1 + y)\left(2 + y + 4\,y^2 + y^3\right).$$

By the rational root test, g has no rational root; therefore, if it is not irreducible, it must factor into two irreducible quadratics. But this would be inconsistent with the factorization of the reduction modulo 5. Hence, g is irreducible.

The discriminant of g is $331776 = (576)^2$, and therefore, the Galois group of g is contained in the alternating group A_4. The resolvent cubic of g is $h(y) = 64 - 48\,y + y^3$. The reduction of h modulo 5 is irreducible over $\mathbb{Z}_5$, so h is irreducible over $\mathbb{Q}$. It follows that the Galois group of g is A_4.

Example 8.8.4. Take $K = \mathbb{Q}$ and $f(x) = x^4 + 16x^2 - 1$. The reduction of f modulo 3 is irreducible over $\mathbb{Z}_3$, so f is irreducible over $\mathbb{Q}$. The discriminant of f is -1081600, so the Galois group G of f is not contained in the alternating group. The resolvent cubic of f is

$$h(x) = 260\,x - 32\,x^2 + x^3 = x\left(260 - 32\,x + x^2\right).$$

Because h is reducible, it follows that G is either Z_4 or D_4. To determine which, we can use the criterion of Lemma 8.8.1.

Note first that $\delta = \sqrt{-1081600} = 1040\sqrt{-1}$, so $\mathbb{Q}(\delta) = \mathbb{Q}(\sqrt{-1})$. The root of the resolvent cubic in $\mathbb{Q}$ is zero, so the polynomial $H(x)$ of Lemma 8.8.1 is $x^2(x^2 - 16x - 1)$. The nonzero roots of this are $8 \pm \sqrt{65}$, so H does not split over $\mathbb{Q}(\delta)$. Therefore, the Galois group is D_4.

The following lemma also implies that the Galois group is D_4 rather than $\mathbb{Z}_4$:

Lemma 8.8.5. *Let K be a subfield of $\mathbb{R}$ and let f be an irreducible quartic polynomial over K whose discriminant is negative. Then the Galois group of f over K is not $\mathbb{Z}_4$.*

Proof. Complex conjugation, which we denote here by γ, is an automorphism of $\mathbb{C}$ that leaves invariant the coefficients of f. The splitting field E of f can be taken to be a subfield of $\mathbb{C}$, and γ induces a K-automorphism of E, since E is Galois over K.

The square root δ of the discriminant δ^2 is always contained in the splitting field. As δ^2 is assumed to be negative, δ is pure imaginary. In particular, E is not contained in the reals, and the restriction of γ to E has order 2.

Suppose that the Galois group G of f is cyclic of order 4. Then G contains a unique subgroup of index 2, which must be the subgroup generated by γ. By by the Galois correspondence, there is a unique intermediate field $K \subseteq L \subseteq E$ of dimension 2 over K, which is the fixed field of γ. But $K(\delta)$ is a quadratic extension of K which is *not* fixed pointwise by γ. This is a contradiction. ■

Example 8.8.6. Take the ground field to be $\mathbb{Q}$ and $f(x) = x^4+5x+5$. The irreducibility of f follows from the Eisenstein criterion. The discriminant of f is 15125, which is not a square in $\mathbb{Q}$; therefore, the Galois group G of f is not contained in the alternating group. The resolvent cubic of f is $25-20\,x+x^3 = (5+x)\left(5-5\,x+x^2\right)$. Because h is reducible, it follows that the G is either Z_4 or D_4, but this time the discriminant is positive, so the criterion of Lemma 8.8.5 fails.

Note that the splitting field of the resolvent cubic is $\mathbb{Q}(\delta) = \mathbb{Q}(\sqrt{5})$, since $\sqrt{\delta} = 55\sqrt{5}$. The one root of the resolvent cubic in $\mathbb{Q}$ is -5. Therefore, the polynomial $H(x)$ discussed in Lemma 8.8.1 is $\left(-5+x^2\right)\left(5-5\,x+x^2\right)$. Because $H(x)$ splits over $\mathbb{Q}(\sqrt{5})$, the Galois group is $\mathbb{Z}_4$.

Example 8.8.7. Set $f(x) = x^4+6x^2+4$. This is a so-called biquadratic polynomial, whose roots are $\pm\sqrt{\alpha}, \pm\sqrt{\beta}$, where α and β are the roots of the quadratic polynomial x^2+6x+4. Because the quadratic polynomial is irreducible over the rationals, so is the quartic f. The discriminant of f is $25600 = 160^2$, so the Galois group is contained in the alternating group. But the resolvent cubic $20\,x - 12\,x^2 + x^3 = (-10+x)\,(-2+x)\,x$ is reducible over $\mathbb{Q}$, so the Galois group is $\mathcal{V}$.

A *Mathematica* notebook **Quartic.nb** for investigation of Galois groups of quartic polynomials can be found on my Web site. This notebook can be used to verify the computations in the examples as well as to help with the Exercises.

Exercises 8.8

8.8.1. Verify Equation (8.8.8).

8.8.2. Show that a linear change of variables $y = x + c$ does not alter the discriminant of a polynomial.

8.8.3. Show how to modify the treatment in this section to deal with a nonmonic quartic polynomial.

8.8.4. Show that the resolvent cubic h of an irreducible quartic polynomial f has the same discriminant as f, $\delta^2(h) = \delta^2(f)$.

8.8.5. Let $f(x)$ be an irreducible quartic polynomial over a field K, and let δ^2 denote the discriminant of f. Show that the splitting field of the resolvent cubic of f equals $K(\delta)$ if, and only if, the resolvent cubic is not irreducible over K.

8.8.6. Show that $x^4 + 3x + 3$ is irreducible over $\mathbb{Q}$, and determine the Galois group.

8.8.7. For p a prime other than 3 and 5, show that x^4+px+p is irreducible with Galois group S_4. The case $p = 3$ is treated in the previous exercise, and the case $p = 5$ is treated in Example 8.8.6.

8.8.8. Find the Galois group of $f(x) = x^4 + 5x^2 + 3$. Compare Example 8.8.7.

8.8.9. Show that the biquadratic $f(x) = x^4 + px^2 + r$ has Galois group $\mathcal{V}$ precisely when r is a square in the ground field K. Show that in general $K(\delta) = K(\sqrt{r})$.

8.8.10. Find examples of the biquadratic $f(x) = x^4 + px^2 + r$ for which the Galois group is $\mathbb{Z}_4$ and examples for which the Galois group is D_4. Find conditions for the Galois group to be cyclic and for the Galois group to be the dihedral group.

8.8.11. Determine the Galois group over $\mathbb{Q}$ for each of the following polynomials. You may need to do computer aided calculations, for example, using the Mathematica notebook **Quartic.nb** on my Web site.

(a) $21 + 6x - 17x^2 + 3x^3 + 21x^4$
(b) $1 - 12x + 36x^2 - 19x^4$
(c) $-33 + 16x - 39x^2 + 26x^3 - 9x^4$
(d) $-17 - 50x - 43x^2 - 2x^3 - 33x^4$
(e) $-8 - 18x^2 - 49x^4$
(f) $40 + 48x + 44x^2 + 12x^3 + x^4$
(g) $82 + 59x + 24x^2 + 3x^3 + x^4$

8.9. Galois Groups of Higher Degree Polynomials

In this section, we shall discuss the computation of Galois groups of polynomials in $\mathbb{Q}[x]$ of degree 5 or more.

If a polynomial $f(x) \in K[x]$ of degree n has irreducible factors of degrees $m_1 \geq m_2 \geq \cdots \geq m_r$, where $\sum_i m_i = n$, we say that $(m_1, m_2, \dots, m_r)$ is the *degree partition* of f. Let $f(x) \in \mathbb{Z}[x]$ and let $\tilde{f}(x) \in \mathbb{Z}_p[x]$ be the reduction of f modulo a prime p; if $\tilde{f}$ has degree partition α, we say that f *has degree partition* α *modulo* p.

The following theorem is fundamental for the computation of Galois groups over $\mathbb{Q}$.

Theorem 8.9.1. *Let f be an irreducible polynomial of degree n with coefficients in $\mathbb{Z}$. Let p be a prime that does not divide the leading coefficient of f nor the discriminant of f. Suppose that f has degree partition α modulo p. Then the Galois group of f contains a permutation of cycle type α.*

If f is an irreducible polynomial of degree n whose Galois group does not contain an n-cycle, then the reduction of f modulo a prime p is *never* irreducible of degree n. If the prime p does not divide the leading coefficient of f or the discriminant, then the reduction of f modulo p factors, according to the theorem. If the prime divides the leading coefficient, then the reduction has degree less than n. Finally, if the prime divides the discriminant, then the discriminant of the reduction is zero; but an irreducible polynomial over $\mathbb{Z}_p$ can never have multiple roots and, therefore, cannot have zero discriminant.

We shall not prove Theorem 8.9.1 here. You can find a proof in B. L. van der Waerden, *Algebra*, Volume I, Frederick Ungar Publishing Co., 1970, Section 8.10 (translation of the 7th German edition, Springer–Verlag, 1966).

This theorem, together with the computation of the discriminant, often suffices to determine the Galois group.

Example 8.9.2. Consider $f(x) = x^5 + 5x^4 + 3x + 2$ over the ground field $\mathbb{Q}$. The discriminant of f is 4557333, which is not a square in $\mathbb{Q}$ and which has prime factors 3, 11, and 138101. The reduction of f modulo 7 is irreducible, and the reduction modulo 41 has degree partition $(2, 1, 1, 1)$. It follows that f is irreducible over $\mathbb{Q}$ and that its Galois group over $\mathbb{Q}$ contains a 5-cycle and a 2-cycle. But a 5-cycle and a 2-cycle generate S_5, so the Galois group of f is S_5.

Example 8.9.3. Consider $f(x) = 4 - 4x + 9x^3 - 5x^4 + x^5$ over the ground field $\mathbb{Q}$. The discriminant of f is $15649936 = 3956^2$, so the Galois group G of f is contained in the alternating group A_5. The prime factors of the discriminant are 2, 23, and 43. The reduction of f modulo 3 is irreducible and the reduction modulo 5 has degree partition $(3, 1, 1)$. Therefore, f is irreducible over $\mathbb{Q}$, and its Galois group over $\mathbb{Q}$ contains a 5-cycle and a 3-cycle. But a 5-cycle and a 3-cycle generate the alternating group A_5, so the Galois group is A_5.

The situation is quite a bit more difficult if the Galois group is not S_n or A_n. In this case, we have to show that certain cycle types *do not* appear. The rest of this section is devoted to a (by no means definitive) discussion of this question.

Example 8.9.4. Consider $f(x) = -3 + 7x + 9x^2 + 8x^3 + 3x^4 + x^5$ over the ground field $\mathbb{Q}$. The discriminant of f is $1306449 = 1143^2$, so the Galois group G of f is contained in the alternating group A_5. The prime factors of the discriminant are 3 and 127. The transitive subgroups of the alternating group A_5 are A_5, D_5, and $\mathbb{Z}_5$, according to Exercise 5.1.20. The reduction of f modulo 2 is irreducible, and the reduction modulo 5 has degree partition $(2, 2, 1)$. Therefore, G contains a 5-cycle and an element of cycle type $(2, 2, 1)$. This eliminates $\mathbb{Z}_5$ as a possibility for the Galois group.

If we compute the factorization of the reduction of f modulo a number of primes, we find no instances of factorizations with degree partition $(3, 1, 1)$. In fact, for the first 1000 primes, the frequencies of degree partitions modulo p are

1^5	$2, 1^4$	$2^2, 1$	$3, 1^2$	$3, 2$	$4, 1$	5
.093	0	.5	0	0	0	.4

These data certainly suggest strongly that the Galois group has no 3-cycles and, therefore, must be D_5 rather than A_5, but the empirical evidence does not yet constitute a proof! I would now like to present some facts that nearly, but not quite, constitute a method for determining that the Galois group is D_5 rather than A_5.

Let us compare our frequency data with the frequencies of various cycle types in the transitive subgroups of S_5. The first table shows the number of elements of each cycle type in the various transitive subgroups of S_5, and the second table displays the frequencies of the cycle types, that is, the number of elements of a given cycle type divided by the order of the group.

	1^5	$2, 1^4$	$2^2, 1$	$3, 1^2$	$3, 2$	$4, 1$	5
$\mathbb{Z}_5$	1	0	0	0	0	0	4
D_5	1	0	5	0	0	0	4
A_5	1	0	15	20	0	0	24
$\mathbb{Z}_4 \ltimes \mathbb{Z}_5$	1	0	5	0	0	10	4
S_5	1	10	15	20	0	30	24

Numbers of elements of various cycle types

	1^5	$2, 1^4$	$2^2, 1$	$3, 1^2$	$3, 2$	$4, 1$	5
$\mathbb{Z}_5$	.2	0	0	0	0	0	.8
D_5	.1	0	.5	0	0	0	.4
A_5	.017	0	.25	.333	0	0	.4
$\mathbb{Z}_4 \ltimes \mathbb{Z}_5$	.05	0	.25	0	0	.5	.2
S_5	.0083	.083	.125	.167	0	.25	.2

Frequencies of various cycle types

Notice that our empirical data for frequencies of degree partitions modulo p for $f(x) = -3 + 7x + 9x^2 + 8x^3 + 3x^4 + x^5$ are remarkably close to the frequencies of cycle types for the group D_5. Let's go back to our previous example, the polynomial $f(x) = 4 - 4x + 9x^3 - 5x^4 + x^5$, whose Galois group is known to be A_5. The frequencies of degree partitions modulo the first 1000 primes are

1^5	$2, 1^4$	$2^2, 1^3$	$3, 1^2$	$3, 2$	$4, 1$	5
.019	0	.25	.32	0	0	.41

These data are, again, remarkably close to the data for the distribution of cycle types in the group A_5. The following theorem (conjectured by Frobenius, and proved by Chebotarev in 1926) asserts that the frequencies of degree partitions modulo primes inevitably approximate the frequencies of cycle types in the Galois group:

Theorem 8.9.5. *(Chebotarev). Let* $f \in \mathbb{Z}[x]$ *be an irreducible polynomial of degree* n*, and let* G *denote the Galois group of* f *over* $\mathbb{Q}$*. For each partition* α *of* n *let* d_α *be the fraction of elements of* G *of cycle type* α*. For each* $N \in \mathbb{N}$*, let* $d_{\alpha,N}$ *be the fraction of primes* p *in the interval* $[1, N]$ *such that* f *has degree partition* α *modulo* p*. Then*

$$\lim_{N \to \infty} d_{\alpha,N} = d_\alpha.$$

Because the distribution of degree partitions modulo primes of the polynomial $f(x) = -3 + 7x + 9x^2 + 8x^3 + 3x^4 + x^5$ for the first 1000 primes is quite close to the distribution of cycle types for the group D_5 and quite far from the distribution of cycle types for A_5, our belief that the Galois group of f is D_5 is encouraged, but still not rigorously confirmed by this theorem. The difficulty is that we do not know for certain that, if we examine yet more primes, the distribution of degree partitions will not shift toward the distribution of cycle types for A_5.

What we would need in order to turn these observations into a method is a practical error estimate in Chebotarev's theorem. In fact, a number of error estimates were published in the 1970s and 1980s, but I have not able to find any description in the literature of a practical method based on these estimates.

The situation is annoying but intriguing. Empirically, the distribution of degree partitions converge rapidly to the distribution of cycle types for the Galois group, so that one can usually identify the Galois group by this method, without having a proof that it is in fact the Galois group. (By the way, there exist pairs of nonisomorphic transitive subgroups of S_n for $n \geq 12$ that have the same distribution of cycle types, so that frequency of degree partitions cannot always determine the Galois group of a polynomial.)

A practical method of computing Galois groups of polynomials of small degree is described in L. Soicher and J. McKay, "Computing Galois Groups over the Rationals," *Journal of Number Theory*, vol. 20 (1985) pp. 273–281. The method is based on the computation and factorization of certain resolvent polynomials. This method has two great advantages: First, it works, and second, it is based on mathematics that you now know. I am not going to describe the method in full here, however. This method, or a similar one, has been implemented in the computer algebra package *Maple*; the command

galois( ) in *Maple* will compute Galois groups for polynomials of degree no more than 7.

Both the nonmethod of Chebotarev and the method of Soicher and McKay are based on having at hand a catalog of potential Galois groups (i.e., transitive subgroups of S_n) together with certain identifying data for these groups. Transitive subgroups of S_n have been cataloged at least for $n \leq 11$; see G. Butler and J. McKay, "The Transitive Subgroups of Degree up to 11," *Communications in Algebra*, vol. 11 (1983), pp. 863–911.

By the way, if you write down a polynomial with integer coefficients at random, the polynomial will probably be irreducible, will probably have Galois group S_n, and you will probably be able to show that the Galois group is S_n by examining the degree partition modulo p for only a few primes. In fact, just writing down polynomials at random, you will have a hard time finding one whose Galois group is *not* S_n. Apparently, not all that much is known about the probability distribution of various Galois groups, aside from the predominant occurrence of the symmetric group.

A major unsolved problem is the so-called *inverse Galois problem*: Which groups can occur as Galois groups of polynomials over the rational numbers? A great deal is known about this problem; for example, it is known that all solvable groups occur as Galois groups over $\mathbb{Q}$. (See Chapter 10 for a discussion of solvability.) However, the definitive solution to the problem is still out of reach.

Exercises 8.9

Find the *probable* Galois groups for each of the following quintic polynomials, by examining reductions of the polynomials modulo primes. In *Mathematica,* the command **Factor[f, Modulus → p]** will compute the factorization of the reduction of a polynomial f modulo a prime p. Using this, you can do a certain amount of computation "by hand." By writing a simple loop in *Mathematica,* you can examine the degree partition modulo p for the first N primes, say for $N = 100$. This will already give you a good idea of the Galois group in most cases. You can find a program for the computation of frequencies of degree partitions on my Web site; consult the notebook **Galois-Groups.nb**.

8.9.1. $x^5 + 2$

8.9.2. $x^5 + 20x + 16$

8.9.3. $x^5 - 5x + 12$

8.9.4. $x^5 + x^4 - 4x^3 - 3x^2 + 3x + 1$

8.9.5. $x^5 - x + 1$

8.9.6. Write down a few random polynomials of various degrees, and show that the Galois group is S_n.

8.9.7. Project: Read the article of Soicher and McKay. Write a program for determining the Galois group of polynomials over $\mathbb{Q}$ of degree ≤ 5 based on the method of Soicher and McKay.

8.9.8. Project: Investigate the probability distribution of Galois groups for polynomials of degree ≤ 4 over the integers. If you have done the previous exercise, extend your investigation to polynomials of degree ≤ 5.

CHAPTER 9

Solvability

9.1. Composition Series and Solvable Groups

This section treats a decomposition of any finite group into simple pieces.

Definition 9.1.1. A group G with no nontrivial proper normal subgroup N (that is, $\{e\} \underset{\neq}{\subset} N \underset{\neq}{\subset} G$) is called *simple*.

We know that a cyclic group of prime order is simple, as it has no proper subgroups at all. In fact, the cyclic groups of prime order comprise all abelian simple groups (see Exercise 9.1.1). In the next section, we will show that the alternating groups A_n for $n \geq 5$ are simple.

One of the heroic achievements of mathematics in this century is the complete classification of finite simple groups, which was finished in 1981. There are a number of infinite families of finite simple groups, namely, the cyclic groups of prime order, alternating groups, and certain groups of matrices over finite fields (simple groups of Lie type). Aside from the infinite families, there are 26 so-called *sporadic simple groups*. In the 1950s not all of the simple groups of Lie type were known, and only five of the sporadic groups were known. In the period from 1960 to 1980, knowledge of the simple groups of Lie type was completed and systematized, and the remaining sporadic groups were discovered. At the same time, classification theorems of increasing strength eventually showed that the known list of finite simple groups covered all possibilities. This achievement was a collaborative effort of many mathematicians.

Now, consider the chain of groups $\{e\} \subsetneq N \subsetneq G$. If N is not simple, then it is possible to interpose a subgroup N', $\{e\} \subsetneq N' \subsetneq N$, with N' normal in N. If G/N is not simple, then there is a proper normal subgroup $\{e\} \subsetneq \bar{N}'' \subsetneq G/N$; then the inverse image $N'' = \{g \in G : gN \in \bar{N}''\}$ is a normal subgroup of G satisfying $N \subsetneq N'' \subsetneq G$. If we continue in this way to interpose intermediate normal subgroups until no further additions are possible, the result is a *composition series*.

Definition 9.1.2. A *composition series* for a group G is a chain of subgroups $\{e\} = G_0 \subsetneq G_1 \subsetneq \cdots \subsetneq G_n = G$ such that for all i, G_i is normal in G_{i+1} and G_{i+1}/G_i is simple.

An induction based on the preceding observations shows that a finite group always has a composition series.

Proposition 9.1.3. *Any finite group has a composition series.*

Proof. A group of order 1 is simple, so has a composition series. Let G be a group of order $n > 1$, and assume inductively that whenever G' is a finite group of order less than n, then G' has a composition series. If G is simple, then $\{e\} \subsetneq G$ is already a composition series. Otherwise, there is a proper normal subgroup $\{e\} \subsetneq N \subsetneq G$. By the induction hypothesis, N has a composition series $\{e\} \subsetneq G_1 \subsetneq \cdots \subsetneq G_r = N$. Likewise, G/N has a composition series $\{e\} \subsetneq N_1 \subsetneq \cdots \subsetneq N_s = G/N$. Let $\pi : G \to G/N$ be the quotient map, and let $G_{r+i} = \pi^{-1}(N_i)$ for $0 \le i \le s$, and finally put $n = r + s$. Then $\{e\} \subsetneq G_1 \subsetneq \cdots \subsetneq G_n = G$ is a composition series for G. ■

For abelian groups we can make a sharper statement: It follows from the structure theory of finite abelian groups that the only simple abelian groups are cyclic of prime order, and that any finite abelian group has a composition series in which the successive quotients are cyclic of prime order (Exercise 9.1.1).

Note that there are many choices to be made in the construction of a composition series. For example, an abelian group of order 45 has compositions series in which successive quotient groups have order $3, 3, 5$, another in which the successive quotient groups have order $3, 5, 3$, and another in which successive quotient groups have order $5, 3, 3$ (Exercise 9.1.2).

A theorem of C. Jordan and O. Hölder (Hölder, 1882, based on earlier work of Jordan) says that the simple groups appearing in any two composition series of a finite group are the same up to order (and isomorphism). I am not going to give a proof of the Jordan-Hölder theorem here, but you can find a proof in any more advanced book on group theory or in many reference works on algebra.

Definition 9.1.4. A finite group is said to be *solvable* if it has a composition series in which the successive quotients are cyclic of prime order.

Proposition 9.1.5. *A finite group* G *is solvable if, and only if, it has a chain of subgroups* $\{e\} \underset{\neq}{\subset} G_1 \underset{\neq}{\subset} \cdots \underset{\neq}{\subset} G_n = G$ *in which each* G_i *is normal in* G_{i+1} *and the quotients* G_{i+1}/G_i *are abelian.*

Proof. Exercise 9.1.3. ■

Exercises 9.1

9.1.1.

(a) Show that any group of order p^n, where p is a prime, has a composition series in which successive quotients are cyclic of order p.

(b) Show that the structure theory for finite abelian groups implies that any finite abelian group has a composition series in which successive quotients are cyclic of prime order. In particular, the only simple abelian groups are the cyclic groups of prime order.

9.1.2. Show that an abelian group G of order 45 has a composition series $\{e\} \subseteq G_1 \subseteq G_2 \subseteq G_3 = G$ in which the successive quotient groups have order $3, 3, 5$, and another in which the successive quotient groups have orders $3, 5, 3$, and yet another in which the successive quotient groups have orders $5, 3, 3$.

9.1.3. Prove Proposition 9.1.5.

9.1.4. Show that the symmetric groups S_n for $n \leq 4$ are solvable.

9.2. Commutators and Solvability

In this section, we develop by means of exercises a description of solvability of groups in terms of *commutators*. This section can be skipped without loss of continuity; however, the final three exercises in the section will be referred to in later proofs.

Definition 9.2.1. The *commutator* of elements x, y in a group is $[x, y] = x^{-1}y^{-1}xy$. The *commutator subgroup* or *derived subgroup* of a group G is the subgroup generated by all commutators of elements of G. The commutator subgroup is denoted $[G, G]$ or G'.

9.2.1. Calculate the commutator subgroup of the dihedral group D_n.

9.2.2. Calculate the commutator subgroup of the symmetric groups S_n for $n = 3, 4$.

9.2.3. For H and K subgroups of a group G, let $[H, K]$ be the subgroup of G generated by commutators $[h, k]$ with $h \in H$ and $k \in K$.

(a) Show that if H and K are both normal, then $[H, K]$ is normal in G, and $[H, K] \subseteq H \cap K$.

(b) Conclude that the commutator subgroup $[G, G]$ is normal in G.

(c) Define $G^{(0)} = G$, $G^{(1)} = [G, G]$, and, in general, $G^{(i+1)} = [G^{(i)}, G^{(i)}]$. Show by induction that $G^{(i)}$ is normal in G for all i and $G^{(i)} \supseteq G^{(i+1)}$.

9.2.4. Show that G/G' is abelian and that G' is the unique smallest subgroup of G such that G/G' is abelian.

Theorem 9.2.2. *For a group G, the following are equivalent:*

(a) *G is solvable.*

(b) *There is a natural number n such that the nth commutator subgroup $G^{(n)}$ is the trivial subgroup $\{e\}$.*

Proof. We use the characterization of solvability in Exercise 9.1.5.

If the condition (b) holds, then

$$\{e\} = G^{(n)} \subseteq G^{(n-1)} \subseteq \cdots \subseteq G^{(1)} \subseteq G$$

is a chain of subgroups, each normal in the next, with abelian quotients. Therefore, G is solvable.

Suppose, conversely, that G is solvable and that

$$\{e\} = H_r \subseteq H_{r-1} \subseteq \cdots \subseteq H_1 \subseteq G$$

is a chain of subgroups, each normal in the next, with abelian quotients. We show by induction that $G^{(k)} \subseteq H_k$ for all k; in particular,

$G^{(r)} = \{e\}$. Since G/H_1 is abelian, it follows from Exercise 9.2.4 that $G^{(1)} \subseteq H_1$.

Assume inductively that $G^{(i)} \subseteq H_i$ for some i ($1 \le i < r$). Since H_i/H_{i+1} is assumed to be abelian, it follows from Exercise 9.2.4 that $[H_i, H_i] \subseteq H_{i+1}$; but then $G^{(i+1)} = [G^{(i)}, G^{(i)}] \subseteq [H_i, H_i] \subseteq H_{i+1}$. This completes the inductive argument. ■

In the following exercises, you can use whatever criterion for solvability appears most convenient.

9.2.5. Show that any subgroup of a solvable group is solvable.

9.2.6. Show that any quotient group of a solvable group is solvable.

9.2.7. Show that if $N \subsetneq G$ is a normal subgroup and both N and G/N are solvable, then also G is solvable.

9.3. Simplicity of the Alternating Groups

In this section, we will prove that the symmetric groups S_n and the alternating groups A_n for $n \ge 5$ are not solvable. In fact, we will see that the alternating group A_n is the unique nontrivial proper normal subgroup of S_n for $n \ge 5$, and moreover that A_n is a simple group for $n \ge 5$.

Recall that conjugacy classes in the symmetric group S_n are determined by cycle structure. Two elements are conjugate precisely when they have the same cycle structure. This fact is used frequently in the following.

Lemma 9.3.1. *For $n \ge 3$, the alternating group A_n is generated by 3–cycles.*

Proof. A product of two 2–cycles in S_n is conjugate to $(12)(12) = e = (123)(132)$, if the two 2–cycles are equal; or to $(12)(23) = (123)$, if the two 2–cycles have one digit in common; or to $(12)(34) = (132)(134)$, if the two 2–cycles have no digits in common. Thus, any product of two 2–cycles can be written as a product of one or two 3–cycles. Any even permutation is a product of an even number of 2–cycles and, therefore, can be written as a product of 3–cycles. ■

Since the center of a group is always a normal subgroup, if we want to show that a nonabelian group is simple, it makes sense to check first that the center is trivial. You are asked to show in Exercise 9.3.1 that for $n \geq 3$ the center of S_n is trivial and in Exercise 9.3.4 that for $n \geq 4$ the center of A_n is trivial.

Theorem 9.3.2. *If $n \geq 5$ and N is a normal subgroup of S_n such that $N \neq \{e\}$, then $N \supseteq A_n$.*

Proof. By Exercise 9.3.1, the center of S_n consists of e alone. Let $\sigma \neq e$ be an element of N; since σ is not central and since 2–cycles generate S_n, there is some 2–cycle τ such that $\sigma\tau \neq \tau\sigma$. Consider the element $\tau\sigma\tau\sigma^{-1}$; writing this as $(\tau\sigma\tau)\sigma^{-1}$ and using the normality of N, we see that this element is in N; on the other hand, writing the element as $\tau(\sigma\tau\sigma^{-1})$, we see that the element is a product of two unequal 2–cycles.

If these two 2–cycles have a digit in common, then the product is a 3–cycle.

If the two 2–cycles have no digit in common, then by normality of N, N contains all elements that are a product of two disjoint 2–cycles. In particular, N contains the elements $(12)(34)$ and $(12)(35)$ and, therefore, the product $(12)(34)(12)(35) = (34)(35) = (435)$. So also in this case, N contains a 3–cycle.

By normality of N, N contains all 3–cycles, and, therefore, by Lemma 9.3.1, $N \supseteq A_n$. ■

Lemma 9.3.3. *If $n \geq 5$, then all 3–cycles are conjugate in A_n.*

Proof. It suffices to show that any 3–cycle is conjugate in A_n to (123). Let σ be a 3–cycle. Since σ and (123) have the same cycle structure, there is an element of $\tau \in S_n$ such that $\tau(123)\tau^{-1} = \sigma$. If τ is even, there is nothing more to do. Otherwise, $\tau' = \tau(45)$ is even, and $\tau'(123)\tau'^{-1} = \sigma$. ■

Theorem 9.3.4. *If $n \geq 5$, then A_n is simple.*

Proof. Suppose $N \neq \{e\}$ is a normal subgroup of A_n and that $\sigma \neq e$ is an element of N. Since the center of A_n is trivial and A_n is generated by 3–cycles, there is a 3–cycle τ that does not commute with σ. Then (as in the proof of the previous theorem) the element $\sigma\tau\sigma^{-1}\tau^{-1}$ is a nonidentity element of N that is a product of two 3–cycles.

The rest of the proof consists of showing that N must contain a 3–cycle. Then by Lemma 9.3.3, N must contain all 3–cycles, and, therefore, $N = A_n$ by Lemma 9.3.1.

The product π of two 3–cycles must be of one of the following types:

1. $(a_1a_2a_3)(a_4a_5a_6)$
2. $(a_1a_2a_3)(a_1a_4a_5) = (a_1a_4a_5a_2a_3)$
3. $(a_1a_2a_3)(a_1a_2a_4) = (a_1a_3)(a_2a_4)$
4. $(a_1a_2a_3)(a_2a_1a_4) = (a_1a_4a_3)$
5. $(a_1a_2a_3)(a_1a_2a_3) = (a_1a_3a_2)$

In either of the last two cases, N contains a 3–cycle.

In case (3), since $n \geq 5$, A_n contains an element $(a_2a_4a_5)$. Compute that $(a_2a_4a_5)\pi(a_2a_4a_5)^{-1} = (a_1a_3)(a_4a_5)$, and the product

$$\pi^{-1}(a_2a_4a_5)\pi(a_2a_4a_5)^{-1} = (a_2a_5a_4)$$

is a 3–cycle in N.

In case (2), compute that $(a_1a_4a_3)\pi(a_1a_4a_3)^{-1} = (a_4a_3a_5a_2a_1)$, and $\pi^{-1}(a_1a_4a_3)\pi(a_1a_4a_3)^{-1} = (a_4a_2a_3)$ is a 3–cycle in N.

Finally, in case (1), compute that

$$(a_1a_2a_4)\pi(a_1a_2a_4)^{-1} = (a_2a_4a_3)(a_1a_5a_6),$$

and the product $\pi^{-1}(a_1a_2a_4)\pi(a_1a_2a_4)^{-1}$ is $(a_1a_4a_6a_2a_3)$, namely, a 5–cycle. But then by case (2), N contains a 3–cycle. ■

These computations may look mysterious and unmotivated. The idea is to take a 3–cycle x and to form the commutator $\pi^{-1}x\pi x^{-1}$. This is a way to get a lot of new elements of N. If we experiment just a little, we can find a 3–cycle x such that the commutator is either a 3–cycle or, at the worst, has one of the cycle structures already dealt with.

Here is an alternative way to finish the proof, which avoids the computations. I learned this method from I. Herstein, *Abstract Algebra*, 3rd edition, Prentice Hall, 1996.

I claim that the simplicity of A_n for $n > 6$ follows from the simplicity of A_6. Suppose $n > 6$ and $N \neq \{e\}$ is a normal subgroup of A_n. By the first part of the proof, N contains a nonidentity element that is a product of two 3–cycles. These two 3–cycles involve at most 6 of the digits $1 \leq k \leq n$; so there is an isomorphic copy of S_6 in S_n such that $N \cap A_6 \neq \{e\}$. Since $N \cap A_6$ is a normal subgroup of A_6, and A_6 is supposed to be simple, it follows that $N \cap A_6 = A_6$, and, in particular, N contains a 3–cycle. But if N contains a 3–cycle, then $N = A_n$, by an argument used in the original proof.

So it suffices to prove that A_5 and A_6 are simple. Take $n = 5$ or 6, and suppose that $N \neq \{e\}$ is a normal subgroup of A_n that is minimal among all such subgroups; that is, if $\{e\} \subseteq K \underset{\neq}{\subset} N$ is a normal subgroup of A_n, then $K = \{e\}$.

Let X be the set of conjugates of N in S_n, and let S_n act on X by conjugation. What is the stabilizer (centralizer) of N? Since N is normal in A_n, $\mathrm{Cent}_{S_n}(N) \supseteq A_n$. There are no subgroups properly between A_n and S_n, so the centralizer is either A_n or S_n. If the centralizer is all of S_n, then N is normal in S_n, so by Theorem 9.3.2, $N = A_n$.

If the centralizer of N is A_n, then X has $2 = [S_n : A_n]$ elements. Let $M \neq N$ be the other conjugate of N in S_n. Then $M = \sigma N \sigma^{-1}$ for any odd permutation σ. Note that $M \cong N$ and, in particular, M and N have the same number of elements.

I leave it as an exercise to show that M is also a normal subgroup of A_n.

Since $N \neq M$, $N \cap M$ is a normal subgroup of A_n properly contained in N. By the assumption of minimality of N, it follows that $N \cap M = \{e\}$. Therefore, MN is a subgroup of A_n, isomorphic to $M \times N$. The group MN is normal in A_n, but if σ is an odd permutation, then $\sigma MN \sigma^{-1} = \sigma M \sigma^{-1} \sigma N \sigma^{-1} = NM = MN$, so MN is normal in S_n. Therefore, by Theorem 9.3.2 again, $MN = A_n$. *In particular, the cardinality of A_n is $|M \times N| = |N|^2$. But neither $5!/2 = 60$ nor $6!/2 = 360$ is a perfect square, so this is impossible.*

This completes the alternative proof of Theorem 9.3.4.

It follows from the simplicity of A_n for $n \geq 5$ that neither S_n nor A_n is solvable for $n \geq 5$ (Exercise 9.3.6).

Exercises 9.3

9.3.1. Show that for $n \geq 3$, the center of S_n is $\{e\}$. Hint: Suppose that $\sigma \in Z(A_n)$. Then the conjugacy class of σ consists of σ alone. What is the cycle structure of σ?

9.3.2. Show that the subgroup $\mathcal{V} = \{e, (12)(34), (13)(24), (14)(23)\}$ of A_4 is normal in S_4.

9.3.3. Show that if A is a normal subgroup of a group G, then the center Z(A) of A is normal in G.

9.3.4. This exercise shows that the center of A_n is trivial for all $n \geq 4$.

(a) Compute that the center of A_4 is $\{e\}$.
(b) Show that for $n \geq 4$, A_n is not abelian.

(c) Use Exercise 9.3.3 and Theorem 9.3.2 to show that if $n \geq 5$, then the center of A_n is either $\{e\}$ or A_n. Since A_n is not abelian, the center is $\{e\}$.

9.3.5. Show that the subgroup M appearing in the alternative proof of 9.3.4 is a normal subgroup of A_n.

9.3.6. Observe that a simple nonabelian group is not solvable, and conclude that neither A_n nor S_n are solvable for $n \geq 5$.

9.4. Cyclotomic Polynomials

In this section, we will study factors of the polynomial $x^n - 1$. Recall that a primitive n^{th} root of unity in a field K is an element $\zeta \in K$ such that $\zeta^n = 1$ but $\zeta^d \neq 1$ for $d < n$. The primitive n^{th} roots of 1 in $\mathbb{C}$ are the numbers $e^{2\pi i r/n}$, where r is relatively prime to n. Thus the number of primitive n^{th} roots is $\varphi(n)$, where φ denotes the the Euler function.

Definition 9.4.1. The nth cyclotomic polynomial $\Psi_n(x)$ is defined by

$$\Psi_n(x) = \prod\{(x - \zeta) : \zeta \text{ a primitive } n^{th} \text{ root of 1 in } \mathbb{C}\}.$$

A priori, $\Psi_n(x)$ is a polynomial in $\mathbb{C}[x]$, but, in fact, it turns out to have integer coefficients. It is evident that $\Psi_n(x)$ is a monic polynomial of degree $\varphi(n)$. Note the factorization:

$$\begin{aligned} x^n - 1 &= \prod \{(x - \zeta) : \zeta \text{ an } n^{th} \text{ root of 1 in } \mathbb{C}\} \\ &= \prod_{d \text{ divides } n} \prod\{(x - \zeta) : \zeta \text{ a primitive } d^{th} \text{ root of 1 in } \mathbb{C}\} \\ &= \prod_{d \text{ divides } n} \Psi_d(x). \end{aligned} \tag{9.4.1}$$

Using this, we can compute the polynomials $\Psi_n(x)$ recursively, beginning with $\Psi_1(x) = x - 1$. See Exercise 9.4.2.

Proposition 9.4.2. *For all n, the cyclotomic polynomial $\Psi_n(x)$ is a monic polynomial of degree $\varphi(n)$ with integer coefficients.*

Proof. The assertion is valid for $n = 1$ by inspection. We proceed by induction on n. Fix $n > 1$, and put

$$f(x) = \prod_{\substack{d<n \\ d \text{ divides } n}} \Psi_d(x),$$

so that $x^n - 1 = f(x)\Psi_n(x)$. By the induction hypothesis, $f(x)$ is a monic polynomial in $\mathbb{Z}[x]$. Therefore,

$$\Psi_n(x) \in \mathbb{Q}(x) \cap \mathbb{C}[x] = \mathbb{Q}[x],$$

by Exercise 9.4.4. Since all the polynomials involved in the factorization $x^n - 1 = f(x)\Psi_n(x)$ are monic, it follows from Gauss's lemma that $\Psi_n(x)$ has integer coefficients. ■

Note that the splitting field of $x^n - 1$ coincides with the splitting field of $\Psi_n(x)$ and is equal to $\mathbb{Q}(\zeta)$, where ζ is any primitive n^{th} root of unity (Exercise 9.4.3). Next, we wish to show that $\Psi_n(x)$ is irreducible over $\mathbb{Q}$, so $\mathbb{Q}(\zeta)$ is a field extension of degree $\varphi(n)$ over $\mathbb{Q}$. First we note the following:

Lemma 9.4.3. *The following statements are equivalent:*

(a) $\Psi_n(x)$ *is irreducible.*

(b) *If ζ is a primitive n^{th} root of unity in $\mathbb{C}$, f is the minimal polynomial of ζ over $\mathbb{Q}$, and p is a prime not dividing n, then ζ^p is a root of f.*

(c) *If ζ is a primitive n^{th} root of unity in $\mathbb{C}$, f is the minimal polynomial of ζ over $\mathbb{Q}$, and r is relatively prime to n, then ζ^r is a root of f.*

(d) *If ζ is a primitive n^{th} root of unity in $\mathbb{C}$, and r is relatively prime to n, then $\zeta \mapsto \zeta^r$ determines an automorphism of $\mathbb{Q}(\zeta)$ over $\mathbb{Q}$.*

Proof. Exercise 9.4.5. ■

Theorem 9.4.4. $\Psi_n(x)$ *is irreducible over* $\mathbb{Q}$.

Proof. Suppose that $\Psi_n(x) = f(x)g(x)$, where $f, g \in \mathbb{Z}[x]$, and f is irreducible. Let ζ be a root of f in $\mathbb{C}$; thus ζ is a primitive n^{th} root of unity and f is the minimal polynomial of ζ over $\mathbb{Q}$. Suppose that p is a prime not dividing n. According to the previous lemma, it suffices to show that ζ^p is a root of f.

Suppose that ζ^p is not a root of f. Then, necessarily, ζ^p is a root of g, or, equivalently, ζ is a root of $g(x^p)$. It follows that $f(x)$ divides

$g(x^p)$, because f is the minimal polynomial of ζ. Reducing all polynomials modulo p, we have that $\tilde{f}(x)$ divides $\tilde{g}(x^p) = (\tilde{g}(x))^p$. In particular, $\tilde{f}(x)$ and $\tilde{g}(x)$ are not relatively prime in $\mathbb{Z}_p[x]$. Now, it follows from $\tilde{\Psi}_n(x) = \tilde{f}(x)\tilde{g}(x)$ that $\tilde{\Psi}_n(x)$ has a multiple root, and, therefore, $x^n - 1$ has a multiple root over $\mathbb{Z}_p$. But this cannot be so, because the derivative of $x^n - 1$ in $\mathbb{Z}_p[x]$, namely $[n]x^{n-1}$, is not identically zero, since p does not divide n, and has no roots in common with $x^n - 1$. This contradiction shows that, in fact, ζ^p is a root of f. ■

Corollary 9.4.5. *The splitting field of $x^n - 1$ over $\mathbb{Q}$ is $\mathbb{Q}(\zeta)$, where ζ is a primitive n^{th} root of unity. The dimension of $\mathbb{Q}(\zeta)$ over $\mathbb{Q}$ is $\varphi(n)$, and the Galois group of $\mathbb{Q}(\zeta)$ over $\mathbb{Q}$ is isomorphic to $\Phi(n)$, the multiplicative group of units in $\mathbb{Z}_n$.*

Proof. The irreducibility of $\Psi_n(x)$ over $\mathbb{Q}$ implies that $\dim_{\mathbb{Q}}(\mathbb{Q}(\zeta)) = \varphi(n)$. We can define an injective homomorphism of $G = \mathrm{Aut}_{\mathbb{Q}}(\mathbb{Q}(\zeta)$ into $\Phi(n)$ by $\sigma \mapsto [r]$ if $\sigma(\zeta) = \zeta^r$. This map is surjective since both groups have the same cardinality. ■

Proposition 9.4.6. *Let K be any field whose characteristic does not divide n. The the Galois group of $x^n - 1$ over K is isomorphic to a subgroup of $\Phi(n)$. In particular, if n is prime, then the Galois group is cyclic of order dividing $n - 1$.*

Proof. Exercise 9.4.6. ■

Exercises 9.4

9.4.1. If ζ is any primitive n^{th} root of unity in $\mathbb{C}$, show that the set of all primitive n^{th} roots of unity in $\mathbb{C}$ is $\{\zeta^r : r \text{ is relatively prime to } n\}$.

9.4.2. Show how to use Equation 9.4.1 to compute the cyclotomic polynomials recursively. Compute the first several cyclotomic polynomials by this method.

9.4.3. Note that the splitting field of $x^n - 1$ coincides with the splitting field of $\Psi_n(x)$ and is equal to $\mathbb{Q}(\zeta)$, where ζ is any primitive n^{th} root of unity.

9.4.4. Show that $\mathbb{Q}(x) \cap \mathbb{C}[x] = \mathbb{Q}[x]$.

9.4.5. Prove Lemma 9.4.3.

9.4.6. Prove Proposition 9.4.6. Hint: Let E be a splitting field of $x^n - 1$ over K. Show that the hypothesis on characteristics implies that E contains a primitive n^{th} root of unity ζ. Define an injective homomorphism from $G = \mathrm{Aut}_K(E)$ into $\Phi(n)$ by $\sigma \mapsto [r]$, where $\sigma(\zeta) = \zeta^r$. Show that this is a well–defined, injective homomorphism of groups. (In particular, check the proof of Corollary 9.4.5.)

9.5. The Equation $x^n - b = 0$

Consider the polynomial $x^n - b \in K[x]$, where $b \neq 0$, and n is relatively prime to the characteristic of K. The polynomial $x^n - b$ has n distinct roots in a splitting field E, and the ratio of any two roots is an nth root of unity, so again E contains n distinct roots of unity and, therefore, a primitive n^{th} root of unity u.

If a is one root of $x^n - b$ in E, then all the roots are of the form $u^j a$, and $E = K(u, a)$. We consider the extension in two stages $K \subseteq K(u) \subseteq E = K(u, a)$.

A $K(u)$-automorphism τ of E is determined by $\tau(a)$, and $\tau(a)$ must be of the form $\tau(a) = u^i a$.

Lemma 9.5.1. *The map $\tau \mapsto \tau(a)a^{-1}$ is an injective group homomorphism from $\mathrm{Aut}_{K(u)}(E)$ into the cyclic group generated by u. In particular, $\mathrm{Aut}_{K(u)}(E)$ is a cyclic group whose order divides n.*

Proof. Exercise 9.5.1. ■

We have proved the following proposition:

Proposition 9.5.2. *If K contains a primitive n^{th} root of unity (where n is necessarily relatively prime to the characteristic) and E is a splitting field of $x^n - b \in K[x]$, then $\mathrm{Aut}_K(E)$ is cyclic. Furthermore, $E = K(a)$, where a is any root in E of $x^n - b$.*

Corollary 9.5.3. *If K is a field, n is relatively prime to the characteristic of K, and E is a splitting field over K of $x^n - b \in K[x]$, then $\mathrm{Aut}_K(E)$ has a cyclic normal subgroup N such that $\mathrm{Aut}_K(E)/N$ is abelian.*

Proof. As shown previously, E contains a primitive n^{th} root of unity u, and if a is one root of $x^n - b$ in E, then $K \subseteq K(u) \subseteq K(u, a) = E$, and the intermediate field $K(u)$ is a Galois over K. By Proposition

9.5.2, $N = \mathrm{Aut}_{K(u)}(E)$ is cyclic, and by the fundamental theorem, N is normal. The quotient $\mathrm{Aut}_K(E)/N \cong \mathrm{Aut}_K(K(u))$ is a subgroup of $\Phi(n)$ and, in particular, abelian, by Proposition 9.4.6. ■

Exercises 9.5

9.5.1. Prove Lemma 9.5.1.

9.5.2. Find the Galois group of $x^{13} - 1$ over $\mathbb{Q}$. Find all intermediate fields in as explicit a form as possible.

9.5.3. Find the Galois group of $x^{13} - 2$ over $\mathbb{Q}$. Find all intermediate fields in as explicit a form as possible.

9.5.4. Let $n = p_1^{m_1} p_2^{m_2} \cdots p_s^{m_s}$. Let ζ be a primitive n^{th} root of unity in $\mathbb{C}$ and let ζ_i be a primitive $p_i^{m_i}$ root of unity. Show that the Galois group of $x^n - 1$ over $\mathbb{Q}$ is isomorphic to the direct product of the Galois groups of $x^{p_i^{m_i}} - 1$, $(1 \le i \le s)$. Show that each $\mathbb{Q}(\zeta_i)$ is a subfield of $\mathbb{Q}(\zeta)$, the intersection of the $\mathbb{Q}(\zeta_i)$ is $\mathbb{Q}$, and the composite of all the $\mathbb{Q}(\zeta_i)$ is $\mathbb{Q}(\zeta)$.

9.6. Solvability by Radicals

Definition 9.6.1. A tower of fields $K = K_0 \subseteq K_1 \subseteq \cdots \subseteq K_r$ is called a *radical tower* if there is a sequence of elements $a_i \in K_i$ such that $K_1 = K_0(a_1), K_2 = K_1(a_2), \ldots, K_r = K_{r-1}(a_r)$, and there is a sequence of natural numbers n_i such that $a_i^{n_i} \in K_{i-1}$ for $1 \le i \le r$.

We call r the *length* of the radical tower. An extension $K \subseteq L$ is called a *radical extension* if there is a radical tower $K = K_0 \subseteq K_1 \subseteq \cdots \subseteq K_r$ with $K_r = L$.

The idea behind this definition is that the elements of a radical extension L can be computed, starting with elements of K, by rational operations—that is, field operations— and by "extracting roots," that is, by solving equations of the form $x^n = b$.

We am aiming for the following result:

Theorem 9.6.2. (Galois) *If* $K \subseteq E \subseteq L$ *are field extensions with* E *Galois over* K *and* L *radical over* K, *then the Galois group* $\mathrm{Aut}_K(E)$ *is a solvable group.*

Lemma 9.6.3. *If* $K \subseteq L$ *is a radical extension of* K*, then there is a radical extension* $L \subseteq \bar{L}$ *such that* $\bar{L}$ *is Galois over* K.

Proof. Suppose there is a radical tower $K = K_0 \subseteq K_1 \subseteq \cdots \subseteq K_r = L$ of length r. We prove the statement by induction on r. If $r = 0$, there is nothing to do. So suppose $r \geq 1$ and there is a radical extension $F \supseteq K_{r-1}$ such that F is Galois over K. There is an $a \in L$ and a natural number n such that $L = K_{r-1}(a)$, and $b = a^n \in K_{r-1}$. Let $p(x) \in K[x]$ be the minimal polynomial of b, and let $f(x) = p(x^n)$. Then a is a root of $f(x)$.

Let $\bar{L}$ be a splitting field of $f(x)$ over F that contains $F(a)$. We have the following inclusions:

$$\begin{array}{ccccccc} & & F & \subseteq & F(a) & \subseteq & \bar{L} \\ & & \cup\!\vert & & \cup\!\vert & & \\ K & \subseteq & K_{r-1} & \subseteq & L & = & K_{r-1}(a). \end{array}$$

If $g(x) \in K[x]$ is a polynomial such that F is a splitting field of $g(x)$ over K, then $\bar{L}$ is a splitting field of $g(x)f(x)$ over K and, hence, $\bar{L}$ is Galois over K.

Finally, for any root α of $f(x)$, α^n is a root of $p(x)$, and, therefore, $\alpha^n \in F$. It follows that $\bar{L}$ is a radical extension of F and, hence, of K. ■

Lemma 9.6.4. *Suppose* $K \subseteq L$ *is a field extension such that* L *is radical and Galois over* K*. Let* $K = K_0 \subseteq K_1 \subseteq \cdots \subseteq K_{r-1} \subseteq K_r = L$ *be a radical tower, and let* $a_i \in K_i$ *and* $n_i \in \mathbb{N}$ *satisfy* $K_i = K_{i-1}(a_i)$ *and* $a_i^{n_i} \in K_{i-1}$ *for* $1 \leq i \leq r$*. Let* $n = n_1 n_2 \cdots n_r$*, and suppose that* K *contains a primitive* n^{th} *root of unity. Then the Galois group* $\mathrm{Aut}_K(L)$ *is solvable.*

Proof. We prove this by induction on the length r of the radical tower. If $r = 0$, then $K = L$ and the Galois group is $\{e\}$. So suppose $r \geq 1$, and suppose that the result holds for radical Galois extensions with radical towers of length less than r. It follows from this inductive hypothesis that the Galois group $G_1 = \mathrm{Aut}_{K_1}(L)$ is solvable.

We have $K_1 = K(a_1)$ and $a_1^{n_1} \in K$. Since n_1 divides n, K contains a primitive n_1^{th} root of unity. Then it follows from Proposition 9.5.2 that K_1 is Galois over K with cyclic Galois group $\mathrm{Aut}_K(K_1)$.

By the fundamental theorem, G_1 is normal in the Galois group $G = \mathrm{Aut}_K(L)$, and $G/G_1 \cong \mathrm{Aut}_K(K_1)$. Since G_1 is solvable and normal in G and the quotient G/G_1 is cyclic, it follows that G is solvable. ■

Proof of Theorem 9.6.2. Let $K \subseteq E \subseteq L$ be field extensions, where E is Galois over K and L is radical over K. By Lemma 9.6.3, we can assume that L is also Galois over K. Since $\mathrm{Aut}_K(E)$ is a quotient group of $\mathrm{Aut}_K(L)$, it suffices to prove that $\mathrm{Aut}_K(L)$ is solvable.

This has been done in Lemma 9.6.4 under the additional assumption that K contains certain roots of unity. The strategy will be to reduce to this case by introducing roots of unity.

Let $K = K_0 \subseteq K_1 \subseteq \cdots \subseteq K_{r-1} \subseteq K_r = L$ be a radical tower, and let $a_i \in K_i$ and $n_i \in \mathbb{N}$ satisfy $K_i = K_{i-1}(a_i)$ and $a_i^{n_i} \in K_{i-1}$ for $1 \le i \le r$. Let $n = n_1 n_2 \cdots n_r$, and let ζ be a primitive n^{th} root of unity in an extension field of L.

Consider the following inclusions:

$$\begin{array}{ccccccc} & & K(\zeta) & \subseteq & L(\zeta) & = & K(\zeta)\cdot L \\ & & \cup\!| & & \cup\!| & & \\ K & \subseteq & K(\zeta)\cap L & \subseteq & L. & & \end{array}$$

By Proposition 8.5.6, $L(\zeta) = K(\zeta) \cdot L$ is Galois over $K(\zeta)$ and, furthermore, $\mathrm{Aut}_{K(\zeta)}(L(\zeta)) \cong \mathrm{Aut}_{K(\zeta)\cap L}(L)$. But $L(\zeta)$ is obtained from $K(\zeta)$ by adjoining the elements a_i, so $L(\zeta)$ is radical over $K(\zeta)$, and, furthermore, Lemma 9.6.4 is applicable to the extension $K(\zeta) \subseteq L(\zeta)$. Hence, $\mathrm{Aut}_{K(\zeta)}(L(\zeta)) \cong \mathrm{Aut}_{K(\zeta)\cap L}(L)$ is solvable.

The extension $K \subseteq K(\zeta)$ is Galois with abelian Galois group, by Proposition 9.4.6. Therefore, every intermediate field is Galois over K with abelian Galois group, by the fundamental theorem. In particular, $K(\zeta) \cap L$ is Galois over K and $\mathrm{Aut}_K(K(\zeta) \cap L)$ is abelian, so solvable.

Write $G = \mathrm{Aut}_K(L)$ and $N = \mathrm{Aut}_{K(\zeta)\cap L}(L)$. Since $K(\zeta) \cap L$ is Galois over K, by the fundamental theorem, N is normal in G and $G/N \cong \mathrm{Aut}_K(K(\zeta) \cap L)$.

Since both N and G/N are solvable, so is G. ■

Definition 9.6.5. Let K be a field and $p(x) \in K[x]$. We say that *p(x) is solvable by radicals* over K if there is a radical extension of K that contains a splitting field of $p(x)$ over K.

The idea of this definition is that the roots of $p(x)$ can be obtained, beginning with elements of K, by rational operations and by extraction of roots.

Corollary 9.6.6. *Let $p(x) \in K[x]$, and let E be a splitting field of $p(x)$ over K. If $p(x)$ is solvable by radicals, then the Galois group $\mathrm{Aut}_K(E)$ is solvable.*

Corollary 9.6.7. (Abel, Galois) *The general equation of degree n over a field K is not solvable by radicals if $n \geq 5$.*

Proof. According to Theorem 8.7.1, the Galois group of the general equation of degree n is S_n, and it follows from Theorem 9.3.4 that S_n is not solvable when $n \geq 5$. ■

This result implies that there can be no analogue of the quadratic (and cubic and quartic) formula for the general polynomial equation of degree 5 or higher. A general method for solving a degree n equation over K is a method of solving the general equation of degree n, of finding the roots u_i in terms of the variables t_i. Such a method can be specialized to any particular equation with coefficients in K. The procedures for solving equations of degrees 2, 3, and 4 are general methods of obtaining the roots in terms of rational operations and extractions of radicals. If such a method existed for equations of degree $n \geq 5$, then the general equation of degree n would have to be solvable by radicals over the field $K(t_1, \dots, t_n)$, which is not so.

9.7. Radical Extensions

This section can be omitted without loss of continuity.

In this section, we will obtain a partial converse to the results of the previous section: If $K \subseteq E$ is a Galois extension with a solvable Galois group, then E is contained in a radical extension. I will prove this only in the case that the ground field has characteristic 0. This restriction is not essential but is made to avoid technicalities. *All the fields in this section are assumed to have characteristic 0.*

We begin with a converse to Proposition 9.5.2.

Proposition 9.7.1. *Suppose the field K contains a primitive n^{th} root of unity. Let E be a Galois extension of K such that $\mathrm{Aut}_K(E)$ is cyclic of order n. Then E is the splitting field of an irreducible polynomial $x^n - b \in K[x]$, and $E = K(a)$, where a is any root of $x^n - b$ in E.*

The proof of this is subtle and requires some preliminary apparatus. Let $K \subseteq E$ be a Galois extension. Recall that E^* denotes the set of nonzero elements of E.

Definition 9.7.2. A function $f : \mathrm{Aut}_K(E) \to E^*$ is called a *multiplicative 1-cocycle* if it satisfies $f(\sigma\tau) = f(\sigma)\sigma(f(\tau))$.

The basic example of a multiplicative 1-cocycle is the following: If a is any nonzero element of E, then the function $g(\sigma) = \sigma(a)a^{-1}$ is a multiplicative 1-cocycle (exercise).

Proposition 9.7.3. *If* $K \subseteq E$ *is a Galois extension, and* $f : \mathrm{Aut}_K(E) \to E^*$ *is a multiplicative 1-cocycle, then there is an element* $a \in E^*$ *such that* $f(\sigma) = \sigma(a)a^{-1}$.

Proof. The elements of the Galois group are linearly independent, by Proposition 8.5.7, so there is an element $b \in E^*$ such that

$$\sum_{\tau \in \mathrm{Aut}_K(E)} f(\tau)\tau(b) \neq 0.$$

Call this nonzero element a^{-1}. In the following computation, the 1-cocycle relation is used in the form $\sigma(f(\tau)) = f(\sigma)^{-1}f(\sigma\tau)$. Now, for any $\sigma \in \mathrm{Aut}_K(E)$,

$$\begin{aligned}
\sigma(a^{-1}) &= \sigma\left(\sum_{\tau \in \mathrm{Aut}_K(E)} f(\tau)\tau(b)\right) \\
&= \sum_{\tau \in \mathrm{Aut}_K(E)} \sigma(f(\tau))\sigma\tau(b) \\
&= \sum_{\tau \in \mathrm{Aut}_K(E)} f(\sigma)^{-1}f(\sigma\tau)\sigma\tau(b) \\
&= f(\sigma)^{-1} \sum_{\tau \in \mathrm{Aut}_K(E)} f(\tau)\tau(b) \\
&= f(\sigma)^{-1}a^{-1}.
\end{aligned}$$

This gives $f(\sigma) = \sigma(a)a^{-1}$. ■

Definition 9.7.4. Let $K \subseteq E$ be a Galois extension. For $a \in E$, the *norm* of a is defined by

$$N(a) = N_K^E(a) = \prod_{\sigma \in \mathrm{Aut}_K(E)} \sigma(a).$$

Note that $N(a)$ is fixed by all $\sigma \in \mathrm{Aut}_K(E)$, so $N(a) \in K$ (because E is a Galois extension). For $a \in K$, $N(a) = a^{\dim_K(E)}$.

Proposition 9.7.5. *(D. Hilbert's theorem 90). Let* $K \subseteq E$ *be a Galois extension with cyclic Galois group. Let* σ *be a generator of* $\mathrm{Aut}_K(E)$*. If* $b \in E^*$ *satisfies* $N(b) = 1$*, then there is an* $a \in E^*$ *such that* $b = \sigma(a)a^{-1}$*.*

Proof. Let n be the order of the cyclic group $\mathrm{Aut}_K(E)$. Define the map $f : \mathrm{Aut}_K(E) \to E^*$ by $f(\mathrm{id}) = 1$, $f(\sigma) = b$, and

$$f(\sigma^i) = \sigma^{i-1}(b) \cdots \sigma(b)b,$$

for $i \geq 1$.

We can check that $N(b) = 1$ implies that $f(\sigma^{i+n}) = f(\sigma^i)$, so f is well–defined on $\mathrm{Aut}_K(E)$ and, furthermore, that f is a multiplicative 1-cocycle (Exercise 9.7.2).

Because f is a multiplicative 1-cocycle, it follows from the previous proposition that there is an $a \in E^*$ such that $b = f(\sigma) = \sigma(a)a^{-1}$. ■

Proof of Proposition 9.7.1. Suppose that $\mathrm{Aut}_K(E)$ is cyclic of order n, that σ is a generator of the Galois group, and that $\zeta \in K$ is a primitive n^{th} root of unity. Because $\zeta \in K$, its norm satisfies $N(\zeta) = \zeta^n = 1$. Hence, by Proposition 9.7.5, there is an element $a \in E^*$ such that $\zeta = \sigma(a)a^{-1}$, or $\sigma(a) = \zeta a$. Let $b = a^n$; we have $\sigma(b) = \sigma(a)^n = (\zeta a)^n = a^n = b$, so b is fixed by $\mathrm{Aut}_K(E)$, and, therefore, $b \in K$, since E is Galois over K. The elements $\sigma^i(a) = \zeta^i a$, $0 \leq i \leq n-1$ are distinct roots of $x^n - b$ in E, and $\mathrm{Aut}_K(E)$ acts transitively on these roots, so $x^n - b = \prod_{i=0}^{n-1}(x - \zeta^i a)$ is irreducible in $K[x]$. (In fact, if f is an irreducible factor of $x^n - b$ in $K[x]$, and $\zeta^i a$ is one root of f, then for all j, we have that $\sigma^{j-i}(\zeta^i a) = \zeta^j a$ is also a root of f; hence, $\deg f \geq n$ and $f(x) = x^n - b$.) Since $\dim_K(K(a)) = n = \dim_K(E)$, we have $K(a) = E$. ■

Theorem 9.7.6. *Suppose* $K \subseteq E$ *is a Galois field extension. If the Galois group* $\mathrm{Aut}_K(E)$ *is solvable, then there is a radical extension* L *of* K *such that* $K \subseteq E \subseteq L$*.*

Proof. Let $n = \dim_K(E)$, and let ζ be a primitive n^{th} root of unity in a field extension of E. Consider the extensions:

$$\begin{array}{ccccc} & & K(\zeta) & \subseteq & E(\zeta) = K(\zeta)\cdot E \\ & & \cup\!| & & \cup\!| \\ K & \subseteq & K(\zeta)\cap E & \subseteq & E. \end{array}$$

By Proposition 8.5.6, $E(\zeta)$ is Galois over $K(\zeta)$ with Galois group

$$\mathrm{Aut}_{K(\zeta)}(E(\zeta)) \cong \mathrm{Aut}_{K(\zeta)\cap E}(E) \subseteq \mathrm{Aut}_K(E).$$

Therefore, $\mathrm{Aut}_{K(\zeta)}(E(\zeta))$ is solvable.

Let $G = G_0 = \mathrm{Aut}_{K(\zeta)}(E(\zeta))$, and let $G_0 \supseteq G_1 \supseteq \cdots \supseteq G_r = \{e\}$ be a composition series of G with cyclic quotients. Define $K_i = \mathrm{Fix}(G_i)$ for $0 \le i \le r$; thus

$$K(\zeta) = K_0 \subseteq K_1 \subseteq \cdots \subseteq K_r = E(\zeta).$$

By the fundamental theorem, each extension $K_{i-1} \subseteq K_i$ is Galois with Galois group $\mathrm{Aut}_{K_{i-1}}(K_i) \cong G_{i-1}/G_i$, which is cyclic. Since K_{i-1} contains a primitive d^{th} root of unity, where $d = \dim_{K_{i-1}}(K_i)$, it follows from Proposition 9.5.2 that $K_i = K_{i-1}(a_i)$, where a_i satisfies an irreducible polynomial $x^d - b_i \in K_{i-1}[x]$.

Therefore, $E(\zeta)$ is a radical extension of $K(\zeta)$. Since also $K(\zeta)$ is a radical extension of K, $E(\zeta)$ is a radical extension of K containing E, as required. ■

Corollary 9.7.7. *If E is the splitting field of a polynomial $p(x) \in K[x]$, and the Galois group $\mathrm{Aut}_K(E)$ is solvable, then $p(x)$ is solvable by radicals.*

Exercises 9.7

9.7.1. If $a \in E^*$, then the function $g(\sigma) = \sigma(a)a^{-1}$ is a multiplicative 1-cocycle.

9.7.2. With notation as in the proof of 9.7.5, check that if $N(b) = 1$, then f is well–defined on $\mathrm{Aut}_K(E)$ and f is a multiplicative 1-cocycle.

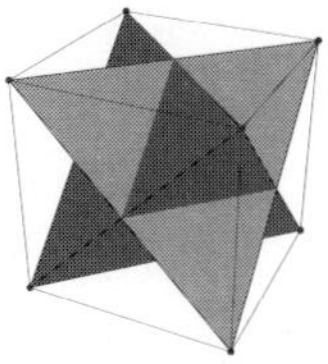

Chapter 10

Isometry Groups

10.1. More on Isometries of Euclidean Space

The goal of this section is to analyze the isometry group of Euclidean space. First, we will show that an isometry of Euclidean space that fixes the origin is actually a linear map. I'm going to choose a somewhat indirect but elegant way to do this, by using a uniqueness result: If an isometry fixes enough points, then it must fix all points, that is, it must be the identity map.

The distance function on $\mathbb{R}^n$ and its subsets used in this section is the standard Euclidean distance function, which is related to the Euclidean norm and the standard inner product by

$$d(\mathbf{a}, \mathbf{b}) = ||\mathbf{a} - \mathbf{b}||,$$

where

$$||\mathbf{a}|| = \langle \mathbf{a}, \mathbf{a} \rangle = \sum_i a_i^2.$$

Recall that an isometry $\tau : R \to \mathbb{R}^n$ defined on a subset $R \subseteq \mathbb{R}^\ell$ is a map that satisfies $d(\tau(\mathbf{a}), \tau(\mathbf{b})) = d(\mathbf{a}, \mathbf{b})$ for all $\mathbf{a}, \mathbf{b} \in R$. Here R is not presumed to be a linear subspace and τ is not presumed to be linear.

Lemma 10.1.1.

(a) *Suppose $R \subseteq \mathbb{R}^\ell$ is a subset containing $\mathbf{0}$, and $\tau : R \to \mathbb{R}^n$ is an isometry such that $\tau(\mathbf{0}) = \mathbf{0}$. Then τ preserves norms and inner products:*

$$||\tau(\mathbf{a})|| = ||\mathbf{a}|| \quad \textit{and} \quad \langle \tau(\mathbf{a}), \tau(\mathbf{b}) \rangle = \langle \mathbf{a}, \mathbf{b} \rangle,$$

for all $\mathbf{a}, \mathbf{b} \in R$.

(b) *In particular, if R is a vector subspace of $\mathbb{R}^\ell$ and $\tau : R \to \mathbb{R}^n$ is a linear isometry, then τ preserves norms and inner products.*

Proof. If $\mathbf{a} \in R$, then $\|\tau(\mathbf{a})\| = d(\mathbf{0}, \tau(\mathbf{a})) = d(\tau(\mathbf{0}), \tau(\mathbf{a})) = d(\mathbf{0}, \mathbf{a}) = \|\mathbf{a}\|$. Thus τ preserves norms.

If $\mathbf{a}, \mathbf{b} \in R$, then

$$d(\mathbf{a}, \mathbf{b})^2 = \|\mathbf{a} - \mathbf{b}\|^2 = \|\mathbf{a}\|^2 + \|\mathbf{b}\|^2 + 2\langle \mathbf{a}, \mathbf{b} \rangle,$$

and likewise

$$d(\tau(\mathbf{a}), \tau(\mathbf{b}))^2 = \|\tau(\mathbf{a}) - \tau(\mathbf{b})\|^2 = \|\tau(\mathbf{a})\|^2 + \|\tau(\mathbf{b})\|^2 + 2\langle \tau(\mathbf{a}), \tau(\mathbf{b}) \rangle.$$

Using that τ preserves both distance and norms, and comparing the last two expressions, we obtain that $\langle \tau(\mathbf{a}), \tau(\mathbf{b}) \rangle = \langle \mathbf{a}, \mathbf{b} \rangle$. This proves part (a), and part (b) follows immediately. ■

Definition 10.1.2. An *affine subspace* of $\mathbb{R}^n$ is a coset (translate) of a linear subspace (i.e., $\mathbf{x}_0 + F$, where F is a linear subspace). The *dimension* of an affine subspace $\mathbf{x}_0 + F$ is the dimension of F. The *affine span* of a subset R of $\mathbb{R}^n$ is the smallest affine subspace containing R.

A *(linear) hyperplane* in $\mathbb{R}^n$ is a linear subspace P of codimension 1; that is, P has dimension $n - 1$. An *affine hyperplane* in $\mathbb{R}^n$ is an affine subspace of dimension $n - 1$.

For example, in $\mathbb{R}^3$, a hyperplane is a two–dimensional plane through the origin. An affine hyperplane is a two–dimensional plane not necessarily passing through the origin.

In $\mathbb{R}^n$, every hyperplane P is the kernel of a *linear functional* $\varphi : \mathbb{R}^n \to \mathbb{R}$. In fact, $\mathbb{R}^n/P$ is one–dimensional, so is isomorphic to $\mathbb{R}$. Composing the canonical projection $\pi : \mathbb{R}^n \to \mathbb{R}^n/P$ with any isomorphism $\tilde{\varphi} : \mathbb{R}^n/P \to \mathbb{R}$ gives a linear functional $\varphi = \tilde{\varphi} \circ \pi : \mathbb{R}^n \to \mathbb{R}$ with kernel P. However, a linear functional $\varphi : \mathbb{R}^n \to \mathbb{R}$ has the form $\varphi(\mathbf{x}) = \langle \mathbf{x}, \boldsymbol{\alpha} \rangle$ for some $\boldsymbol{\alpha} \in \mathbb{R}^n$. Thus a linear hyperplane P is always of the form $\{\mathbf{x} \in \mathbb{R}^n : \langle \mathbf{x}, \boldsymbol{\alpha} \rangle = 0\}$ for some $\boldsymbol{\alpha} \in \mathbb{R}^n$. Given $\boldsymbol{\alpha} \in \mathbb{R}^n$, write P_{α} for the linear hyperplane $\{\mathbf{x} \in \mathbb{R}^n : \langle \mathbf{x}, \boldsymbol{\alpha} \rangle = 0\}$. The affine hyperplane $\mathbf{x}_0 + P_{\alpha}$ is then characterized as the set of points $\mathbf{x}$ such that $\langle \mathbf{x}, \boldsymbol{\alpha} \rangle = \langle \mathbf{x}_0, \boldsymbol{\alpha} \rangle$ (Exercise 10.1.2). We can show that any n points in $\mathbb{R}^n$ lie on some affine hyperplane (Exercise 10.1.3).

The following statement generalizes the well–known theorem of plane geometry that says that the set of points equidistant to two given points is the perpendicular bisector of the segment joining the two given points.

Proposition 10.1.3. *Let* $\mathbf{a}$ *and* $\mathbf{b}$ *be distinct points in* $\mathbb{R}^n$*. Then the set of points that are equidistant to* $\mathbf{a}$ *and* $\mathbf{b}$ *is the affine hyperplane* $\mathbf{x}_0 + P_{\boldsymbol{\alpha}}$*, where* $\mathbf{x}_0 = (\mathbf{a}+\mathbf{b})/2$ *and* $\boldsymbol{\alpha} = \mathbf{b} - \mathbf{a}$*.*

Proof. Exercise 10.1.4. ■

Corollary 10.1.4. *Let* $\mathbf{a}_i$ $(1 \le i \le n+1)$ *be* $n+1$ *points in* $\mathbb{R}^n$ *that do not lie on any affine hyperplane. If* $\mathbf{a}$ *and* $\mathbf{b}$ *satisfy* $d(\mathbf{a}, \mathbf{a}_i) = d(\mathbf{b}, \mathbf{a}_i)$ *for* $1 \le i \le n+1$*, then* $\mathbf{a} = \mathbf{b}$*.*

Proof. This is just the contrapositive of Proposition 10.1.3. ■

Corollary 10.1.5. *Let* R *be a subset of* $\mathbb{R}^n$*, and suppose* $\mathbf{a}_i$ $(1 \le i \le n+1)$ *are* $n+1$ *points in* R *that do not lie on any affine hyperplane. If an isometry* $\tau : R \to \mathbb{R}^n$ *satisfies* $\tau(\mathbf{a}_i) = \mathbf{a}_i$ *for* $i \le 1 \le n+1$*, then* $\tau(\mathbf{a}) = \mathbf{a}$ *for all* $\mathbf{a} \in R$*.*

Proof. For any point $\mathbf{a} \in \mathbb{R}^n$, $d(\tau(\mathbf{a}), \mathbf{a}_i) = d(\tau(\mathbf{a}), \tau(\mathbf{a}_i)) = d(\mathbf{a}, \mathbf{a}_i)$ for $1 \le i \le n+1$. By Corollary 10.1.4, $\tau(\mathbf{a}) = \mathbf{a}$. ■

Theorem 10.1.6.

(a) *Let* R *be any nonempty subset of* $\mathbb{R}^n$*, and let* $\tau : R \to \mathbb{R}^n$ *be an isometry. Then there exists an affine isometry* $T : \mathbb{R}^n \to \mathbb{R}^n$ *that extends* τ*.*

(b) *If the affine span of* R *is equal to* $\mathbb{R}^n$*, then the extension in part (a) is unique.*

(c) *If* $\mathbf{0} \in R$ *and* $\tau(0) = 0$*, then every extension in part (a) is linear.*

(d) *Any isometry of* $\mathbb{R}^n$ *is affine. Any isometry fixing the origin is linear.*

Proof. Consider first the special case that R contains $\mathbf{0}$ and $\tau(\mathbf{0}) = \mathbf{0}$. It follows from Lemma 10.1.1. (a) that τ preserves norms and inner products.

Let $V = \text{span}(R)$ and let $\{\mathbf{a}_1, \dots, \mathbf{a}_k\}$ be a basis of V contained in R. Let $T_1 : V \to \mathbb{R}^n$ be the unique linear map satisfying $T_1(\mathbf{a}_i) = \tau(\mathbf{a}_i)$ for $1 \le i \le k$. I claim that T_1 is isometric. In fact, for $\mathbf{x} =$

$\sum_i \alpha_i \mathbf{a}_i \in V$, we have

$$\begin{aligned}\|T_1(\mathbf{x})\|^2 &= \|\sum_i \alpha_i \tau(\mathbf{a}_i)\|^2 = \langle \sum_i \alpha_i \tau(\mathbf{a}_i), \sum_j \alpha_j \tau(\mathbf{a}_j)\rangle \\ &= \sum_{i,j} \alpha_i \alpha_j \langle \tau(\mathbf{a}_i), \tau(\mathbf{a}_j)\rangle = \sum_{i,j} \alpha_i \alpha_j \langle \mathbf{a}_i, \mathbf{a}_j\rangle \\ &= \langle \sum_i \alpha_i \mathbf{a}_i, \sum_j \alpha_j \mathbf{a}_j\rangle = \|\sum_i \alpha_i \mathbf{a}_i\|^2 = \|\mathbf{x}\|^2,\end{aligned}$$

which proves the claim.

We now have an isometric linear map $T_1 : V \to T_1(V) \subseteq \mathbb{R}^n$. Let $V^\perp$ and $T_1(V)^\perp$ denote the orthogonal complements of V and $T_1(V)$ in $\mathbb{R}^n$. These linear subspaces are both $n-k$–dimensional subspaces of $\mathbb{R}^n$, so there exists a linear isometry T_2 from $V^\perp$ to $T(V)^\perp$. Define $T(\mathbf{a}_1 + \mathbf{a}_2) = T_1(\mathbf{a}_1) + T_2(\mathbf{a}_2)$ for $\mathbf{a}_1 \in V$ and $\mathbf{a}_2 \in V^\perp$. It is easy to check that $T : \mathbb{R}^n \to \mathbb{R}^n$ is a linear isometry. By construction, $T(\mathbf{a}_i) = T_1(\mathbf{a}_i) = \tau(\mathbf{a}_i)$ for $1 \le i \le k$.

The composed isometry $T^{-1} \circ \tau : R \to V$ fixes every element of $\{\mathbf{0}, \mathbf{a}_1, \dots, \mathbf{a}_k\}$. Since this is a set of $k+1$ vectors in the k–dimensional Euclidean space V that does not lie on any $k-1$–dimensional affine subspace, it follows from Corollary 10.1.5 that $T^{-1} \circ \tau(\mathbf{a}) = \mathbf{a}$ for all $\mathbf{a} \in R$; that is $\tau(\mathbf{a}) = T(\mathbf{a})$, for all $\mathbf{a} \in R$.

This completes the proof of part (a) under our special auxiliary hypothesis.

For the general case of part (a), choose any element $\mathbf{a}_0 \in R$. Write $\mathbf{b}_0 = \tau(\mathbf{a}_0)$. Define $R' = R - \mathbf{a}_0 = \{\mathbf{a} - \mathbf{a}_0 : \mathbf{a} \in R\}$. Let $\tau'(\mathbf{x}) = \tau(\mathbf{x} + \mathbf{a}_0) - \mathbf{b}_0$ for $\mathbf{x} \in R'$. Then $\tau' : R' \to \mathbb{R}^n$ is an isometry, $\mathbf{0} \in R'$, and $\tau'(\mathbf{0}) = \mathbf{0}$. Therefore, by the special case already considered, there exists a linear isometry $T' : \mathbb{R}^n \to \mathbb{R}^n$ that extends τ'. Then $T : \mathbf{x} \mapsto T'(\mathbf{x} - \mathbf{a}_o) + \mathbf{b}_0$ is an isometric affine map on $\mathbb{R}^n$ that extends τ. This completes the proof of part (a).

Retaining the notation of the last paragraph, the affine span of R equals $\text{span}(R') + \mathbf{a}_0$. The affine span of R equals $\mathbb{R}^n$ if, and only if, the span of R' equals $\mathbb{R}^n$. In this case, there can be only one linear extension of τ'. But linear extensions of τ' correspond one-to-one with affine extensions of τ, so there can be only one affine extension of τ. This proves (b).

For part (c), suppose that $\mathbf{0} \in R$ and $\tau(\mathbf{0}) = \mathbf{0}$. Suppose L is a linear map, and $T : \mathbf{x} \mapsto L(\mathbf{x}) + \mathbf{b}_0$ is an affine extension of τ. Then $\mathbf{b}_0 = T(\mathbf{0}) = \tau(\mathbf{0}) = \mathbf{0}$, so $T = L$.

Finally, part (d) follows by taking $R = \mathbb{R}^n$. ■

The next result considers the problem of expressing an isometry as a product of reflections.

Theorem 10.1.7.

(a) *Any linear isometry of* $\mathbb{R}^n$ *is a product of at most* n *orthogonal reflections.*

(b) *Any element of* $O(n, \mathbb{R})$ *is a product of at most* n *reflection matrices.*

Proof. Let τ be a linear isometry of $\mathbb{R}^n$, let $\mathbb{E} = \{\hat{\mathbf{e}}_i\}$ denote the standard orthonormal basis of $\mathbb{R}^n$, and write $\mathbf{f}_i = \tau(\hat{\mathbf{e}}_i)$ for $i \le i \le n$. We show by induction that, for $1 \le p \le n$. there is a product ρ_p of at most p orthogonal reflections satisfying $\tau_p(\hat{\mathbf{e}}_i) = \mathbf{f}_i$ for $1 \le i \le p$. If $\hat{\mathbf{e}}_1 = \mathbf{f}_1$, put $\rho_1 = \mathrm{id}$; otherwise, $\mathbf{0}$ lies on the hyperplane consisting of points equidistant to $\hat{\mathbf{e}}_1$ and $\mathbf{f}_1$, and the reflection in this hyperplane maps $\hat{\mathbf{e}}_1$ to $\mathbf{f}_1$. In this case, let ρ_1 be the orthogonal reflection in this hyperplane.

Now, suppose $\rho = \rho_{p-1}$ is a product of at most $p-1$ orthogonal reflections that maps $\hat{\mathbf{e}}_i$ to $\mathbf{f}_i$ for $1 \le i \le p-1$. If also $\rho(\hat{\mathbf{e}}_p) = \mathbf{f}_p$, then put $\rho_p = \rho$. Otherwise, observe that $\{\mathbf{f}_1, \dots, \mathbf{f}_{p-1}, \mathbf{f}_p\}$, and

$$\{\mathbf{f}_1, \dots, \mathbf{f}_{p-1}, \rho(\hat{\mathbf{e}}_p)\} = \{\rho(\hat{\mathbf{e}}_1), \dots, \rho(\hat{\mathbf{e}}_{p-1}), \rho(\hat{\mathbf{e}}_p)\}$$

are both orthonormal sets, hence, $\{\mathbf{f}_1, \dots, \mathbf{f}_{p-1}\} \cup \{\mathbf{0}\}$ lies on the hyperplane of points equidistant to $\mathbf{f}_p$ and $\rho(\hat{\mathbf{e}}_p)$. If σ is the orthogonal reflection in this hyperplane, then $\rho_p = \sigma \circ \rho$ has the desired properties. This completes the induction. Now $\tau = \rho_n$ by application of Corollary 10.1.5 to $\rho_n^{-1} \circ \tau$. ∎

Corollary 10.1.8.

(a) *An element of* $O(n, \mathbb{R})$ *has determinant equal to* 1 *if, and only if, it is a product of an even number of reflection matrices.*

(b) *An element of* $O(n, \mathbb{R})$ *has determinant equal to* -1 *if, and only if, it is a product of an odd number of reflection matrices.*

This should remind you of even and odd permutations; recall that a permutation is even if, and only if, it is a product of an even number of 2-cycles, and odd if, and only if, it is a product of an odd number of 2-cycles. The similarity is no coincidence; see Exercise 10.1.7.

We now work out the structure of the group of all isometries of $\mathbb{R}^n$.

For $\mathbf{b} \in \mathbb{R}^n$, define the translation $\tau_{\mathbf{b}}$ by $\tau_{\mathbf{b}}(\mathbf{x}) = \mathbf{x} + \mathbf{b}$.

Lemma 10.1.9. *The set of translations of* $\mathbb{R}^n$ *is a normal subgroup of the group of isometries of* $\mathbb{R}^n$*, and is isomorphic to the additive group* $\mathbb{R}^n$*. For any isometry* σ*, and any* $\mathbf{b} \in \mathbb{R}^n$*, we have* $\sigma\tau_{\mathbf{b}}\sigma^{-1} = \tau_{\sigma(\mathbf{b})}$.

Proof. Exercise 10.1.8. ■

Let τ be an isometry of $\mathbb{R}^n$, and let $\mathbf{b} = \tau(\mathbf{0})$. Then $\sigma = \tau_{-\mathbf{b}}\tau$ is an isometry satisfying $\sigma(\mathbf{0}) = \mathbf{0}$, so σ is linear by Theorem 10.1.6. Thus, $\tau = \tau_{\mathbf{b}}\sigma$ is a product of a linear isometry and a translation.

Theorem 10.1.10. *The group* $\mathrm{Isom}(n)$ *of isometries of* $\mathbb{R}^n$ *is the semidirect product of the group of linear isometries and the translation group. Thus,* $\mathrm{Isom}(n) \cong O(n, \mathbb{R}) \ltimes \mathbb{R}^n$.

Proof. We have just observed that the product of the group of translations and the group of linear isometries is the entire group of isometries. The intersection of these two subgroups is trivial, and the group of translations is normal. Therefore, the isometry group is a semidirect product of the two subgroups. ■

The remainder of this section contains some results on the geometric classification of isometries of $\mathbb{R}^2$, which we will use in Section 10.4 for classification of two–dimensional crystal groups.

An *affine reflection* in the hyperplane through a point $\mathbf{x}_0$ and perpendicular to a unit vector $\boldsymbol{\alpha}$ is given by $\mathbf{x} \mapsto \mathbf{x} - 2\langle \mathbf{x} - \mathbf{x}_0, \boldsymbol{\alpha}\rangle\boldsymbol{\alpha}$.

A *glide–reflection* is the product $\tau_{\mathbf{a}}\sigma$, where σ is an affine reflection and $\mathbf{a}$ is parallel to the hyperplane of the reflection σ.

An *affine rotation* with center $\mathbf{x}_0$ is given by $\tau_{\mathbf{x}_0}R\tau_{-\mathbf{x}_0}$, where R is a rotation.

Proposition 10.1.11. *Every isometry of* $\mathbb{R}^2$ *is a translation, a glide–reflection, an affine reflection, or an affine rotation.*

Proof. Every isometry can be written uniquely as a product $\tau_{\mathbf{h}}B$, where B is a linear isometry. If $B = E$, then the isometry is a translation.

If $B = R_\theta$ is a rotation through an angle $0 < \theta < 2\pi$, then $E - R_\theta$ is invertible, so we can solve the equation $\mathbf{h} = \mathbf{x} - R_\theta\mathbf{x}$. Then $\tau_{\mathbf{h}}R_\theta = \tau_{\mathbf{x}}\tau_{-R_\theta(\mathbf{x})}R_\theta = \tau_{\mathbf{x}}R_\theta\tau_{-\mathbf{x}}$, an affine rotation.

If $B = j_{\boldsymbol{\alpha}}$ is a reflection (where $\boldsymbol{\alpha}$ is a unit vector), write $\mathbf{h} = s\boldsymbol{\alpha} + t\boldsymbol{\beta}$, where $\boldsymbol{\beta}$ is a unit vector perpendicular to $\boldsymbol{\alpha}$. If $t = 0$, then $\mathbf{x} \mapsto \tau_{\mathbf{h}}\, j_{\boldsymbol{\alpha}}(\mathbf{x}) = \tau_{s\boldsymbol{\alpha}}\, j_{\boldsymbol{\alpha}}(\mathbf{x}) = \mathbf{x} - 2\langle \mathbf{x} - (s/2)\boldsymbol{\alpha}, \boldsymbol{\alpha}\rangle\boldsymbol{\alpha}$ is an affine reflection. Otherwise $\tau_{\mathbf{h}}\, j_{\boldsymbol{\alpha}} = \tau_{t\boldsymbol{\beta}}(\tau_{s\boldsymbol{\alpha}}j_{\boldsymbol{\alpha}})$ is a glide–reflection. ■

Exercises 10.1

10.1.1. Suppose A, B, C, D are four noncoplanar points in $\mathbb{R}^3$. and $\alpha, \beta, \gamma, \delta$ are positive numbers. Show by elementary geometry that there is an most one point P such that $d(P, A) = \alpha$, $d(P, B) = \beta$, $d(P, C) = \gamma$, and $d(P, D) = \delta$. Now show that an isometry of $\mathbb{R}^3$ that fixes A, B, C, and D is the identity map.

10.1.2. Show that $\mathbf{x}_0 + P_{\boldsymbol{\alpha}} = \{\mathbf{x} \in \mathbb{R}^n : \langle \mathbf{x}, \boldsymbol{\alpha} \rangle = \langle \mathbf{x}_0, \boldsymbol{\alpha} \rangle\}$.

10.1.3. Show that any n points in $\mathbb{R}^n$ lie on some affine hyperplane.

10.1.4. Prove the Proposition 10.1.3 as follows: Show that $||\mathbf{x} - \mathbf{a}|| = ||\mathbf{x} - \mathbf{b}||$ is equivalent to $2\langle \mathbf{x}, \mathbf{b} - \mathbf{a} \rangle = ||b||^2 - ||a||^2 = 2\langle \mathbf{x}_0, \mathbf{b} - \mathbf{a} \rangle$, where $\mathbf{x}_0 = (\mathbf{a} + \mathbf{b})/2$.

10.1.5. Show that a set $\{\mathbf{a}_0, \mathbf{a}_1, \ldots, \mathbf{a}_n\}$ does not lie on any affine hyperplane if, and only if, the set $\{\mathbf{a}_1 - \mathbf{a}_0, \mathbf{a}_2 - \mathbf{a}_0, \ldots, \mathbf{a}_n - \mathbf{a}_0\}$ is linearly independent.

10.1.6. Prove Theorem 10.1.6 more directly as follows: Using that τ preserves inner products, show that $||\tau(\mathbf{a}+\mathbf{b}) - \tau(\mathbf{a}) - \tau(\mathbf{b})|| = 0$, hence, τ is additive, $\tau(\mathbf{a} + \mathbf{b}) = \tau(\mathbf{a}) + \tau(\mathbf{b})$. It remains to show homogeneity, $\tau(s\mathbf{a}) = s\tau(\mathbf{a})$. Show that this follows for rational s from the additivity of τ, and use a continuity argument for irrational s.

10.1.7.

(a) Show that the homomorphism $T : S_n \longrightarrow GL(n, \mathbb{R})$ described in Exercise 2.4.13 has range in $O(n, \mathbb{R})$.
(b) Show that for any 2-cycle (a, b), $T((a, b))$ is the orthogonal reflection in the hyperplane $\{\mathbf{x} : x_a = x_b\}$.
(c) $\pi \in S_n$ is even if, and only if, $\det(T(\pi)) = 1$.
(d) Show that any element of S_n is a product of at most n 2-cycles.

10.1.8. Prove Lemma 10.1.9.

10.1.9. $\text{Isom}(n)$ is the subgroup of the affine group $\text{Aff}(n)$ consisting of those affine transformations $T_{A,\mathbf{b}}$ such that A is orthogonal. Show that $\text{Isom}(n)$ is isomorphic to the group of $(n+1)$-by-$(n+1)$ matrices $\begin{bmatrix} A & b \\ 0 & 1 \end{bmatrix}$ such that $A \in O(n, \mathbb{R})$.

10.2. Euler's Theorem

In this section we discuss Euler's theorem on convex polyhedra and show that there are only five regular polyhedra.

Theorem 10.2.1. *Let* v, e, *and* f *be, respectively, the number of vertices, edges, and faces of a convex polyhedron. Then* $v - e + f = 2$.

Let us assume this result for the moment, and see how it leads quickly to a classification of regular polyhedra. A regular polyhedron is one whose faces are mutually congruent regular polygons, and at each of whose vertices the same number of edges meet. Let p denote the number of edges on each face of a regular polyhedron, and q the valence of each vertex (i.e. the number of edges meeting at the vertex). For the known regular polyhedra we have the following data:

polyhedron	v	e	f	p	q
tetrahedron	4	6	4	3	3
cube	8	12	6	4	3
octahedron	6	12	8	3	4
dodecahedron	20	30	12	5	3
icosahedron	12	30	20	3	5

Since each edge is common to two faces and to two vertices, we have:

$$e = fp/2 = vq/2.$$

Lemma 10.2.2.

(a) *Solving the equations*

$$e = fp/2 = vq/2$$

together with

$$2 = v - e + f$$

for v, e, f *in terms of* p *and* q *gives*

$$v = \frac{4p}{2p + 2q - qp}, \qquad e = \frac{2pq}{2p + 2q - qp}, \qquad f = \frac{4q}{2p + 2q - qp}.$$

(b) *It follows that that* $2p + 2q > qp$.

(c) *The only pairs* (p, q) *satisfying this inequality (as well as* $p \geq 3$, $q \geq 3$*) are* $(3, 3)$, $(3, 4)$, $(4, 3)$, $(3, 5)$, *and* $(5, 3)$.

Proof. Exercise 10.2.1. ■

Corollary 10.2.3. *There are only five regular convex polyhedra.*

Next, we will take a brief side trip to prove Euler's theorem, which we will interpret as a result of *graph theory.*

An *embedded graph* in $\mathbb{R}^3$ is a set of the form $V \cup E_1 \cup \cdots \cup E_n$, where V is a finite set of distinguished points (called *vertices*), and the E_i are smooth curves (called *edges*) such that

(a) Each curve E_i begins and ends at a vertex.
(b) The curves E_i have no self-intersections.
(c) Two curves E_i and E_j can intersect only at vertices, $E_i \cap E_j \subseteq V$.

The smoothness of the curves E_i and conditions (a) and (b) mean that for each i, there is a smooth injective function $\varphi_i : [0,1] \to \mathbb{R}^3$ whose image is E_i such that $\varphi_i(0), \varphi_i(1) \in V$.

Figure 10.2.1 shows a graph.

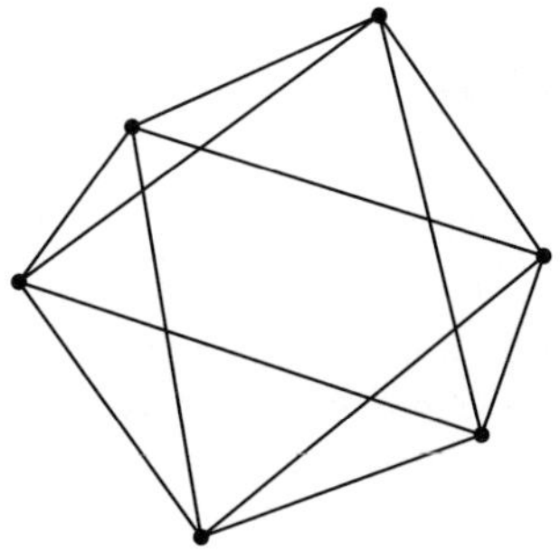

Figure 10.2.1. A graph.

Definition 10.2.4. The *valence* of a vertex on a graph is the number of edges containing that vertex as an endpoint.

I leave it to you to formulate precisely the notions of a *path* on a graph; a *cycle* is a path that begins and ends at the same vertex.

Definition 10.2.5. A graph that admits no cycles is called a *tree.*

Figure 10.2.2 shows a tree.

The following is a graph theoretic version of Euler's theorem:

Theorem 10.2.6. *Let $\mathcal{G}$ be a connected graph on the sphere $S = \{\mathbf{x} \in \mathbb{R}^3 : \|\mathbf{x}\| = 1\}$. Let e and v be the number of edges and vertices of $\mathcal{G}$, and let f be the number of connected components of $S \setminus \mathcal{G}$. Then $\varepsilon(\mathcal{G}) = v - e + f = 2$.*

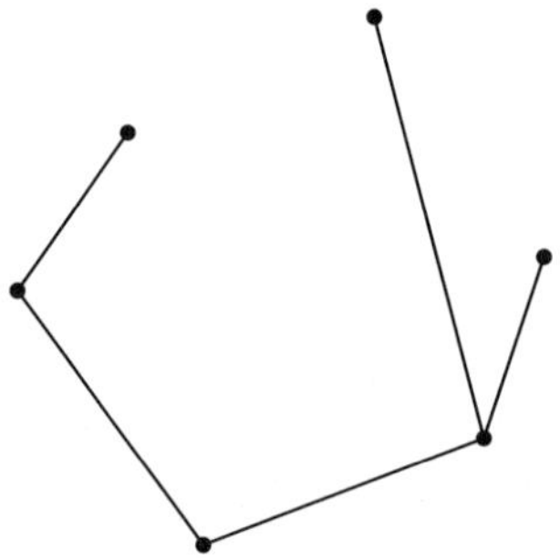

Figure 10.2.2. A tree.

Proof. If $\mathcal{G}$ has exactly one edge, then $e = 1$, $v = 2$ and $f = 1$, so the formula is valid. So suppose that $\mathcal{G}$ has at least two edges, and that the result holds for all connected graphs on the sphere with fewer edges. If $\mathcal{G}$ admits a cycle, then choose some edge belonging to a cycle, and let $\mathcal{G}'$ be the graph resulting from deleting the chosen edge (but not the endpoints of the edge) from $\mathcal{G}$. Then $\mathcal{G}'$ remains connected. Furthermore, $v(\mathcal{G}') = v$, $e(\mathcal{G}') = e - 1$, and $f(\mathcal{G}') = f - 1$. Hence, $\varepsilon(\mathcal{G}) = \varepsilon(\mathcal{G}') = 2$. If $\mathcal{G}$ is a tree, let $\mathcal{G}'$ be the graph resulting from deleting a vertex with valence 1 together with the edge containing that vertex (but not the other endpoint of the edge). The $v(\mathcal{G}') = v - 1$, $e(\mathcal{G}') = e - 1$, and $f(\mathcal{G}') = f = 1$. Hence, $\varepsilon(\mathcal{G}) = \varepsilon(\mathcal{G}') = 2$. This completes the proof. ■

Proof of Euler's theorem: We can suppose without loss of generality that $\mathbf{0}$ is contained in the interior of the convex polyhedron, which is in turn contained in the interior of the unit sphere S. The projection $\mathbf{x} \mapsto \mathbf{x}/\|\mathbf{x}\|$ onto the sphere maps the polyhedron bijectively onto the sphere. The image of the edges and vertices is a connected graph $\mathcal{G}$ on the sphere with e edges and v vertices, and each face of the polyhedron projects onto a connected component in $S \setminus \mathcal{G}$. Hence, Euler's theorem for polyhedra follows from Theorem 10.2.6. ■

The projection of a dodecahedron onto the sphere is shown in Figure 10.2.3.

Figure 10.2.3. Dodecahedron projected on the sphere.

Exercises 10.2

10.2.1. Prove Lemma 10.2.2.

10.2.2. Explain how the data $p \geq 3$, $q = 2$ can be sensibly interpreted, in the context of Lemma 10.2.2.

10.2.3. Determine by case-by-case inspection that the symmetry groups of the five regular polyhedra satisfy

$$|G| = vq = fp.$$

Find an explanation for these formulas. Hint: Consider actions of G on vertices and on faces.

10.2.4. Show that a tree always has a vertex with valence 1, and that in a connected tree, there is exactly one path between any two vertices.

10.3. Finite Rotation Groups

In this section we will classify the finite subgroups of $SO(3, \mathbb{R})$ and $O(3, \mathbb{R})$, largely by means of exercises.

It's easy to obtain the corresponding result for two dimensions: A finite subgroup of $SO(2, \mathbb{R})$ is cyclic. A finite subgroup of $O(2, \mathbb{R})$ is either cyclic or a dihedral group D_n for $n \geq 1$ (Exercise 10.3.1).

The classification for $SO(3, \mathbb{R})$ proceeds by analyzing the action of the finite group on the set of points left fixed by some element of the group. Let G be a finite subgroup of the rotation group $SO(3, \mathbb{R})$. Each nonidentity element $g \in G$ is a rotation about some axis; thus, g has two fixed points on the unit sphere S, called the *poles* of g. The group G acts on the set $\mathcal{P}$ of poles, since if $\mathbf{x}$ is a pole of g and $h \in G$, then $h\mathbf{x}$ is a pole of hgh^{-1}. You are asked to show in the Exercises that the stabilizer of a pole in G is a nontrivial cyclic group.

Let M denote the number of orbits of G acting on $\mathcal{P}$. Applying Burnside's Lemma 5.2.2 to this action gives

$$M = \frac{1}{|G|}(|\mathcal{P}| + 2(|G| - 1)),$$

because the identity of G fixes all elements of $\mathcal{P}$, while each non-identity element fixes exactly two elements. We can rewrite this equation as

$$(M - 2)\,|G| = |\mathcal{P}| - 2, \qquad (10.3.1)$$

which shows us that

$$M \geq 2.$$

We have the following data for several known finite subgroups of the rotation group:

group	$\lvert G\rvert$	*orbits*	*orbit sizes*	*stabilizer sizes*
$\mathbb{Z}_n$	n	2	$1, 1$	n, n
D_n	$2n$	3	$2, n, n$	$n, 2, 2$
tetrahedron	12	3	$6, 4, 4$	$2, 3, 3$
cube/octahedron	24	3	$12, 8, 6$	$2, 3, 4$
dodec/icosahedron	60	3	$30, 20, 12$	$2, 3, 5$

Theorem 10.3.1. *The finite subgroups of* $SO(3, \mathbb{R})$ *are the cyclic groups* $\mathbb{Z}_n$, *the dihedral groups* D_n, *and the rotation groups of the regular polyhedra.*

Proof. Let $\mathbf{x}_1, \dots, \mathbf{x}_M$ be representatives of the M orbits. We rewrite

$$|\mathcal{P}|/|G| = \sum_{i=1}^{M} \frac{|\mathcal{O}(\mathbf{x}_i)|}{|G|} = \sum_{i=1}^{M} \frac{1}{|\mathrm{Stab}(\mathbf{x}_i)|}.$$

Putting this into the orbit counting equation, writing M as $\sum_{i=1}^{M} 1$, and rearranging gives

$$\sum_{i=1}^{M} \left(1 - \frac{1}{|\mathrm{Stab}(\mathbf{x}_i)|}\right) = 2\left(1 - \frac{1}{|G|}\right). \qquad (10.3.2)$$

Now, the right-hand side is strictly less than 2, and each term on the left is at least 1/2, since the size of each stabilizer is at least 2, so we have

$$2 \leq M \leq 3.$$

If $M = 2$, we obtain from Equation (10.3.1) or (10.3.2) that there are exactly two poles, and thus two orbits of size 1. Hence, there is only one rotation axis for the elements of G, and G must be a finite cyclic group.

If $M = 3$, write $2 \leq a \leq b \leq c$ for the sizes of the stabilizers for the three orbits. Equation (10.3.2) rearranges to

$$\frac{1}{a} + \frac{1}{b} + \frac{1}{c} = 1 + \frac{2}{|G|}. \qquad (10.3.3)$$

Since the right side is greater than 1 and no more than 2, the only solutions are $(2, 2, n)$ for $n \geq 2$, $(2, 3, 3)$, $(2, 3, 4)$, and $(2, 3, 5)$.

The rest of the proof consists of showing that the only groups that can realize these data are the dihedral group D_n acting as rotations of the regular n–gon, and the rotation groups of the tetrahedron, the octahedron, and the icosahedron. The idea is to construct the geometric figures from the data on the orbits of poles.

It's a little tedious to read the details of the four cases but fun to work out the details for yourself. The Exercises provide a guide. ■

Exercises 10.3

10.3.1. Show that the finite subgroup of $SO(2, \mathbb{R})$ is cyclic and, consequently, a finite subgroup of $SO(3, \mathbb{R})$ that consists of rotations about a single axis is cyclic. Show that a finite subgroup of $O(2, \mathbb{R})$ is either cyclic or a dihedral group D_n for $n \geq 1$.

10.3.2. Show that the stabilizer of any pole is a cyclic group of order at least 2. Observe that the stabilizer of a pole $\mathbf{x}$ is the same as the stabilizer of the pole $-\mathbf{x}$, so the orbits of $\mathbf{x}$ and of $-\mathbf{x}$ have the same size. Consider the example of the rotation group G of the tetrahedron. What are the orbits of G acting on the set of poles of G? For which poles $\mathbf{x}$ is it true that $\mathbf{x}$ and $-\mathbf{x}$ belong to the same orbit?

10.3.3. Let G be a finite subgroup of $SO(3, \mathbb{R})$ with the data $M = 3$, (a, b, c)
$= (2, 2, n)$ with $n \geq 2$. Show that $|G| = 2n$, and the sizes of the three orbits of G acting on $\mathcal{P}$ are n, n, and 2. The case $n = 2$ is a bit

special, so consider first the case $n \geq 3$. There is one orbit of size 2, which must consist of a pair of poles $\{\mathbf{x}, -\mathbf{x}\}$; and the stabilizer of this pair is a cyclic group of rotations about the axis determined by $\{\mathbf{x}, -\mathbf{x}\}$. Show that there is an n-gon in the plane through the origin perpendicular to $\mathbf{x}$, whose n vertices consist of an orbit of poles of G, and show that G is the rotation group of this n-gon.

10.3.4. Extend the analysis of the last exercise to the case $n = 2$. Show that G must be $D_2 \cong \mathbb{Z}_2 \times \mathbb{Z}_2$, acting as symmetries of a rectangular card.

10.3.5. Let G be a finite subgroup of $SO(3, \mathbb{R})$ with the data $M = 3$, (a, b, c)
$= (2, 3, 3)$. Show that $|G| = 12$, and the size of the three orbits of G acting on $\mathcal{P}$ are $6, 4$, and 4. Consider an orbit $\mathcal{O} \subseteq \mathcal{P}$ of size 4, and let $\mathbf{u} \in \mathcal{O}$. Choose a vector $\mathbf{v} \in \mathcal{O} \setminus \{\mathbf{u}, -\mathbf{u}\}$. Let $g \in G$ generate the stabilizer of $\mathbf{u}$ [so $o(g) = 3$].

Conclude that $\{\mathbf{u}, \mathbf{v}, g\mathbf{v}, g^2\mathbf{v}\} = \mathcal{O}$. Deduce that $\{\mathbf{v}, g\mathbf{v}, g^2\mathbf{v}\}$ are the three vertices of an equilateral triangle, which lies in a plane perpendicular to $\mathbf{u}$ and that $\|\mathbf{u} - \mathbf{v}\| = \|\mathbf{u} - g\mathbf{v}\| = \|\mathbf{u} - g^2\mathbf{v}\|$.

Now, let $\mathbf{v}$ play the role of $\mathbf{u}$, and conclude that the four points of $\mathcal{O}$ are equidistant and are, therefore, the four vertices of a regular tetrahedron $\mathcal{T}$. Hence, G acts as symmetries of $\mathcal{T}$; since $|G| = 12$, conclude that G is the rotation group of $\mathcal{T}$.

10.3.6. Let G be a finite subgroup of $SO(3, \mathbb{R})$ with the data $M = 3$, (a, b, c)
$= (2, 3, 4)$. Show that $|G| = 24$, and the size of the three orbits of G acting on $\mathcal{P}$ are $12, 8$, and 6.

Consider the orbit $\mathcal{O} \subseteq \mathcal{P}$ of size 6. Since there is only one such orbit, $\mathcal{O}$ must contain together with any of its elements $\mathbf{x}$ the opposite vector $-\mathbf{x}$.

Let $\mathbf{u} \in \mathcal{O}$. Choose a vector $\mathbf{v} \in \mathcal{O} \setminus \{\mathbf{u}, -\mathbf{u}\}$. The stabilizer of $\{\mathbf{u}, -\mathbf{u}\}$ is cyclic of order 4; let g denote a generator of this cyclic group.

Show that $\{\mathbf{v}, g\mathbf{v}, g^2\mathbf{v}, g^3\mathbf{v}\}$ is the set of vertices of a square that lies in a plane perpendicular to $\mathbf{u}$. Show that $-\mathbf{v} = g^2\mathbf{v}$ and that the plane of the square bisects the segment $[\mathbf{u}, -\mathbf{u}]$.

Using rotations about the axis through $\mathbf{v}, -\mathbf{v}$, show that $\mathcal{O}$ consists of the 6 vertices of a regular octahedron. Show that G is the rotation group of this octahedron.

10.3.7. Let G be a finite subgroup of $SO(3, \mathbb{R})$ with the data $M = 3$, (a, b, c)
$= (2, 3, 5)$.Show that $|G| = 60$, and the size of the three orbits of G acting on $\mathcal{P}$ are $30, 20$, and 12.

Consider the orbit $\mathcal{O} \subseteq \mathcal{P}$ of size 12. Since there is only one such orbit, $\mathcal{O}$ must contain together with any of its elements $\mathbf{x}$ the opposite vector $-\mathbf{x}$.

Let $\mathbf{u} \in \mathcal{O}$ and let $\mathbf{v} \in \mathcal{O}$ satisfy $\|\mathbf{u}-\mathbf{v}\| \leq \|\mathbf{u}-\mathbf{y}\|$ for all $\mathbf{y} \in \mathcal{O} \setminus \{\mathbf{u}\}$.

Let g be a generator of the stabilizer of $\{\mathbf{u}, -\mathbf{u}\}$, $o(g) = 5$. Show that the 5 points $\{g^i\mathbf{v} : 0 \leq i \leq 4\}$ are the vertices of a regular pentagon that lies on a plane perpendicular to $\mathbf{u}$.

Show that the plane of the pentagon cannot bisect the segment $[\mathbf{u}, -\mathbf{u}]$, that $\mathbf{u}$ and the vertices of the pentagon lie all to one side of the bisector of $[\mathbf{u}, -\mathbf{u}]$, and finally that the 12 points $\{\pm\mathbf{u}, \pm g^i\mathbf{v} : 0 \leq i \leq 4\}$ comprise the orbit $\mathcal{O}$. Show that these 12 points are the vertices of a regular icosahedron and that G is the rotation group of this icosahedron.

It is not very much work to extend our classification results to finite subgroups of $O(3, \mathbb{R})$. If G is a finite subgroup of $O(3, \mathbb{R})$, then $H = G \cap SO(3, \mathbb{R})$ is a finite rotation group, so it is on the list of Theorem 10.3.1.

10.3.8. If G is a finite subgroup of $O(3, \mathbb{R})$ and the inversion $i : \mathbf{x} \mapsto -\mathbf{x}$ is an element of G, then $G = H \cup iH \cong H \times \mathbb{Z}_2$, where $H = G \cap SO(3, \mathbb{R})$. Conversely, for any H on the list of Theorem 10.3.1, $G = H \cup iH \cong H \times \mathbb{Z}_2$ is a subgroup of $O(3, \mathbb{R})$.

10.3.9.

(a) Suppose G is a finite subgroup of $O(3, \mathbb{R})$, that G is not contained in $SO(3, \mathbb{R})$, and that the inversion is not an element of G. Let $H = G \cap SO(3, \mathbb{R})$. Show that $\psi : a \mapsto \det(a)a$ is an isomorphism of G onto a subgroup $\tilde{G}$ of $SO(3, \mathbb{R})$, and $H \subseteq \tilde{G}$ is an index 2 subgroup.

(b) Conversely, if $H \subseteq \tilde{G}$ is an index 2 pair of subgroups of $SO(3, \mathbb{R})$, let $R \in \tilde{G} \setminus H$, and define $G = H \cup (-R)H \subseteq O(3, \mathbb{R})$. Show that G is a subgroup of $O(3, \mathbb{R})$, G is not contained in $SO(3, \mathbb{R})$, and the inversion is not an element of G. Furthermore, $\psi(G) = \tilde{G}$.

(c) Show that the complete list of index 2 pairs in $SO(3, \mathbb{R})$ is
   (i) $\mathbb{Z}_n \subseteq \mathbb{Z}_{2n}$, $n \geq 1$
   (ii) $\mathbb{Z}_n \subseteq D_n$, $n \geq 2$
   (iii) $D_n \subseteq D_{2n}$, $n \geq 2$
   (iv) The rotation group of the tetrahedron contained in the rotation group of the cube.

This exercise completes the classification of finite subgroups of $O(3, \mathbb{R})$. Note that this is more than a classification up to group isomorphism; the groups are classified by their mode of action. For example, the abstract group Z_{2n} acts as a rotation group but also as a group of rotations and reflection–rotations, due to the pair $\mathbb{Z}_n \subseteq \mathbb{Z}_{2n}$

on the list of the previous exercise. More precisely, the classification is *up to conjugacy*: Recall that two subgroups A and B of $O(3, \mathbb{R})$ are conjugate if there is a $g \in O(3, \mathbb{R})$ such that $A = gBg^{-1}$; conjugacy is an equivalence relation on subgroups. Our results classify the conjugacy classes of subgroups of $O(3, \mathbb{R})$.

10.4. Crystals

In this section, we shall investigate crystals in two and three dimensions, with the goal of analyzing their symmetry groups. We will encounter several new phenomena in group theory in the course of this discussion: Let's first say what we mean by a crystal.

Definition 10.4.1. A *lattice* L in a real vector space V (for us, $V = \mathbb{R}^2$ or $V = \mathbb{R}^3$) is the set of integer linear combinations of some basis of V.

For example, take the basis $\mathbf{a} = \begin{bmatrix} 1 \\ 0 \end{bmatrix}, \mathbf{b} = \begin{bmatrix} 1/2 \\ \sqrt{3}/2 \end{bmatrix}$ in $\mathbb{R}^2$. Figure 10.4.1 shows (part of) the lattice generated by $\{\mathbf{a}, \mathbf{b}\}$.

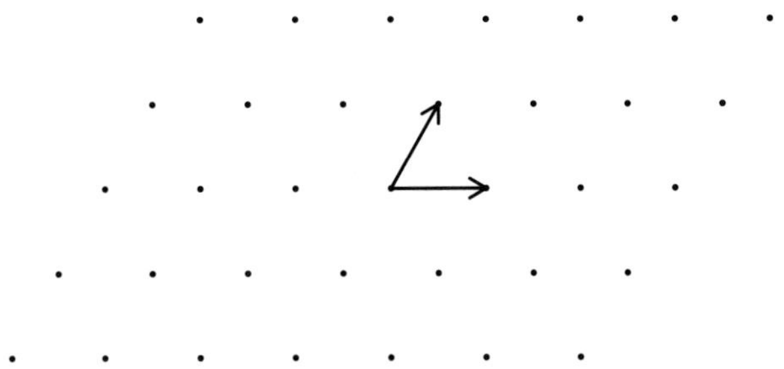

Figure 10.4.1. Hexagonal lattice.

Definition 10.4.2. A *fundamental domain* for a lattice L in V is a closed region D in V such that

(a) $\bigcup_{\mathbf{x} \in L} \mathbf{x} + D = V$

(b) For $\mathbf{x} \neq \mathbf{y}$ in L, $(\mathbf{x} + D) \cap (\mathbf{y} + D)$ is contained in the boundary of $(\mathbf{x} + D)$.

That is, the translates of D by elements of L cover V, and two different translates of D can intersect only in their boundary. (I am not interested in using complicated sets for D; convex polygons in $\mathbb{R}^2$ and convex polyhedra in $\mathbb{R}^3$ will be general enough.)

For the lattice displayed in Figure 10.4.1, two different fundamental domains are

1. The parallelepiped spanned by $\{\mathbf{a}, \mathbf{b}\}$, namely,

$$\{s\mathbf{a} + t\mathbf{b} : 0 \leq s, t \leq 1\}$$

2. A hexagon centered at the origin, two of whose vertices are at $(0, \pm 1/\sqrt{3})x$

Definition 10.4.3. A *crystal* consists of some geometric figure in a fundamental domain of a lattice together with all translates of this figure by elements of the lattice.

For example, take the hexagonal fundamental domain for the lattice L described previously, and the geometric figure in the fundamental domain displayed in Figure 10.4.2. The crystal generated by translations of this pattern is shown in Figure 10.4.3.

Figure 10.4.2. Pattern in fundamental domain.

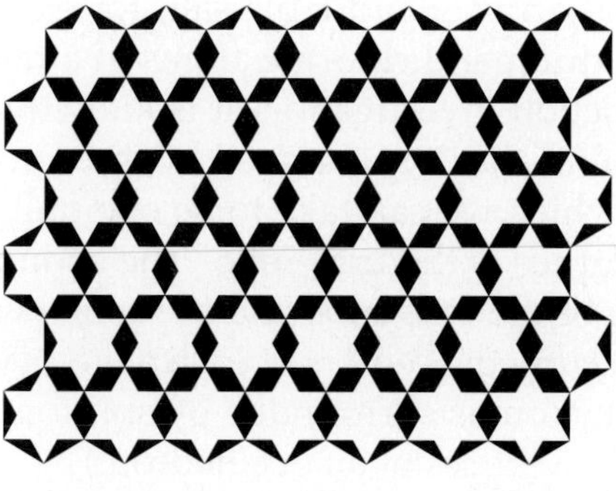

Figure 10.4.3. Hexagonal crystal.

Crystals in two dimensions (in the form of fabrics, wallpapers, carpets, quilts, tilework, basket weaves, and lattice work) are a mainstay of decorative art. See Figure 10.4.4. You can find many examples by browsing in your library under "design" and related topics. For some remarkable examples created by "chaotic" dynamical systems, see M. Field and M. Golubitsky, *Symmetry in Chaos*, Oxford University Press, 1992.

Figure 10.4.4. "Crystal" in decorative art.

What we have defined as a crystal in three dimensions is an idealization of physical crystals. Many solid substances are crystalline, consisting of arrangements of atoms that are repeated in a three–dimensional array. Solid table salt, for example, consists of huge arrays of sodium and chlorine atoms in a ratio of one to one. From x-ray diffraction investigation, it is known that the lattice of a salt crystal is cubic, that is, generated by three orthogonal vectors of equal length, which we can take to be one unit. A fundamental domain for this lattice is the unit cube. The sodium atoms occupy the eight vertices of the unit cube and the centers of the faces. The chlorine atoms occupy the centers of each edge as well as the center of the cube. (Each atom is surrounded by six atoms of the opposite sort, arrayed at the vertices of an octahedron.)

We turn to the main concern of this section: What is the group of isometries of a crystal? Of course, the symmetry group of a crystal built on a lattice L will include at least L, acting as translations. I will make the standing assumption that enough stuff was put into the pattern in the fundamental domain so that the crystal has no translational symmetries *other* than those in L.

Assumption: L *is the group of translations of the crystal.*

Now I want to define a "linear part" of the group of symmetries of a crystal, the so-called point group. One could be tempted by such examples as that in Figure 10.4.3 to define the point group of a crystal to be intersection of the symmetry group of the crystal with the orthogonal group. This is pretty close to being the right concept, but it is not quite right in general, as shown by the following example: Let L be the square lattice generated by $\{\hat{\mathbf{e}}_1, \hat{\mathbf{e}}_2\}$. Take as a fundamental domain the square with sides of length 1, centered at the origin. This square is divided into quarters by the coordinate axes. In the southeast quarter, put a small copy of the front page of today's *New York Times*. In the northwest corner, put a mirror image copy of the front page. The crystal generated by this data has no rotation or reflection symmetries. (See Figure 10.4.5.) Nevertheless, its symmetry group is larger than L, namely, $\tau_{\hat{\mathbf{e}}_2/2}\sigma$ is a symmetry, where σ is the reflection in the y–axis.

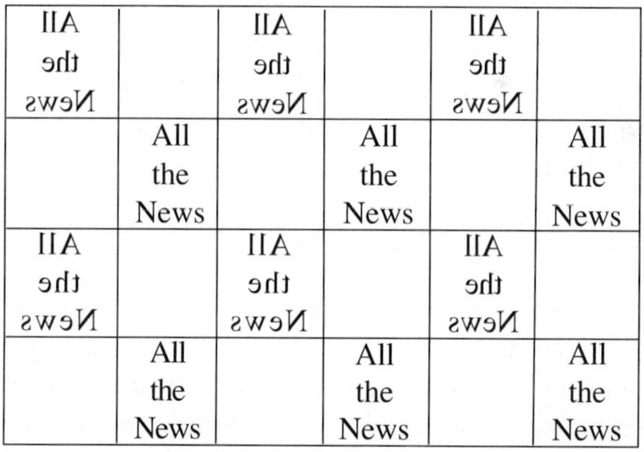

Figure 10.4.5. *New York Times* crystal.

This example points us to the proper definition of the point group.

The symmetry group G of a crystal is a subgroup of Isom(V) (where $V = \mathbb{R}^2$ or $\mathbb{R}^3$). We know from Theorem 10.1.10 that Isom(V) is the semidirect product of the group V of translations and the orthogonal group O(V). The lattice L is the intersection of G with the group of translations and is a normal subgroup of G. The quotient group G/L is isomorphic to the image of G in $\mathrm{Isom}(V)/V \cong O(V)$, by Proposition 2.7.18.

Definition 10.4.4. The point group G^0 of a crystal is G/L.

Recall that the quotient map of $\mathrm{Isom}(V)$ onto $O(V)$ is given as follows: Each isometry of V can be written uniquely as a product $\tau\sigma$, where τ is a translation and σ is a linear isometry. The image of $\tau\sigma$ in $O(V)$ is σ. For the *New York Times* crystal, the symmetry group is generated by $\tau_{\hat{\mathbf{e}}_1}$, $\tau_{\hat{\mathbf{e}}_2}$, and $\tau_{\hat{\mathbf{e}}_2/2}\sigma$, where σ is the reflection in the y–axis. Therefore, the point group is the two element group generated by σ in $O(2, \mathbb{R})$.

Of course, there are examples where $G^0 \cong G \cap O(V)$; this happens when G is the semidirect product of L and $G \cap O(V)$ (Exercise 10.4.2).

Now, consider $\mathbf{x} \in L$ and $A \in G^0$. By definition of G^0, there is an $\mathbf{h} \in V$ such that $\tau_{\mathbf{h}}A \in G$. Then $(\tau_{\mathbf{h}}A)\tau_{\mathbf{x}}(\tau_{\mathbf{h}}A)^{-1} = \tau_{\mathbf{h}}\tau_{A\mathbf{x}}\tau_{-\mathbf{h}} = \tau_{A\mathbf{x}} \in G$. It follows from our assumption that $A\mathbf{x} \in L$. Thus, we have proved the following:

Lemma 10.4.5. L *is invariant under* G^0.

Let $\mathbb{F}$ be a basis of L, that is, a basis of the ambient vector space V $(= \mathbb{R}^2$ or $\mathbb{R}^3)$ such that L consists of *integer* linear combinations of $\mathbb{F}$. Since for $A \in G^0$ and $\mathbf{a} \in \mathbb{F}$, $A\mathbf{a} \in L$, it follows that the matrix of A with respect to the basis $\mathbb{F}$ is integer valued. This observation immediately yields a strong restriction on G^0:

Proposition 10.4.6. *Any rotation in* G^0 *is of order* 2, 3, 4, *or* 6.

Proof. On the one hand, a rotation $A \in G^0$ has an integer valued matrix with respect to $\mathbb{F}$ so, in particular, the trace of A is an integer. On the other hand, with respect to a suitable orthonormal basis of V, A has the matrix

$$\begin{bmatrix} \cos(\theta) & -\sin(\theta) & 0 \\ \sin(\theta) & \cos(\theta) & 0 \\ 0 & 0 & 1 \end{bmatrix}$$

in the three–dimensional case or

$$\begin{bmatrix} \cos(\theta) & -\sin(\theta) \\ \sin(\theta) & \cos(\theta) \end{bmatrix}$$

in the two–dimensional case. So it follows that $2\cos(\theta)$ is an integer, and, therefore, $\theta = \pm 2\pi/k$ for $k \in \{2, 3, 4, 6\}$. ■

Corollary 10.4.7. *The point group of a two–dimensional crystal is one of the following ten groups (up to conjugacy in* $O(2,\mathbb{R})$*):* $\mathbb{Z}_n$ *or* D_n *for* $n = 1, 2, 3, 4, 6$*.*

Proof. This follows from Exercise 10.3.1 and Proposition 10.4.6. ■

We will call the classes of point groups, up to conjugacy in $O(2,\mathbb{R})$, the *geometric point groups.*

Corollary 10.4.8.
(a) *The elements of infinite order in a two–dimensional crystal group are either translations or glide–reflections.*
(b) *An element of infinite order is a translation if, and only if, it commutes with the square of each element of infinite order.*

Proof. The first part follows from the classification of isometries of $\mathbb{R}^2$, together with the fact that rotations in a two–dimensional crystal group are of finite order. Since the square of a glide–reflection or of a translation is a translation, translations commute with the square of each element of infinite order. On the other hand, if τ is a glide–reflection, and σ is a translation in a direction not parallel to the line of reflection of τ, then τ does not commute with σ^2. ■

Lemma 10.4.9. *Let* L *be a two–dimensional lattice. Let* $\mathbf{a}$ *be a vector of minimal length in* L*, and let* $\mathbf{b}$ *be a vector of minimal length in* $L \setminus \mathbb{R}\mathbf{a}$*. Then* $\{\mathbf{a}, \mathbf{b}\}$ *is a basis for* L*; that is, the integer linear span of* $\{\mathbf{a}, \mathbf{b}\}$ *is* L*.*

Proof. Exercise 10.4.4. ■

Lemma 10.4.10. *Suppose the point group of a two–dimensional crystal contains a rotation* R *of order 3,4, or 6. Then the lattice* L *must be*
(a) *the hexagonal lattice spanned by two vectors of equal length at an angle of* $\pi/3$*, if* R *has order 3 or 6* or
(b) *the square lattice spanned by two orthogonal vectors of equal length, if* R *has order 4.*

Proof. Exercise 10.4.5. ■

Lemma 10.4.11. *Let G_i be symmetry groups of two–dimensional crystals, with lattices L_i and point groups G_i^0, for $i = 1, 2$. Suppose $\varphi : G_1 \to G_2$ is a group isomorphism. Then*

(a) $\varphi(L_1) = L_2$.

(b) *There is a $\Phi \in GL(\mathbb{R}^2)$ such that $\varphi(\tau_{\mathbf{x}}) = \tau_{\Phi(\mathbf{x})}$ for $\mathbf{x} \in L_1$. The matrix of Φ with respect to bases of L_1 and L_2 is integer valued, and the inverse of this matrix is integer valued.*

(c) *φ induces an isomorphism $\tilde{\varphi} : G_1^0 \to G_2^0$. For $B \in G_1^0$, we have $\tilde{\varphi}(B) = \Phi B \Phi^{-1}$.*

(d) *φ maps affine rotations to affine rotations, affine reflections to affine reflections, translations to translations, and glide–reflections to glide–reflections.*

Proof. If τ is a translation in G_1, then τ commutes with the square of every element of infinite order in G_1. It follows that $\varphi(\tau)$ is an element of infinite order in G_2 with the same property, so $\varphi(\tau)$ is a translation, by Corollary 10.4.8. This proves part (a).

The isomorphism $\varphi : L_1 \to L_2$ extends to a linear isomorphism $\Phi : \mathbb{R}^2 \to \mathbb{R}^2$, whose matrix A with respect to bases of L_1 and L_2 is integer valued; applying the same reasoning to φ^{-1} shows that A^{-1} is also integer valued. This proves part (b).

If π_i denotes the quotient map from G_i to $G_i^0 = G_i/L_i$, for $i = 1, 2$, then $\pi_2 \circ \varphi : G_1 \to G_2^0$ is a surjective homomorphism with kernel L_1; therefore, there is an induced isomorphism $\tilde{\varphi} : G_1^0 = G_1/L_1 \to G_2^0$ such that $\tilde{\varphi} \circ \pi_1 = \pi_2 \circ \varphi$, by the Homomorphism Theorem 2.7.6.

On the other hand, G_i^0 can be identified with a finite subgroup of $O(\mathbb{R}^2)$: G_i^0 is the set of $B \in O(\mathbb{R}^2)$ such that there exists a $\mathbf{h} \in \mathbb{R}^2$ such that $\tau_{\mathbf{h}} B \in G_i$. So let $B \in G_1^0$ and let $\mathbf{h} \in \mathbb{R}^2$ satisfy $\tau_{\mathbf{h}} B \in G_1$. Then $\varphi(\tau_{\mathbf{h}} B) = \tau_{\mathbf{k}} \tilde{\varphi}(B)$ for some $\mathbf{k} \in \mathbb{R}^2$. Compute that $\tau_{B\mathbf{x}} = (\tau_{\mathbf{h}} B)\tau_{\mathbf{x}}(\tau_{\mathbf{h}} B)^{-1}$ for $\mathbf{x} \in L_1$. Applying φ to both sides gives $\tau_{\Phi(B\mathbf{x})} = \tau_{\mathbf{k}} \tilde{\varphi}(B) \tau_{\Phi(\mathbf{x})} \tilde{\varphi}(B)^{-1} \tau_{-\mathbf{k}} = \tau_{\tilde{\varphi}(B)\Phi(\mathbf{x})}$. Therefore, $\tilde{\varphi}(B) = \Phi B \Phi^{-1}$, which completes the proof of (c).

The isomorphism φ maps translations to translations, by part (a). The glide–reflections are the elements of infinite order that are not translations, so φ also maps glide–reflections to glide–reflections. The affine rotations in G_1 are elements of the form $g = \tau_{\mathbf{h}} B$, where B is a rotation in G_1^0; for such an element, $\varphi(g) = \tau_{\mathbf{k}} \Phi B \Phi^{-1}$ is also a rotation. Use a similar argument for affine reflections, or use the fact that affine reflections are the elements of finite order that are not affine rotations. ∎

Remark 10.4.12. We can show that any finite subgroup of $GL(\mathbb{R}^n)$ is conjugate in $GL(R^n)$ to a subgroup of $O(\mathbb{R}^n)$, and subgroups of $O(\mathbb{R}^n)$ that are conjugate in $GL(\mathbb{R}^n)$ are also conjugate in $O(\mathbb{R}^n)$. In particular, isomorphic two–dimensional crystal groups have point groups belonging to the same geometric class. These statements will be verified in the Exercises.

Theorem 10.4.13. *There are exactly* 17 *isomorphism classes of two–dimensional crystal groups. They are distributed among the geometric point group classes, as in Table* 10.4.1.

geometric class	*number of isomorphism classes*
$\mathbb{Z}_1$	1
D_1	3
$\mathbb{Z}_2$	1
D_2	4
$\mathbb{Z}_3$	1
D_3	2
$\mathbb{Z}_4$	1
D_4	2
$\mathbb{Z}_6$	1
D_6	1

Table 10.4.1. Distribution of classes of crystal groups.

Proof. The strategy of the proof is to go through the possible geometric point group classes and produce for each a list of possible crystal groups; these are then shown to be mutually nonisomorphic by using Lemma 10.4.11. In the proof, G will always denote the crystal group, G^0 the point group, and L the lattice. The main difficulties are already met in the case $G^0 = D_1$.

Point group $\mathbb{Z}_1$. The lattice L is the entire symmetry groups. Any two lattices in $\mathbb{R}^2$ are isomorphic as groups. A crystal with trivial point group is displayed in Figure 10.4.6.

Point group D_1. The point group is generated by a single reflection σ in a line A through the origin. Let B be an orthogonal line through the origin. If $\mathbf{v}$ is any vector in L that is not in $A \cup B$, then $\mathbf{v} + \sigma(\mathbf{v})$ is in $L \cap A$ and $\mathbf{v} - \sigma(\mathbf{v})$ is in $L \cap B$. Let L_0 be the lattice generated by $L \cap (A \cup B)$.

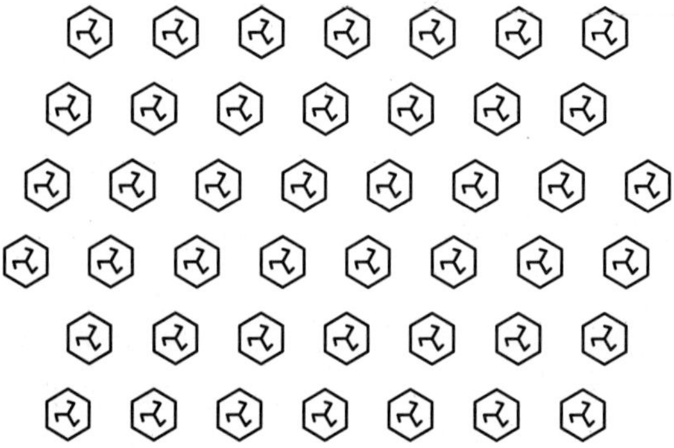

Figure 10.4.6. Crystal with trivial point group.

Case: $L_0 = L$. In this case the lattice L is generated by orthogonal vectors $\mathbf{a} \in A$ and $\mathbf{b} \in B$.

If $\sigma \in G$, then G is the semidirect product of L and D_1. This isomorphism class is denoted D_{1m}.

If $\sigma \notin G$, then G contains a glide–reflection $\tau_h\sigma$ in a line parallel to A; without loss of generality, we can suppose that the origin is on this line and that $g = \tau_{s\mathbf{a}}\sigma$, with $0 < |s| \le 1/2$. Since $g^2 = \tau_{2s\mathbf{a}} \in G$, we have $|s| = 1/2$. Then G is generated by L and the glide–reflection $\tau_{\mathbf{a}/2}\sigma$. This isomorphism class is denoted by D_{1g}.

Case: $L_0 \neq L$. Let $\mathbf{u}$ be a vector of shortest length in $L \setminus L_0$; we can assume without loss of generality that $\mathbf{u}$ makes an acute angle with both the generators $\mathbf{a}$ and $\mathbf{b}$ of L_0. Put $\mathbf{v} = \sigma(\mathbf{u})$. We can now show that $\mathbf{u}$ and $\mathbf{v}$ generate L, $\mathbf{a} = \mathbf{u}+\mathbf{v}$, $\mathbf{b} = \mathbf{u}-\mathbf{v}$ (Exercise 10.4.8).

As in the previous case, if G contains a glide–reflection, then we can assume without loss of generality that it contains the glide–reflection $\tau_{\mathbf{a}/2}\sigma = \tau_{(1/2)(\mathbf{u}+\mathbf{v})}\sigma$. But then G also contains the reflection

$$\tau_{-\mathbf{v}}\tau_{(1/2)(\mathbf{u}+\mathbf{v})}\sigma = \tau_{(1/2)(\mathbf{u}-\mathbf{v})}\sigma.$$

Thus, G is a semidirect product of L and D_1. This isomorphism class is labeled D_{1c}.

The classes D_{1m}, D_{1g}, and D_{1c} are mutually nonisomorphic: D_{1m} and D_{1c} are both semidirect products of L and D_1, while D_{1g} is not. In D_{1m}, the reflection σ is represented by the matrix $\begin{bmatrix} 1 & 0 \\ 0 & -1 \end{bmatrix}$ with respect to a basis of L, and in D_{1g}, σ is represented by $\begin{bmatrix} 0 & 1 \\ 1 & 0 \end{bmatrix}$. These matrices are not conjugate in $GL(2, \mathbb{Z})$, so by Lemma 10.4.11, the crystal groups are not isomorphic.

The *New York Times* crystal is of type D_{1g}. Figure 10.4.7 displays crystals of types D_{1m} and D_{1c}.

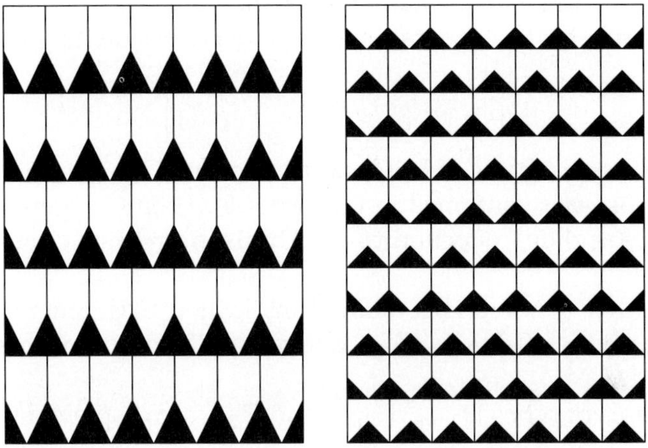

Figure 10.4.7. Crystals with point group D_1.

Point group $\mathbb{Z}_2$. The point group is generated by the half-turn $-E$, so G contains an element $\tau_h(-E)$, which is also a half-turn about some point. The group G is, thus, a semidirect product of L and $\mathbb{Z}_2$; there is no restriction on the lattice. A crystal with symmetry type Z_2 is displayed in Figure 10.4.8.

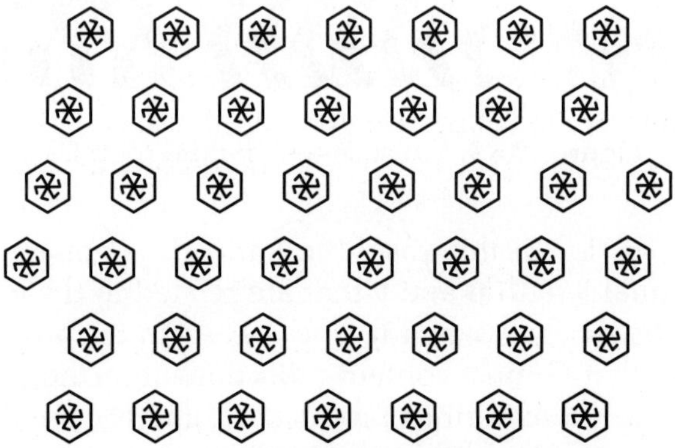

Figure 10.4.8. Crystal of symmetry type Z_2.

Point group D_2. Here G^0 is generated by reflections in two orthogonal lines A and B. As for the point group D_1, we have to consider two cases. Let L_0 be the lattice generated by $L \cap (A \cup B)$.

Case: $L_0 = L$. The lattice L is generated by orthogonal vectors $\mathbf{a} \in A$ and $\mathbf{b} \in B$. Here there are three further possibilities: Let α and β denote the reflections in the lines A and B.

If both α and β are contained in G, then G is the semidirect product of L and D_2. The generators of D_2 are represented by matrices

$$\begin{bmatrix} 1 & 0 \\ 0 & -1 \end{bmatrix} \quad \text{and} \quad \begin{bmatrix} -1 & 0 \\ 0 & 1 \end{bmatrix}$$

with respect to a basis of L. This class is called D_{2mm}.

Suppose α is contained in G but not β. Then G is generated by L, α, and the glide–reflection $\tau_{\mathbf{b}/2}\beta$. This isomorphism class is called D_{2mg}.

If neither α nor β is contained in G, then G is generated by L and two glide–reflections $\tau_{\mathbf{a}/2}\alpha$ and $\tau_{\mathbf{b}/2}\beta$. This isomorphism class is called D_{2gg}.

Crystals of types D_{2mm}, D_{2mg}, and D_{2gg} are displayed in Figure 10.4.9.

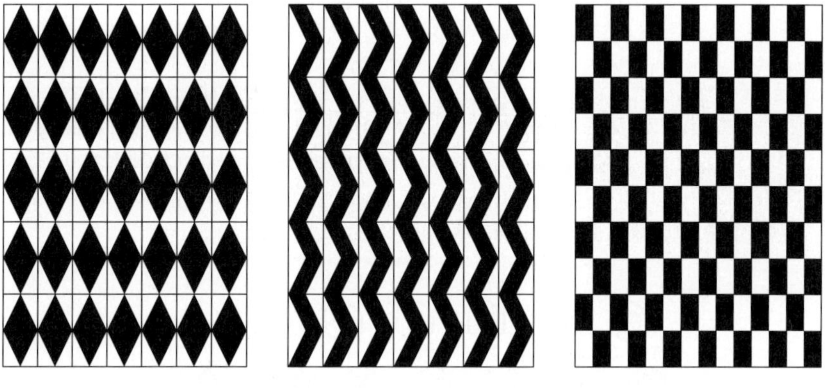

Figure 10.4.9. Crystals with point group D_2.

Case: $L_0 \neq L$. In this case, the lattice L is generated by two vectors of equal length $\mathbf{u}$ and $\mathbf{v}$ that are related to the generators $\mathbf{a}$ and $\mathbf{b}$ of L_0 by $\mathbf{a} = \mathbf{u} + \mathbf{v}$ and $\mathbf{b} = \mathbf{u} - \mathbf{v}$. As in the analysis of case D_{1c}, we find that G must contain reflections in orthogonal lines A and B. Thus, G is a semidirect product of L and D_2, but the matrices of the generators of D_2 with respect to a basis of L are

$$\begin{bmatrix} 0 & 1 \\ 1 & 0 \end{bmatrix} \quad \text{and} \begin{bmatrix} 0 & 1 \\ -1 & 0 \end{bmatrix}.$$

This isomorphism class is called D_{2c}.

The four classes D_{2mm}, D_{2mg}, D_{2gg}, and D_{2c} are mutually non-isomorphic: The groups D_{2mm} and D_{2c} are semidirect products of

L and D_2, but the other two are not. The groups D_{2mm} and D_{2c} are nonisomorphic because they have matrix representations that are not conjugate in $GL(2, \mathbb{Z})$. D_{2mg}, D_{2gg} are nonisomorphic because the former contains an element of order 2 while the latter does not.

A crystal of type D_{2c} is shown in Figure 10.4.10.

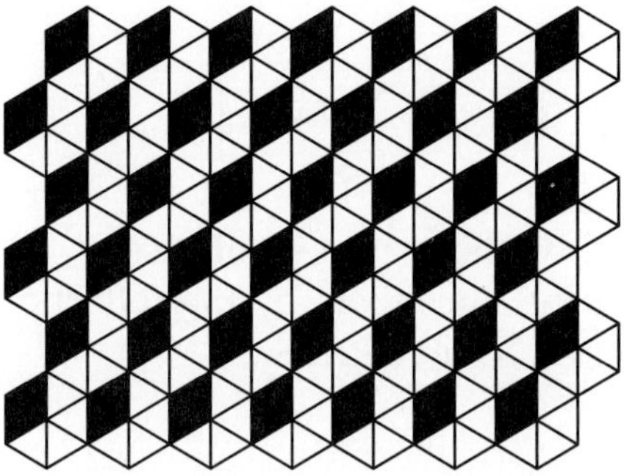

Figure 10.4.10. Crystal of type D_{2c}.

Point group $\mathbb{Z}_3$: By Exercise 10.4.5, the lattice is necessarily the hexagonal lattice generated by two vectors of equal length at an angle of $\pi/3$. G must contain an affine rotation of order 3 about some point, so G is a semidirect product of L and $\mathbb{Z}_3$. A crystal with symmetry type Z_3 is displayed in Figure 10.4.11.

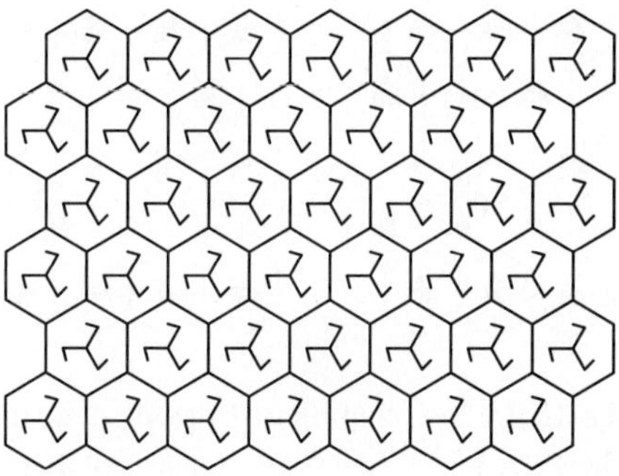

Figure 10.4.11. Crystal of symmetry type Z_3.

Point group D_3: As in the previous case, the lattice is necessarily the hexagonal lattice generated by two vectors of equal length at an angle of $\pi/3$.

Now, G^0 contains reflections in two lines forming an angle of $\pi/3$. We can argue as in case D_{1c} that G actually contains reflections in lines parallel to these lines. These two reflections generate a copy of D_3 in G, and hence G must be a semidirect product of L and D_3.

There are still two possibilities for the action of D_3 on L: Let A and B be the lines fixed by two reflections in $D_3 \subseteq G$. As in the case D_1, the lattice L must contain points on the lines A and B. Let L_0 be the lattice generated by $L \cap (A \cup B)$.

If $L_0 = L$, then D_3 is generated by reflections in the lines containing the six shortest vectors in the lattice. Generators for D_3 have matrices

$$\begin{bmatrix} 1 & 1 \\ 0 & -1 \end{bmatrix} \quad \text{and} \quad \begin{bmatrix} -1 & 0 \\ 1 & 1 \end{bmatrix}$$

with respect to a basis of L.

If $L_0 \neq L$, then, by the argument of case D_{1c}, L is generated by vectors that bisect the lines fixed by the reflections in D_3. In this case, generators of D_3 have matrices

$$\begin{bmatrix} 0 & 1 \\ 1 & 0 \end{bmatrix} \quad \text{and} \quad \begin{bmatrix} -1 & -1 \\ 0 & 1 \end{bmatrix}$$

with respect to a basis of L.

The two groups are nonisomorphic because the matrix representations are not conjugate in $GL(2, \mathbb{Z})$.

Crystals of symmetry types D_{3m} and D_{3c} are displayed in Figure 10.4.12.

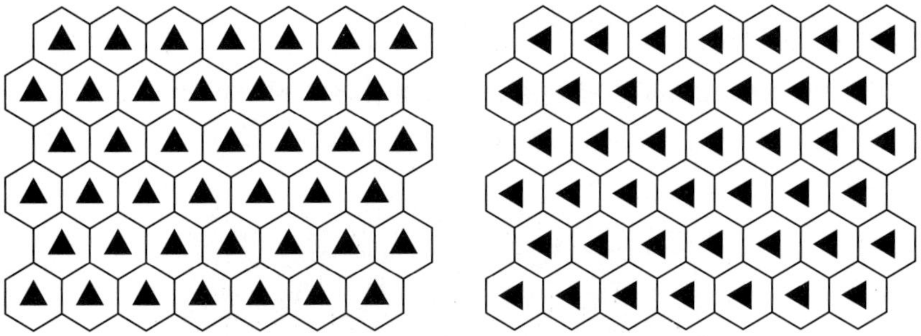

Figure 10.4.12. Crystals with point group D_3.

Point group $\mathbb{Z}_4$: The group G must contain an affine rotation of order 4 about some point, so G is a semidirect product of L and Z_4. The lattice is necessarily square, by Lemma 10.4.6. A crystal of symmetry type Z_4 is shown in Figure 10.4.13.

Point group D_4: The lattice is necessarily square, generated by orthogonal vectors of equal length $\mathbf{a}$ and $\mathbf{b}$. G contains a rotation of order 4. The point group G^0 contains reflections in the lines spanned by $\mathbf{a}$, $\mathbf{b}$, $\mathbf{a}+\mathbf{b}$, and $\mathbf{a}-\mathbf{b}$. By the argument for the case D_{1c}, G must actually contain reflections in the lines parallel to $\mathbf{a}+\mathbf{b}$ and $\mathbf{a}-\mathbf{b}$.

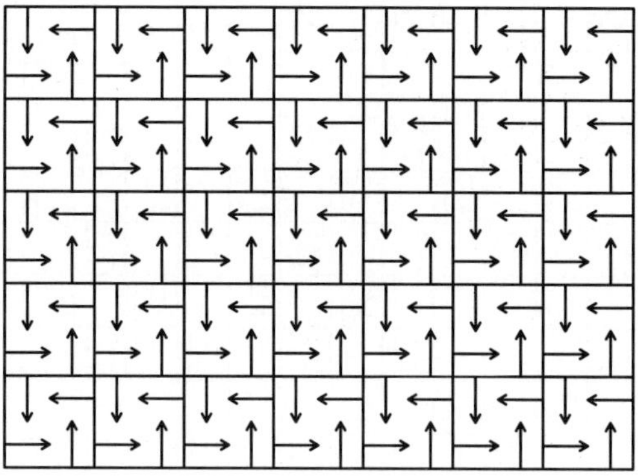

Figure 10.4.13. Crystal of type Z_4.

Without loss of generality, we can suppose that the origin is the intersection of these two lines. There are still two possibilities:

Case: G contains reflections in the lines spanned by $\mathbf{a}$ and $\mathbf{b}$. Then G is a semidirect product of L and D_4. This case is called D_{4m}.

Case: G does not contain a reflection in the line spanned by $\mathbf{a}$, but contains a glide–reflection $\tau_{\mathbf{a}/2}\sigma$, where σ is the reflection in the line spanned by $\mathbf{a}$. This case is called D_{4g}.

The two groups are nonisomorphic because one is a semidirect product of L and D_4, and the other is not.

Crystals of symmetry types D_{4m} and D_{4g} are displayed in Figures 10.4.14 and 10.4.15.

Figure 10.4.14. Crystal with point group D_{4m}.

Point group $\mathbb{Z}_6$: The lattice is hexagonal and the group G contains a rotation of order 6. G is a semidirect product of L and $\mathbb{Z}_6$. A crystal of symmetry type Z_6 is displayed in Figure 10.4.16.

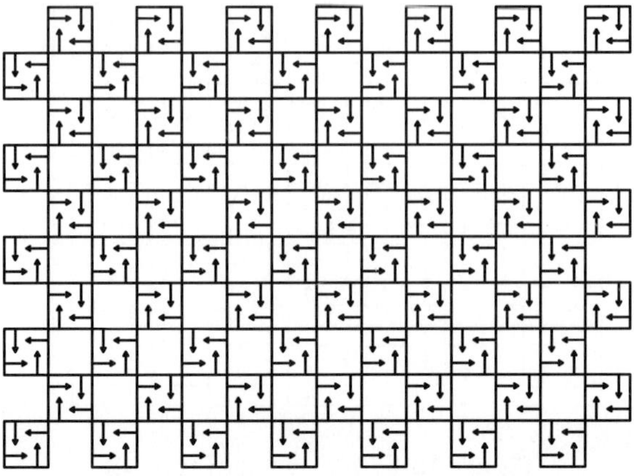

Figure 10.4.15. Crystal with point group D_{4g}.

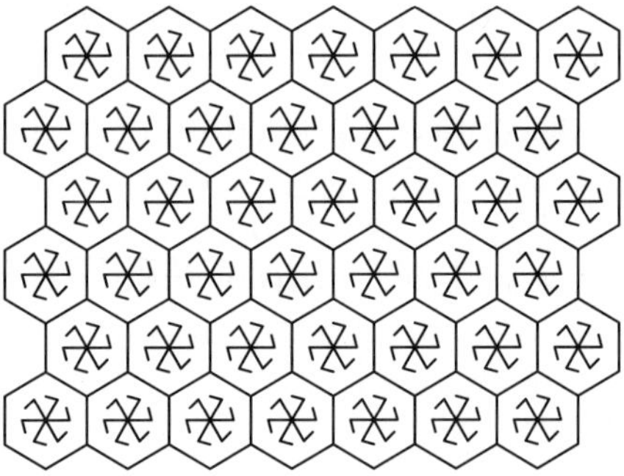

Figure 10.4.16. Crystal of symmetry type Z_6.

Point group D_6: The lattice is hexagonal and the group G contains a rotation of order 6. G also contains reflections in the lines spanned by the six shortest vectors in L, and in the lines spanned by sums and differences of these vectors. Observe that the example in Figure 10.4.3 has symmetry type D_6.

This completes the proof of Theorem 10.4.13. ■

It is possible to classify three–dimensional crystals using similar principles. This was accomplished in 1890 by the crystallographer Fedorov (at a time when the atomic hypothesis was still definitely hypothetical). There are 32 geometric crystal classes [conjugacy classes in $O(3, \mathbb{R})$ of finite groups that act on lattices]; 73 arithmetic crystal classes [conjugacy classes in $GL(3, \mathbb{Z})$ of finite groups that act on lattices]; and 230 isomorphism classes of crystal groups.

About two decades after the theoretical classification of crystals by their symmetry type was achieved, developments in technology made experimental measurements of crystal structure possible. Namely, it was hypothesized by von Laue that the then newly discovered x-rays would have wave lengths comparable to the distance between atoms in crystalline substances and could be diffracted by crystal surfaces. Practical experimental implementation of this idea was developed by W. H. and W. L. Bragg (father and son) in 1912 and 1913.

Exercises 10.4

10.4.1. Devise examples in which the group of translation symmetries of a crystal built on a lattice L is strictly larger than L. (This can only happen if there is not enough "information" in a fundamental domain to prevent further translational symmetries.)

10.4.2. Show that if G is the semidirect product of L and $G \cap O(V)$, then $G^0 \cong G \cap O(V)$.

10.4.3. For several of the groups listed in Corollary 10.4.7, construct a two–dimensional crystal with that point group.

10.4.4. Let L be a two–dimensional lattice. Let $\mathbf{a}$ be a vector of minimal length in L, and let $\mathbf{b}$ be a vector of minimal length in $L \setminus \mathbb{R}\mathbf{a}$.

(a) By using $\|\mathbf{a} \pm \mathbf{b}\| \geq \|\mathbf{b}\|$, show that $|\langle \mathbf{a}, \mathbf{b} \rangle| \leq \|\mathbf{a}\|^2/2$.

(b) Let L_0 be the integer linear span of $\{\mathbf{a}, \mathbf{b}\}$. If $L_0 \neq L$, show that there is a nonzero vector $\mathbf{v} = s\mathbf{a} + t\mathbf{b} \in L \setminus L_0$ with $|s|, |t| \leq 1/2$.

(c) Using part (a), show that such a vector would satisfy

$$\|\mathbf{v}\|^2 \leq (3/4)\|b\|^2.$$

Show that this implies $\mathbf{v} = \mathbf{0}$.

(d) Conclude that $\{\mathbf{a}, \mathbf{b}\}$ is a basis for L; that is, the integer linear span of $\{\mathbf{a}, \mathbf{b}\}$ is L.

10.4.5. Prove Lemma 10.4.10. Hint: Let $\mathbf{a}$ be a vector of minimal length in L; apply the previous exercise to $\{\mathbf{a}, R\mathbf{a}\}$.

10.4.6. The purpose of this exercise and the next is to verify the assertions made in Remark 10.4.12. Let G be a finite subgroup of $GL(\mathbb{R}^n)$. For $\mathbf{x}, \mathbf{y} \in \mathbb{R}^n$ define $\langle\langle \mathbf{x}|\mathbf{y} \rangle\rangle = \sum_{g \in G} \langle g\mathbf{x}|g\mathbf{y} \rangle$.

(a) Show that $(\mathbf{x}, \mathbf{y}) \mapsto \langle\langle \mathbf{x}|\mathbf{y} \rangle\rangle$ defines an inner product on $\mathbb{R}^n$. That is, this is a positive definite, bilinear function.

(b) Show that if $g \in G$, then $\langle\langle g\mathbf{x}|g\mathbf{y} \rangle\rangle = \langle\langle \mathbf{x}|\mathbf{y} \rangle\rangle$.

(c) Conclude that the matrix of g with respect to an orthonormal basis for the inner product $\langle\langle\ |\ \rangle\rangle$ is orthogonal.

(d) Conclude that there is a matrix A (the change of basis matrix) such that AgA^{-1} is orthogonal, for all $g \in G$.

10.4.7. Let T be an element of $GL(n, \mathbb{R})$. It is known that T has a polar decomposition $T = U\sqrt{T^*T}$, where U is an orthogonal matrix and $\sqrt{T^*T}$ is a self-adjoint matrix that commutes with any matrix commuting with T^*T. Refer to a text on linear algebra for a discussion of these ideas.

(a) Suppose G_1 and G_2 are subgroups of $O(n, \mathbb{R})$ and that T is an element of $GL(n, \mathbb{R})$ such that $TG_1T^{-1} = G_2$. Define $\varphi(g) = TgT^{-1}$ for $g \in G_1$, and verify that φ is an isomorphism of G_1 onto G_2.

(b) Use that fact that $G_i \subseteq O(n, \mathbb{R})$ to show that $\varphi(g^*) = \varphi(g)^*$ for $g \in G_1$. Here A^* is used to denote the transpose of the matrix A.

(c) Verify that $Tg = \varphi(g)T$ for $g \in G_1$. Apply the transpose operation to both sides of this equation, and use that $g^* = g^{-1} \in G_1$ for all $g \in G_1$ to conclude that also $gT^* = T^*\varphi(g)$ for $g \in G_1$.

(d) Deduce that T^*T commutes with all elements of G_1 and, therefore, so does $\sqrt{T^*T}$. Conclude that $\varphi(g) = UgU^* = UgU^{-1}$. Thus, G_1 and G_2 are conjugate in $O(n, \mathbb{R})$.

10.4.8. Consider the case of point group D_1 and $L_0 \neq L$. Let $\mathbf{u}$ be a vector of shortest length in $L \setminus L_0$; assume without loss of generality that $\mathbf{u}$ makes an acute angle with both the generators $\mathbf{a}$ and $\mathbf{b}$ of L_0. Put $\mathbf{v} = \sigma(\mathbf{u})$. Show that $\mathbf{u}$ and $\mathbf{v}$ generate L, $\mathbf{a} = \mathbf{u} + \mathbf{v}$, $\mathbf{b} = \mathbf{u} - \mathbf{v}$.

10.4.9. Fill in the details of the analysis for D_3. Give examples of crystals of both symmetry types D_{3m} and D_{3c}.

10.4.10. Fill in the details of the analysis of point group D_4, and give examples of crystals of both D_4 symmetry types.

10.4.11. Fill in the details for the case D_6.

10.4.12. Of what symmetry types are the following crystals?

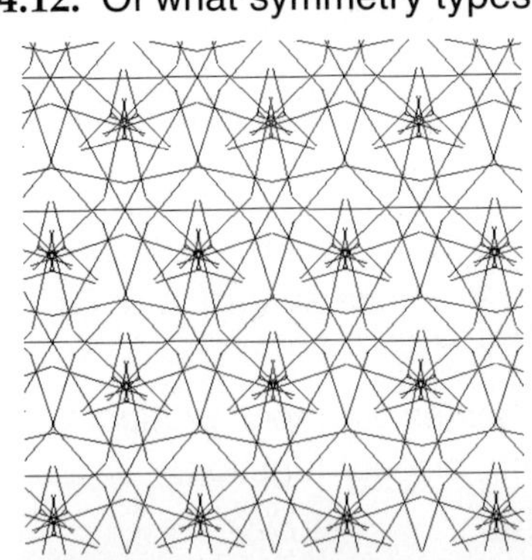

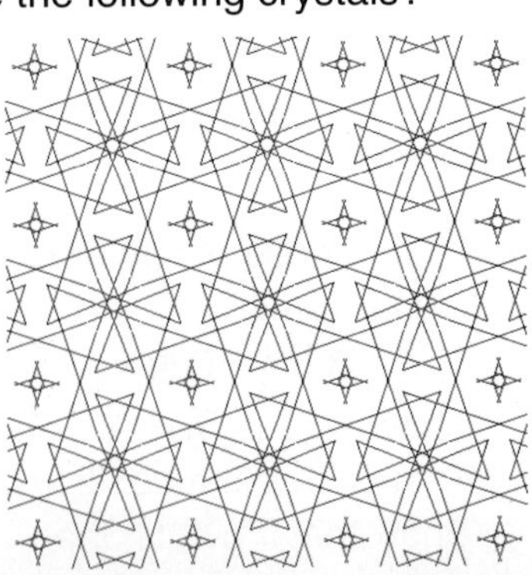

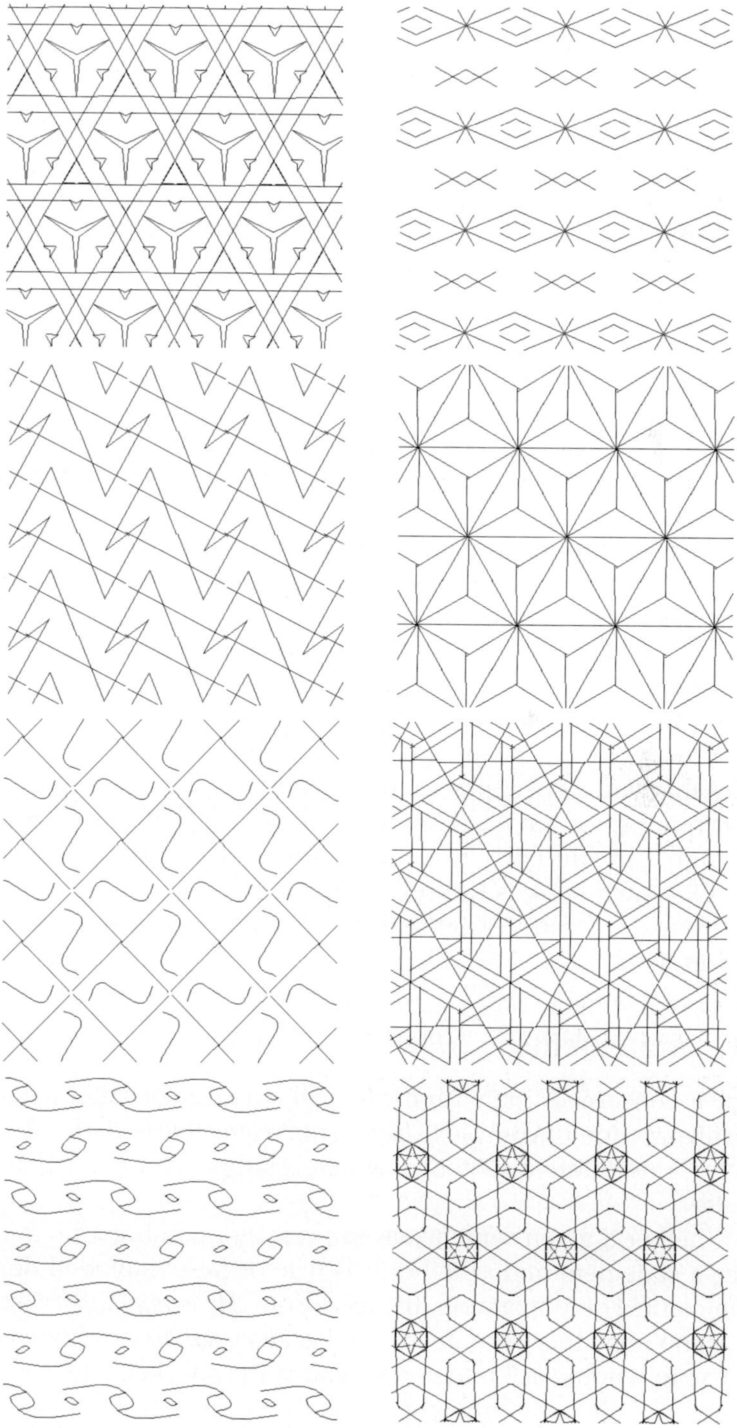

APPENDIX A

Almost Enough about Logic

The purpose of this text is to help you develop an appreciation for, and a facility in, the *practice* of mathematics. We will need to pay some attention to the language of mathematical discourse, so I include informal essays on logic and on the language of sets. These are intended to observe and describe the use of logic and set language in everyday mathematics, rather than to examine the logical or set theoretic foundations of mathematics.

To practice mathematics, you need only a very little bit of set theory or logic; but need to use that little bit accurately. This requires that you respect precise usage of mathematical language.

A number of auxiliary texts are available that offer a a more thorough, but still fairly brief and informal treatment of logic and sets. One such book is Keith Devlin, *Sets, Functions, and Logic, An Introduction to Abstract Mathematics*, 2nd ed., Chapman and Hall, London, 1992.

A.1. Statements

Logic is concerned first with the logical structure of statements and the construction of complex statements from simple parts. A statement is a declarative sentence, which is supposed to be either true or false.

Whether a given sentence is sufficiently unambiguous to qualify as a statement and whether it is true or false may well depend upon context. For example, the assertion *"He is my brother"* standing alone is neither true nor false, because we do not know from context to whom *"he"* refers, nor who is the speaker. The sentence becomes a statement only when both the speaker and the *"he"* have been identified. The same thing happens frequently in mathematical writing, with sentences that contain variables, for example, the sentence

$$x > 0.$$

Such a sentence, which is capable of becoming a statement when the variables have been adequately identified, is called a *predicate* or, perhaps less pretentiously, a *statement with variables*. Such a sentence is neither true nor false until the variable has been identified.

It is the job of every writer of mathematics (you, for example!) to strive to abolish ambiguity. In particular, it is never permissible to introduce a symbol without declaring what it should stand for, unless the symbol has a conventional meaning (e.g., $\mathbb{R}$ for the set of real numbers). Otherwise, what you write will be meaningless or incomprehensible.

A.2. Logical Connectives

Statements can be combined or modified by means of *logical connectives* to form new statements; the validity of such a composite statement depends only on the validity of its components. The basic logical connectives are *and, or, not,* and *if ... then.* We consider these in turn.

A.2.1. The Conjunction *And*

For statements A and B, the statement "A and B" is true exactly when both A and B are true. This is conventionally illustrated by a *truth table*:

A	B	A and B
t	t	t
t	f	f
f	t	f
f	f	f

The table contains one row for each of the four possible combinations of truth values of A and B; the last entry of the row is the corresponding truth value of "A and B." (The logical connective *and* thus defines a function from ordered pairs truth values to truth values, i.e., from $\{t, f\} \times \{t, f\}$ to $\{t, f\}$.)

A.2.2. The Disjunction *Or*

For statements A and B, the statement "A or B" is true when at least one of the component statements is true.

A	B	A or B
t	t	t
t	f	t
f	t	t
f	f	f

In everyday speech, or sometimes is taken to mean "one or the other, but not both," but in mathematics the universal convention is that or means "one or the other or both."

A.2.3. The Negation *Not*

The negation "not(A)" of a statement A is true when A is false and false when A is true.

A	not(A)
t	f
f	t

Of course, given an actual statement A, we do not generally negate it by writing "not(A)." Instead, we employ one of various means afforded by our natural language. The negation of

- *I am satisfied with your explanation.*

is

- *I am not satisfied with your explanation.*

The statement

- *All of the committee members supported the decision.*

has various negations:

- *Not all of the committee members supported the decision.*
- *At least one of the committee members did not support the decision.*
- *At least one of the committee members opposed the decision.*

The following is not a correct negation. Why not?

- *All of the committee members did not support the decision.*

At this point we might try to combine the negation not with the conjunction and or the disjunction or. We compute the truth table of "not(A and B)," as follows:

A	B	A and B	not(A and B)
t	t	t	f
t	f	f	t
f	t	f	t
f	f	f	t

Next, we observe that "not(A) or not(B)" has the same truth table as "not(A and B)" (i.e., defines the same function from $\{t, f\} \times \{t, f\}$ to $\{t, f\}$).

A	B	not(A)	not(B)	not(A) or not(B)
t	t	f	f	f
t	f	f	t	t
f	t	t	f	t
f	f	t	t	t

We say that two *statement formulas* such as "not(A and B)" and "not(A) or not(B)" are *logically equivalent* if they have the same truth table; when we substitute actual statements for A and B in the logically equivalent statement formulas, we end up with two composite statements with exactly the same meaning.

Exercise A.1. Check similarly that "not(A or B)" is logically equivalent to "not(A) and not(B)." Also verify that "not(not(A))" is equivalent to A.

A.2.4. The Implication *If . . . Then*

Next, we consider the implication

- *If A, then B.*

or

- *A implies B.*

We define "if A, then B" to mean "not(A and not(B))," or, equivalently, "not(A) or B"; this is fair enough, since we want "if A, then B" to mean that one cannot have A without also having B. The negation of "A implies B" is thus "A and not(B)."

Exercise A.2. Write out the truth table for "A implies B" and for its negation.

Exercise A.3. Sometimes students jump to the conclusion that "A implies B" is equivalent to one or another of the following: "A and B," "B implies A," or "not(A) implies not(B)." Check that in fact "A implies B" is not equivalent to any of these by writing out the truth tables and noticing the differences.

Exercise A.4. However, "A implies B" is equivalent to its contrapositive "not(B) implies not(A)." Write out the truth tables to verify this.

Exercise A.5. Verify that "A implies (B implies C)" is logically equivalent to "(A and B) implies C."

Exercise A.6. Verify that "A or B" is equivalent to "not(A) implies B."

Often a statement of the form "A or B" is most conveniently proved by assuming A does not hold and proving B.

The use of the connectives *and, or,* and *not* in logic and mathematics coincides with their use in everyday language, and their meaning is clear. Since "if ... then" has been defined in terms of these other connectives, its meaning ought to be just as clear. However, the use of "if ... then" in everyday language often differs (somewhat subtly) from that prescribed here, and we ought to clarify this by looking at an example. Sentences using "if ... then" in everyday speech frequently concern the uncertain future, for example, the sentence

(∗) *If it rains tomorrow, our picnic will be ruined.*

At first glance, "if ... then" seems to be used here with the prescribed meaning:

- *It is not true that it will rain tomorrow without our picnic being ruined.*

However, we notice something amiss when we form the negation. (When we are trying to understand an assertion, it is often helpful to consider the negation.) According to our prescription, the negation ought to be

- *It will rain tomorrow, and our picnic will not be ruined.*

However, the actual negation of the sentence (∗) ought to comment on the consequences of the weather without predicting the weather:

(∗∗) *It is possible that it will rain tomorrow, and our picnic will not be ruined.*

What is going on here? Any sentence about the future must at least implicitly take account of uncertainty; the purpose of the original sentence (∗) is to deny uncertainty, by issuing an absolute prediction:

- *Under all circumstances, if it rains tomorrow, our picnic will be ruined.*

The negation (∗∗) readmits uncertainty.

The preceding example is distinctly nonmathematical, but actually something rather like this also occurs in mathematical usage of "if ... then". Very frequently, "if ... then" sentences in mathematics also involve the *universal quantifier* "for every."

- *For every* x, *if* $x \neq 0$, *then* $x^2 > 0$.

Often the quantifier is only implicit; we write instead

- *If* $x \neq 0$, *then* $x^2 > 0$.

The negation of this is not

- $x \neq 0$ *and* $x^2 \leq 0$,

as we would expect if we ignored the (implicit) quantifier. Because of the quantifier, the negation is actually

- *There exists an* x *such that* $x \neq 0$ *and* $x^2 \leq 0$.

Quantifiers and the negation of quantified statements are discussed more thoroughly in the next section.

Here are a few commonly used logical expressions:

- "A if B" means "B implies A."
- "A only if B" means "A implies B."
- "A if, and only if, B" means "A implies B, and B implies A."
- *Unless* means "if not," but *if not* is equivalent to "or." (Check this!)
- Sometimes *but* is used instead of *and* for emphasis.

A.3. Quantifiers

We now turn to a more systematic discussion of quantifiers. Frequently, statements in mathematics assert that all objects of a certain type have a property, or that there exists at least one object with a certain property.

A.3.1. Universal Quantifier

A statement with a *universal quantifier* typically has the form

- *For all* x, $P(x)$,

where $P(x)$ is a predicate containing the variable x. Here are some examples:

- *For every* x, *if* $x \neq 0$, *then* $x^2 > 0$.
- *For each* f, *if* f *is a differentiable real valued function on* $\mathbb{R}$, *then* f *is continuous.*

I have already mentioned that it is not unusual to omit the important introductory phrase "for all." When the variable x has not already been specified, a sentence such as

- *If* $x \neq 0$, *then* $x^2 > 0$.

is conventionally interpreted as containing the universal quantifier. Another usual practice is to include part of the hypothesis in the introductory phrase, thus limiting the scope of the variable:

- *For every nonzero* x, *the quantity* x^2 *is positive.*
- *For every* $f : \mathbb{R} \to \mathbb{R}$, *if* f *is differentiable, then* f *is continuous.*

It is actually preferable style not to use a variable at all, when this is possible:

- *The square of a nonzero real number is positive.*

Sometimes the validity of a quantified statement may depend on the context. For example,

- *For every* x, *if* $x \neq 0$, *then* $x^2 > 0$.

is a true statement if it has been established by context that x is a real number, but false if x is a complex number. The statement is meaningless if no context has been established for x.

A.3.2. Existential Quantifier

Statements containing the *existential quantifier* typically have the form

- *There exists an* x *such that* $P(x)$,

where $P(x)$ is a predicate containing the variable x. Here are some examples.

- *There exists an* x *such that* x *is positive and* $x^2 = 2$.
- *There exists a continuous function* $f : (0,1) \to \mathbb{R}$ *that is not bounded.*

A.3.3. Negation of Quantified Statements

Let us consider how to form the negation of sentences containing quantifiers. The negation of the assertion that every x has a certain property is that *some* x does not have this property; thus the negation of

- *For every* x, $P(x)$.

is

- *There exists an* x *such that not* $P(x)$.

Consider the preceding examples of (true) statements containing universal quantifiers; their (false) negations are

- *There exists a nonzero* x *such that* $x^2 \leq 0$.
- *There exists a function* $f : \mathbb{R} \to \mathbb{R}$ *such that* f *is differentiable and* f *is not continuous.*

Similarly, the negation of a statement

- *There exists an* x *such that* $P(x)$.

is

- *For every* x, *not* $P(x)$.

Consider the preceding examples of (true) statements containing existential quantifiers; their (false) negations are

- *For every* x, $x \leq 0$ *or* $x^2 \neq 2$.
- *For every function* $f : (0,1) \to \mathbb{R}$, *if* f *is continuous, then* f *is bounded.*

These examples require careful thought; you should think about them until they become absolutely clear.

The various logical elements that we have discussed can be combined as necessary to express complex concepts. An example that is familiar to you from your calculus course is the definition of "the function f is continuous at the point y":

- *For every* $\varepsilon > 0$*, there exists a* $\delta > 0$ *such that (for all* x*) if* $|x - y| < \delta$*, then* $|f(x) - f(y)| < \varepsilon$.

Exercise A.7. Form the negation of this statement.

A.3.4. Order of Quantifiers

It is important to realize that the order of universal and existential quantifiers cannot be changed without utterly changing the meaning of the sentence. This is a true sentence:

- *For every integer* n*, there exists an integer* m *such that* $m > n$.

This is a false sentence, obtained by reversing the order of the quantifiers:

- *There exists an integer* m *such that for every integer* n*,* $m > n$.

A.4. Deductions

Logic concerns not only statements but also deductions. Basically there is only one rule of deduction:

- *If A, then* B*. A. Therefore* B.

For quantified statements this takes the form

- *For all* x*, if* $A(x)$*, then* $B(x)$*.* $A(\alpha)$*. Therefore* $B(\alpha)$.

Here is an example:

- *Every subgroup of an abelian group is normal.* $\mathbb{Z}$ *is an abelian group, and* $3\mathbb{Z}$ *is a subgroup. Therefore,* $3\mathbb{Z}$ *is a normal subgroup of* $\mathbb{Z}$.

If you don't know (yet) what this means, it doesn't matter: You don't *have* to know what it means in order to appreciate its form. Here is another example of exactly the same form:

- *Every car will eventually end up as a pile of rust. Kathryn's Miata is a car. Therefore, it will eventually end up as a pile of rust.*

As you begin to read proofs, you should look out for the verbal variations that this one form of deduction takes and make note of them for your own use.

Most statements requiring proof are "if ... then" statements. To prove "if A, then B," we have to assume A, and prove B under this assumption. To prove "For all x, $A(x)$ implies $B(x)$," we assume that $A(\alpha)$ holds for a particular (but arbitrary) α, and prove $B(\alpha)$ for this particular α. There are many examples of such proofs in the main text.

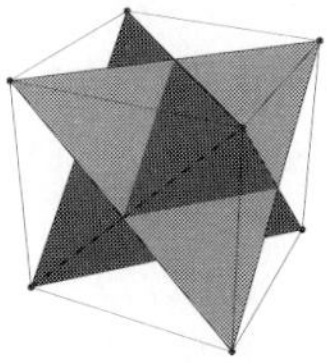

APPENDIX B

Almost Enough about Sets

This is an essay on the language of sets. As with logic, we treat the subject of sets in an intuitive fashion rather than as an axiomatic theory.

A *set* is a collection of (mathematical) objects. The objects contained in a set are called its *elements*. We write $x \in A$ if x is an element of the set A. Very small sets can be specified by simply listing their elements, for example, $A = \{1, 5, 7\}$. For sets A and B, we say that A is *contained* in B, and we write $A \subseteq B$ if each element of A is also an element of B. That is, if $x \in A$, then $x \in B$. (Because of the implicit universal quantifier, the negation of this is that there exists an element of A that is not an element of B.)

Two sets are *equal* if they contain exactly the same elements. This might seem like a quite stupid thing to mention, but in practice, we often have two quite different descriptions of the same set, and we have to do a lot of work to show that the two sets contain the same elements. To do this, it is often convenient to show that each is contained in the other. That is, $A = B$ if, and only if, $A \subseteq B$ and $B \subseteq A$.

Subsets of a given set are frequently specified by a property or predicate; for example, $\{x \in \mathbb{R} : 1 \leq x \leq 4\}$ denotes the set of all real numbers between 1 and 4. Note that set containment is related to logical implication in the following fashion: If a property $P(x)$ implies a property $Q(x)$, then the set corresponding to $P(x)$ is contained in the set corresponding to $Q(x)$. For example, $x < -2$ implies that $x^2 > 4$, so $\{x \in \mathbb{R} : x < -2\} \subseteq \{x \in \mathbb{R} : x^2 > 4\}$.

The *intersection* of two sets A and B, written $A \cap B$, is the set of elements contained in both sets. $A \cap B = \{x : x \in A \text{ and } x \in B\}$. Note the relation between intersection and the logical conjunction.

If $A = \{x \in C : P(x)\}$ and $B = \{x \in C : Q(x)\}$, then $A \cap B = \{x \in C : P(x) \text{ and } Q(x)\}$.

The *union* of two sets A and B, written $A \cup B$, is the set of elements contained in at least one of the two sets. $A \cup B = \{x : x \in A \quad \text{or} \quad x \in B\}$. Set union and the logical disjunction are related as are set intersection and logical conjunction. If $A = \{x \in C : P(x)\}$ and $B = \{x \in C : Q(x)\}$, then $A \cup B = \{x \in C : P(x) \quad \text{or} \quad Q(x)\}$.

Given finitely many sets—for example, five sets A, B, C, D, E—we similarly define their intersection $A \cap B \cap C \cap D \cap E$ to consist of those elements that are in all of the sets, and the union $A \cup B \cup C \cup D \cup E$ to consist of those elements that are in at least one of the sets.

There is a unique set with no elements at all, called the *empty set*, or the *null set*, and usually denoted $\emptyset$.

Proposition B.1. *The empty set is a subset of every set.*

Proof. Given an arbitrary set A, we have to show that $\emptyset \subseteq A$; that is, for every element $x \in \emptyset$, we have $x \in A$. The negation of this statement is that there exists an element $x \in \emptyset$ such that $x \notin A$. But this negation is false, because there are no elements at all in $\emptyset$! So the original statement is true. ■

If the intersection of two sets is the empty set, we say that the sets are *disjoint*, or *nonintersecting*.

Here is a small theorem concerning the properties of set operations.

Proposition B.2. *For all sets* A, B, C,

(a) $A \cup A = A$, *and* $A \cap A = A$.
(b) $A \cup B = B \cup A$, *and* $A \cap B = B \cap A$.
(c) $(A \cup B) \cup C = A \cup B \cup C = A \cup (B \cup C)$, *and* $(A \cap B) \cap C = A \cap B \cap C = A \cap (B \cap C)$.
(d) $A \cap (B \cup C) = (A \cap B) \cup (A \cap C)$, *and* $A \cup (B \cap C) = (A \cup B) \cap (A \cup C)$.

The proofs are just a matter of checking definitions.

Given two sets A and B, we define the *relative complement* of B in A, denoted $A \setminus B$, to be the elements of A that are not contained in B. That is, $A \setminus B = \{x \in A : x \notin B\}$.

In general, all the sets appearing in some particular mathematical discussion are subsets of some "universal" set U; for example, we might be discussing only subsets of the real numbers $\mathbb{R}$. (However, there is no universal set once and for all, for all mathematical discussions; the assumption of a "set of all sets" leads to contradictions.) It is customary and convenient to use some special notation such as $\mathcal{C}(B)$ for the complement of B relative to U, and to refer to $\mathcal{C}(B) = \mathcal{U} \setminus B$ simply as *the complement of* B. [The notation $\mathcal{C}(B)$ is not standard.]

Exercise B.1. Show that the sets $A \cap B$ and $A \setminus B$ are disjoint and have union equal to A.

Exercise B.2 (de Morgan's laws). For any sets A and B, show that

$$\mathcal{C}(A \cup B) = \mathcal{C}(A) \cap \mathcal{C}(B),$$

and

$$\mathcal{C}(A \cap B) = \mathcal{C}(A) \cup \mathcal{C}(B).$$

Exercise B.3. For any sets A and B, show that $A \setminus B = A \cap \mathcal{C}(B)$.

Exercise B.4. For any sets A and B, show that

$$(A \cup B) \setminus (A \cap B) = (A \setminus B) \cup (B \setminus A).$$

Another important construction with sets is the Cartesian product. For any two sets A and B, an *ordered pair* of elements (a, b) is just a list with two items, the first from A and the second from B. The Cartesian product $A \times B$ is the set of all such pairs. Order counts, and $A \times B$ is not the same as $B \times A$, unless of course $A = B$. (The Cartesian product is named after Descartes, who realized the possibility of coordinatizing the plane as $\mathbb{R} \times \mathbb{R}$.)

We recall the notion of a *function from* A *to* B and some terminology regarding functions that is standard throughout mathematics. A function f from A to B is a rule that gives for each element of $a \in A$ an "outcome" in $f(a) \in B$. More formally, a function is a subset of $A \times B$ that contains for each element $a \in A$ exactly one pair (a, b); the subset contains (a, b) if, and only if, $b = f(a)$. A is called the *domain* of the function, B the *codomain*, $f(a)$ is called the *value* of the function at a, and the set of all values, $\{f(a) : a \in A\}$, is called the *range* of the function. In general, the range is only a subset of B; a function is said to be *surjective*, or *onto*, if its range is all of B; that is, for each $b \in B$, there exists an $a \in A$, such that $f(a) = b$. Figure B.1 exhibits a surjective function.

A function f is said to be *injective*, or *one to one*, if for each two distinct elements a and a' in A, we have $f(a) \neq f(a')$. Equivalently, $f(a) = f(a')$ implies that $a = a'$. Figure B.2 displays an injective and a noninjective function.

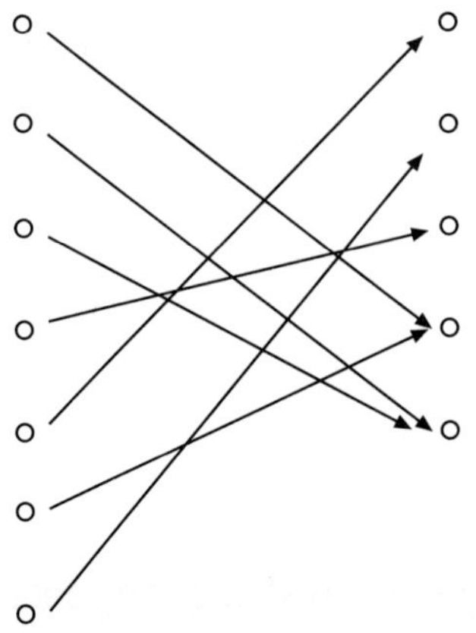

Figure B.1. A surjection.

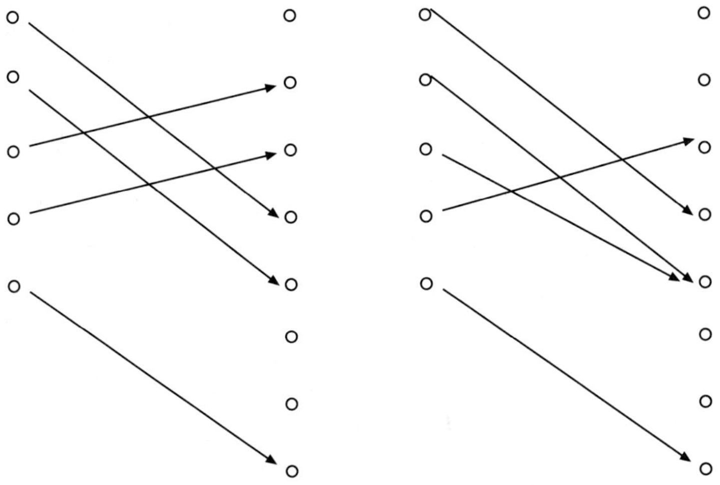

Figure B.2. Injective and noninjective functions.

Finally, f is said to be *bijective* if it is both injective and surjective. A bijective function (or *bijection*) is also said to be a *one to one correspondence* between A and B, since it matches up the elements of the two sets one to one. When f is bijective, there is an *inverse function* f^{-1} defined by $f^{-1}(b) = a$ if, and only if, $f(a) = b$. Figure B.3 displays a bijective function.

If $f : X \to Y$ is a function and A is a subset of X, we write $f(A)$ for $\{f(a) : a \in A\} = \{y \in Y : \text{ there exists } a \in A \text{ such that } y = f(a)\}$. We refer to $f(A)$ as the *image of* A *under* f. If B is a subset of Y, we write $f^{-1}(B)$ for $\{x \in X : f(x) \in B\}$. We refer to $f^{-1}(B)$ as the *preimage of* B *under* f.

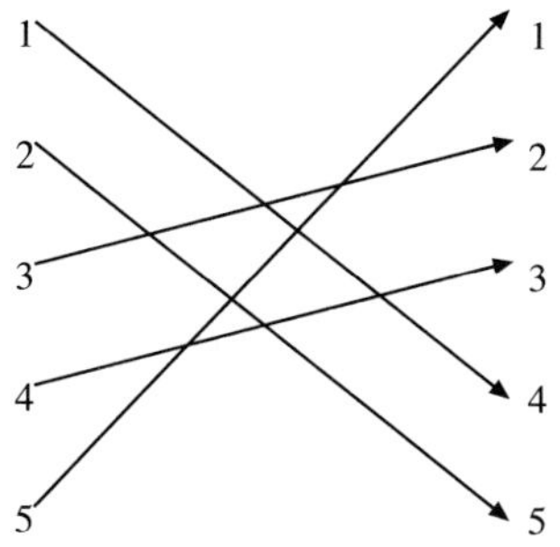

Figure B.3. A bijection.

Exercise B.5. Let $f : X \to Y$ be a function, and let E and F be subsets of X. Show that $f(E) \cup f(F) = f(E \cup F)$. Also, show that $f(E) \cap f(F) \supseteq f(E \cap F)$; give an example to show that we can have strict containment.

Exercise B.6. Let $f : X \to Y$ be a function, and let E and F be subsets of Y. Show that $f^{-1}(E) \cup f^{-1}(F) = f^{-1}(E \cup F)$. Also, show that $f^{-1}(E) \cap f^{-1}(F) = f^{-1}(E \cap F)$. Finally, show that $f^{-1}(E \setminus F) = f^{-1}(E) \setminus f^{-1}(F)$.

B.1. Families of Sets; Unions and Intersections

The elements of a set can themselves be sets! This is not at all an unusual situation in mathematics. For example, for each natural number n we could take the set $X_n = \{x : 0 \le x \le n\}$ and consider the collection of all of these sets X_n, which we call $\mathcal{F}$. Thus, $\mathcal{F}$ is a set whose members are sets. In order to at least try not to be confused, we often use words such as collection or family in referring to a set whose elements are sets.

Given a family $\mathcal{F}$ of sets, we define the union of $\mathcal{F}$, denoted $\cup\mathcal{F}$, to be the set of elements that are contained in at least one of the members of $\mathcal{F}$; that is,

$$\cup\mathcal{F} = \{x : \text{ there exists } A \in \mathcal{F} \text{ such that } x \in A\}.$$

Similarly, the intersection of the family $\mathcal{F}$, denoted $\cap\mathcal{F}$, is the set of elements that are contained in every member of $\mathcal{F}$; that is,

$$\cap\mathcal{F} = \{x : \text{ for every } A \in \mathcal{F}, \quad x \in A\}.$$

In the example we have chosen, $\mathcal{F}$ is an *indexed family of sets*, indexed by the natural numbers; that is, $\mathcal{F}$ consists of one set X_n for each natural number n. For indexed families we often denote the union and intersection as follows:

$$\cup\mathcal{F} = \bigcup_{n\in\mathbb{N}} X_n = \bigcup_{n=1}^{\infty} X_n$$

and

$$\cap \mathcal{F} = \bigcap_{n \in \mathbb{N}} X_n = \bigcap_{n=1}^{\infty} X_n.$$

It is a fact, perhaps obvious to you, that $\bigcup_{n \in \mathbb{N}} X_n$ is the set of all nonnegative real numbers, whereas $\bigcap_{n \in \mathbb{N}} X_n = X_1$.

B.2. Finite and Infinite Sets

I include here a brief discussion of finite and infinite sets. I assume an intuitive familiarity with the natural numbers $\mathbb{N} = \{1, 2, 3, \ldots\}$. A set S is said to be *finite* if there is some natural number n and a bijection between S and the set $\{1, 2, \ldots, n\}$. A set is *infinite* otherwise. A set S is said to be *countably infinite* if there is a bijection between S and the set of natural numbers. A set is said to be *countable* if it is either finite or countably infinite. *Enumerable* or *denumerable* are synonyms for *countable.* (Some authors prefer to make a distinction between *denumerable,* which they take to mean "countably infinite," and *countable,* which they use as we have.) It is a bit of a surprise to most people that infinite sets come in different sizes; in particular, there are infinite sets that are not countable. For example, using the completeness of the real numbers, we can show that the set of real numbers is not countable.

It is clear that every set is either finite or infinite, but it is not always possible to determine which. For example, it is pretty easy to show that there are infinitely many prime numbers, but it is *unknown* at present whether there are infinitely many twin primes, that is, successive odd numbers, such as 17 and 19, both of which are prime. Let us observe that although the even natural numbers $2\mathbb{N} = \{2, 4, 6, \ldots\}$, the natural numbers $\mathbb{N}$, and the integers $\mathbb{Z} = \{0, \pm 1, \pm 2, \ldots\}$ satisfy

$$2\mathbb{N} \subseteq \mathbb{N} \subseteq \mathbb{Z},$$

and no two of the sets are equal, they all are countable sets, so all have the same size or cardinality. Two sets are said to have the *same cardinality* if there is a bijection between them. One characterization of infinite sets is that a set is infinite if, and only if, it has a proper subset with the same cardinality.

If there exists a bijection from a set S to $\{1, 2, \ldots, n\}$, we say the cardinality of S equals n, and write $|S| = n$. (It cannot happen that there exist bijections from S to both $\{1, 2, \ldots, n\}$ and $\{1, 2, \ldots, m\}$ for $n \neq m$.)

Here are some theorems about finite and infinite sets that we will not prove here.

Theorem B.3. *A subset of a finite set is finite. A subset of a countable set is countable. A set S is countable if, and only if, there is an injective function with domain S and range contained in $\mathbb{N}$.*

Theorem B.4. *A union of finitely many finite sets is finite. A union of countably many countable sets is countable.*

Using this result, we see that the rational numbers $\mathbb{Q} = \{a/b : a, b \in \mathbb{Z}, b \neq 0\}$ is a countable set. Namely, $\mathbb{Q}$ is the union of the sets

$$\begin{aligned} A_1 &= \mathbb{Z} \\ A_2 &= (1/2)\mathbb{Z} = \{0, \pm 1/2, \pm 2/2, \pm 3/2 \dots\} \\ &\dots \\ A_n &= (1/n)\mathbb{Z} = \{0, \pm 1/n, \pm 2/n, \pm 3/n \dots\} \\ &\dots, \end{aligned}$$

each of which is countably infinite.

APPENDIX C

Induction

C.1. Proof by Induction

Suppose you need to climb a ladder. If you are able to reach the first rung of the ladder and you are also able to get from any one rung to the next, then there is nothing to stop you from climbing the whole ladder. This is called *the principle of mathematical induction.*

Mathematical induction is often used to prove statements about the natural numbers, or about families of objects indexed by the natural numbers. Suppose that you need to prove a statement of the form

- *For all* $n \in \mathbb{N}$, $P(n)$,

where $P(n)$ is a predicate. Examples of such statements are

- *For all* $n \in \mathbb{N}$, $1 + 2 + \cdots + n = (n)(n+1)/2$.
- *For all* n *and for all permutations* $\pi \in S_n$, π *has a unique decomposition as a product of disjoint cycles.* (See Section 1.5.)

To prove that $P(n)$ holds for all $n \in \mathbb{N}$, it suffices to show that $P(1)$ holds (you can reach the first rung) and that whenever $P(k)$ holds, then also $P(k+1)$ holds (you can get from any one rung to the next). Then $P(n)$ holds for all n (you can climb the whole ladder).

Principle of Mathematical Induction. *For the statement "For all* $n \in \mathbb{N}$, $P(n)$*" to be valid, it suffices that*

1. $P(1)$*, and*
2. *For all* $k \in \mathbb{N}$, $P(k)$ *implies* $P(k+1)$.

To prove that "*For all* $k \in \mathbb{N}$, $P(k)$ *implies* $P(k+1)$," you have to assume $P(k)$ for a fixed but arbitrary value of k and prove $P(k+1)$ under this assumption. This sometimes seems like a big cheat to beginners, for we seem to be assuming what we want to prove, namely, that $P(n)$ holds. But it is not a cheat at all; we are just showing that it is possible to get from one rung to the next.

As an example we prove the identity

$$P(n) : 1 + 2 + \cdots + n = (n)(n+1)/2$$

by induction on n. The statement $P(1)$ reads

$$1 = (1)(2)/2,$$

which is evidently true. Now, we assume $P(k)$ holds for some k, that is,

$$1 + 2 + \cdots + k = (k)(k+1)/2,$$

and prove that $P(k+1)$ also holds. The assumption of $P(k)$ is called the *induction hypothesis*. Using the induction hypothesis, we have

$$1 + 2 + \cdots + k + (k+1) = (k)(k+1)/2 + (k+1) = \frac{(k+1)}{2}(k+2),$$

which is $P(k+1)$. This completes the proof of the identity.

The principle of mathematical induction is equivalent to the following principle:

Well–Ordering Principle. *Every nonempty subset of the natural numbers has a least element.*

Another form of the principle of mathematical induction is the following:

Principle of Mathematical Induction, 2nd Form. *For the statement "For all $n \in \mathbb{N}$, $P(n)$" to be valid, it suffices that:*

1. $P(1)$, *and*
2. *For all $k \in \mathbb{N}$, if $P(r)$ for all $r \leq k$, then also $P(k+1)$.*

The two forms of the principle of mathematical induction and the well–ordering principle are all equivalent statements about the natural numbers. That is, assuming any one of these principles, we can prove the other two. The proof of the equivalence is somewhat more abstract than the actual subject matter of this course, so I prefer to omit it. When you have more experience with doing proofs, you may wish to provide your own proof of the equivalence.

C.2. Definitions by Induction

It is frequently necessary or convenient to define some sequence of objects (numbers, sets, functions, ...) *inductively* or *recursively*. That means the nth object is defined in terms of the first $n-1$ objects (i.e., in terms of all of the objects preceding the nth), instead of there being a formula or procedure which tells you once and for all how to define the nth object. For example, the sequence of Fibonacci numbers is defined by the recursive rule:

$$f_1 = f_2 = 1 \quad f_n = f_{n-1} + f_{n-2} \quad \text{for } n \geq 3.$$

The well–ordering principle, or the principle of mathematical induction, implies that such a rule suffices to define f_n for all natural numbers n. For f_1 and f_2 are defined by an explicit formula (we can get to the first rung), and if $f_1, \dots, f_k$ have been defined for some k, then the recursive rule $f_{k+1} = f_k + f_{k-1}$ also defines f_{k+1} (we can get from one rung to the next).

Principle of Inductive Definition. To define a sequence of objects $A_1, A_2, \dots$ it suffices to have

1. A definition of A_1
2. For each $k \in \mathbb{N}$, a definition of A_{k+1} in terms of $\{A_1, \dots, A_k\}$

Here is an example relevant to this course: Suppose we are working in a system with an associative multiplication (perhaps a group, perhaps a ring, perhaps a field). Then, for an element a, we can define a^n for $n \in \mathbb{N}$ by the recursive rule: $a^1 = a$ and for all $k \in \mathbb{N}$, $a^{k+1} = a^k a$.

Here is another example where the objects being defined are intervals in the real numbers. The goal is to compute an accurate approximation to $\sqrt{7}$. We define a sequence of intervals $A_n = [a_n, b_n]$ with the properties

1. $b_n - a_n = 6/2^n$,
2. $A_{n+1} \subseteq A_n$ for all $n \in \mathbb{N}$, and
3. $a_n^2 < 7$ and $b_n^2 > 7$ for all $n \in \mathbb{N}$.

Define $A_1 = [1, 7]$. If $A_1, \dots, A_k$ have been defined, let $c_k = (a_k + b_k)/2$. If $c_k^2 < 7$, then define $A_{k+1} = [c_k, b_k]$. Otherwise, define $A_k = [a_k, c_k]$. (Remark that all the numbers a_k, b_k, c_k are rational, so it is never true that $c_k^2 = 7$.) You should do a proof (by induction) that the sets A_n defined by this procedure do satisfy the properties just listed. This example can easily be transformed into a computer program for calculating the square root of 7.

C.3. Multiple Induction

Let a be an element of an algebraic system with an associative multiplication (a group, a ring, a field). Consider the problem of showing that, for all natural numbers m and n, $a^m a^n = a^{m+n}$. It would seem that some sort of inductive procedure is appropriate, but two integer variables have to be involved in the induction. How is this to be done? Let $P(m, n)$ denote the predicate "$a^m a^n = a^{m+n}$." To establish that for all $m, n \in \mathbb{N}, P(m, n)$, I claim that it suffices to show that

(1) $P(1, 1)$.
(2) For all $r, s \in \mathbb{N}$ if $P(r, s)$ holds, then $P(r + 1, s)$ holds.
(3) For all $r, s \in \mathbb{N}$ if $P(r, s)$ holds, then $P(r, s + 1)$ holds.

To justify this intuitively, we have to replace our image of a ladder with the grid of integer points in the quarter-plane, $\{(m, n) : m, n \in \mathbb{N}\}$. Showing $P(1, 1)$ gets us onto the grid. Showing (2) allows us to get from any point on the grid to the adjacent point to the right, and showing (3) allows us to get from any point on the grid to the adjacent point above. By taking steps to the right and up from $(1, 1)$, we can reach any point on the grid, so eventually $P(m, n)$ can be established for any (m, n).

As intuitive as this picture may be, it is not entirely satisfactory, because it seems to require a new principle of induction on two variables. And if we needed to do an induction on three variables, we would have to invent yet another principle. It is more satisfactory to justify the procedure by the principle of induction on one integer variable.

To do this, it is useful to consider the following very general situation. Suppose we want to prove a proposition of the form "for all $t \in T, P(t)$," where T is some set and P is some predicate. Suppose T can be written as a union of sets $T = \bigcup_1^\infty T_k$, where (T_k) is an increasing sequence of subsets $T_1 \subseteq T_2 \subseteq T_3 \dots$. According to the usual principle of induction, it suffices to show

(a) For all $t \in T_1$, $P(t)$, and
(b) For all $k \in \mathbb{N}$, if $P(t)$ holds for all $t \in T_k$, then also $P(t)$ holds for all $t \in T_{k+1}$.

In fact, this suffices to show that for all $n \in \mathbb{N}$, and for all $t \in T_n$, $P(t)$. But since each $t \in T$ belongs to some T_n, $P(t)$ holds for all $t \in T$.

Now, to apply this general principle to the situation of induction on two integer variable, we take T to be the set $\{(m, n) : m, n \in \mathbb{N}\}$. There are various ways in which we could choose the sequence of subsets T_k; for example we can take $T_k = \{(m, n) : m, n \in \mathbb{N} \text{ and } m \leq k\}$. Now, suppose we have a predicate $P(m, n)$ for which we can prove the following:

(1) $P(1, 1)$
(2) For all $r, s \in \mathbb{N}$ if $P(r, s)$ holds then $P(r + 1, s)$ holds.
(3) For all $r, s \in \mathbb{N}$ if $P(r, s)$ holds then $P(r, s + 1)$ holds.

Using (1) and (3) and induction on one variable, we can conclude that $P(1, n)$ holds for all $n \in \mathbb{N}$; that is, $P(m, n)$ holds for all $(m, n) \in T_1$. Now, fix $k \in \mathbb{N}$, and suppose that $P(m, n)$ holds for all $(m, n) \in T_k$, that is, whenever $m \leq k$. Then, in particular, $P(k, 1)$ holds. It then follows from (2) that $P(k+1, 1)$ holds. Now, from this and (3) and induction on one variable, it follows that $P(k+1, n)$ holds for all $n \in \mathbb{N}$. But then $P(m, n)$ holds for all (m, n) in $T_k \cup \{(k+1, n) : n \in \mathbb{N}\} = T_{k+1}$. Thus, we can prove (a) and (b) for our sequence T_k.

Exercise C.1. Define $T_k = \{(m, n) : m, n \in \mathbb{N}, m \leq k, \text{ and } n \leq k\}$. Show how statements (1) through (3) imply (a) and (b) for this choice of T_k.

Exercise C.2. Let a be a real number (or an element of a group, or an element of a ring). Show by induction on two variables that $a^m a^n = a^{m+n}$ for all $m, n \in \mathbb{N}$.

Appendix D

Complex Numbers

In this appendix, we review the construction of the complex numbers from the real numbers. As you know, the equation $x^2 + 1 = 0$ has no solution in the real numbers. However, it is possible to construct a field containing $\mathbb{R}$ by appending to $\mathbb{R}$ a solution of this equation.

To begin with, consider the set $\mathbb{C}$ of all formal sums $a+bi$, where a and b are in $\mathbb{R}$ and i is just a symbol. We give this set the structure of a two–dimensional real vector space in the obvious way: Addition is defined by $(a + bi) + (a' + b'i) = (a + a') + (b + b')i$ and multiplication with real scalars by $\alpha(a + bi) = \alpha a + \alpha bi$.

Next, we *try* to define a multiplication on $\mathbb{C}$ in such a way that the distributive law holds and also $i^2 = -1$, and $i\alpha = \alpha i$. These requirements force the definition: $(a + bi)(c + di) = (ac - bd) + (ad + bc)i$. Now it is completely straightforward to check that this multiplication is commutative and associative, and that the distributive law does indeed hold. Moreover, $(a + 0i)(c + di) = ac + adi$, so multiplication by $(a + 0i)$ coincides with scalar multiplication by $a \in \mathbb{R}$. In particular, $1 = 1 + 0i$ is the multiplicative identity in $\mathbb{C}$, and we can identify $\mathbb{R}$ with the set of elements $a + 0i$ in $\mathbb{C}$.

To show that $\mathbb{C}$ is a field, it remains only to check that nonzero elements have multiplicative inverses. It is straightforward to compute that

$$(a + bi)(a - bi) = a^2 + b^2 \in \mathbb{R}.$$

Hence, if not both a and b are zero, then

$$(a + bi)(\frac{a}{(a^2 + b^2)} - \frac{b}{(a^2 + b^2)}i) = 1.$$

It is a remarkable fact that every polynomial with complex coefficients has a complete set of roots in $\mathbb{C}$; that is, every polynomial with complex coefficients is a product of linear factors $x - \alpha$. This theorem is due to Gauss and is often know as the fundamental theorem of algebra. All proofs of this theorem contain some analysis, and the most straightforward proofs involve some complex analysis; you can find a proof in any text on complex analysis. In general, a field K with the property that every polynomial with coefficients in K has a complete set of roots in K is called *algebraically closed*.

For any complex number $z = a + bi \in \mathbb{C}$, we define the complex conjugate of z by $\bar{z} = a - bi$, and the modulus of z by $|z| = \sqrt{a^2 + b^2}$. Note that $z\bar{z} = |z|^2$, and $z^{-1} = \bar{z}/|z|^2$. The *real part* of z, denoted $\Re z$, is $a = (z + \bar{z})/2$, and the *imaginary part* of z, denoted $\Im z$ is $b = (z - \bar{z})/2i$. (Note that the imaginary part of z is a real number, and $z = \Re z + i\Im z$.)

If $z = a + bi$ is a complex number with modulus 1 (that is, $a^2 + b^2 = 1$), then there is a real number t such that $a = \cos t$ and $b = \sin t$; t is determined up to addition of an integer multiple of 2π.

Exercise D.1. Consider two complex numbers of modulus 1, namely, $\cos t + i \sin t$ and $\cos s + i \sin s$. Use trigonometric identities to verify that

$$(\cos t + i \sin t)(\cos s + i \sin s) = \cos(s + t) + i \sin(s + t).$$

We introduce the notation $e^{it} = \cos t + i \sin t$; then the result of the previous exercise is $e^{it}e^{is} = e^{i(t+s)}$.

For any complex number z, $z/|z|$ has modulus 1, so is of the form e^{it} for some t. Therefore, z itself can be written in the form $z = |z|(z/|z|) = |z|e^{it}$.

Exercise D.2. Write $5 - 3i$ in the form re^{it}, where $r > 0$ and $t \in \mathbb{R}$.

The form $z = re^{it}$ for a complex number is called the *polar form*. Multiplication of two complex numbers written in polar form is particularly easy to compute: $r_1e^{it}r_2e^{is} = r_1r_2e^{i(s+t)}$. It follows from this that for complex numbers z_1 and z_2, we have $|z_1z_2| = |z_1||z_2|$.

For a complex number $z = re^{it}$, we have $z^n = r^ne^{int}$.

Exercise D.3. Show that a complex number z satisfies $z^n = 1$ if, and only if, $z \in \{e^{i2\pi k/n} : 0 \le k \le n - 1\}$. Such a complex number is called an nth root of unity.

Appendix E

Review of Linear Algebra

E.1. Linear algebra in K^n

This appendix provides a quick review of linear algebra. We let K denote one of the fields $\mathbb{Q}$, $\mathbb{R}$, or $\mathbb{C}$. K^n is the set of n-tuples of elements of K, viewed as column vectors $\mathbf{x} = \begin{bmatrix} x_1 \\ x_2 \\ \vdots \\ x_n \end{bmatrix}$. The set K^n has two operations, component-by-component addition of vectors, and multiplication of a vector by a "scalar" in K,

$$\begin{bmatrix} x_1 \\ x_2 \\ \vdots \\ x_n \end{bmatrix} + \begin{bmatrix} y_1 \\ y_2 \\ \vdots \\ y_n \end{bmatrix} = \begin{bmatrix} x_1 + y_1 \\ x_2 + y_2 \\ \vdots \\ x_n + y_n \end{bmatrix}, \quad \text{and} \quad \alpha \begin{bmatrix} x_1 \\ x_2 \\ \vdots \\ x_n \end{bmatrix} = \begin{bmatrix} \alpha x_1 \\ \alpha x_2 \\ \vdots \\ \alpha x_n \end{bmatrix}.$$

The zero vector $\mathbf{0}$, with all components equal to zero, is the identity for addition of vectors.

Definition E.1. A *linear combination* of set S of vectors is any vector of the form $\alpha_1 \mathbf{v}_1 + \alpha_2 \mathbf{v}_2 + \cdots + \alpha_s \mathbf{v}_s$, where for all i, $\alpha_i \in K$ and $\mathbf{v}_i \in S$. The *span* of S is the set of all linear combinations of S. We denote the span of S by $\text{span}(S)$.

The span of the empty set is the set containing only the zero vector $\{\mathbf{0}\}$.

Definition E.2. A set S vectors is *linearly independent* if for all natural numbers s, for all $\boldsymbol{\alpha} = \begin{bmatrix} \alpha_1 \\ \vdots \\ \alpha_s \end{bmatrix} \in K^s$, and for all sequences $(\mathbf{v}_1, \ldots \mathbf{v}_s)$ of *distinct* vectors in S, if $\alpha_1\mathbf{v}_1 + \alpha_2\mathbf{v}_2 + \cdots + \alpha_s\mathbf{v}_s = \mathbf{0}$, then $\boldsymbol{\alpha} = \mathbf{0}$. Otherwise, S is *linearly dependent*.

Note that a linear independent set cannot contain the zero vector. The empty set is linearly independent, since there are no sequences of its elements.

Definition E.3. A *vector subspace* or *linear subspace* V of K^n is a subset that is closed under taking linear combinations. That is V is its own span.

Note that $\{\mathbf{0}\}$ and K^n are vector subspaces.

Definition E.4. Let V be a vector subspace of K^n. A subset of V is called a *basis* of V if the set is linearly independent and has span equal to V.

The *standard basis* of K^n is the set

$$\hat{\mathbf{e}}_1 = \begin{bmatrix} 1 \\ 0 \\ \vdots \\ 0 \end{bmatrix}, \hat{\mathbf{e}}_2 = \begin{bmatrix} 0 \\ 1 \\ \vdots \\ 0 \end{bmatrix}, \ldots, \hat{\mathbf{e}}_n = \begin{bmatrix} 0 \\ 0 \\ \vdots \\ 1 \end{bmatrix}.$$

Example E.5. Show that $\{ \begin{bmatrix} 1 \\ 1 \\ 1 \end{bmatrix}, \begin{bmatrix} 1 \\ 2 \\ 5 \end{bmatrix}, \begin{bmatrix} 4 \\ 2 \\ 7 \end{bmatrix} \}$ is a basis of $\mathbb{R}^3$. We have to show two things: that the set is linearly independent and that its span is $\mathbb{R}^3$. (Actually, as we will observe a little later, it would suffice to prove only one of these statements.) For linear independence, we have to show that if

$$\alpha_1 \begin{bmatrix} 1 \\ 1 \\ 1 \end{bmatrix} + \alpha_2 \begin{bmatrix} 1 \\ 2 \\ 5 \end{bmatrix} + \alpha_3 \begin{bmatrix} 4 \\ 2 \\ 7 \end{bmatrix} = \begin{bmatrix} 0 \\ 0 \\ 0 \end{bmatrix}, \tag{E.1}$$

then all the α_i equal zero. The vector equation (E.1) is equivalent to the matrix equation

$$\begin{bmatrix} 1 & 1 & 4 \\ 1 & 2 & 2 \\ 1 & 5 & 7 \end{bmatrix} \begin{bmatrix} \alpha_1 \\ \alpha_2 \\ \alpha_3 \end{bmatrix} = \begin{bmatrix} 0 \\ 0 \\ 0. \end{bmatrix} \tag{E.2}$$

The matrix *row reduces* to the identity matrix $\begin{bmatrix} 1 & 0 & 0 \\ 0 & 1 & 0 \\ 0 & 0 & 1 \end{bmatrix}$, so the solution set of (E.2) is the same as the solution set of

$$\begin{bmatrix} 1 & 0 & 0 \\ 0 & 1 & 0 \\ 0 & 0 & 1 \end{bmatrix} \begin{bmatrix} \alpha_1 \\ \alpha_2 \\ \alpha_3 \end{bmatrix} = \begin{bmatrix} 0 \\ 0 \\ 0. \end{bmatrix} \tag{E.3}$$

But this equation is equivalent to $\alpha_1 = 0$, $\alpha_2 = 0$, $\alpha_3 = 0$. This shows that the given set of three vectors is linearly independent.

To show that the span of the given set of vectors is all of $\mathbb{R}^3$, we must show that for every $\begin{bmatrix} x \\ y \\ z \end{bmatrix}$, there exist coefficients $\alpha_1, \alpha_2, \alpha_3$ such that

$$\begin{bmatrix} x \\ y \\ z \end{bmatrix} = \alpha_1 \begin{bmatrix} 1 \\ 1 \\ 1 \end{bmatrix} + \alpha_2 \begin{bmatrix} 1 \\ 2 \\ 5 \end{bmatrix} + \alpha_3 \begin{bmatrix} 4 \\ 2 \\ 7 \end{bmatrix}. \tag{E.4}$$

The vector equation (E.4) is equivalent to the matrix equation

$$\begin{bmatrix} x \\ y \\ z \end{bmatrix} = \begin{bmatrix} 1 & 1 & 4 \\ 1 & 2 & 2 \\ 1 & 5 & 7 \end{bmatrix} \begin{bmatrix} \alpha_1 \\ \alpha_2 \\ \alpha_3 \end{bmatrix}. \tag{E.5}$$

It suffices to solve this equation for each of the standard basis vectors $\hat{\mathbf{e}}_1, \hat{\mathbf{e}}_2, \hat{\mathbf{e}}_3$, for if each of these vectors is in the span, then all of $\mathbb{R}^3$ is contained in the span. For example, solving the equation

$$\begin{bmatrix} 1 \\ 0 \\ 0 \end{bmatrix} = \begin{bmatrix} 1 & 1 & 4 \\ 1 & 2 & 2 \\ 1 & 5 & 7 \end{bmatrix} \begin{bmatrix} \alpha_1 \\ \alpha_2 \\ \alpha_3 \end{bmatrix} \tag{E.6}$$

by row reduction gives $\begin{bmatrix} \alpha_1 \\ \alpha_2 \\ \alpha_3 \end{bmatrix} = (1/11) \begin{bmatrix} 4 \\ -5 \\ 3 \end{bmatrix}$. We proceed similarly for $\hat{\mathbf{e}}_2, \hat{\mathbf{e}}_3$.

If $\{\mathbf{v}_1, \dots, \mathbf{v}_s\}$ is a basis of V, then every element of V can be written as a linear combination of $(\mathbf{v}_1, \dots, \mathbf{v}_s)$ in one and only one way. In general, to find the coefficients of a vector with respect to a basis, we have to solve a linear equation; for example, to write $\begin{bmatrix} 2 \\ -7 \\ 13 \end{bmatrix}$ in terms of the basis $\left\{ \begin{bmatrix} 1 \\ 1 \\ 1 \end{bmatrix}, \begin{bmatrix} 1 \\ 2 \\ 5 \end{bmatrix}, \begin{bmatrix} 4 \\ 2 \\ 7 \end{bmatrix} \right\}$ of the previous example, we have to solve the equation

$$\begin{bmatrix} 2 \\ -7 \\ 13 \end{bmatrix} = \alpha_1 \begin{bmatrix} 1 \\ 1 \\ 1 \end{bmatrix} + \alpha_2 \begin{bmatrix} 1 \\ 2 \\ 5 \end{bmatrix} + \alpha_3 \begin{bmatrix} 4 \\ 2 \\ 7 \end{bmatrix},$$

which is equivalent to the matrix equation

$$\begin{bmatrix} 2 \\ -7 \\ 13 \end{bmatrix} = \begin{bmatrix} 1 & 1 & 4 \\ 1 & 2 & 2 \\ 1 & 5 & 7 \end{bmatrix} \begin{bmatrix} \alpha_1 \\ \alpha_2 \\ \alpha_3 \end{bmatrix}.$$

Definition E.6. A *linear transformation* or *linear map* from a vector subspace $V \subseteq K^n$ to K^m is a map T satisfying $T(\mathbf{x}+\mathbf{y}) = T(\mathbf{x}) + T(\mathbf{y})$ for all $\mathbf{x}, \mathbf{y} \in V$ and $T(\alpha\mathbf{x}) = \alpha T(\mathbf{x})$ for all $\alpha \in K$ and $\mathbf{x} \in V$.

The typical example of a linear transformation is given by matrix multiplication: Let M be an m-by-n matrix (m rows, n columns); then $\mathbf{x} \mapsto T_M(\mathbf{x}) = M\mathbf{x}$ is a linear transformation from K^n to K^m. If $M^1, \dots M^n$ denote the columns of M, then

$$M \begin{bmatrix} x_1 \\ x_2 \\ \vdots \\ x_n \end{bmatrix} = x_1 M^1 + \cdots x_n M^n.$$

Thus $M\mathbf{x}$ is a linear combination of the columns of M. In particular, $M\hat{\mathbf{e}}_j = M^j$ for $1 \le j \le n$.

To any linear transformation $T : K^n \to K^m$, we can associate its *standard matrix* M_T, which is the m-by-n matrix whose columns are $T(\hat{\mathbf{e}}_1), \dots, T(\hat{\mathbf{e}}_n)$. Then for any vector $\mathbf{x} = \begin{bmatrix} x_1 \\ \vdots \\ x_n \end{bmatrix} = \sum x_i \hat{\mathbf{e}}_i$, we have

$$T(\mathbf{x}) = T(\sum x_i \hat{\mathbf{e}}_i) = \sum x_i T(\hat{\mathbf{e}}_i) = \sum x_i M_T^i = M_T \mathbf{x}. \qquad \text{(E.7)}$$

Denote the set of linear transformations from K^n to K^m by

$$\mathrm{Hom}(K^n, K^m)$$

and the set of m-by-n matrices over K by

$$\mathrm{Mat}_{m,n}(K).$$

The correspondence between linear transformations and matrices has the following important algebraic properties:

Proposition E.7.

(a) *For any* m, n, *the correspondence* $T \mapsto M_T$ *is a bijection between* $\mathrm{Hom}(K^n, K^m)$ *and* $\mathrm{Mat}_{m,n}(K)$.

(b) $M_{S+T} = M_S + M_T$, *and* $M_{\alpha T} = \alpha M_T$, *for* $S, T \in \mathrm{Hom}(K^n, K^m)$ *and* $\alpha \in K$.

(c) *The standard matrix of the identity transformation on* K^n *is the* n*-by-*n *identity matrix* E_n, *with 1's on the diagonal and 0's elsewhere.*

(d) *If* $S \in \mathrm{Hom}(K^n, K^m)$ *and* $T \in \mathrm{Hom}(K^m, K^\ell)$, *then* $M_{T \circ S} = M_T M_S$.

Proof. Assertions (a) through (c) are left to the reader. For (d), note that $T \circ S(\mathbf{x}) = T(S(\mathbf{x})) = M_T S(\mathbf{x}) = M_T M_S \, \mathbf{x}$ and on the other hand, $T \circ S(\mathbf{x}) = M_{T \circ S} \, \mathbf{x}$, for all $\mathbf{x} \in K^n$, by Equation (E.7). ■

Example E.8. Suppose a linear transformation T of $\mathbb{R}^2$ satisfies $T(\begin{bmatrix} 1 \\ 2 \end{bmatrix}) = \begin{bmatrix} 5 \\ 7 \end{bmatrix}$ and $T(\begin{bmatrix} 2 \\ 3 \end{bmatrix}) = \begin{bmatrix} 1 \\ 4 \end{bmatrix}$. Let's compute the standard matrix of T. To do so, we have to compute $T(\hat{\mathbf{e}}_1)$, and $T(\hat{\mathbf{e}}_2)$. This can be done by expressing $\hat{\mathbf{e}}_i$ as linear combinations of $\begin{bmatrix} 1 \\ 2 \end{bmatrix}$ and $\begin{bmatrix} 2 \\ 3 \end{bmatrix}$. Thus we have to solve

$$\hat{\mathbf{e}}_1 = \alpha_1 \begin{bmatrix} 1 \\ 2 \end{bmatrix} + \alpha_2 \begin{bmatrix} 2 \\ 3 \end{bmatrix} = \begin{bmatrix} 1 & 2 \\ 2 & 3 \end{bmatrix} \begin{bmatrix} \alpha_1 \\ \alpha_2 \end{bmatrix},$$

and

$$\hat{\mathbf{e}}_2 = \beta_1 \begin{bmatrix} 1 \\ 2 \end{bmatrix} + \beta_2 \begin{bmatrix} 2 \\ 3 \end{bmatrix} = \begin{bmatrix} 1 & 2 \\ 2 & 3 \end{bmatrix} \begin{bmatrix} \beta_1 \\ \beta_2 \end{bmatrix}.$$

Solving these equations by row reduction gives

$$\begin{bmatrix} \alpha_1 \\ \alpha_2 \end{bmatrix} = \begin{bmatrix} -3 \\ 2 \end{bmatrix} \quad \text{and} \quad \begin{bmatrix} \beta_1 \\ \beta_2 \end{bmatrix} = \begin{bmatrix} 2 \\ -1 \end{bmatrix}.$$

Therefore,

$$T(\hat{\mathbf{e}}_1) = -3\,T(\begin{bmatrix} 1 \\ 2 \end{bmatrix}) + 2\,T(\begin{bmatrix} 2 \\ 3 \end{bmatrix}) = -3 \begin{bmatrix} 5 \\ 7 \end{bmatrix} + 2 \begin{bmatrix} 1 \\ 4 \end{bmatrix} = \begin{bmatrix} -13 \\ -13 \end{bmatrix}$$

and

$$T(\hat{\mathbf{e}}_2) = 2\,T(\begin{bmatrix}1\\2\end{bmatrix}) - 1\,T(\begin{bmatrix}2\\3\end{bmatrix}) = 2\begin{bmatrix}5\\7\end{bmatrix} - 1\begin{bmatrix}1\\4\end{bmatrix} = \begin{bmatrix}9\\10\end{bmatrix}.$$

Hence, the standard matrix of T is $M = \begin{bmatrix}-13 & 9\\-13 & 10\end{bmatrix}$. As a check, you should verify that $M\begin{bmatrix}1\\2\end{bmatrix} = T(\begin{bmatrix}1\\2\end{bmatrix}) = \begin{bmatrix}5\\7\end{bmatrix}$ and $M\begin{bmatrix}2\\3\end{bmatrix} = T(\begin{bmatrix}2\\3\end{bmatrix}) = \begin{bmatrix}1\\4\end{bmatrix}$.

Definition E.9. The *range* of a linear transformation $T : V \to K^m$ is $\{T(\mathbf{x}) : \mathbf{x} \in K^n\}$.

The range of a linear transformation $T : V \to K^m$ is a vector subspace of K^m (exercise).

If $T : K^n \to K^m$ is linear, the range of T is the span of $\{T(\hat{\mathbf{e}}_1), \dots, T(\hat{\mathbf{e}}_n)\}$. For any $\mathbf{x} \in K^n$ can be written uniquely as a linear combination $\mathbf{x} = \sum \alpha_i \hat{\mathbf{e}}_i$, and then we have $T(\mathbf{x}) = \sum \alpha_i T(\mathbf{e}_i)$. If M is the standard matrix of T, then the range of T is the span of the columns of M.

Definition E.10. Let V be a vector subspace of K^n and let $T : V \to K^m$ be a linear transformation. The *kernel* of T is $\{\mathbf{x} \in V : T(\mathbf{x}) = \mathbf{0}\}$.

For V a vector subspace of K^n, the kernel of a linear transformation $T : V \to K^m$ is a vector subspace of K^n (Exercise E.2).

Exercise E.1. Complete the proof of Proposition E.7.

Exercise E.2. Let $T : V \to K^m$ be a linear transformation from a vector subspace V of K^n to K^m. Show that the kernel of T is a vector subspace of K^n and that the range of T is a vector subspace of K^m.

E.2. Bases and Dimension

The basic fact about finite–dimensional linear algebra is as follows:

Theorem E.11. *If* $m < n$ *and* $T : K^n \to K^m$ *is a linear transformation, then* $\ker(T) \neq \{\mathbf{0}\}$. *That is, the kernel contains nonzero vectors.*

Sketch of the proof. Consider the standard matrix M of T, which is m-by-n. The kernel of T is the same as the set of solutions to the matrix equation $M\mathbf{x} = 0$. Solving this equation by row reduction, and taking into account that the number or rows of the matrix is less than the number of columns, we find that there exist nonzero solutions. ■

Corollary E.12. *If* S *is a subset of* K^n *of cardinality greater than* n*, then* S *is linearly dependent.*

Proof. Let $(\mathbf{v}_1, \ldots, \mathbf{v}_{n+1})$ be a sequence of distinct vectors in S. Let A be the n-by-$n+1$ matrix with columns $(\mathbf{v}_1, \ldots, \mathbf{v}_{n+1})$. Since the number of columns of A is greater than the number of rows, the linear transformation $T_A : K^s \to K^n$ has a nonzero kernel. Equivalently, the matrix equation $A\mathbf{a} = 0$ has nonzero solutions. But that means that the columns of A are linearly dependent. ■

Corollary E.13. *If* $s < n$*, then any set of* s *vectors in* K^n *has span properly contained in* K^n*.*

Proof. Let $\{\mathbf{v}_1, \ldots, \mathbf{v}_s\}$ be a set of s vectors. Form the matrix M that has columns $(\mathbf{v}_1, \ldots, \mathbf{v}_s)$. The span of $\{\mathbf{v}_1, \ldots, \mathbf{v}_s\}$ is the same as the set $\{M\mathbf{x} : \mathbf{x} \in K^s\}$. Suppose that the span is all of K^n.

Then, in particular, each of the standard basis elements $\hat{\mathbf{e}}_j$ ($1 \le j \le n$) is contained in the span, so there exists a solution $\mathbf{a}^j$ to the matrix equation $\hat{\mathbf{e}}_j = M\mathbf{a}^j$. Let A be the matrix whose columns are $(\mathbf{a}^1, \ldots, \mathbf{a}^n)$, and let $E = E_n$ be the identity matrix, whose columns are $(\hat{\mathbf{e}}_1, \ldots, \hat{\mathbf{e}}_n)$. Then we have the matrix equation $E = MA$. For all $\mathbf{x} \in K^n$, we have $E\mathbf{x} = \mathbf{x}$. Suppose $\mathbf{x} \in \ker(T_A)$. Then $\mathbf{x} = E\mathbf{x} = M(A\mathbf{x}) = \mathbf{0}$. Thus the kernel of T_A is $\{\mathbf{0}\}$.

But A is s-by-n, and $T_A : K^n \to K^s$, with $s < n$, so by Theorem E.11, $\ker(T_A) \neq \{\mathbf{0}\}$. This is a contradiction, which resulted from the assumption that the span of $\{\mathbf{v}_1, \ldots, \mathbf{v}_s\}$ is equal to K^n. ■

Corollary E.14. *Every basis of* K^n *has exactly* n *elements.*

Proof. Since the basis is linearly independent, it must have no more than n elements. And since its span is all of K^n, it must have no fewer than n elements. ■

Lemma E.15. *Let V be a vector subspace of K^n, and let $S = \{\mathbf{v}_1, \dots, \mathbf{v}_s\}$ be a set of s distinct vectors in V.*

(a) *If S is linearly independent and the span of S is properly contained in V, then there exists a vector $\mathbf{v} \in V$ such that $S \cup \{\mathbf{v}\}$ is linearly independent.*

(b) *If the span of S is V, but S is linearly dependent, then there is a j such that the set $S \setminus \{\mathbf{v}_j\}$ has span equal to V.*

Proof. For part (a): Since the span of S is properly contained in V, there is a vector $\mathbf{v} \in V$ that is not in the span. I claim that $S \cup \{\mathbf{v}\}$ is linearly independent. Let $\alpha_1, \dots, \alpha_s, \beta$ satisfy $(\sum_{i=1}^{s} \alpha_i \mathbf{v}_i) + \beta \mathbf{v} = 0$.

If $\beta \neq 0$, then we could solve for $\mathbf{v}$, namely,

$$\mathbf{v} = \sum_{i=1}^{s} (-\alpha_i/\beta)\mathbf{v}_i,$$

which would contradict that $\mathbf{v}$ is not in the span of S. Hence $\beta = 0$. But then we have $\sum_{i=1}^{s} \alpha_i \mathbf{v}_i = 0$, which implies that the remaining α_i are zero by the linear independence of S.

For part (b): If the set is linearly dependent, then for some j, the vector $\mathbf{v}_j$ can be written as a linear combination of the remaining $\mathbf{v}_i$; in fact, if $\sum_i \alpha_i \mathbf{v}_i = \mathbf{0}$, and $\alpha_j \neq 0$, then $\mathbf{v}_j = \sum_{i \neq j}(-\alpha_i/\alpha_j)\mathbf{v}_i$.

Then if a vector $\mathbf{x}$ is written as linear combination of $(\mathbf{v}_1, \dots, \mathbf{v}_s)$, we can substitute for $\mathbf{v}_j$ the expression $\sum_{i \neq j}(-\alpha_i/\alpha_j)\mathbf{v}_i$, and thus express $\mathbf{x}$ as a linear combination of the sequence with $\mathbf{v}_j$ removed. Hence the span of $S \setminus \{\mathbf{v}_j\}$ removed is the same as the span of S. ∎

Call a set of vectors in V *maximal linearly independent* if it is linearly independent and is not contained in a larger linearly independent set. Since any linear independent set in V has no more than n elements, *any linear independent set is contained in a maximal linearly independent set.* (If a given linear independent set S is not already maximal, it is contained in a larger linearly independent set S_1. If S_1 is not maximal, it is contained in a larger linearly independent set S_2, and so forth. But the process of enlarging the linearly independent sets must stop after at most n stages, as all of the S_i have size no greater than n.)

According to part (a) of the lemma, *a maximal linearly independent set in V has span equal to V, and hence is a basis of V.*

Call a set of vectors in V *spanning* if its span is equal to V. Call it *minimal spanning* if its span is V and if no proper subset is spanning. Note that every spanning set contains a finite spanning set. In fact, if S is spanning, and $\{\mathbf{v}_1, \dots, \mathbf{v}_t\}$ is a basis of V, then for each j, $\mathbf{v}_j$ can be written as a linear combination of finitely many elements of S; that is, $\mathbf{v}_j$ is in the span of a finite subset S_j of S. Then the finite set $\cup_j S_j \subseteq S$ has span containing all of the $\mathbf{v}_j$, hence equal to V. It follows that *a minimal spanning set is finite*. Now, according to part (b) of the lemma, *a minimal spanning sequence in V is linearly independent, and hence a basis of V.*

We have proved the following statement:

Corollary E.16. *Let V be a linear subspace of K^n.*

(a) *V has a basis, and any basis of V has no more than n elements.*
(b) *Any linearly independent set in V is contained in a basis of V.*
(c) *Any spanning set in V has a subset that is a basis of V.*

Now let V be a nonzero vector subspace of K^n, and let $\{\mathbf{v}_1, \dots, \mathbf{v}_s\}$ be a basis of V. We define a linear map T from K^s to $V \subseteq K^n$ by $\begin{bmatrix} \alpha_1 \\ \vdots \\ \alpha_s \end{bmatrix} \mapsto \sum_i \alpha_i \mathbf{v}_i$. This map is one to one with range equal to V, by the definition of a basis. Let T^{-1} be the inverse map, from V to K^s. You can check that a set $\{\mathbf{x}_1, \dots, \mathbf{x}_r\}$ is a basis of V if, and only if, $\{T^{-1}(\mathbf{x}_1), \dots, T^{-1}(\mathbf{x}_r)\}$ is a basis of K^s (Exercise E.3).

Corollary E.17. *Any two bases of a vector subspace V of K^n have the same cardinality. The cardinality of a basis is called the dimension of V.*

Proof. If V is the zero subspace, then its unique basis is the empty set. Thus $\{\mathbf{0}\}$ has dimension 0.

Assume that V is not the zero subspace, so any basis of V has cardinality greater than 0. If V has two bases of cardinality s and t, then there exist invertible linear maps $S : K^s \to V$ and $T : K^t \to V$. Then $T^{-1}S : K^s \to K^t$ is an invertible linear map, which entails that $s = t$. ■

In the following, that a linear transformation $T : K^n \to K^m$ has a left inverse means that there is a linear transformation $L : K^m \to K^n$ such that $L \circ T = \mathrm{id}_{K^n}$. Similarly, that T has a right inverse means that there is a linear transformation $R : K^m \to K^n$ such that $T \circ R = \mathrm{id}_{K^m}$. A matrix has a left inverse if there is a matrix A such that $AM = E$, where E denotes the identity matrix of the appropriate size. Similarly, M has a right inverse if there is a matrix B such that $MB = E$.

Proposition E.18. *Let* M *be an* n*-by-*n *matrix over* K*, and let* T *be the linear transformation of* K^n *given by multiplication by* M*. The following conditions are equivalent:*

(a) *The columns of* M *are linearly independent.*
(b) *The columns of* M *have span equal to* K^n.
(c) $\ker(T) = \{\mathbf{0}\}$.
(d) $\text{range}(T) = K^n$.
(e) T *is a linear isomorphism.*
(f) T *has a left inverse.*
(g) T *has a right inverse.*
(h) M *is invertible.*
(i) M *has a left inverse.*
(j) M *has a right inverse.*
(k) M^t *is invertible.*
(l) *The rows of* M *are linearly independent.*
(m) *The rows of* M *have span equal to* K^n.

Proof. The equivalence of (a) and (b) follows from Corollaries E.14 and E.16. For $\mathbf{x} \in K^n$, $T(\mathbf{x}) = Mx$ is a linear combination of the columns of M; the columns are linearly independent if, and only if, $T(\mathbf{x}) = M\mathbf{x} \neq \mathbf{0}$ for $\mathbf{x} \neq \mathbf{0}$; this gives the equivalence of (a) and (c). Points (b) and (d) are equivalent since the range of T is the span of the columns of M. Thus we have (a) $\iff$ (b) $\iff$ (c) $\iff$ (d).

Points (c) and (d) together imply (e), and (e) implies each of (c) and (d), so (c) $\iff$ (d) $\iff$ (e).

If T has a left inverse, then T is injective. And if T is invertible, then it has a left inverse. So we have (f) $\Rightarrow$(c) $\Rightarrow$(e) $\Rightarrow$(f). Similarly (g) $\Rightarrow$(d) $\Rightarrow$(e) $\Rightarrow$(g).

We have (e) $\iff$ (h), (f) $\iff$ (i), and (g) $\iff$ (j), by Proposition E.7.

Recall that the transpose M^t of a matrix M is the matrix with rows and columns switched. Transposition satisfies the identity $(MN)^t = N^tM^t$, as follows by computation. Thus $NM = E \iff M^tN^t = E^t = E$, which gives (h) $\iff$ (k). Finally, the implications (k) $\iff$ (l) $\iff$ (m) amount to the implications (h) $\iff$ (a) $\iff$ (b) applied to M^t in place of M. ■

Exercise E.3. Let V and W be linear subspaces of K^n and K^m, and let $T : V \to W$ be an invertible linear map. Let S be a set of vectors in V.

(a) Show that the inverse map $T^{-1} : W \to V$ is also linear.
(b) Show that S is linearly independent if, and only if, its image $T(S)$ is linearly independent.

(c) Show that S has span equal to V if, and only if, $T(S)$ has span equal to W.
(d) Show that S is a basis of V if, and only if, $T(S)$ is a basis of W.

Exercise E.4. Show that if $T : K^s \to K^t$ is an invertible linear map, then $s = t$.

Exercise E.5. Verify that the empty set is the unique basis of $\{\mathbf{0}\}$.

Exercise E.6. Let $T : K^n \to K^m$ be a linear map.
(a) Show that if T is injective, then $m \geq n$.
(b) Show that the dimension of the range of T is $\leq n$.
(c) Show that if T is surjective, then $n \geq m$.

E.3. Inner Product and Orthonormal Bases

The standard *inner product* on $\mathbb{R}^n$ is $\langle \mathbf{x}, \mathbf{y} \rangle = \sum_i x_i y_i$. The inner product has the following properties:

1. For fixed $\mathbf{y}$, the map $\mathbf{x} \mapsto \langle \mathbf{x}, \mathbf{y} \rangle$ is linear; that is,
$$\langle \alpha_1 \mathbf{x}_1 + \alpha_2 \mathbf{x}_2, \mathbf{y} \rangle = \alpha_1 \langle \mathbf{x}_1, \mathbf{y} \rangle + \alpha_2 \langle \mathbf{x}_2, \mathbf{y} \rangle$$
for all $\alpha_1, \alpha_2 \in \mathbb{R}$ and $\mathbf{x}_1, \mathbf{x}_2, \mathbf{y} \in \mathbb{R}^n$.
2. $\langle \mathbf{x}, \mathbf{y} \rangle = \langle \mathbf{y}, \mathbf{x} \rangle$ for all $\mathbf{x}, \mathbf{y} \in \mathbb{R}^n$.
3. $\langle \mathbf{x}, \mathbf{x} \rangle \geq 0$ and $\langle \mathbf{x}, \mathbf{x} \rangle = 0$ if, and only if $\mathbf{x} = \mathbf{0}$.

We define the *norm* of $\mathbf{x} \in \mathbb{R}^n$ by $\|\mathbf{x}\| = \langle \mathbf{x}, \mathbf{x} \rangle^{1/2} = (\sum_i x_i^2)^{1/2}$. The inner product satisfies the important *Cauchy-Schwartz inequality*, $|\langle \mathbf{x}, \mathbf{y} \rangle| \leq \|\mathbf{x}\| \, \|\mathbf{y}\|$. The standard *Euclidean distance function* on $\mathbb{R}^n$ is defined by $d(\mathbf{x}, \mathbf{y}) = \|\mathbf{x} - \mathbf{y}\| = (\sum_i (x_i - y_i)^2)^{1/2}$.

Exercise E.7. Prove the Cauchy-Schwartz inequality, using the hint in the text. Hint: Use the inequality
$$\langle \mathbf{x} + t\mathbf{y}, \mathbf{x} + t\mathbf{y} \rangle \geq 0,$$
and minimize with respect to the real variable t, using calculus.)

We say that two vectors are *orthogonal* if their inner product is zero. A set of vectors is said to be *orthonormal* if each vector in the set has norm equal to one, and distinct vectors in the set are orthogonal. An orthonormal set is always linearly independent. For if $\mathbf{v}_i, \dots, \mathbf{v}_s$ are elements of an orthonormal set and $\sum_i \alpha_i \mathbf{v}_i = \mathbf{0}$, then for each j, we have $0 = \langle \mathbf{0}, \mathbf{v}_j \rangle = \langle \sum_i \alpha_i \mathbf{v}_i, \mathbf{v}_j \rangle = \sum_i \alpha_i \langle \mathbf{v}_i, \mathbf{v}_j \rangle = \alpha_j$. Hence an orthonormal set in $\mathbb{R}^n$ never has more than n elements, and an orthonormal set with n elements is a basis of $\mathbb{R}^n$.

If S is any set of vectors, then the set of vectors orthogonal to every vector in S, denoted $S^\perp$ is a subspace of $\mathbb{R}^n$.

It is very easy to find the expansion of a vector with respect to an orthonormal basis. In fact, if $\{\mathbf{v}_i, \dots, \mathbf{v}_n\}$ is an orthonormal basis of $\mathbb{R}^n$, then for every $\mathbf{x} \in \mathbb{R}^n$, we have $\mathbf{x} = \sum_i \langle \mathbf{x}, \mathbf{v}_i \rangle \mathbf{v}_i$. To see this, note that $\mathbf{y} = \mathbf{x} - \sum_i \langle \mathbf{x}, \mathbf{v}_i \rangle \mathbf{v}_i$ is orthogonal to each $\mathbf{v}_j$, hence orthogonal to every linear combination of the $\mathbf{v}_j$'s, hence orthogonal to every vector in $\mathbb{R}^n$, since $\{\mathbf{v}_i, \dots, \mathbf{v}_n\}$ is a basis of $\mathbb{R}^n$. But then $\mathbf{y}$ is orthogonal to itself, hence zero.

There are plenty of orthonormal bases of $\mathbb{R}^n$. In fact, let $S = \{\mathbf{a}_1, \dots, \mathbf{a}_s\}$ be any finite set of vectors. Let $S_j = \{\mathbf{a}_1, \dots, \mathbf{a}_j\}$ for $j \leq s$. Then we can inductively define orthonormal sets B_j such that $\text{span}(B_j) = \text{span}(S_j)$ for each j. In particular, if S has span equal to $\mathbb{R}^n$, then $B = B_s$ is an orthonormal basis of $\mathbb{R}^n$. The inductive procedure (the Gram-Schmidt procedure) goes as follows. First, we can assume that the $\mathbf{a}_i$ are all nonzero. Put $B_1 = \{\mathbf{a}_1 / ||\mathbf{a}_1||\}$. If B_j is already defined, then put $\mathbf{w}_j = \mathbf{a}_{j+1} - \sum_{\mathbf{v} \in B_j} \langle \mathbf{a}_{j+1}, \mathbf{v} \rangle \mathbf{v}$. Then $\mathbf{w}_j$ is orthogonal to to B_j. If $\mathbf{w}_j = \mathbf{0}$, then put $B_{j+1} = B_j$. Otherwise, put $B_{j+1} = B_j \cup \{\mathbf{w}_j / ||\mathbf{w}_j||\}$. In any case, B_{j+1} is orthonormal. We can check without difficulty that $\text{span}(B_{j+1}) = \text{span}(S_{j+1})$, assuming that $\text{span}(B_j) = \text{span}(S_j)$.

Exercise E.8. For any subset S of $\mathbb{R}^n$, show that $S^\perp$ is a subspace of $\mathbb{R}^n$. Show that $(S^\perp)^\perp = \text{span}(S)$.

Exercise E.9. Say that $\mathbb{R}^n$ is the direct sum of vector subspaces M and N, $\mathbb{R}^n = M \oplus N$, if $M + N = \mathbb{R}^n$ and $M \cap N = \{\mathbf{0}\}$. Show that for any subspace M of $\mathbb{R}^n$, we have $\mathbb{R}^n = M \oplus M^\perp$. Show that every vector $\mathbf{x} \in \mathbb{R}^n$ has a unique decomposition $\mathbf{x} = \mathbf{x}_1 + \mathbf{x}_2$, where $\mathbf{x}_1 \in M$ and $\mathbf{x}_2 \in M^\perp$. Show that the map $P_M : \mathbf{x} \mapsto \mathbf{x}_1$ is well defined and linear, with range M and with kernel $M^\perp$, and that $P_M = P_M \circ P_M$.

Exercise E.10. Let M be a subspace of R^n, and let $\{\mathbf{f}_1, \dots, \mathbf{f}_s\}$ be an orthonormal basis of M. Show that $P_M(\mathbf{x}) = \sum_i \langle \mathbf{x}, \mathbf{f}_i \rangle \mathbf{f}_i$.

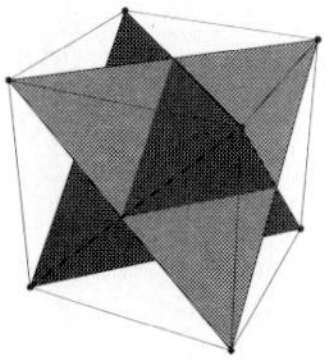

APPENDIX F

Models of Regular Polyhedra

To make the models of the regular polyhedra, copy the patterns on the following pages onto heavy card stock (65– to 70–pound stock), using as much magnification as possible. Then cut out the patterns (including the grey glue tabs), and score them along the internal lines with a utility knife. Glue them (using a slow-drying glue in order to give yourself time to adjust the position). The patterns for the icosahedron and dodecahedron come in two pieces, each of which makes one "hemisphere" of the polyhedron, which you have to glue together.

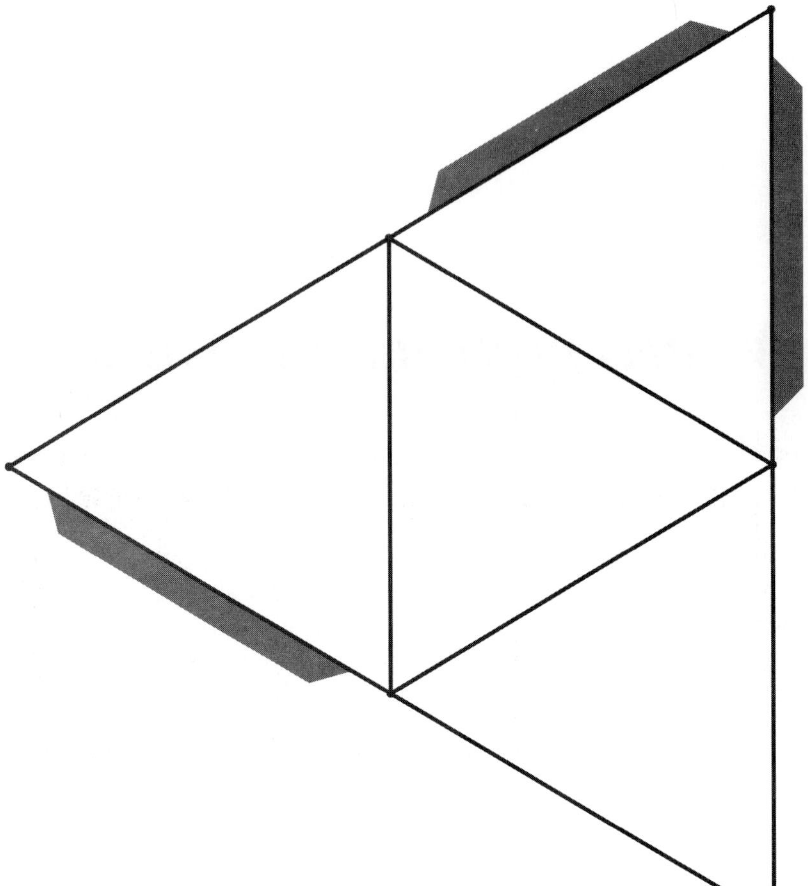

Figure F.1. Tetrahedron pattern.

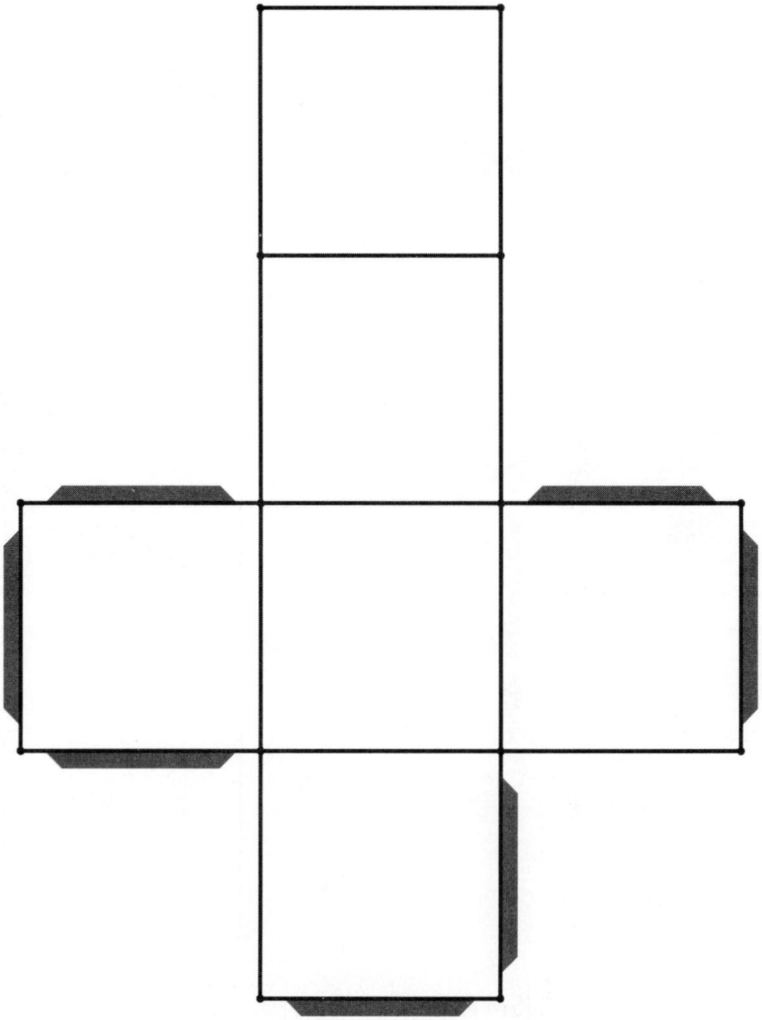

Figure F.2. Cube pattern.

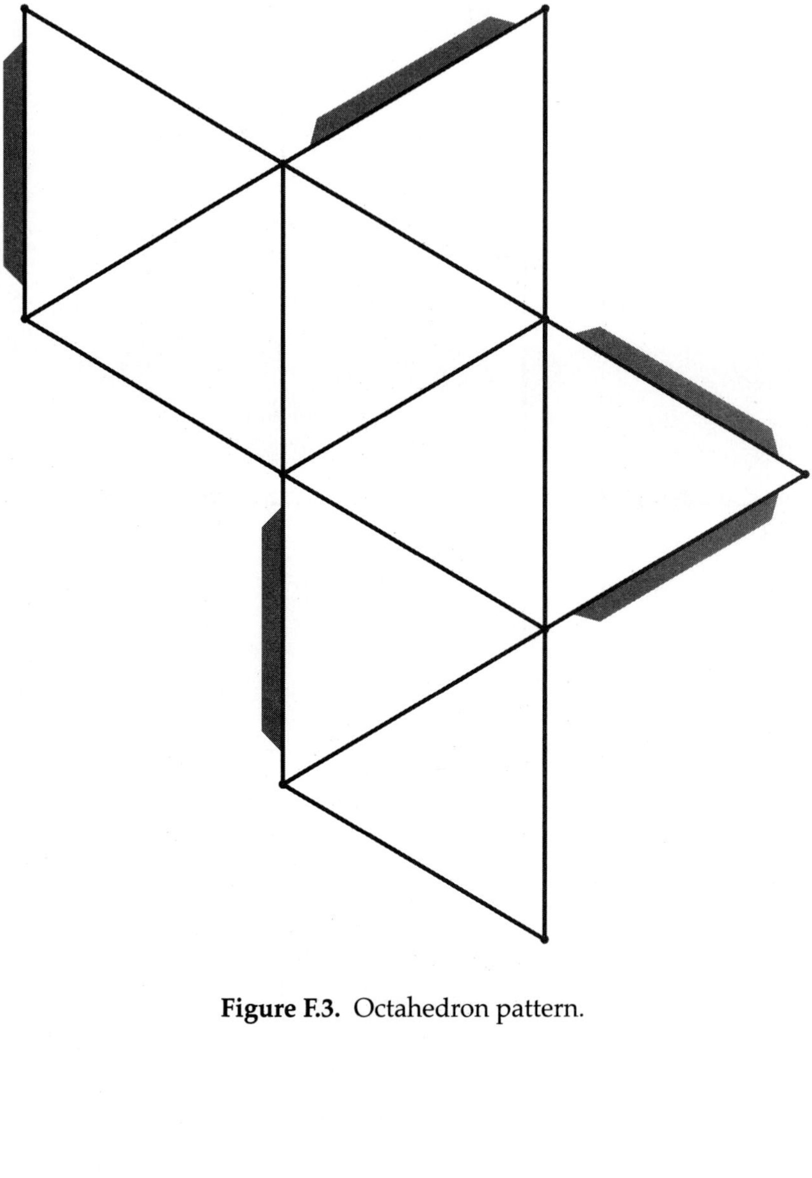

Figure F.3. Octahedron pattern.

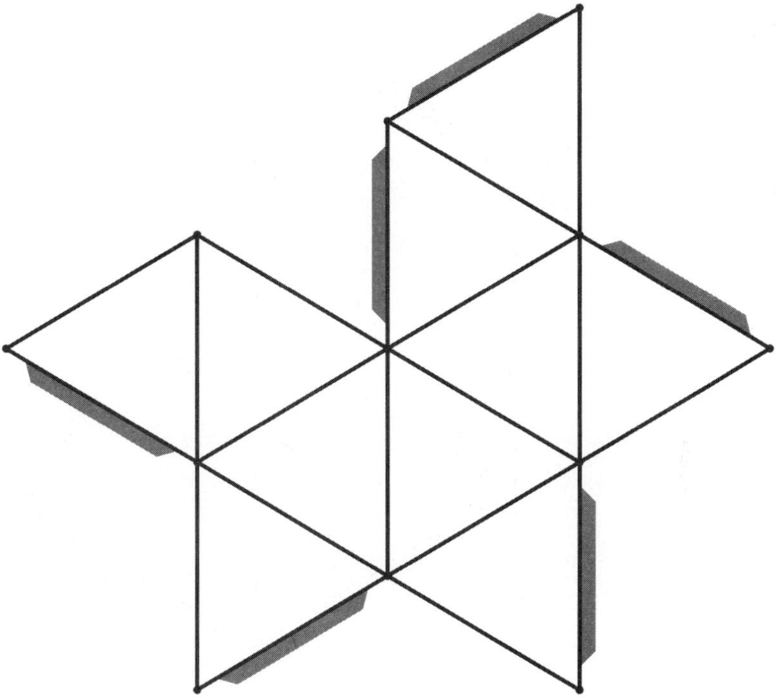

Figure F.4. Icosahedron top pattern.

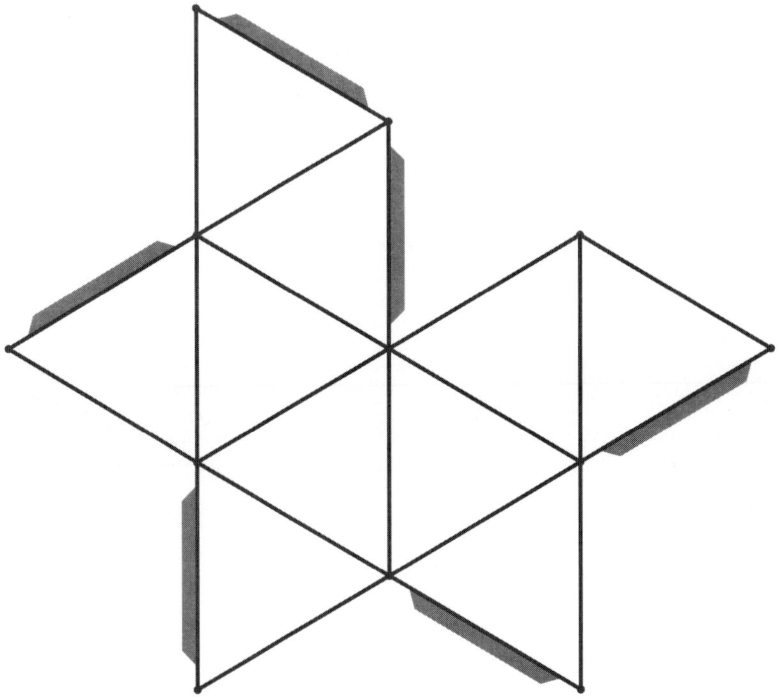

Figure F.5. Icosahedron bottom pattern.

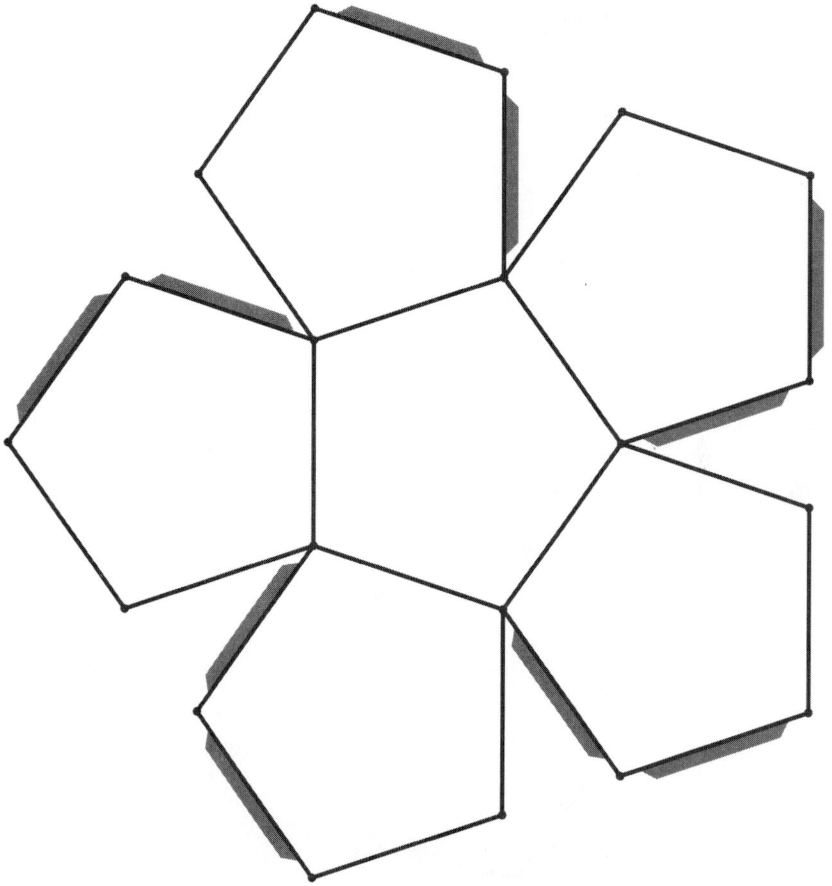

Figure F.6. Dodecahedron top pattern.

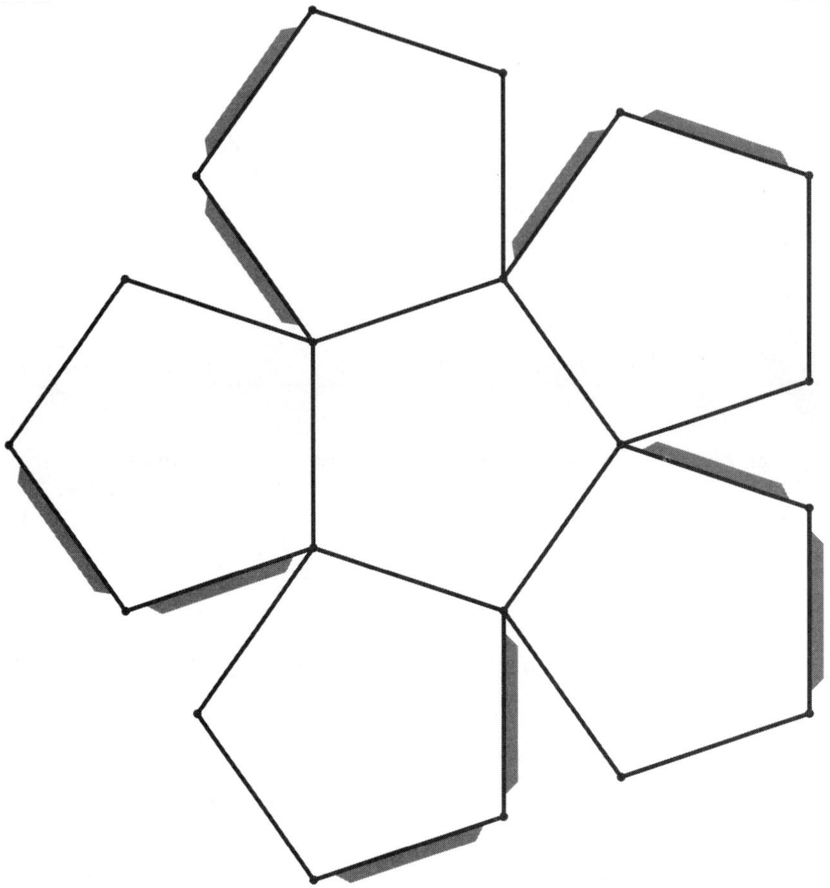

Figure F.7. Dodecahedron bottom pattern.

APPENDIX G

Suggestions for Further Study

The student who has worked his or her way through a substantial portion of this text stands on the threshold of modern mathematics and has access to many related topics.

Here are a few suggestions for further study.

First let me mention some algebra texts that you might use for collateral study, as a source of additional exercises, or for additional topics not discussed here:

Elementary:

- I. N. Herstein, *Abstract Algebra,* 3rd edition, Prentice Hall, 1996.
- T. Shifrin, *Abstract Algebra* , Prentice Hall, 1995.

More advanced:

- D. S. Dummit and R. M. Foote, Abstract Algebra, Prentice Hall, 1991.
- I. N. Herstein, *Topics in Algebra,* 2nd edition, John Wiley, 1975.
- M. Artin, *Algebra,* Prentice Hall, 1995.

Linear (and multilinear) algebra is one of the most useful topics in algebra. For further study of mathematics and for applications of mathematics, eventually you need a better knowledge of linear algebra than you gained in your first course. For this you can look to

- S. Axler, *Linear Algebra Done Right,* Springer–Verlag, 1995.
- K. M. Hoffman and R. Kunze, *Linear Algebra,* 2nd edition, Prentice Hall, 1971. 1971
- S. Lang, *Linear Algebra,* Springer–Verlag, 1987.

Both group theory and field theory have substantial contact with number theory. For an introduction to number theory, see

- G. Andrews, *Number Theory*, Dover Publications, 1994. (Original edition, Saunders, 1971.)

For a computational approach to polynomial rings in several variables and algebraic geometry, there is a beautiful and accessible text:

- D. Cox, J. Little, and D. O'Shea, *Ideals, Varieties and Algorithms*, Springer-Verlag, 1992.

For further study of group theory, my own preference is for the theory of representations and applications. I recommend

- W. Fulton and J. Harris, *Representation Theory, A First Course*, Springer-Verlag, 1991.
- B. Simon, *Representations of Finite and Compact Groups*, American Mathematical Society, 1996.
- S. Sternberg, *Group Theory and Physics*, Cambridge University Press, 1994.

These books are quite challenging, but they are accessible with a knowledge of this course, linear algebra, and undergraduate analysis.

Index

Symbols

$A \setminus B$, 410
$Ax + b$ group, 135
D_n, 152
e^{it}, 422
$K(x)$, 233
$K[x, x^{-1}]$, 79
$K[x]$, 44
$N_G(H)$, 205
$o(a)$, 96
$Phi(n)$, 215
  structure of, 164
$R \oplus S$, 77, 232
$R \times S$, 77, 232
S_n, 20
$[a]$, 38
$Aut(G)$, 214
$Aut(L)$, 307
$Aut_F(L)$, 286
$Aut_K(L)$, 307
$\bar{z}$, 422
$\mathbb{C}$, 44, 421
$Cent(g)$, 206
$\cap$, 409
$\cup$, 410
$dim(V)$, 170
$dim_K(L)$, 277
$End(V)$, 175
$\emptyset$, 410
$Fix(H)$, 287, 309
$Hom(V, W)$, 174
$\Im z$, 422
$Int(G)$, 214
$\in$, 409
$\mathbb{N}$, 26, 414
$\Phi(n)$, 72
  structure of, 163
$\mathbb{Q}$, 44, 415
$\mathbb{R}$, 44
$\Re z$, 422
$SL(n, \mathbb{R})$, 115, 138
$SO(n, \mathbb{R})$, 197
$\subseteq$, 409
φ, 65
$\mathbb{Z}$, 26, 414
$\mathbb{Z}_n$, 39, 115, 133
((S)), 266
New York Times
  crystal, 385

A

Aardvarks, 148
Abel, N. H., 272
Abelian group, 87, 119, 144
  Sylow subgroups, 223
Affine group, 119, 135, 152, 154, 373
Affine reflection, 372
Affine rotation, 372

Affine span, 368
Affine subspace, 368
Affine transformation, 12
Algebraic element, 278
Algebraic number, 278
Algebraically closed field, 421
Algorithm
for g.c.d of integers, 31, 36
for g.c.d. of polynomials, 55
Alternating group, 115, 125, 352–356
And, 401
Associates, 253
Automorphism, 118
Automorphism group, 144, 214–216
of $(\mathbb{Z}_n)^k$, 216
of S_3, 144
of S_n, 214
of $\mathbb{Q}$, 215
of $\mathbb{Z}$, 214
of $\mathbb{Z}^2$, 215
of $\mathbb{Z}_2 \times \mathbb{Z}_2$, 216
of $\mathbb{Z}_n \times \mathbb{Z}_n$, 216
of $\mathbb{Z}_n$, 215
Automorphism of a field, 307
Ax + b group, 119, 152, 154, 373
Axis of symmetry, 177

B

Basis, 166, 424
ordered, 171
Bijective, 411
Binomial coefficients, 56
Bragg, W. H., 397
Bragg, W. L., 397
Burnside's lemma, 211
Butler, G., 346

C

Canonical projection, 130
Cardinality, 23, 414
Cartesian product, 411
of rings, 77
Cartesian product of rings, 232
Cauchy's theorem, 218
Cauchy, A., 272
Center, 122, 124
Centralizer, 206, 216
Characteristic, 250
Characteristic function, 62
Chebotarev, 345
Chinese remainder theorem, 42, 44, 77, 247, 260
Chirality, 200
Class equation, 216
Clock arithmetic, 36
Codomain, 411
Commutator, 351
Commutator subgroup, 144, 351
Complement, 410
relative, 410
Complement of a subspace, 171
Complex conjugate, 422
Complex numbers, 44, 421–422
Composition series, 349
Congruence class, 38
Congruence modulo n, 37
Conjugacy, 131, 204
Conjugacy class, 131, 204, 209, 216
Conjugate elements, 117
Conjugate subgroups, 204
Conjunction, 401
Contrapositive, 403
Convex polyhedra, 374
Convex set, 16
Coset, 119
Countable, 414
Counting
colorings, 212
formulas, 207
necklaces, 211
Cryptography, 80
Crystal, 383
Crystal groups
classification, 389, 396
Cube
full symmetry group, 201
rotation group, 180
Cycle, 23, 375
Cyclic group, 94, 155
Cyclic groups, 133
Cyclic subgroup, 94
Cyclotomic polynomial, 356–359

D

de Morgan's laws, 411
Deduction, 407
Definition by induction, 417

Degree of a polynomial, 46
Denumerable, 414
Derived subgroup, 351
Determinant
  as homomorphism, 111
Devlin, 400
Dihedral group, 104–109, 152, 154
Dimension, 170
  of an affine subspace, 368
Direct product, 145, 148
Direct sum
  internal, 171
  of rings, 77
  of vector spaces, 171
Direct sum of rings, 232
Discriminant, 329
  computation of, 329
  criterion for Galois group, 329, 340
  expressed as a resultant, 333
Disjoint, 410
Disjunction, 401
Divisibility of integers, 26–36
Divisibility of polynomials, 44–54
Division with remainder
  for integers, 29
Division with remainder for polynomials, 49
Dodecahedron
  full symmetry group, 201
  rotation group, 185
Domain, 411
Double coset, 124
Dual polyhedron, 183

E

Eisenstein's criterion, 269
Elementary symmetric functions, 320
Empty set, 410
Enumerable, 414
Equivalence class, 127
Equivalence relation, 125
Euclidean domain, 253
Euler φ function, 65
Euler's theorem, 374
Euler's theorem (number theory), 66
Even permutation, 115, 125
Existence of subgroups
  Cauchy's theorem, 218
  Sylow's theorem, 219
Existential quantifier, 406
Exponential function as homomorphism, 137

F

F–automorphism, 286
Fedorov, 396
Fermat's little theorem (number theory), 61
Fiber, 129
Field
  definition, 78, 228
  of complex numbers, 421
Field extension, 275
  Galois, 312
  separable, 310
Field of fractions, 248
Field, M., 383
Finite abelian group, 155–164
Finite set, 414
Finite subgroups of $O(3, \mathbb{R})$, 381
Finite subgroups of $SO(3, \mathbb{R})$, 377
Fixed field, 309
Fourier, 273
Fractional linear transformations, 143
Full symmetry group, 200
Function, 411
Fundamental domain, 382
Fundamental theorem of algebra, 421
Fundamental theorem of Galois theory, 315

G

g.c.d, 50
Galois, 273
Galois correspondence, 286, 314–319
Galois extension, 294, 312
Galois group, 309
  discriminant criterion, 329
  of a cubic polynomial, 330, 333
  of a quartic polynomial, 330, 335–342
  of the general polynomial, 328
Gauss, 272
Gauss's lemma, 262
Gaussian integers, 251, 253
General equation of degree n, 327
General polynomial of degree n, 327

Generator of a cyclic group, 94
Generators for a group, 93
Generators for a ring, 230
Geometric point group, 386
Glide–reflection, 372
Golubitsky, M., 383
Graphs, 375
Greatest common divisor, 30, 50, 253
Group
  abelian, 87, 119, 144
  affine, 135
  cyclic, 133
  definition, 70
  dihedral, 104–109
  finite abelian, 155–164
  of automorphisms, 144
  of inner automorphisms, 144
  of invertible maps, 19
  of order p^n, 216
  of prime order, 122
  order pq, 222
  order p^2, 217
  order p^3, 222
  projective linear, 138
  simple, 348
  solvable, 350
  special linear, 138
Group action, 203
Group algebra, 234
Group ring, 234
Groups of small order, 86
  classification, 223

H

Hausdorff maximal principal, 173
Hilbert, D., 364
Homogeneous polynomial, 319
Homomorphism, 109
  from $\mathbb{Z}$ to $\mathbb{Z}_n$, 112
  from $\mathbb{Z}$ to a cyclic subgroup, 112
  kernel, 114
  of rings, 77, 234
  unital, 235
Homomorphism theorems
  for groups, 136–143
  for rings, 242–245
Hyperplane
  affine, 368
  linear, 368
Hölder, O., 350

I

Icosahedron
  full symmetry group, 201
  rotation group, 185
Ideal, 237
  generated by a subset, 239
  maximal, 245
  principal, 238, 241
Idempotent, 241, 251
Identity element, 84
Image, 412
Imaginary part of a complex number, 422
Implication, 403
Index of a subgroup, 121
Induction, 416
  on a totally ordered set, 325
Infinite set, 414
Infinitude of primes, 28
Injective, 411
Inner automorphism, 118
Inner automorphism group, 144
Integers, 26, 414
Integral domain, 248
Intersection, 409
Inverse, 84
Inverse function, 412
Inversion symmetry, 190
Invertible element, 41, 76
Irreducibility criteria, 268–271
Irreducible element, 253
Isom(n), 372
Isometries of $\mathbb{R}^2$, 372
Isometry, 195
Isometry group, 195, 367
  semidirect product structure, 372
Isomorphism, 86
  of rings, 77, 232
Isomorphism of crystal groups, 387

J

Jordan, C., 350
Jordan-Hölder theorem, 350

K

Kappe, L., 337
Kernel, 114, 237, 241

L

Lagrange's theorem, 121
Lagrange, J-L., 272
Lattice, 382
  of subgroups, 93
Lattice (partial order), 93
Laurent polynomial, 79
Leading coefficient, 47
Leading term, 47
Line segment, 16
Linear combination, 165
Linear dependence, 166
Linear independence, 166, 423
Linear isometry, 196, 369
  classification, 199
  product of reflections, 200, 370
Linear map, 167, 426
Linear subspace, 424
Linear transformation, 167, 426
  as homomorphism, 112
Logical connectives, 401
Logical equivalence, 403

M

Matrices of symmetries, 11
Matrix representation, 11, 14
Matrix rings
  simplicity, 239
Maximal ideal, 245
McKay, J., 345
Minimal polynomial, 279
Modular arithmetic, 36
Modulus of a complex number, 422
Monic monomial, 319
Monic polynomial, 47
Monomial, 319
Monomial symmetric function, 323
Multiple induction, 418
Multiple roots, 305
Multiplication of symmetries, 5
Multiplication table, 5, 7, 85
Multiplicative 1-cocycle, 363
Multiplicative inverse, 41
multiplicative inverse, 76
Multiplicity of a root, 305

N

n-fold axis of symmetry, 177
Natural numbers, 26, 414
Negation, 402
Negation of quantified statements, 406
Nilpotent element, 248
Normal subgroup, 114, 124
Normalizer of a subgroup, 205

O

Octahedron
  full symmetry group, 201
  rotation group, 183
Odd permutation, 115
One to one, 411
Onto, 411
Or, 401
Orbit counting lemma, 211
Orbit of an action, 203
Order of an element, 96
Order of quantifiers, 407
Ordered basis, 171
Orthogonal group, 91
Orthogonal matrix, 91, 196

P

p-group, 216
  center, 217
  composition series, 217
Partial order, 93
  by set inclusion, 93
Partition
  of an integer, 160, 322
Partition of a set, 125
Path, 375
Permutation group, 19, 117–119, 155
Permutations, 16
  cycle structure, 204
Point group, 386
Poisson, 273
Polar form of a complex number, 422
Polynomial
  cubic, 274–277

Galois group, 330, 333
splitting field, 284–292
minimal, 279
quartic
Galois group, 330, 335–342
separable, 310
Polynomial ring, 44
Predicate, 401
Preimage, 412
primitive element, 261
Principal ideal, 238, 241
Projective linear group, 138
Proper factor, 253
Proper factorization, 253
Public key cryptography, 80

Q

Quantifier
existential, 406
universal, 404, 405
Quotient group, 132, 133
Quotient homomorphism
of groups, 133
Quotient map, 130
for vector spaces, 168
Quotient vector space, 168

R

Range, 411
Rational functions, 233
Rational numbers, 44, 415
Rational root test, 270
Real numbers, 44
Real part of a complex number, 422
Reflection-rotations, 193
Reflections, 189
Regular polyhedra, 176
classification, 374
Relative complement, 410
Relatively prime elements, 253
Relatively prime integers, 33
Relatively prime polynomials, 53
Residue class, 38
Resolvent cubic, 336
Resultant, 331
Ring
definition, 75, 228
examples, 78
simple, 239
Root of a polynomial, 54
Roots of unity, 95, 422
RSA method, 80
Ruffini, P., 272

S

Semidirect product, 153
Separable field extension, 310
Separable polynomial, 310
Set, 409
countable, 414
finite, 414
infinite, 414
Sign homomorphism, 115
Simple group, 348
classification, 348
Simple ring, 239
Simple roots, 305
Simplicity
of matrix rings, 239
Simplicity of matrix rings, 241
Soicher, L., 345
Solution by radicals, 272
Solvable group, 350
Span, 165
Special linear group, 115, 138
Special orthogonal group, 197
Splitting field, 303
Stabilizer subgroup, 204
Standard matrix, 426
Subfield, 275
Subgroup, 91
index 2, 124, 125
normal, 114, 124
Subgroup generated by S, 93
Subring, 230
Subring generated by S, 230
Subspace, 167
Surjective, 411
Surjective maps and equivalence relations, 129
Sylow subgroup, 219
Sylow theorems, 219
Symmetric functions, 319
Symmetric group, 19, 117–119, 125, 155, 352–356
Conjugacy classes, 204, 209
Symmetric polynomials, 319

Symmetries, 2
 of a brick, 190
 of a square tile, 191
 of geometric figures, 3

T

Tetrahedron
 full symmetry group, 202
 rotation group, 177
Total degree, 319
Transcendental element, 278
Transcendental number, 278
Transitive action, 204
Tree, 375
Trigonometric polynomials, 79
Truth table, 401
Twin primes, 414

U

Union, 410
Unique factorization domain, 254
Unique factorization theorem
 for K[x], 53
 for the integers, 34
Unit, 76
Unital ring homomorphism, 235
Units in $\mathbb{Z}_n$
 structure of, 163
Universal quantifier, 404, 405
 implicit, 404
Universal set, 410

V

Valence, 375
van der Waerden, 342
Vector space, 165
Vector subspace, 167, 424
von Laue, 397

W

Warren, B., 337
Well–ordering principle, 417

Z

Zero divisor, 41
Zorn's lemma, 173